Quantum Concepts in Physics

Written for advanced undergraduates, physicists, and historians and philosophers of physics, this book tells the story of the development of our understanding of quantum phenomena through the extraordinary years of the first three decades of the twentieth century.

Rather than following the standard axiomatic approach, this book adopts a historical perspective, explaining clearly and authoritatively how pioneers such as Heisenberg, Schrödinger, Pauli and Dirac developed the fundamentals of quantum mechanics and merged them into a coherent theory, and why the mathematical infrastructure of quantum mechanics has to be as complex as it is. The author creates a compelling narrative, providing a remarkable example of how physics and mathematics work in practice. The book encourages an enhanced appreciation of the interactions between mathematics, theory and experiment, helping the reader gain a deeper understanding of the development and content of quantum mechanics than with any other text at this level.

Malcolm Longair is Emeritus Jacksonian Professor of Natural Philosophy and Director of Development at the Cavendish Laboratory, University of Cambridge. He has held many highly respected positions within physics and astronomy, and he has served on and chaired many international committees, boards and panels, working with both NASA and the European Space Agency. He has received much recognition for his work, including the Pilkington Prize of the University of Cambridge for Excellence in Teaching and a CBE in the millennium honours list for his services to astronomy and cosmology. His previous well-received books for Cambridge University Press include *Theoretical Concepts in Physics* (2003), *The Cosmic Century: A History of Astrophysics and Cosmology* (2005) and *High Energy Astrophysics* (2011).

Quantum Concepts in Physics

An Alternative Approach to the Understanding of Quantum Mechanics

MALCOLM LONGAIR

Cavendish Laboratory, University of Cambridge

CAMBRIDGE
UNIVERSITY PRESS

University Printing House, Cambridge CB2 8BS, United Kingdom

One Liberty Plaza, 20th Floor, New York, NY 10006, USA

477 Williamstown Road, Port Melbourne, VIC 3207, Australia

314-321, 3rd Floor, Plot 3, Splendor Forum, Jasola District Centre, New Delhi - 110025, India

79 Anson Road, #06-04/06, Singapore 079906

Cambridge University Press is part of the University of Cambridge.

It furthers the University's mission by disseminating knowledge in the pursuit of education, learning and research at the highest international levels of excellence.

www.cambridge.org
Information on this title: www.cambridge.org/9781107017092

First published 2013
Reprinted 2014

A catalogue record for this publication is available from the British Library

ISBN 978-1-107-01709-2 Hardback

For Deborah

Contents

Part II The Old Quantum Theory

Preface

How this book came about

This book is the outcome of a long cherished ambition to write a follow-up to my book *Theoretical Concepts in Physics* (*TCP2*) (Longair, 2003). In that book, I took the story of the development of theoretical concepts in physics up to the discovery of quanta and the acceptance by the physics community that quanta and quantisation are essential features of the new physics of the early twentieth century. There was neither space nor scope to take that story further – it was just too complicated and would have required more advanced mathematics than I wished to include in that volume.

This book is my attempt to do for quantum mechanics what I did for classical physics and relativity in *TCP2*. The objective is to try to reconstruct as closely as possible the way in which quantum mechanics was created out of a mass of diverse experimental data and mathematical analyses through the period from about 1900 to 1930. In my view, quantisation and quanta are the greatest discoveries in the physics of the twentieth century. The phenomena of quantum mechanics have no direct impact upon our consciousness which to all intents and purposes is a world dominated by classical physics. But quantum mechanics underlies all the phenomena of matter and radiation and is the basis of essentially all aspects of civilisation in the twenty-first century.

There is no lack of excellent books on quantum mechanics which is one of the staples of all courses in undergraduate physics. Most of the successful texts adopt an axiomatic approach in which quantum mechanics is derived from a set of basic axioms, the consequences of which are elucidated in the subsequent mathematical elaboration. The first complete exposition of this approach was Dirac's classic book *Principles of Quantum Mechanics* of 1930 which may be thought of as the ultimate goal of this book (Dirac, 1930a). But how did it all come about? Can we understand why the theory has to be as complex as it is and how did the interpretation of the formalism come about?

Just as the core of *TCP2* was inspired by the essays of Martin J. Klein (1967), so this book was inspired long ago by the book *Sources of Quantum Mechanics* edited by B. L. van der Waerden (1967). I had an ambition to use van der Waerden's book as the basis of the equivalent of *TCP2* for the development of quantum mechanics. This was reinforced by the appearance of the massive six-volume series *The Historical Development of Quantum Theory* by Jagdish Mehra and Helmut Rechenberg which provides a very thorough, authoritative survey of the history of quantum mechanics and which were published between 1982 and 2001 (Mehra and Rechenberg, 1982a,b,c,d, 1987, 2000, 2001). Equally inspiring was *The Conceptual Development of Quantum Mechanics* by Max Jammer which covers similar ground in a single volume (Jammer, 1989). Another inspiration was the book *Inward Bound* by Abraham Pais (1985) which sets the development of quantum mechanics and quantum phenomena in a much longer time-frame. In my view, these truly excellent books are quite hard work and can only be readily appreciated by those who already have a strong foundation in classical and quantum physics. They are quite a challenge for those seeking more readily accessible enlightenment.

The historical approach and level of presentation

The experience of teaching and writing a number of books convinced me of the value of rethinking the foundations of physics from a somewhat historical perspective, at the same time making as few assumptions as is reasonable about the mathematical sophistication of the reader. As in *TCP2*, I assume some fluency in physics and mathematics, but nothing that would be beyond the first couple of years of the typical course in physics. It is useful to restate some of the objectives of *TCP2* which apply equally to the approach adopted in this book, in contrast to the standard way in which the subject is tackled.

The origin of *TCP2* can be traced to discussions in the Cavendish Laboratory in the mid-1970s among those of us who were involved in teaching theoretically biased undergraduate courses. There was a feeling that the syllabuses lacked coherence from the theoretical perspective and that the students were not quite clear about the scope of *physics* as opposed to *theoretical physics*. As our ideas evolved, it became apparent that a discussion of these ideas would be of value for all final-year students. The course entitled *Theoretical Concepts in Physics* was therefore designed to be given in the summer term in July and August to undergraduates entering their final year. It was to be strictly non-examinable and entirely optional. Students obtained no credit from having attended the course beyond an increased appreciation of physics and theoretical physics. I was invited to give this course of lectures for the first time, with the considerable challenge of attracting students to 9.00 am lectures on Mondays, Wednesdays and Fridays during the most glorious summer months in Cambridge.

The course was designed to contain the following elements:

(a) *The interaction between experiment and theory*. Particular stress would be laid upon the importance of experiment and, in particular, the role of advanced technology in leading to theoretical insights.

(b) The importance of having available the *appropriate mathematical tools for tackling theoretical problems*.

(c) *The theoretical background to the basic concepts of modern physics*, emphasising underlying themes such as *symmetry, conservation, invariance*, and so on.

(d) *The role of approximations and models in physics*.

(e) *The analysis of real scientific papers in theoretical physics*, providing insight into how professional physicists tackle real problems.

(f) *The consolidation and revision* of many of the basic physical concepts which all final-year undergraduates can reasonably be expected to have at their fingertips.

(g) Finally, *to convey my own personal enthusiasm for physics and theoretical physics*. My own research has been in high energy astrophysics and astrophysical cosmology, but I remain a physicist at heart. My own view is that astronomy, astrophysics and cosmology are no more than subsets of physics, but applied to the Universe on the large scale. I am one of the very lucky generation who began research in astrophysics in the early 1960s and who have witnessed the astonishing revolutions which have taken place in our understanding of all aspects of the physics of the Universe. But, the same can be said of all areas of physics. The subject is not a dead, pedagogic discipline, the only object of which is to provide examination questions for students. It is an active, extensive subject in a robust state of good health.

My objective in writing *Quantum Concepts in Physics* has been to adopt the same user-friendly approach as in *TCP2* but now applied to the discovery of quantum mechanics. I should emphasise that this is a *personal approach* to the understanding of quantum mechanics, but it has the great virtue of forcing the writer and reader to think hard about the issues at stake at each stage in the development through one of the most dramatic periods in the evolution of our understanding of fundamental processes in physics. One of the differences as compared with *TCP2* is that somewhat more advanced mathematical tools have to be introduced to appreciate the full essence of the story. I have tried to lay out the necessary mathematics in as simple a form as I could devise, without sacrificing rigour. In my view, final-year undergraduates and their teachers should have little trouble in coping with these requirements.

Let me also emphasise that this book is *not* a textbook on quantum mechanics. It is certainly *not* a substitute for the systematic development of these topics through the standard axiomatic approach to the discipline. You should regard this book as a supplement to the standard courses, but one which I hope will enhance your understanding, appreciation and enjoyment of the physics. Certainly, I have learned a huge amount about quantum mechanics through studying the works of genius of the pioneers of the subject.

The challenge

Let me make it clear at the outset that the amount of material which has to be condensed into a single manageable volume is immense. Some impression of the magnitude of the task can be appreciated from the almost 4500 pages of the magnificent series by Mehra and Rechenberg. In addition, the history of physics literature is vast. As a result, I have had to be selective, and although the course is tortuous, I have had to streamline the story to reach my goal in a finite space. For further enlightenment, which I thoroughly recommend, there is no alternative but to delve into the writings of Mehra, Rechenberg, Jammer, Pais and the many other authors cited in the text.

I should also confess that, although I have taught numerous courses on quantum physics, I do not regard myself as a 'black-belt' quantum physicist. This has the advantage that I am embarking on a voyage of personal intellectual discovery as well. I like very much the splendid remark of Fitzgerald,

> 'A Briton wants emotion in his science, something to raise enthusiasm, something with human interest.' (Fitzgerald, 1902)

I confess to belonging to that school. I hope you will enjoy this adventure as much as I do.

Malcolm Longair
Cambridge and Venice, 2012

Acknowledgements

I have greatly benefitted from interactions with many colleagues in the Cavendish Laboratory over the years while this project has been mulled over in my mind. These began in the 1970s when the hot debates in the Cavendish Teaching Committee, which I chaired for a number of years, gave fascinating insights into what my distinguished colleagues considered to be the central core of physical thinking. The stimulus of giving parallel courses in undergraduate physics with my colleagues undoubtedly influenced my understanding. These colleagues include John Waldram, David Green and Paul Alexander, whose insights I much appreciate. I also benefitted from the lecture notes in quantum mechanics prepared by Michael Payne and Howard Hughes, which I have used as reference points in my thinking to ensure that I was to end up in the right place by the end of the book. Once the writing was completed, I was most grateful to receive the advice of Malcolm Perry and Anna Żytkow, who generously scrutinised parts of the text where I felt deeper mathematical insights would improve the clarity of the exposition – I am most grateful to them for the changes they recommended and which I implemented.

As in my earlier books, I have greatly benefitted from the advice of David Green on the subtleties of *LaTeX* coding. His expert advice has greatly improved the appearance of the text and the mathematics. I am most grateful to the librarians in the Rayleigh Library, Nevenka Huntic and Helen Suddaby, who have been unfailingly helpful in tracking down many of the rarer books and papers referred to in the text. Similar thanks go to Mark Hurn, Librarian at the Institute of Astronomy, for uncovering various little-known treasures in that library.

I would emphasise that *Quantum Concepts in Physics* is a much more personal voyage of discovery than my previous books. I knew what I wanted to achieve at the outset, but the working out and research were carried out in parallel with the writing. As such, the usual disclaimers that I am solely responsible for errors of content and judgement are even more apposite than usual. This is a story which can be told in many different ways, with different emphases according to the inclination of the writer. It is all the more important that the reader should consult the many texts referenced in this book to obtain the fuller picture.

As in all my work, the love, support, encouragement and understanding of my family, Deborah, Mark and Sarah, means more to me than can ever be expressed in words.

Figure credits

Most of the authors of the papers which include figures reproduced in this book are deceased. I am most grateful to the publishers who have been most helpful in giving permission for the use of the figures from journals, books and other media for which they now hold the copyright. Every effort has been made to track down the copyright owners of all the pictures, but some of them have proved to be beyond what I consider to be reasonable effort. The origins of all the figures are given in the figure captions. In addition, the following list includes the original publishers of the figures and the publishers who now hold the copyright, as well as the specific forms of acknowledgement requested by them.

Annalen der Physik. Reproduced by kind permission of the Annalen der Physik. Figs 3.2 and 14.1. Also, with kind permission from John Wiley and Sons.

Annales de Chimie et de Physique. Reproduced by kind permission of the Annales de Chimie et de Physique. Fig. 3.1.

Bickerstaff, R., Manchester. Reproduced by kind permission of R. Bickerstaff, Manchester. Fig. 1.1.

Cavendish Laboratory. Reproduced by kind permission of the Cavendish Laboratory. Figs 1.3b, 2.2, 4.2b, 16.3, 18.4, 18.5 and 18.6.

Chapman and Hall, Ltd., London. Reproduced by kind permission of Chapman and Hall, Ltd., London. Figs 8.6a and b. Also, with kind permission from Springer Science and Business Media.

Chicago University Press. Reproduced by kind permission of Chicago University Press. Figs 1.5a and b.

Creative Commons. Reproduced by kind permission of Creative Commons. Figs 7.6b (Diagram drawn for wikipedia by Theresa Knott) and 18.3b (http://labspace.open.ac.uk/mod/resource/view.php?id=431626).

Dover publications. Reproduced by kind permission of Dover publications. Fig. 16.1.

Edinburgh Philosophical Journal. Reproduced by kind permission of Edinburgh Philosophical Journal. Fig. 1.6.

Gyldendal: London and Copenhagen. Reproduced by kind permission of Gyldendal: London and Copenhagen. Figs 8.2, 8.4 and 8.5.

Institut International de Physique Solvay. Reproduced courtesy of the Institut International de Physique Solvay: Photograph by Benjamin Couprie. Fig. 3.3.

Journal of the Franklin Institute. Reproduced by kind permission of Journal of the Franklin Institute. Fig. 7.3b. Also with kind permission from Elsevier.

MacMillan and Company. Reproduced by kind permission of MacMillan and Company. Figs 1.4 and 2.4a and b.

Methuen: London. Reproduced by kind permission of Methuen: London. Figs 4.4, 5.3, 5.4 and 7.4a and b.

Museum Boerhaave, Leiden. Reproduced by kind permission of the Museum Boerhaave, Leiden. Fig. 4.1.

Naturwissenschaften, Die. Reproduced by kind permission of Die Naturwissenschaften. Fig. 14.2*a* and *b*. Also, with kind permission from Springer Science and Business Media.

Niels Bohr Archive. Reproduced by kind permission of the Niels Bohr Archive of the Niels Bohr Institute. Fig. 7.7.

Nuovo Cimento, Il. Reproduced with kind permission of the Società Italiana di Fisica. Fig. 15.1.

Oversigt over det Kgl. Danske Videnskabernes. Reproduced by kind permission of the Oversigt over det Kgl. Danske Videnskabernes. Fig. 8.1.

Pergamon Press. Reproduced by kind permission of the Pergamon Press. Figs 6.1*a* and *b* and 6.2.

Philosophical Magazine. Reproduced by kind permission of the Philosophical Magazine. Figs 4.2*a*, 4.3, 4.5 and 8.7.

Philosophical Transactions of the Royal Society of London. Reproduced by kind permission of the Royal Society of London. Fig. 1.3*b*.

Pic Du Midi Observatory. Reproduced by kind permission of Drs Rozelot, Desnoux and Buil of the Pic Du Midi Observatory. Figs 7.1*a*, *b* and *c*.

Physical Review. Reproduced by kind permission of the American Physical Society. Figs 9.1, 9.2 and 6.2. Copyright 1923, 1927 and 1933 by the American Physical Society.

Proceedings of the Royal Society of London. Reproduced by kind permission of the Royal Society of London. Figs 9.3 and 18.7.

Verhandlungen der Deutschen Physikalischen Gesellschaft. Reproduced by kind permission of the Deutschen Physikalischen Gesellschaft. Fig. 4.6.

Wolski, Andrzej. Reproduced by kind permission of Andrzej Wolski from his lecture notes. Fig. 1 of Notes (page 396).

Zhurnal Russkoe Fiziko-Khimicheskoe Obshchestvo. Reproduced by kind permission of the Zhurnal Russkoe Fiziko-Khimicheskoe Obshchestvo. Fig. 1.2.

Zeitschrift für Physik. Reproduced by kind permission of the Zeitschrift für Physik. Figs 7.6*a*, 8.3, 18.1*a*, 18.1*b*, 18.2 and 18.3*a*. Also, with kind permission from Springer Science and Business Media.

PART I

THE DISCOVERY OF QUANTA

Physics and theoretical physics in 1895

1.1 The triumph of nineteenth century physics

The nineteenth century was an era of unprecedented advance in the understanding of the laws of physics. In mechanics and dynamics, more and more powerful mathematical tools had been developed to enable complex dynamical problems to be solved. In thermodynamics, the first and second laws were firmly established, through the efforts of Rudolf Clausius and William Thomson (Lord Kelvin), and the full ramifications of the concept of entropy for classical thermodynamics were being elaborated. James Clerk Maxwell had derived the equations of electromagnetism which were convincingly validated by Heinrich Hertz's experiments of 1887 to 1889. Light and electromagnetic waves were the same thing, thus providing a firm theoretical foundation for the wave theory of light which could account for virtually all the known phenomena of optics.

Sometimes the impression is given that experimental and theoretical physicists of the 1890s believed that the combination of thermodynamics, electromagnetism and classical mechanics could account for all known physical phenomena and that all that remained was to work out the consequences of these recently won achievements. As remarked by Brian Pippard in his survey of physics in 1900,[1] Albert Michelson's famous remark that

> 'Our future discoveries must be looked for in the sixth place of decimals.' (Michelson, 1903)

has often been quoted out of context and is better viewed in the light of Maxwell's words in his inaugural lecture as the first Cavendish Professor of Experimental Physics in 1871:

> 'I might bring forward instances gathered from every branch of science, showing how the labour of careful measurement has been rewarded by the discovery of new fields of research, and by the development of new scientific ideas.' (Maxwell, 1890)

Maxwell's prescient words were the battle-cry for the extraordinary events which were to take place over the succeeding decades. In fact, the late nineteenth century was a period of ferment in the physical sciences when many awkward fundamental problems remained to be solved. These exercised the minds of the greatest physicists of the period. Ultimately, the resolution of these problems was to revolutionise the foundations of physics with the discovery of the wholly different world of quantum mechanics. Let us begin by reviewing some of the issues which led to the crisis of early twentieth century physics.[2]

1.2 Atoms and molecules in the nineteenth century

The origin of the modern concept of atoms and molecules can be traced to the understanding of the laws of chemistry in the early years of the nineteenth century. In the late eighteenth century Antoine-Laurent de Lavoisier established the *law of conservation of mass* in chemical reactions. Then, in the period between 1798 and 1804, Joseph Louis Proust established his *law of definite proportions*, according to which:

> 'A given chemical compound contains the same elements united in the same fixed proportions by mass.'[3]

For example, oxygen makes up 8/9 of the mass of any sample of pure water, while hydrogen makes up the remaining 1/9. In 1803, John Dalton followed up this law with his *law of multiple proportions* according to which:

> 'When two elements combine together to form more than one compound, the weights of one element which unites with a given weight of the other are in simple multiple proportion.'

Next, in 1808 Joseph-Louis Gay-Lussac published his *law of combining volumes of gases*, which states:

> 'The volumes of gases taking part in a chemical change either as reagents or as products, bear a simple numerical relation to one another if all measurements are made under the same conditions of temperature and pressure.'

For example, two volumes of hydrogen react with one volume of oxygen in forming two volumes of water vapour.

These concepts were synthesised and taken much further by Dalton in his influential treatise *A New System of Chemical Philosophy* (Dalton, 1808). He asserted that the ultimate particles, or atoms, of a chemically homogeneous substance all had the same weight and shape and drew up a table of the relative weights of the atoms of a number of simple substances (Fig. 1.1). According to his hypothesis, the atoms are particles of matter which cannot be subdivided into more primitive forms by chemical processes. Originally, Dalton and Berzelius considered that equal volumes of gases under identical physical conditions contain the same number of atoms, but this concept did not agree with the observed relations of the volumes of different combining gases. The solution was provided by Amadeo Avogadro who in 1811 realised that the physical unit was not a single atom but a cluster of a small number of atoms, which he defined as *molecules*, namely the smallest particle of the gas which moves about as a whole. *Avogardo's hypothesis* then states that:

> 'Equal volumes of all gases under the same conditions of temperature and pressure contain the same numbers of molecules.'

A central role in what follows concerned the molecular, or atomic, weight of a substance. This is defined to be the weight of the particle on a scale in which the oxygen atom has weight 16 units. Correspondingly, the gram-molecular weight was defined as the weight of the particle on a scale in which the weight of oxygen was 16 gram.[4] Consequently,

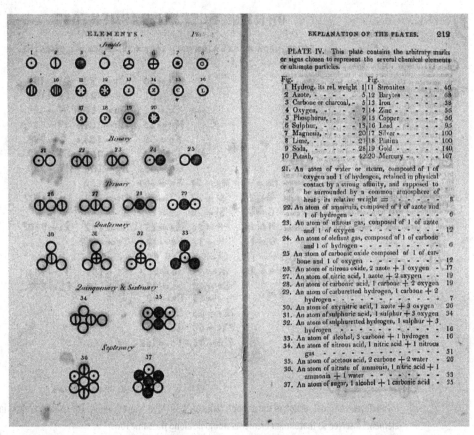

Fig. 1.1   Dalton's symbols for the atoms of various elements and their compounds (Dalton, 1808).

the gram-molecular weights of all substances contain the same numbers of molecules. Avogadro's hypothesis was disregarded by chemists for almost 50 years until 1858 when Stanislao Cannizzaro convinced the leading chemists of the truth of the hypothesis.

A major preoccupation was putting some order into the properties of the chemical elements. In 1789, Lavoisier published a list of 33 chemical elements and grouped them into gases, metals, non-metals and earths. The search was on for a more precise classification scheme. It was known that certain groups of elements had similar chemical properties – for example, the alkali metals sodium, potassium and rubidium and the halogens, chlorine, bromine and iodine. Dmitri Mendeleyev in 1869 and Julius Meyer in 1870 independently published what became known as the *periodic table* of the elements. The tables were constructed by listing the elements in order of increasing atomic weight and then starting a new column when similar classes of elements appeared (Fig. 1.2). Mendeleyev left gaps in the table if an appropriate element had not yet been discovered and then used the trends to predict the properties of the missing elements. Examples included scandium, gallium and germanium. On occasion, he altered the order by atomic weight in order to match similar elements in different rows.

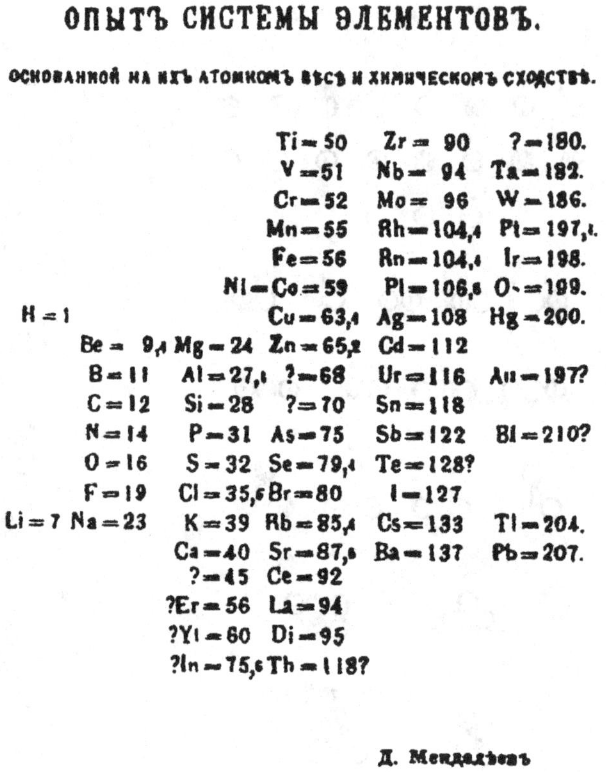

ОПЫТЪ СИСТЕМЫ ЭЛЕМЕНТОВЪ.

ОСНОВАННОЙ НА ИХЪ АТОМНОМЪ ВѢСѢ И ХИМИЧЕСКОМЪ СХОДСТВѢ.

```
                        Ti = 50    Zr = 90    ? = 180.
                        V = 51     Nb = 94    Ta = 182.
                        Cr = 52    Mo = 96    W = 186.
                        Mn = 55    Rh = 104,4  Pt = 197,4.
                        Fe = 56    Rn = 104,4  Ir = 198.
                      Ni = Co = 59  Pl = 106,6  O- = 199.
        H = 1           Cu = 63,4  Ag = 108   Hg = 200.
             Be = 9,4 Mg = 24 Zn = 65,2  Cd = 112
             B = 11   Al = 27,4 ? = 68   Ur = 116   Au = 197?
             C = 12   Si = 28  ? = 70    Sn = 118
             N = 14   P = 31   As = 75   Sb = 122   Bi = 210?
             O = 16   S = 32   Se = 79,4 Te = 128?
             F = 19   Cl = 35,6 Br = 80   I = 127
        Li = 7 Na = 23  K = 39  Rb = 85,4  Cs = 133   Tl = 204.
                       Ca = 40  Sr = 87,6  Ba = 137   Pb = 207.
                       ? = 45   Ce = 92
                     ?Er = 56   La = 94
                     ?Yt = 60   Di = 95
                     ?In = 75,6 Th = 118?
```

Д. Менделѣевъ

Fig. 1.2 Mendeleyev's original version of the periodic table of 1869 (Mendeleyev, 1869). The question marks indicate unknown elements inserted so that similar elements would lie along the same row.

The process of filling in the elements in the periodic table continued throughout the nineteenth century. The understanding of the physics of the atoms of different elements was to be a major concern of Niels Bohr as he struggled to incorporate them into the old quantum theory in the early 1920s. The chemists did not really need the atomic hypothesis to make progress, but rather the empirical rules described above were sufficient to enable remarkable progress to be made in the understanding of chemical processes. As the interest of the chemists waned, however, the physicists took up the reins with the need to provide a microscopic interpretation of the laws of thermodynamics.

1.3 The kinetic theory of gases and Boltzmann's statistical mechanics

The discovery of the first and second laws of thermodynamics in the early 1850s placed thermodynamics on a firm theoretical foundation and these laws were to be elucidated by the next generation of theoretical physicists. Among the challenges was the physical interpretation of the law of increase of entropy which exercised the greatest minds of the subsequent period, including Clausius, Maxwell, Boltzmann and Planck.

1.3.1 The kinetic theory of gases

The laws of thermodynamics describe the properties of matter in bulk. In fact, the theory denies that there is any microscopic structure, its great merit being that it provides general relations between the macroscopic properties of material systems. Nonetheless, Clausius and Maxwell had no hesitation in developing the kinetic theory of gases, considering them to consist of vast numbers of particles making continuous elastic collisions with each other and the walls of the containing vessel. Clausius provided the first systematic account of the theory in 1857 in his paper entitled *On the nature of the motion, which we call heat* (Clausius, 1857). He succeeded in deriving the equation of state of a monatomic gas by working in terms of the mean velocities of the particles. Whilst accounting for the perfect gas law admirably, it did not give good agreement with the known values of the ratios of their specific heat capacities, $\gamma = C_p/C_V$ for molecular gases, where C_p and C_V are the specific heat capacities at constant pressure and constant volume respectively. From experiment, γ was found to be 1.4 for molecular gases, whereas the kinetic theory predicted $\gamma = 1.67$. In the last sentence of his paper, Clausius recognised the important point that there must therefore exist other means of storing kinetic energy within molecular gases which can increase their internal energy per molecule.

One feature of Clausius's work was of particular significance for Maxwell. From the kinetic theory, Clausius worked out the mean velocities of air molecules from his formula $RT = \frac{1}{3}NM\overline{u^2}$. For oxygen and nitrogen, he deduced velocities of 461 and 492 m s^{-1} respectively. The Dutch meteorologist Christoph Buys Ballot criticised this aspect of the theory, since it is well known that pungent odours take minutes to permeate a room. Clausius's response was that the air molecules collide with each other and therefore diffuse from one part of a volume to another, rather than propagate in straight lines. In his paper, Clausius introduced the concept of the *mean free path* of the atoms and molecules of gases for the first time (Clausius, 1858). Thus, in the kinetic theory of gases, it must be supposed that there are continually collisions between the molecules.

Both papers by Clausius were known to Maxwell when he turned to the problem of the kinetic theory of gases in 1859 and 1860. His work was published in 1860 in a characteristically novel and profound series of papers entitled *Illustrations of the dynamical theory of gases* (Maxwell, 1860a,b,c). In a few brief paragraphs,[5] he derived the formula for the velocity distribution $f(u)$ of the particles of the gas and introduced statistical concepts into the kinetic theory of gases and thermodynamics,

$$f(u)\,\mathrm{d}u = 4\pi \left(\frac{m}{2\pi kT}\right)^{3/2} u^2 \exp\left(-\frac{mu^2}{2kT}\right)\mathrm{d}u\,. \tag{1.1}$$

Maxwell immediately noted

> 'that the velocities are distributed among the particles according to the same law as the errors are distributed among the observations in the theory of the *method of least squares*.'

Francis Everitt has written that this derivation of *Maxwell's velocity distribution* marks the beginning of a new epoch in physics (Everitt, 1975). The statistical nature of the laws of thermodynamics and the modern theory of statistical mechanics follow directly from his analysis.

Maxwell, however, ran up against exactly the same problem as Clausius. If only the translational degrees of freedom are taken into account, the value of γ should be 1.67. Maxwell also considered the case in which the rotational degrees of freedom of non-spherical molecules were taken into account as well as their translational motions, but this calculation resulted in a ratio of specific heat capacities $\gamma = 1.33$, again inconsistent with the value 1.4 observed in the common molecular gases. In the last sentence of his great paper, he makes the discouraging remark:

> 'Finally, by establishing a necessary relation between the motions of translation and rotation of all particles not spherical, we proved that a system of such particles could not possibly satisfy the known relation between the two specific heats of all gases.'

1.3.2 The viscosity of gases

Despite this difficulty, Maxwell immediately applied the kinetic theory of gases to their transport properties – diffusion, thermal conductivity and viscosity. His calculation of the coefficient of viscosity of gases was of special importance.[6] Specifically, he worked out how the coefficient of *dynamic* or *absolute viscosity* η is expected to change with pressure and temperature. He found the result

$$\eta = \tfrac{1}{3}\lambda\bar{u}nm = \tfrac{1}{3}\frac{m\bar{u}}{\sigma}, \qquad (1.2)$$

where λ is the mean free path of the molecules, $\bar{u}$ is their mean velocity, n their number density and σ the collision cross-section of the molecules – λ and σ are related by $\lambda = 1/n\sigma$. Maxwell was surprised to find that the coefficient of viscosity is *independent of the pressure*, since there is no dependence upon number density n in (1.2). The reason is that, although there are fewer molecules per unit volume as n decreases, the mean free path increases as n^{-1}, enabling the increment of momentum transfer to take place over greater distances. Furthermore, as the temperature of the gas increases, $\bar{u}$ increases as $T^{1/2}$. Therefore, the viscosity of a gas should increase with temperature, unlike the behaviour of liquids. This somewhat counter-intuitive result was the subject of a brilliant set of experiments carried out by Maxwell from 1863 to 1865 (Fig. 1.3). He confirmed the prediction of the kinetic theory that the viscosity of gases is independent of the pressure. He expected to discover the $T^{1/2}$ law as well, but in fact found a stronger dependence, $\eta \propto T$.

In his great paper of 1867, he interpreted this result as indicating that there must be a repulsive force between the molecules which varied with distance r as r^{-5}. This was a profound discovery since it meant that there was no longer any need to consider the molecules to be 'elastic spheres of definite radius' (Maxwell, 1867). The repulsive force,

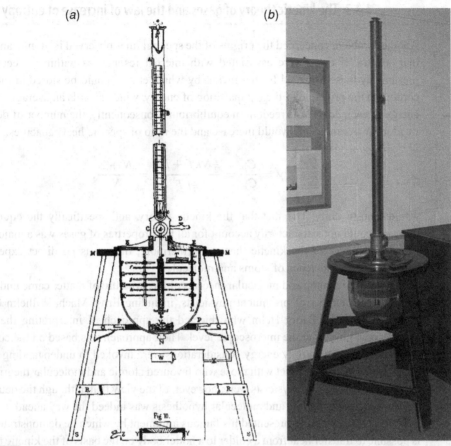

Fig. 1.3   (*a*) Maxwell's apparatus for measuring the viscosity of gases. The gas fills the chamber and the glass discs oscillate as a torsion pendulum. The viscosity of the gas is found by measuring the rate of decay of the oscillations of the torsion balance. The oscillations of the torsion balance were measured by reflecting a light beam from the mirror attached to the suspension. The pressure and temperature of the gas could be varied. The oscillations were started magnetically since the volume of the chamber had to be perfectly sealed. (*b*) Maxwell's apparatus on display in the Cavendish Laboratory.

proportional to r^{-5}, meant that encounters between molecules would take the form of deflections through different angles, depending upon the impact parameter. Maxwell showed that it was more appropriate to think in terms of a *relaxation time*, roughly the time it would take a molecule to be deflected through $90°$, as a result of random encounters with other molecules. According to Maxwell's analysis, molecules could be replaced by centres of repulsion, or, in his words, 'mere points, or pure centres of force endowed with inertia' – it was no longer necessary to make any special assumption about molecules as hard, elastic spheres. To express this insight more provocatively, the concept of collisions between particles was replaced by interactions between fields of force.

1.3.3 The kinetic theory of gases and the law of increase of entropy

Another problem concerned the origins of the spectral lines observed in atomic and molecular spectra. If these were associated with internal resonances within molecules, then presumably these provided further means by which energy could be stored in the gas according to the principle of the equipartition of energy, which awards an average of $\frac{1}{2}kT$ of energy to each degree of freedom in equilibrium. Consequently, the number of degrees of freedom N per molecule would increase and the ratio of specific heat capacities,

$$\gamma = \frac{C_p}{C_V} = \frac{\frac{1}{2}NkT + kT}{\frac{1}{2}kT} = \frac{N+2}{N}, \qquad (1.3)$$

would tend to unity. The fact that the kinetic theory, and specifically the equipartition theorem, could not satisfactorily account for all the properties of gases was a major barrier to the acceptance of the kinetic theory. Furthermore, there was no direct experimental evidence for the existence of atoms and molecules.

The status of atomic and molecular theories of the structure of matter came under attack from a small number of prominent physicists, including Ernst Mach, Wilhelm Ostwald, Pierre Duhem and Georg Helm who rejected the approach of interpreting the laws of macroscopic physics at the microscopic level. Their approach was based on the concept of 'energetics', in which only energy considerations were invoked in understanding physical phenomena, in clear conflict with those who favoured atomic and molecular theories. Most late nineteenth century physicists were, however, of the view that, although the details were not quite right, the atomic and molecular hypothesis was indeed the way ahead.

In 1867, Maxwell first presented his famous argument by which he demonstrated how it is possible to transfer heat from a colder to a hotter body on the basis of the kinetic theory of gases, in violation of the strict application of the law of increase of entropy (Maxwell, 1867). This argument is commonly referred to as involving 'Maxwell's demon'.[7] The Maxwell velocity distribution describes the range of velocities which inevitably must be present in a gas in thermal equilibrium at temperature T. He considered a vessel divided into two halves, A and B, the gas in A being hotter than that in B with a small hole drilled in the partition between them. Whenever a fast molecule moves from B to A, heat is transferred from the colder to the hotter body without the influence of any external agency. It is overwhelmingly more likely that hot molecules move from A to B and in this process heat flows from the hotter to the colder body with the consequence that the entropy of the whole system increases. According to the kinetic theory of gases, however, there is a very small but finite probability that the reverse will happen spontaneously and entropy will decrease in this natural process.

In the late 1860s, Clausius and Boltzmann attempted to derive the second law of thermodynamics from mechanics, an approach known as the *dynamical interpretation of the second law*. The dynamics of individual particles were followed in the hope that they would ultimately lead to an understanding of the origin of the second law. Maxwell rejected this approach as a matter of principle because of the simple but compelling argument that Newton's laws of motion and Maxwell's equations for the electromagnetic field are time reversible and consequently the irreversibility implicit in the second law cannot be explained by a dynamical theory. The second law could only be understood as a statement about the statistical behaviour of an immense number of particles.

Eventually, Boltzmann accepted Maxwell's doctrine concerning the statistical nature of the second law and set about working out the formal relationship between entropy and probability,

$$S = k \ln p, \tag{1.4}$$

where S is the entropy, p the probability of that state and k is a universal constant. In his analysis, the value of the constant k, Boltzmann's constant, was not known. Boltzmann's analysis is of considerable mathematical complexity and, indeed, this was one of the problems which stood in the way of the scientists of his day fully appreciating the deep significance of what he had achieved. Among those who did was Josiah Willard Gibbs who, in his fundamental text *Elementary Principles of Statistical Mechanics*, demonstrated that systems of huge numbers of particles did indeed tend to thermodynamic equilibrium, although there were necessarily fluctuations about the average properties of the gas at every stage of the evolution towards equilibrium (Gibbs, 1902).

1.4 Maxwell's equations for the electromagnetic field

One of the unquestioned triumphs of nineteenth century physics was Maxwell's discovery of the equations for the electromagnetic field.[8] Building on the brilliant experimental investigations of Michael Faraday, Maxwell discovered the equations for the electromagnetic field:[9]

$$\text{curl } \boldsymbol{E} = -\frac{\partial \boldsymbol{B}}{\partial t}, \tag{1.5}$$

$$\text{curl } \boldsymbol{H} = \boldsymbol{J} + \frac{\partial \boldsymbol{D}}{\partial t}, \tag{1.6}$$

$$\text{div } \boldsymbol{D} = \rho_e, \tag{1.7}$$

$$\text{div } \boldsymbol{B} = 0. \tag{1.8}$$

In this modern notation, $\boldsymbol{D}$ and $\boldsymbol{B}$ are the electric and magnetic flux densities, $\boldsymbol{E}$ and $\boldsymbol{H}$ are the electric and magnetic field strengths and $\boldsymbol{J}$ is the current density. The term $\partial \boldsymbol{D}/\partial t$ is the famous *displacement current*, originally introduced on the basis of Maxwell's mechanical model for material media, or vacua (Maxwell, 1861a,b, 1862a,b). All mention of the mechanical origins of his model disappears from the final version of the theory in his great paper *A dynamical theory of the electromagnetic field* (Maxwell, 1865).

Maxwell's was only one of a number of theories of the electromagnetic field and it took some time before the full validation of his theory came about. The remarkable result of Maxwell's calculations was that the speed of light in a vacuum depended only upon the fundamental constants of electrostatics ϵ_0, the permittivity of free space, and magnetostatics μ_0, the permeability of free space, through the relation $c = (\epsilon_0 \mu_0)^{-1/2}$. The determination

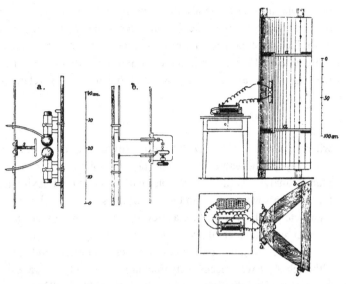

Hertz's apparatus for the generation and detection of electromagnetic radiation. The emitter *a* produced electromagnetic radiation in discharges between the spherical conductors. The detector *b* consisted of a similar device with the jaws of the detector placed as close together as possible to achieve maximum sensitivity. The emitter was placed at the focus of a cylindrical paraboloid reflector to produce a directed beam of radiation (Hertz, 1893).

of these quantities by laboratory experiments was a major concern of Maxwell and his successor as Cavendish Professor, John William Strutt, Lord Rayleigh.

The final validation of the equations was obtained through the brilliant experiments of Hertz in the period 1887–1889, almost a decade after Maxwell's death (Hertz, 1893). Hertz demonstrated that electromagnetic disturbances are propagated at the speed of light in free space (Fig. 1.4). In addition, these waves behaved in all respects exactly like light, the subject headings of his great book being rectilinear propagation, polarisation, reflection, and refraction. This was the final proof of the validity of Maxwell's equations. Ironically, in the same experiments which provided confirmation of Maxwell's theory, Hertz discovered the *photoelectric effect*, the liberation of cathode rays by ultraviolet and optical radiation, which was to prove to be evidence for the fact that, in some circumstances, radiation behaves like particles.

Maxwell resigned from his post at King's College London in 1865 in order to look after the family estate at Glenlair in the Scottish borders and to continue his studies into his many areas of scientific interest. During the succeeding years, he put an enormous effort into the writing of his *Treatise on Electricity and Magnetism* (Maxwell, 1873). The Treatise is unlike many of the other great texts such as Newton's *Principia Mathematica* in that it is not a systematic presentation of the subject but a work in progress, reflecting Maxwell's extensive approach to research. In a later conversation, Maxwell remarked that the aim of the Treatise was not to expound his theory finally to the world, but to educate himself by presenting a view of the stage he had reached. Maxwell's somewhat disconcerting advice was to read the four parts of the Treatise in parallel rather than in sequence.

One of the most important results appears for the first time in Part 4, Sect. 792 of Volume 2 in which Maxwell worked out the pressure which radiation exerts on a conductor. This profound result provides the relation between the pressure p and the energy density ε of a 'gas' of electromagnetic radiation $p = \frac{1}{3}\varepsilon$, derived entirely from Maxwell's electromagnetic theory.[10] This key result was used by Boltzmann in his paper of 1884 in which he derived the Stefan–Boltzmann law from classical thermodynamics (Sect. 1.7.1).

But more than experimental validation of the equations was needed. A new and different perspective had to be adopted to reveal the full power of Maxwell's equations for the electromagnetic field. As Freeman Dyson has written,

> 'Maxwell's theory had to wait for the next generation of physicists, Hertz and Lorentz and Einstein, to reveal its power and clarify its concepts. The next generation grew up with Maxwell's equations and was at home with a Universe built out of fields. The primacy of fields was as natural to Einstein as the primacy of mechanical structures had been for Maxwell.' (Dyson, 1999)

1.5 The Michelson–Morley experiment and the theory of relativity

The unification of light and electromagnetism encouraged experimental physicists to find evidence for the medium through which electromagnetic phenomena are propagated, the *aether*. The most famous experiment was the pioneering interferometric measurements of Michelson (Fig. 1.5a). The final result published by Michelson and Edward Morley (1887) showed no evidence for the drift of the aether relative to the arms of the interferometer, the level of significance of this famous null result being quite enormous (Fig. 1.5b).

One of the concerns about Maxwell's theory of the electromagnetic field was that the forms of the equations are not invariant with respect to the Galilean transformations of Newtonian physics. For this reason, Maxwell's equations were regarded as being 'non-relativistic' – the equations would only hold good in a single preferred frame of reference and in all others there would be additional terms.[11] Woldmar Voigt (1887) was aware of the null result of Michelson's experiments and in 1887 published an analysis of the transformations between inertial frames of reference with the assumption that the speed of propagation of the waves was the same in all inertial frames. To express his insight in different terms, he was the first physicist to work out the set of transformations which would leave the wave equation form invariant. He found the following transformations:

$$
\begin{cases}
t' = t - \dfrac{Vx}{c^2}, \\
x' = x - ct, \\
y' = y/\gamma, \\
z' = z/\gamma,
\end{cases}
\qquad
\gamma = \left(1 - \dfrac{V^2}{c^2}\right)^{-1/2},
\tag{1.9}
$$

where V is the relative velocity of two inertial frames of reference. These transformations are almost identical to the standard Lorentz transformations of special relativity, but little

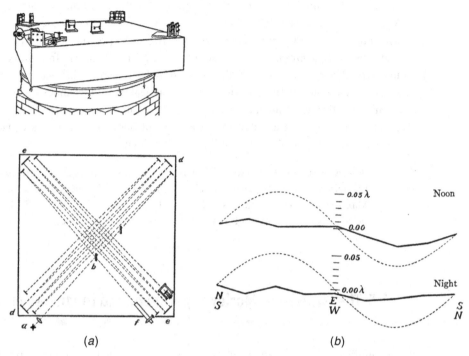

Fig. 1.5 (a) The Michelson–Morley experiment of 1887. (b) The null result of the Michelson–Morely experiment. The solid lines show the average movement of the central fringe as the apparatus was rotated through 360°. The broken line shows one eighth of the sinusoidal variation expected if the Earth moved through a stationary aether at 30 km s^{-1} (Michelson, 1927).

attention was paid to his analysis.[12] Voigt showed that these transformations reduce to the standard expression for the Doppler shift in the limit $V \ll c$.

Explanations of the null result of the Michelson–Morley experiment were proposed independently by Fitzgerald and Hendrik Lorentz. In his brief note of 1889, Fitzgerald suggested that a contraction of the length of the arm of the interferometer by a factor $\gamma = (1 - V^2/c^2)^{-1/2}$ in the direction of motion through the aether could account for the null result (Fitzgerald, 1889). Lorentz had agonised about the null result of the Michelson–Morley experiment and came to the same conclusion that it could be explained by a physical contraction of dimensions in the direction of motion of the apparatus through the aether (Lorentz, 1892a). Over the succeeding decade, he built the Fitzgerald–Lorentz contraction into his theory of the electron. One of the suggestive results of the electrodynamics of moving charges was that, according to Maxwell's theory of the electromagnetic field, the field lines would be squashed perpendicular to the direction of motion[13] by the same factor γ. Eventually in 1904 Lorentz arrived at the complete expressions for the Lorentz transformations by a somewhat tortuous route[14] (Lorentz, 1904), in stark contrast to the much deeper and simpler approach of Einstein in his great paper of 1905, *On the electrodynamics of moving bodies* (Einstein, 1905c). The Lorentz transformations established by Lorentz and Einstein were

$$
\begin{cases}
t' = \gamma \left(t - \dfrac{Vx}{c^2} \right), \\
x' = \gamma(x - ct), \\
y' = y, \\
z' = z,
\end{cases}
\qquad
\gamma = \left(1 - \dfrac{V^2}{c^2} \right)^{-1/2}. \qquad (1.10)
$$

These transformations were quickly adopted by the physics community and were to play a key role in unravelling the physics of the spectra of atoms and molecules over the coming decades. Notice that the reason for the similarity of Voigt's and Lorentz's transformations is the scale invariance of the wave equation

$$
\left(\nabla^2 - \frac{1}{c^2} \frac{\partial^2}{\partial t^2} \right) \phi = 0, \qquad (1.11)
$$

under scaling relations $dx \to \kappa\, dx$, $dy \to \kappa\, dy$, $dz \to \kappa\, dz$ and $dt \to \kappa\, dt$.

1.6 The origin of spectral lines

The first decades of the nineteenth century marked the beginnings of quantitative experimental spectroscopy. The breakthrough resulted from the pioneering experiments and theoretical understanding of the laws of interference and diffraction of waves by Thomas Young. In his Bakerian Lecture of 1801 to the Royal Society of London, *On the theory of light and colours*, he used the wave theory of light of Christiaan Huygens to account for the results of interference experiments, such as his famous double-slit experiment (Young, 1802). Among the most striking achievements of this paper was the measurement of the wavelengths of light of different colours using a diffraction grating with 500 grooves per inch. From this time onwards, wavelengths were used to characterise the colours in the spectrum.

In 1802, William Wollaston made spectroscopic observations of sunlight and discovered five strong dark lines, as well as two fainter lines in the spectrum (Wollaston, 1802). The full significance of these observations only became apparent following the remarkable experiments of Joseph Fraunhofer. Fraunhofer's motivation for studying the solar spectrum was his realisation that accurate measurements of the refractive indices of glasses should be made using monochromatic light. In his spectroscopic observations of the Sun, he rediscovered the narrow dark lines which would provide precisely defined wavelength standards. In his words,

> 'I wanted to find out whether in the colour-image (that is, spectrum) of sunlight, a similar bright stripe was to be seen, as in the colour-image of lamplight. But instead of this, I found with the telescope almost countless strong and weak vertical lines, which however are darker than the remaining part of the colour-image; some seem to be completely black.'

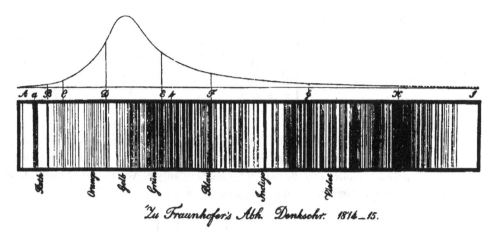

Fig. 1.6 Fraunhofer's solar spectrum of 1814 showing the vast numbers of dark absorption lines. The colours of the various regions of the spectrum are labelled, as well as the letters A, a, B, C, D, E, b, F, G and H, indicating the most prominent absorption lines. The continuous line above the spectrum shows the approximate solar continuum intensity, as estimated by Fraunhofer (Fraunhofer, 1817a,b).

He labelled the 10 strongest lines in the solar spectrum A, a, B, C, D, E, b, F, G and H and recorded 574 fainter lines between the B and H lines (Fig. 1.6) (Fraunhofer, 1817a,b), the notation still used today. From the technical point of view, a major advance was the invention of the *spectroscope* with which the deflection of light passing through the prism could be measured precisely. To achieve this, he placed a theodolite on its side and observed the spectrum through a telescope mounted on the rotating ring.

The understanding of the dark lines in the solar spectrum had to await developments in laboratory spectroscopy. In his papers of 1817, Fraunhofer noted that the dark D lines coincided with the bright double line seen in lamplight. In 1849, Léon Foucault performed a key experiment in which sunlight was passed through a sodium arc so that the two spectra could be compared precisely. To his surprise, the solar spectrum displayed even darker D lines when passed though the arc than without the arc present (Foucault, 1849). He followed up this observation with an experiment in which the continuum spectrum of light from glowing charcoal was passed through the arc and the dark D lines of sodium were found to be imprinted on the transmitted spectrum.

Ten years later, the experiment was repeated by Gustav Kirchhoff who made the further crucial observation that, to observe an absorption feature, the source of the light had to be hotter than the absorbing flame. These results were immediately followed up in 1859 by his understanding of the relation between the emissive and absorptive properties of any substance, what is now known as Kirchhoff's *law of emission and absorption of radiation* (Kirchhoff, 1859).[15] This states that, in thermal equilibrium, the radiant energy emitted by a body at any frequency is precisely equal to the radiant energy absorbed at the same wavelength. Specifically, for isotropic radiation, the *monochromatic emission coefficient* j_ν is defined such that the increment in intensity dI_ν radiated into the solid angle $d\Omega$ from the cylindrical volume dV of area dA and length dl is

$$dI_\nu \, dA \, d\Omega = j_\nu \, dV \, d\Omega, \tag{1.12}$$

where j_ν has units $W\,m^{-3}\,Hz^{-1}\,sr^{-1}$. Since the emission is assumed to be isotropic, the *volume emissivity* of the medium is $\varepsilon_\nu = 4\pi j_\nu$. We can write the volume of the cylinder $dV = dA\,dl$ and so

$$dI_\nu = j_\nu\,dl\,. \qquad (1.13)$$

The *monochromatic absorption coefficient* α_ν is defined by the relation

$$dI_\nu\,dA\,d\Omega = -\alpha_\nu\,I_\nu\,dA\,d\Omega\,dl\,. \qquad (1.14)$$

Kirchhoff showed that in thermodynamic equilibrium

$$\alpha_\nu B_\nu(T) = j_\nu\,. \qquad (1.15)$$

In other words, the emission and absorption coefficients for any physical process are related by the unknown spectrum of equilibrium radiation $B_\nu(T)$. This expression enabled Kirchhoff to understand the relation between the emission and absorption properties of flames, arcs, sparks and the solar atmosphere. In 1859, very little was known about the form of $B_\nu(T)$. As Kirchhoff remarked,

'It is a highly important task to find this function.'

This was one of the great experimental challenges for the remaining decades of the nineteenth century. Kirchhoff's profound insight was the beginning of a long and tortuous story which was to lead to Planck's discovery of the formula for black-body radiation over 40 years later.

Throughout the 1850s, there was considerable effort in Europe and in the USA aimed at identifying the emission and absorption lines produced by different substances in flame, spark and arc spectra. Different elements and compounds possessed distinctive patterns of spectral lines and attempts were made to relate these to the lines observed in the solar spectrum. In 1859, for example, Julius Plücker identified the Fraunhofer F line with the bright Hβ line of hydrogen and the C line was more or less coincident with Hα, demonstrating the presence of hydrogen in the solar atmosphere.

The most important work resulted from the studies of Robert Bunsen and Kirchhoff. In Kirchhoff's great papers of 1861–1863 entitled *Investigations of the solar spectrum and the spectra of the chemical elements*, the solar spectrum was compared with the spark spectra of 30 elements using a four-prism arrangement with which it was possible to view both the spectrum of the element and the solar spectrum simultaneously (Kirchhoff, 1861, 1862, 1863). Kirchhoff concluded that the cool, outer regions of the solar atmosphere contained iron, calcium, magnesium, sodium, nickel and chromium and probably cobalt, barium, copper and zinc as well.

While the patterns of spectral lines provided the fingerprints for the presence of different elements in stars, a major challenge was to discover formulae which could describe the wavelengths of the lines in the spectra of different elements. The first and most important success was the spectrum of hydrogen. In his remarkable papers of 1885, the Swiss schoolmaster Johann Jakob Balmer used laboratory and astronomical spectra to describe the wavelengths λ of the lines in the spectrum of hydrogen by the expression:

$$\lambda = \lambda_0 \, \frac{m^2}{m^2 - 4}, \tag{1.16}$$

where $m = 3, 4, 5, \ldots$ and $\lambda_0 = 3645$ Å. The constant λ_0 corresponds to the wavelength limit of the Balmer series. In terms of frequencies ν or wavenumbers $n = \lambda^{-1}$, Balmer's formula can be written

$$\nu = \nu_0 \left(1 - \frac{2^2}{m^2} \right) \quad \text{or} \quad \frac{1}{\lambda} = n = R_\infty \left(\frac{1}{2^2} - \frac{1}{m^2} \right), \tag{1.17}$$

where R_∞ is known as the *Rydberg constant*. With the aid of subsequent astronomical observations, Balmer was able to test his formula up to $m = 16$ and found precise agreement with (1.16) (Balmer, 1885).[16] This was the first quantum mechanical formula to be discovered. Balmer died 15 years before the deep significance of his numerological discovery was appreciated by Niels Bohr. A similar regularity appeared in Pickering's observation of 1896 of the star ζ Puppis. He discovered a sequence of absorption lines resembling the Balmer series, which became known as the *Pickering series* (Pickering, 1896). He showed that the lines could be described by Balmer's formula provided half-integral values of the principal quantum number m were used.

Of the numerous formulae proposed to describe the lines in the spectra of different elements, two are of special significance. In 1889, Johannes Rydberg presented a memoir to the Royal Swedish Academy in which he proposed that the spectral lines of all series in atomic spectra could be described by the formula

$$n = n_0 - \frac{R_\infty}{(m + \mu)^2}, \tag{1.18}$$

where m is a positive integer and R_∞, the same *Rydberg constant* introduced in (1.17), was to be a constant for all series. The empirical constant μ is known as the *quantum defect*. The spectroscopists had found various regularities in the spectra of the elements, the strongest lines forming the *principal series*, while other regularities were found for broader lines which were known as the *diffuse series* and a *sharp series* of lines which were so named. For each series of each element, n_0 and μ were estimated from the experimentally measured wavelengths, the quantum defect μ lying in the range $0 < \mu < 1$. If $\mu = 0$, the formula reduces to Balmer's formula. The formula was applied successfully to the principal, diffuse and sharp series of the lines in the spectra of sodium, potassium, magnesium, calcium and zinc. This formula was later generalised to the following forms for the different series:

$$\text{Principal series} \quad n_p = R_\infty \left[(1 + s)^{-2} - (m + p)^{-2} \right], \tag{1.19}$$

$$\text{Diffuse series} \quad n_d = R_\infty \left[(2 + p)^{-2} - (m + d)^{-2} \right], \tag{1.20}$$

$$\text{Sharp series} \quad n_s = R_\infty \left[(2 + p)^{-2} - (m + s)^{-2} \right], \tag{1.21}$$

where $m = 3, 4, 5, \ldots$ and the constants s, p and d are chosen to fit the observed spectra for each series.

In 1908, these formulae were generalised further by Walther Ritz (1908) who formulated his *combination principle* according to which every spectral line could be expressed as the difference of two so-called *spectral terms*, each of which depended upon an integer m as well as the constants appearing in (1.19)–(1.21). Thus, Ritz wrote:

$$\nu = (2, p, \pi) - (m, s, \sigma), \tag{1.22}$$

where the terms in brackets are generalisations of the terms in (1.18). The inference is that the spectral lines of any element include frequencies that are either the sum or the difference of the frequencies of two other lines. Thus, if the individual terms in the Ritz formula are A, B and C and the observed lines are $(A - B)$ and $(B - C)$, by adding the frequencies we find $(A - B) + (B - C) = (A - C)$ and by subtracting them $(A - B) - (A - C) = (C - B)$. This proved to be a powerful tool for disentangling complex spectra of elements into their different constituent series. But, even more importantly, Ritz's combination principle was to prove to lie at the very heart of the revolution which was to lead to the reformulation of the laws of physics at the atomic level (Sect. 11.3).

1.7 The spectrum of black-body radiation

Kirchhoff's exhortation about the central importance of determining the equilibrium distribution of radiation $B(T)$ stimulated theoretical and experimental physicists to take up the challenge. The function $B(T)$ is known as the black-body spectrum because it is the equilibrium spectrum of a perfect radiator and absorber in thermal equilibrium. The next task is to understand the preliminaries which were to lead to Planck's epochal discovery in 1900. The first steps involved determining the total amount of radiation emitted by a black-body, the Stefan–Boltzmann law.

1.7.1 The derivation of the Stefan–Boltzmann law

Josef Stefan deduced the law which bears his name empirically from experimental data published by Tyndall in 1865 on the radiation of a platinum strip heated electrically to different temperatures. In 1879, he published a paper in which he stated:

> 'From weak red heat (about 525 C) to complete white heat (about 1200 C) the intensity of radiation increases from 10.4 to 122, thus nearly twelvefold (more precisely 11.7). This observation caused me to take the heat radiation as proportional to the fourth power of the absolute temperature. The ratio of the absolute temperature $273 + 1200$ and $273 + 525$ raised to the fourth power gives 11.6.' (Stefan, 1879)

This relation refers to the total radiation emitted by the heated body,

$$-\left(\frac{\mathrm{d}E}{\mathrm{d}t}\right) = \text{total radiant energy per second} \propto T^4. \tag{1.23}$$

In 1884, Boltzmann, Stefan's pupil, deduced this law theoretically from considerations of classical thermodynamics (Boltzmann, 1884). Suppose a closed volume is filled only with electromagnetic radiation and that it contains a piston so that the 'gas' of radiation can be compressed or expanded. If the heat dQ is added to the system, the total internal energy is increased by dU and work is done on the piston as it is pushed outwards slightly so that the volume increases by dV. By conservation of energy,

$$dQ = dU + p\, dV. \tag{1.24}$$

Now rearrange this relation by introducing the associated increase in entropy $dS = dQ/T$:

$$T\, dS = dU + p\, dV. \tag{1.25}$$

This relation is converted into a partial differential equation by dividing through by dV at constant T

$$T\left(\frac{\partial S}{\partial V}\right)_T = \left(\frac{\partial U}{\partial V}\right)_T + p. \tag{1.26}$$

We now use the Maxwell's relation[17]

$$\left(\frac{\partial p}{\partial T}\right)_V = \left(\frac{\partial S}{\partial V}\right)_T,$$

to recast this relation as

$$T\left(\frac{\partial p}{\partial T}\right)_V = \left(\frac{\partial U}{\partial V}\right)_T + p. \tag{1.27}$$

Thus, if we know the relation between U and T, that is, the equation of state for the gas, the relation between temperature and energy density can be found. This relation for a 'gas' of electromagnetic radiation was derived by Maxwell in his *Treatise on Electricity and Magnetism*, $p = \frac{1}{3}\varepsilon$, where ε is the energy density of radiation (see Sect. 1.4).

Therefore, $U = \varepsilon V$ and $p = \frac{1}{3}\varepsilon$ and so, from equation (1.11), we find

$$T\left(\frac{\partial(\frac{1}{3}\varepsilon)}{\partial T}\right)_V = \left(\frac{\partial(\varepsilon V)}{\partial V}\right)_T + \frac{1}{3}\varepsilon; \qquad \frac{1}{3}T\left(\frac{\partial \varepsilon}{\partial T}\right)_V = \varepsilon + \frac{1}{3}\varepsilon = \frac{4}{3}\varepsilon.$$

Therefore,

$$\frac{d\varepsilon}{\varepsilon} = 4\frac{dT}{T}; \quad \ln \varepsilon = 4 \ln T; \quad \varepsilon \propto T^4. \tag{1.28}$$

This is Boltzmann's contribution to the Stefan–Boltzmann law. In modern form, the law is written $I = \sigma T^4$, where I is the radiant energy emitted per unit area per second from the surface of a black-body at temperature T. The experimental evidence for the Stefan–Boltzmann law was not particularly convincing in 1884 and it was not until 1897 that Lummer and Pringsheim undertook their very careful experiments which showed that the law was indeed correct with high precision.

1.7.2 Wien's displacement law and the spectrum of black-body radiation

The spectrum of black-body radiation was poorly known until the turn of the century but there had already been some important work done on the theoretical form which the radiation law should have. In 1894, Wilhelm (Willy) Wien published his derivation of his *displacement law* by a combination of electromagnetism and thermodynamics and this was to prove to be of central importance in the development of the theory of black-body radiation (Wien, 1894).

First, consider the adiabatic expansion of a 'gas' of electromagnetic radiation. The flow of heat in or out of the system $dQ = dU + p\,dV$ and in an adiabatic expansion, $dQ = 0$. Further, since $U = \varepsilon V$ and $p = \frac{1}{3}\varepsilon$,

$$d(\varepsilon V) + \tfrac{1}{3}\varepsilon\,dV = 0; \quad V\,d\varepsilon + \varepsilon\,dV + \tfrac{1}{3}\varepsilon\,dV; \quad \frac{d\varepsilon}{\varepsilon} = -\frac{4}{3}\frac{dV}{V}. \tag{1.29}$$

Integrating,

$$\varepsilon = \text{constant} \times V^{-4/3}. \tag{1.30}$$

But $\varepsilon = aT^4$ where $a = 4\sigma/c$ and hence

$$TV^{1/3} = \text{constant}. \tag{1.31}$$

Since V is proportional to the cube of the dimension of a spherical volume, $V \propto r^3$ and so $T \propto r^{-1}$.

The next step is to work out the relation between the wavelength of the radiation and the volume of the enclosure. This is no more than the Doppler shift formula for radiation enclosed within the expanding volume V, $\lambda \propto r$, that is, the wavelength of the radiation increases linearly proportionally to the size of the spherical volume.[18] We now combine this result with the relation $T \propto r^{-1}$ to find

$$T \propto \lambda^{-1}. \tag{1.32}$$

This is one aspect of *Wien's displacement law*. If radiation is adiabatically expanded, the wavelength of radiation changes inversely with temperature if we follow a particular set of waves. In other words, the wavelength of the radiation is 'displaced' as the temperature changes. In particular, if we follow the maximum of the radiation spectrum it should follow the T^{-1} law. This was found to be in agreement with experiment.

Wien now went further and combined the Stefan–Boltzmann law with the law $T \propto \lambda^{-1}$ to set constraints on what the spectral form of the radiation had to be. My own version of the argument is as follows. The first step is to note that if any system of bodies is enclosed in a perfectly reflecting enclosure, eventually they will all come to the same temperature because of the emission and absorption of radiation. At the microscopic level, the system comes into thermal equilibrium, whatever the properties of the walls, as a result of the principle of detailed balance. The radiation comes into equilibrium with the objects in the enclosure so that as much energy is radiated as absorbed by the bodies per unit time and the equilibrium spectrum is a black-body spectrum. The radiation will be isotropic and so the only parameters which can characterise the radiation are the temperature T of the enclosure and the wavelength of the radiation λ.

If the black-body radiation is initially at temperature T_1 and the enclosure is expanded adiabatically, then since an adiabatic expansion takes place infinitely slowly, the radiation takes up an equilibrium spectrum at all stages in the expansion until it reaches temperature T_2. The crucial point is that, in an adiabatic expansion, the radiation spectrum has black-body form at all stages in the expansion from T_1 to T_2. The unknown law for the radiation spectrum must therefore scale appropriately with temperature.

Consider the radiation in the wavelength interval λ_1 to $\lambda_1 + d\lambda_1$, and let its energy density be $\varepsilon = u(\lambda_1)\,d\lambda_1$. Then, in the expansion, according to Boltzmann's analysis, the energy associated with the radiation of any particular set of waves decreases as T^4 and hence

$$\frac{u(\lambda_1)\,d\lambda_1}{u(\lambda_2)\,d\lambda_2} = \left(\frac{T_1}{T_2}\right)^4 . \tag{1.33}$$

But $\lambda_1 T_1 = \lambda_2 T_2$ and hence $d\lambda_1 = (T_2/T_1)\,d\lambda_2$. Therefore,

$$\frac{u(\lambda_1)}{T_1^5} = \frac{u(\lambda_2)}{T_2^5} , \tag{1.34}$$

that is,

$$\frac{u(\lambda)}{T^5} = \text{constant} , \tag{1.35}$$

and, since $\lambda T = \text{constant}$, this can be rewritten as

$$u(\lambda)\lambda^5 = \text{constant} . \tag{1.36}$$

Notice that $u(\lambda)$ is the energy density per unit wavelength interval in the radiation spectrum. Now the only combination of T and λ which is a constant is the product λT and hence we can write that, in general, the constant on the right-hand side of (1.36) can only be constructed out of functions involving λT. Therefore the radiation law must have the form

$$u(\lambda)\lambda^5 = f(\lambda T) , \tag{1.37}$$

or

$$u(\lambda)\,d\lambda = \lambda^{-5} f(\lambda T)\,d\lambda . \tag{1.38}$$

This is *Wien's displacement law* in its entirety and it sets constraints on the form of the radiation law for black-body radiation. In terms of frequency, the relation can be written

$$u(\lambda)\,d\lambda = u(\nu)\,d\nu; \quad \lambda = c/\nu, \quad d\lambda = -\frac{c}{\nu^2}\,d\nu ,$$

and hence

$$u(\nu)\,d\nu = \left(\frac{c}{\nu}\right)^{-5} f\left(\frac{\nu}{T}\right) \left(-\frac{c}{\nu^2}\,d\nu\right) , \tag{1.39}$$

that is,

$$u(\nu)\,d\nu = \nu^3 f\left(\frac{\nu}{T}\right) d\nu . \tag{1.40}$$

This brilliant argument shows how far Wien was able to get using only rather general thermodynamic arguments. This work was new in 1894 when Planck first became interested in the problem of the spectrum of black-body radiation.

1.8 The gathering storm

The triumphs of nineteenth century physics were hard-won and their full implications were still being understood. To the thorny problems described in the sections of this chapter were about to be added a whole new set of discoveries which were indeed to rock the foundations of the subject. Most physicists clung to the hope that the challenges described in this chapter would eventually be resolved within the context of classical physics, but even that hope was to prove illusory. Physics was about to enter a period when the foundations were cut away and it would be 30 years before the theory of quantum mechanics was discovered. Through this period of uncertainty, some of the greatest minds in physics pitted themselves against almost insuperable problems with courage and imagination.

2 Planck and black-body radiation

2.1 The key role of experimental technique

1895 was not an arbitrary choice as the cut-off year for the history recounted in Chap. 1. Over the following years a number of experimental discoveries were made which were to change dramatically the face of physics. The origins of these discoveries can be traced to the need for increased precision in experimental physics. The industrial revolution and the widespread availability of electricity and electrical communication required more exact understanding of the physical properties of materials and also the establishment of international standards. These necessitated a much more professional approach to the teaching of experimental physics and its associated theory. As part of that movement the Clarendon Laboratory was founded in Oxford in 1868 and the Cavendish Laboratory in Cambridge in 1874. Of particular significance for this chapter was the foundation of the Physikalisch-Technische Reichsanstaldt in Berlin in 1887 with the task of providing precise measurements of the physical properties of materials which would be of importance to industry. There was an expectation that these laboratories would develop new techniques for undertaking precision measurements.

New experimental techniques were developed thanks to the development of new technologies. To mention only a few of the more important of these, better vacuums were available to physicists through the invention of the Geissler pump, invented by Johann Heinrich Wilhelm Geissler, a brilliant inventor and glass-blower, in about 1855. The vacuum was produced by trapping air in a mercury column and forcing it down the column by the force of gravity. Typically pressures of 0.1 mm of mercury could be obtained within his vacuum tubes. Higher voltages could be placed across discharge tubes with the invention of the Rühmkorff coil which consisted of a transformer with a small number of windings on the primary and a very large number on the secondary. High precision spectroscopy was revolutionised by the superb diffraction gratings produced by Henry Rowland, who pioneered the construction of high precision ruling engines. Samuel Pierpoint Langley perfected the techniques of infrared spectrometry through the invention of the platinum bolometer with which temperature changes as small as 10^{-4} K could be measured. Although Langley's interests were primarily astronomical, the techniques were used to great advantage in the determination of the spectrum of black-body radiation in the final decade of the nineteenth century. The combination of these and many other advances in technology and experimental technique were about to produce unexpected fruits.

2.2 1895–1900: The changing landscape of experimental physics[1]

The vacuum tube was to play a central role in the discoveries of the 1890s. The basic apparatus consisted of a thin-walled glass vacuum tube with positive and negative electrodes between which a high voltage could be maintained. It was realised in the early nineteenth century that gases are poor conductors of electricity since they are electrically neutral. At very low pressures and high voltages, however, electrical discharges were observed in vacuum tubes. Geissler tubes became popular scientific toys and revealed a remarkable range of coloured discharges. William Crookes began his systematic studies of the conduction of electricity through gases in 1879. The appearance of the discharge in the vacuum tube changed as the pressure of the gas decreased. The colourful displays disappeared as the pressure decreased but a current continued to flow. At low enough pressures, the walls of the tube glowed green due to phosphorescence. It was found that objects placed in the tube cast a shadow on the walls of the tube opposite the cathode. It was inferred that a stream of *cathode rays* must be dragged off the cathode causing the walls of the tube to glow. By 1895, it was established that the cathode rays were negatively charged particles which could be deflected by magnetic fields. In 1895 Jean-Baptiste Perrin collected the cathode rays inside a vacuum tube and found directly that they had negative electric charge.

2.2.1 The discovery of X-rays

In 1895, Wilhelm Conrad Röntgen discovered, by accident, that wrapped unexposed photographic plates left close to discharge tubes were darkened. In addition, if a discharge tube were completely surrounded by thin black cardboard, fluorescent materials left close to it glowed in the dark. Röntgen came to the correct conclusion that both phenomena were associated with some new form of radiation emitted by the discharge tube, and he named these X-rays (Röntgen, 1895). His startling X-ray image of his wife's hand, showing the bones of the hand and her massive ring, had an immediate impact (Fig. 2.1*a*). Overnight, X-rays became a matter of the greatest public interest and were very rapidly incorporated into the armoury of the doctor's surgery. In 1896, there were already more than 1000 articles on X-rays. The X-rays were more penetrating than cathode rays, since they could blacken photographic plates at a considerable distance from the hot spot on the discharge tube, which was known to be their source. Their identification with 'ultra-ultraviolet' radiation was only convincingly demonstrated in 1906, when Charles Barkla found that the X-radiation was polarised (Barkla, 1906) and in 1912, when Max von Laue had the inspiration of looking for their diffraction by crystals (Friedrich *et al.*, 1912; Laue, 1912), in the process opening up the new field of X-ray crystallography. Measurements of the interaction of X-rays with matter was to prove to be a central tool in disentangling the structure of atoms.

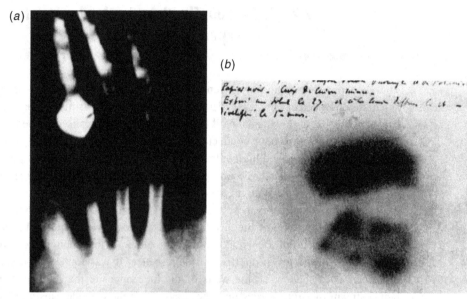

Fig. 2.1   (*a*) The first X-ray picture of Röntgen's wife's hand, illustrating the ability of X-rays to penetrate through soft tissues and reveal the bone structure. (*b*) Becquerel's developed plate showing a strong image, despite the fact that the radioactive salt had not been exposed to sunlight.

2.2.2 The discovery of radioactivity

The association of X-rays with fluorescent materials led to the search for other sources of X-radiation. In 1896, Henri Becquerel, who came from a distinguished family of French physicists, tested several known fluorescent substances before investigating samples of potassium uranyl disulphate. The photographic plates were wrapped in several sheets of black paper, the phosphorescent material was exposed to sunlight and then the plate developed to find if it had been darkened by X-rays. Becquerel's remarkable discovery was that the plates were darkened even when the phosphorescent material was not exposed to light (Fig. 2.1*b*). This was the discovery of natural radioactivity (Becquerel, 1896). In further experiments carried out in the same year, Becquerel showed that the amount of radioactivity was proportional to the amount of uranium in the substance and that the radioactive flux of radiation was constant in time. Another important discovery was that the radiation from the uranium compounds discharged electroscopes. Pierre Curie and Marie Skłodowska-Curie repeated Becquerel's experiments in 1897 and showed that the intensity of radioactivity was proportional to the amount of uranium in different samples. Other radioactive substances were soon identified. Thorium was discovered in 1898 (Schmidt, 1898). By concentrating uranium residues, the Curies discovered the new element polonium, named after Marie's native land. The new element decreased in radioactivity exponentially – radioactive substances have a certain half-life. In September 1898, very strong radioactivity was found in the barium group of residues. Sufficient of the radioactive substance was isolated to show that a new element had been discovered, radium (Curie and Skłodowska-Curie, 1898; Curie *et al.*, 1898).

As soon as the discovery of radioactivity was announced, Ernest Rutherford took up the study of radioactivity. In his first publication on the subject, Rutherford established that there are at least two separate types of radiation emitted by radioactive substances (Rutherford, 1899). He called the component which is most easily absorbed α-radiation (or α-rays) and the much more penetrating component β-radiation (or β-rays). It took another 10 years before Rutherford conclusively demonstrated that the α-radiation consisted of the nuclei of helium atoms (Rutherford and Royds, 1909). In contrast, the β-rays were convincingly shown by Walter Kaufmann to have the same mass-to-charge ratio as the recently discovered electron (Kaufmann, 1902). γ-radiation was discovered in 1900 by Paul Villard as an extremely penetrating form of radiation emitted in radioactive decays – the γ-rays were undeflected by a magnetic field (Villard, 1900a,b). The γ-rays were conclusively identified as electromagnetic waves 14 years later when Rutherford and Edward Andrade observed the reflection of γ-rays from crystal surfaces (Rutherford and Andrade, 1913).

The α-, β- and γ-rays were the only known radiations which could cause the ionisation of air. The characteristic property which distinguished them was their penetrating power. In quantitative terms, these were:

- The α-particles ejected in radioactive decays produce a dense stream of ions and are stopped in air within about 0.05 m. This is called the *range* of the particles.
- The β-particles have greater ranges, but there is not a well-defined value for any particular radioactive decay.
- The γ-rays were found to have by far the longest ranges, a few centimetres of lead being necessary to reduce their intensity by a factor of 10.

2.2.3 The discovery of the electron

With Röntgen's announcement of the discovery of X-rays, John Joseph (JJ) Thomson immediately directed his research efforts to the study of electrical discharges in gases. Experimenting with cathode rays was difficult in Thomson's time because a very good vacuum was required. Otherwise, the cathode rays collided with air molecules, ionising the gas and making it a conductor which shielded the rays. Thomson recognised the necessity of extremely low pressures. He and his assistant Ebeneezar Everett, who built all his apparatus and carried out the experiments, took great pains to secure excellent vacua and to eliminate the effects of outgassing from the materials of the tube (Fig. 2.2).

The discovery of the electron in 1897 is traditionally attributed to Thomson on the basis of his famous series of experiments, in which he established that the charge-to-mass ratio, e/m_e, of cathode rays is about two-thousand times that of hydrogen ions. At the same time, several physicists were hot on the trail.

- In 1896, Pieter Zeeman discovered the broadening of spectral lines when a sodium flame is placed between the poles of a strong electromagnet. Lorentz interpreted this result in terms of the splitting of the spectral lines due to the motion of the 'ions' in the atoms about the magnetic field direction – he found a lower limit of 1000 for the value of e/m_e of the 'ions'.

Fig. 2.2   The vacuum tube with which J. J. Thomson measured the charge-to-mass ratio of cathode rays.

- In January 1897, Emil Wiechert used the magnetic deflection technique to obtain a measurement of e/m_e for cathode rays and concluded that these particles had mass between 2000 and 4000 times smaller than that of hydrogen, assuming their electric charge was the same as that of hydrogen ions. He obtained only an upper limit to the speed of the particles since it was assumed that the kinetic energy of the cathode rays was $E_{kin} = eV$, where V is the accelerating voltage of the discharge tube.
- Walter Kaufmann's experiment was similar to Thomson's. He found the same values of e/m_e, no matter which gas filled the discharge tube, a result which puzzled him. He found a value of e/m_e 1000 times greater than that of hydrogen ions. He concluded

> '... that the hypothesis of cathode rays as emitted particles is by itself inadequate for a satisfactory explanation of the regularities I have observed.'

Thomson was the first of these pioneers to interpret the experiments in terms of a sub-atomic particle. In his words, cathode rays constituted

> '... a new state, in which the subdivision of matter is carried very much further than in the ordinary gaseous state.' (Thomson, 1897)

Thomson further showed that the particles ejected in the photoelectric effect, discovered by Hertz in 1887, were also identical with electrons. In 1898, Thomson modified one of C. T. R. Wilson's early cloud chambers to measure the charge of the electron. He counted the total number of droplets formed and their total charge. From these, he estimated $e = 2.2 \times 10^{-19}$ C, compared with the present standard value of 1.602×10^{-19} C. This experiment was the precursor of the famous Millikan oil drop experiment, in which the water-vapour droplets were replaced by fine drops of a heavy oil, which did not evaporate during the course of the experiment. Thus, Thomson pursued a more sustained and detailed campaign than the other physicists in establishing the universality of what became known as *electrons*, the name coined for cathode rays by Johnstone Stoney in 1891 (Stoney, 1891).

2.3 Planck and the spectrum of black-body radiation

The discoveries in experimental physics summarised in the last section form the background to Planck's research into the nature of black-body radiation – these were the cutting-edge areas of research which had uncovered a whole new range of phenomena which required theoretical explanation. Up to 1894, Planck's researches had been principally into elucidating the nature of classical thermodynamics and the law of increase of entropy. Following the death of Kirchhoff in 1889, he succeeded to Kirchhoff's chair at the University of Berlin. This was a propitious advance since Berlin was one of the most active centres of physics research in the world.

In 1894, Planck turned his attention to the problem of the spectrum of black-body radiation. It is likely that his interest in this problem was stimulated by Wien's important paper of that year, which was discussed in Sect. 1.7.2 (Wien, 1894). Wien's analysis has a strong thermodynamic flavour, which must have appealed to Planck.

In 1895, Planck published the first results of his work on the resonant scattering of plane electromagnetic waves by an oscillating dipole. This was the first of Planck's papers in which he diverged from his previous areas of interest in that it appeared to be about electromagnetism rather than entropy. In the last words of the paper, however, Planck made it clear that he regarded this as a first step towards tackling the problem of the spectrum of black-body radiation. His aim was to set up a system of oscillators in an enclosed cavity which would radiate and interact with the radiation produced so that after a long time the system would come into equilibrium. He could then apply the laws of thermodynamics to black-body radiation with a view to understanding the origin of its spectrum. He explained why this approach offered the prospect of providing insights into basic thermodynamic processes. When energy is lost by an oscillator by radiation, it is not lost as heat, but as electromagnetic waves. The process can be considered 'conservative' because, if the radiation is enclosed in a box with perfectly reflecting walls, it can then react back on the oscillator.[2] Furthermore, the process is independent of the nature of the oscillator. In Planck's words,

> 'The study of conservative damping seems to me to be of great importance, since it opens up the prospect of a possible general explanation of irreversible processes by means of conservative forces – a problem that confronts research in theoretical physics more urgently every day.'[3]

Planck believed that, by studying classically the interaction of oscillators with electromagnetic radiation, he would be able to show that entropy increases absolutely for a system consisting of matter and radiation. These ideas, which were elaborated in a series of five papers, did not work, as was pointed out by Boltzmann. One cannot obtain a monotonic approach to equilibrium without some statistical assumptions about the way in which the system approaches the equilibrium state. This can be understood from Maxwell's simple but compelling argument concerning the time reversibility of the laws of mechanics, dynamics and electromagnetism.

Finally, Planck conceded that statistical assumptions were necessary and introduced the concept of 'natural radiation', which corresponded to Boltzmann's assumption of 'molecular chaos'. Once the assumption was made that there exists a state of 'natural radiation', Planck was able to make considerable progress with his programme. The first thing he did was to relate the energy density of radiation in an enclosure to the average energy of the oscillators within it. This is a very important result and Planck derived it entirely from classical arguments concerning the emission and absorption of radiation by a set of oscillators.

Why did Planck treat oscillators rather than, say, atoms, molecules, lumps of rock, and so on? The reason is that, in thermal equilibrium, everything is in equilibrium with everything else – rocks are in equilibrium with atoms and oscillators and therefore there is no advantage in treating complicated objects. The advantage of considering simple harmonic oscillators is that the radiation and absorption laws can be calculated exactly. To express this important point in another way, Kirchhoff's law (1.15), which relates the emission and absorption coefficients, j_ν and α_ν respectively,

$$\alpha_\nu B_\nu(T) = j_\nu \, ,$$

shows that, if we can determine these coefficients for any process, we can find the universal equilibrium spectrum $B_\nu(T)$.

I have given a detailed derivation of Planck's relation between the mean energy of the oscillator and the spectrum of radiation in thermal equilibrium with it in Chap. 12 of *TCP2*. Here I summarise some of the key results and aspects of the analysis which will prove important in later parts of the story.

2.3.1 The rate of radiation of an oscillator

Charged particles emit electromagnetic radiation when they are accelerated. The total rate of loss of energy by an accelerated electron is

$$-\left(\frac{\mathrm{d}E}{\mathrm{d}t}\right)_{\mathrm{rad}} = \frac{|\ddot{\boldsymbol{p}}|^2}{6\pi\epsilon_0 c^3} = \frac{e^2|\ddot{\boldsymbol{r}}|^2}{6\pi\epsilon_0 c^3} \, , \tag{2.1}$$

where $\boldsymbol{p} = e\,\boldsymbol{r}$ is the dipole moment with respect to some origin and e the electric charge of the electron – note that the radiation depends only upon the instantaneous acceleration $\ddot{\boldsymbol{r}}$ of the electron. This result is often referred to as *Larmor's formula*. The three essential properties of the radiation of an accelerated charged particle are:

(a) The total radiation rate is given by Larmor's formula (2.1) in which the acceleration is the *proper acceleration* of the charged particle and the radiation loss rate is measured in the *instantaneous rest frame* of the particle.

(b) The *polar diagram* of the radiation is of *dipolar* form, that is, the electric field strength varies as $\sin\theta$ and the power radiated per unit solid angle varies as $\sin^2\theta$, where θ is the angle with respect to the acceleration vector of the particle. There is no radiation along the direction of the acceleration vector and the field strength is greatest at right angles to it.

(c) The radiation is *polarised* with the electric field vector lying in the direction of the acceleration vector of the particle, as projected onto a sphere at distance r.

We now apply the result (2.1) to the case of an oscillator performing simple harmonic oscillations with amplitude x_0 at angular frequency ω_0, $x = |x_0| \exp(i\omega t)$. Therefore, $\ddot{x} = -\omega_0^2 |x_0| \exp(i\omega t)$, or, taking the real part, $\ddot{x} = -\omega_0^2 |x_0| \cos \omega t$. The instantaneous rate of loss of energy by radiation is therefore

$$-\left(\frac{dE}{dt}\right) = \frac{\omega_0^4 e^2 |x_0|^2}{6\pi \epsilon_0 c^3} \cos^2 \omega_0 t. \tag{2.2}$$

The average value of $\cos^2 \omega t$ is $\frac{1}{2}$ and hence the average rate of loss of energy in the form of electromagnetic radiation is

$$-\left(\frac{dE}{dt}\right)_{\text{average}} = \frac{\omega_0^4 e^2 |x_0|^2}{12\pi \epsilon_0 c^3}. \tag{2.3}$$

As shown below, the energy E of the oscillator is $E = \frac{1}{2} m |x_0|^2 \omega_0^2$ and so

$$-\frac{dE}{dt} = \gamma E, \tag{2.4}$$

where $\gamma = \omega_0^2 e^2 / 6\pi \epsilon_0 c^3 m$. This expression can be rewritten in terms of the *classical electron radius* $r_e = e^2 / 4\pi \epsilon_0 m_e c^2$. Thus $\gamma = 2r_e \omega_0^2 / 3c$, where m_e is the mass of the electron.

2.3.2 Radiation damping of an oscillator

The equation of damped simple harmonic motion can be written in the form

$$m\ddot{x} + a\dot{x} + kx = 0,$$

where m is the (reduced) mass, k the spring constant and $a\dot{x}$ the damping force. The energies associated with each of these terms is found by multiplying through by $\dot{x}$ and integrating with respect to time.

$$\frac{1}{2}m \int_0^t d(\dot{x}^2) + \int_0^t a\dot{x}^2 \, dt + \frac{1}{2}k \int_0^t d(x^2) = 0. \tag{2.5}$$

We identify the terms in (2.5) with the kinetic energy, the damping energy loss and the potential energy of the oscillator respectively. For simple harmonic motion of the form $x = |x_0| \cos \omega t$, the average kinetic and potential energies are:

average kinetic energy $= \frac{1}{4} m |x_0|^2 \omega_0^2$;   average potential energy $= \frac{1}{4} k |x_0|^2$.

If the damping is very small, the natural frequency of oscillation of the oscillator is $\omega_0^2 = k/m$. Thus, the average kinetic and potential energies of the oscillator are equal and the total energy is the sum of the two,

$$E = \frac{1}{2} m |x_0|^2 \omega_0^2. \tag{2.6}$$

Inspection of the second term on the left-hand side of (2.5) shows that the average rate of loss of energy by radiation from the oscillator is

$$-\left(\frac{dE}{dt}\right)_{rad} = \tfrac{1}{2}a|x_0|^2\omega_0^2 = \frac{a}{m}E . \tag{2.7}$$

This relation can now be compared with the expression for the loss rate by radiation of the oscillator (2.4). We obtain the correct expression for the decay in amplitude of the oscillator by identifying γ with a/m in (2.7). This phenomenon is known as the *radiation damping of an oscillator*.

We can appreciate now why Planck believed this was a fruitful way of attacking the problem. The radiation loss does not go into heat, but into electromagnetic radiation and the constant γ depends only upon fundamental constants, if we take the oscillator to be an oscillating electron. In contrast, in frictional damping, the energy goes into heat and the loss rate formula contains constants appropriate to the material. Furthermore, in the case of electromagnetic waves, if the oscillators and waves are confined within an enclosure with perfectly reflecting walls, energy is not lost from the system and the waves can react back on the oscillators returning energy to them.[4] This is why Planck called the damping *conservative damping*.

2.3.3 The equilibrium radiation spectrum of a harmonic oscillator

The expressions for the dynamics of an oscillator undergoing radiation, or natural, damping which have just been derived are:

$$m\ddot{x} + a\dot{x} + kx = 0; \quad \ddot{x} + \gamma\dot{x} + \omega_0^2 x = 0 . \tag{2.8}$$

If an electromagnetic wave is incident on the oscillator, energy can be transferred to it and a forcing term is added to (2.8),

$$\ddot{x} + \gamma\dot{x} + \omega_0^2 x = \frac{F}{m} . \tag{2.9}$$

If the oscillator is accelerated by the E field of an incident wave, $F = eE_0 \exp(i\omega t)$. To find the response of the oscillator, a trial solution for x of the form $x = |x_0| \exp(i\omega t)$ is adopted. Then

$$|x_0| = \frac{eE_0}{m\left(\omega_0^2 - \omega^2 + i\gamma\omega\right)} . \tag{2.10}$$

Notice that there is a complex factor in the denominator and this means that the oscillator does not vibrate in phase with the incident wave. This does not matter for our calculation. We are only interested in the square of the modulus of the amplitude.

Now, let us work out the rate of radiation of the oscillator under the influence of the incident radiation field. If we set this equal to the 'natural' radiation of the oscillator, we will have found the equilibrium spectrum – the work done by the incident radiation field is just enough to supply the energy loss per second by the oscillator. From (2.2), the rate of radiation of the oscillator is

$$-\left(\frac{dE}{dt}\right) = \frac{\omega^4 e^2 |x_0|^2}{6\pi \epsilon_0 c^3} \cos^2 \omega t,$$

where we now use the value of $|x_0|$ derived in (2.10). We need the square of the modulus of x_0 and this is found by multiplying x_0 by its complex conjugate, that is,

$$|x_0|^2 = \frac{e^2 E_0^2}{m^2 \left[\left(\omega_0^2 - \omega^2\right)^2 + \gamma^2 \omega^2\right]}.$$

Therefore, the radiation rate is

$$-\left(\frac{dE}{dt}\right) = \frac{\omega^4 e^4 E_0^2}{12\pi \epsilon_0 c^3 m^2 \left[\left(\omega_0^2 - \omega^2\right)^2 + \gamma^2 \omega^2\right]}, \tag{2.11}$$

where we have taken averages over time, that is, $\langle \cos^2 \omega t \rangle = \frac{1}{2}$.

We next replace the value of E_0^2 in our formula by the sum over all the waves of angular frequency ω incident upon the oscillator to find the total *average* reradiated power,

$$-\left(\frac{dE}{dt}\right) = \frac{\omega^4 e^4 \frac{1}{2} \sum_i E_{0i}^2}{6\pi \epsilon_0 c^3 m^2 \left[\left(\omega_0^2 - \omega^2\right)^2 + \gamma^2 \omega^2\right]}. \tag{2.12}$$

The next step is to note that the loss rate (2.12) is part of a continuum intensity distribution and so this sum can be written as an incident intensity[5] in the frequency band ω to $\omega + d\omega$, that is,

$$I(\omega)\, d\omega = \frac{1}{2}\epsilon_0 c \sum_i E_{0i}^2. \tag{2.13}$$

Therefore, the total average radiation loss rate is

$$-\left(\frac{dE}{dt}\right) = \frac{\omega^4 e^4}{6\pi \epsilon_0^2 c^4 m^2} \frac{I(\omega)\, d\omega}{\left[\left(\omega_0^2 - \omega^2\right)^2 + \gamma^2 \omega^2\right]} = \frac{8\pi r_e^2}{3} \frac{\omega^4 I(\omega)\, d\omega}{\left[\left(\omega_0^2 - \omega^2\right)^2 + \gamma^2 \omega^2\right]}, \tag{2.14}$$

where we have introduced the classical electron radius, $r_e = e^2/4\pi \epsilon_0 m_e c^2$.

Now the response curve of the oscillator is very sharply peaked about the value ω_0, because the radiation rate is very small in comparison with the total energy of the oscillator, that is, $\gamma \ll 1$ (see Fig. 2.3). We can therefore make some simplifying approximations. If ω appears on its own, $\omega \to \omega_0$ and $(\omega_0^2 - \omega^2) = (\omega_0 + \omega)(\omega_0 - \omega) \approx 2\omega_0(\omega_0 - \omega)$. Therefore,

$$-\left(\frac{dE}{dt}\right) = \frac{2\pi r_e^2}{3} \frac{\omega_0^2 I(\omega)\, d\omega}{\left[(\omega - \omega_0)^2 + (\gamma^2/4)\right]}. \tag{2.15}$$

Finally, we expect $I(\omega)$ to be a slowly varying function in comparison with the sharpness of the response curve of the oscillator and so we can set $I(\omega)$ equal to a constant over the range of values of ω of interest,

$$-\left(\frac{dE}{dt}\right) = \frac{2\pi r_e^2}{3} I(\omega_0) \int_0^\infty \frac{\omega_0^2\, d\omega}{\left[(\omega - \omega_0)^2 + (\gamma^2/4)\right]}. \tag{2.16}$$

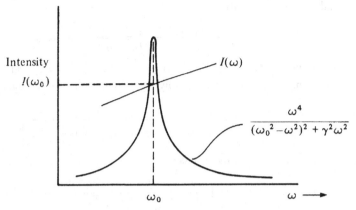

Fig. 2.3 Illustrating the response curve of the oscillator to waves of different frequencies.

The integral is easy if we set the lower limit equal to minus infinity which is perfectly permissible since the function tends very rapidly to zero at frequencies away from that of the peak intensity. Using

$$\int_{-\infty}^{\infty} \frac{dx}{x^2 + a^2} = \frac{\pi}{a},$$

the integral (2.16) becomes $2\pi/\gamma$ and so

$$-\left(\frac{dE}{dt}\right) = \frac{2\pi \omega_0^2 r_e^2}{3} \frac{2\pi}{\gamma} I(\omega_0) = \frac{4\pi^2 \omega_0^2 r_e^2}{3\gamma} I(\omega_0). \qquad (2.17)$$

The rate of radiation should now be set equal to the spontaneous radiation rate of the oscillator. There is only one complication. We have assumed that the oscillator can respond to incident radiation arriving from any direction. For a single oscillator with axis in, say, the x-direction, there are directions of incidence in which the oscillator does not respond to the incident wave – this occurs if a component of the incident electric field is perpendicular to the dipole axis of the oscillator. We get round this problem by the argument used by Richard Feynman. Suppose three oscillators are mutually at right angles to each other. Then this system can respond like a completely free oscillator and follow any incident electric field. Therefore, we obtain the correct result if we suppose that (2.17) is the radiation which would be emitted by *three mutually perpendicular oscillators*, each oscillating at frequency ω_0. Equating (2.17) to three times (2.4), we find the rather remarkable result:

$$\frac{4\pi^2 \omega_0^2 r_e^2}{3\gamma} I(\omega_0) = 3\gamma E \qquad \text{where} \qquad \gamma = \frac{2r_e \omega_0^2}{3c},$$

and hence

$$I(\omega_0) = \frac{\omega_0^2}{\pi^2 c^2} E. \qquad (2.18)$$

Writing (2.18) in terms of the spectral energy density $u(\omega_0)$,[6]

$$u(\omega_0) = \frac{I(\omega_0)}{c} = \frac{\omega_0^2}{\pi^2 c^3} E \,. \tag{2.19}$$

We can now drop the subscript zero on ω_0 since this result applies to all frequencies in equilibrium. In terms of frequencies:

$$u(\omega)\,d\omega = u(\nu)\,d\nu = \frac{\omega^2}{\pi^2 c^3} E\,d\omega \,,$$

that is,

$$u(\nu) = \frac{8\pi\nu^2}{c^3} E \,. \tag{2.20}$$

This is the remarkable result published by Planck in June 1899 (Planck, 1899). All information about the nature of the oscillator has completely disappeared from the problem. All that remains is the average energy of the oscillator. The meaning behind the relation is obviously very deep and fundamental in a thermodynamic sense. The whole analysis has proceeded through a study of the electrodynamics of oscillators, and yet the final result contains no trace of the means by which we arrived at the answer. One can appreciate how excited Planck must have been when he discovered this basic result – if we can work out the average energy of an oscillator of frequency ν in an enclosure at temperature T, we can find immediately the spectrum of black-body radiation.

2.3.4 Rayleigh and the spectrum of black-body radiation

In fact, the same relation (2.20) between the average energy of an oscillator and the intensity of radiation in thermal equilibrium was also worked out by a completely different route by John William Strutt, Baron Rayleigh in 1900.[7] Rayleigh was the author of the famous book *The Theory of Sound* (1894) and so approached the problem from the point of view of the equilibrium distribution of waves in a box (Rayleigh, 1900). Suppose the box is a cube with sides of length L. Inside the box, all possible wave modes consistent with the boundary conditions are allowed to come into thermodynamic equilibrium at temperature T. The wave equation for the waves in the box is

$$\nabla^2 \psi = \frac{\partial^2 \psi}{\partial x^2} + \frac{\partial^2 \psi}{\partial y^2} + \frac{\partial^2 \psi}{\partial z^2} = \frac{1}{c_s^2}\frac{\partial^2 \psi}{\partial t^2} \,, \tag{2.21}$$

where c_s is the speed of the waves. The walls are fixed and so the waves must have zero amplitude $\psi = 0$ at $x, y, z = 0$ and $x, y, z = L$. The solution of this problem is well-known:

$$\psi = C \mathrm{e}^{-i\omega t} \sin\frac{l\pi x}{L} \sin\frac{m\pi y}{L} \sin\frac{n\pi z}{L} \,, \tag{2.22}$$

corresponding to standing waves which fit perfectly into the box in three dimensions, provided l, m and n are integers. Each combination of l, m and n is called a *mode of oscillation* of the waves in the box. These modes are *complete*, *independent* and *orthogonal* and so represent all the ways in which the medium can oscillate within the box.

We now substitute (2.22) into the wave equation (2.21) and find the relation between the values of l, m, n and the angular frequency of the waves, ω,

$$\frac{\omega^2}{c^2} = \frac{\pi^2}{L^2}(l^2 + m^2 + n^2) = \frac{\pi^2 p^2}{L^2}, \qquad (2.23)$$

where $p^2 = l^2 + m^2 + n^2$. We thus find a relation between the modes as parameterised by $p^2 = l^2 + m^2 + n^2$ and their angular frequency ω. According to the Maxwell–Boltzmann equipartition theorem, energy is shared equally among each independent mode and so we need to know the number of modes in the range p to $p + \mathrm{d}p$. We find this by drawing a three-dimensional lattice in l, m, n space and evaluating the number of modes in the octant of a sphere in (l, m, n) space. If p is large, the number of modes in an octant of a spherical shell of radius p and width $\mathrm{d}p$ is,

$$n(p)\,\mathrm{d}p = \frac{L^3 \omega^2}{2\pi^2 c^3}\,\mathrm{d}\omega. \qquad (2.24)$$

The waves are electromagnetic waves and so there are two independent linear polarisations for any given mode. Therefore, there are twice as many modes as given by (2.24).

According to the Maxwell–Boltzmann doctrine of the equipartition of energy, each mode of oscillation has average energy $\overline{E}$. Then, the energy density of electromagnetic radiation in the box is:

$$u(\nu)\,\mathrm{d}\nu\, L^3 = \overline{E} n(p)\,\mathrm{d}p = \frac{L^3 \omega^2 \overline{E}}{\pi^2 c^3}\,\mathrm{d}\omega, \qquad (2.25)$$

that is,

$$u(\nu) = \frac{8\pi \nu^2}{c^3}\overline{E}, \qquad (2.26)$$

exactly the same result as that derived by Planck from electrodynamics (2.20).

A number of features of Rayleigh's analysis are worth noting. First of all, Rayleigh deals directly with the electromagnetic waves themselves, rather than with the oscillators which are the source of the waves and which are in equilibrium with them. Second, central to the result is the doctrine of the equipartition of energy. What the Maxwell–Boltzmann doctrine states is that, if the system is left long enough, irregularities in the energy distribution among the oscillations are smoothed out by various energy interchange mechanisms. In many natural phenomena, the equilibrium distributions are set up very quickly.

2.4 Towards the spectrum of black-body radiation

Planck's next step is at first slightly surprising. Classically, in thermodynamic equilibrium, each degree of freedom is allocated $\frac{1}{2}kT$ of energy and hence the mean energy of a harmonic oscillator should be $\overline{E} = kT$, because it has two degrees of freedom, those associated with the squared terms $\dot{x}^2$ and x^2 in the expression for its energy. Setting $\overline{E} = kT$, we find

$$u(\nu) = \frac{8\pi \nu^2}{c^3} kT \, , \tag{2.27}$$

exactly the result derived by Rayleigh. This turned out to be the correct expression for the black-body radiation law at low frequencies, the Rayleigh–Jeans law. Why did Planck not do this? First of all, the equipartition theorem of Maxwell and Boltzmann is a result of statistical thermodynamics and this was the point of view which Planck had rejected. As we discussed in Sect. 1.3, it was far from clear in 1899 how secure the equipartition theorem really was. Planck had already had his fingers burned by Boltzmann when he failed to note the necessity of statistical assumptions in order to derive the equilibrium state of black-body radiation (Sect. 2.3). He therefore adopted a rather different approach. To quote his words:

'I had no alternative than to tackle the problem once again – this time from the opposite side – namely from the side of thermodynamics, my own home territory where I felt myself to be on safer ground. In fact, my previous studies of the Second Law of Thermodynamics came to stand me in good stead now, for at the very outset I hit upon the idea of correlating not the temperature of the oscillator but its entropy with its energy . . . While a host of outstanding physicists worked on the problem of the spectral energy distribution both from the experimental and theoretical aspect, every one of them directed his efforts solely towards exhibiting the dependence of the intensity of radiation on the temperature. On the other hand, I suspected that the fundamental connection lies in the dependence of entropy upon energy. . . . Nobody paid any attention to the method which I adopted and I could work out my calculations completely at my leisure, with absolute thoroughness, without fear of interference or competition.' (Planck, 1950)

In March 1900, Planck worked out the following relation for the change in entropy of a system which is not in equilibrium (Planck, 1900a):

$$\Delta S = \frac{3}{5} \frac{\partial^2 S}{\partial E^2} \Delta E \, dE \, . \tag{2.28}$$

This equation applies to an oscillator the entropy of which deviates from the maximum value in that an individual resonator deviates by an amount ΔE from the equilibrium energy E. The entropy change occurs when the energy of the oscillator is changed by dE. Thus, if ΔE and dE have opposite signs, so that the system tends to return towards equilibrium, the entropy change is positive and the function $\partial^2 S/\partial E^2$ must necessarily have a negative value. A negative value of $\partial^2 S/\partial E^2$ means that there is an entropy maximum and thus if ΔE and dE are of opposite signs, the system must approach equilibrium.

To appreciate what Planck did next, we need to review the dramatic developments in the experimental determinations of the properties of black-body radiation. Otto Lummer had been appointed to the Physikalisch-Technische Reichsanstaldt in Berlin in 1887 and led the group involved in establishing standards of luminosity. There was no intention initially to determine the spectrum of black-body radiation, but this came about through the need to develop precision methods of measurement of optical and infrared intensities. He was joined by Ferdinand Kurlbaum in 1891 at the Reichsanstaldt and collaborated with Ernst Pringsheim, who was appointed to a titular professorship at the University of Berlin in 1896,

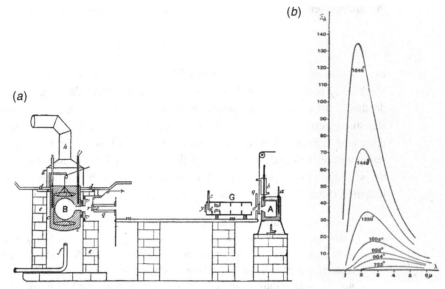

Fig. 2.4 (*a*) Lummer and Pringsheim's apparatus of 1897 for determining experimentally the Stefan–Boltzmann law with high precision (Allen and Maxwell, 1952). (*b*) The spectrum of black-body radiation plotted on linear scales of intensity and wavelength as determined by Lummer and Pringsheim in 1899 for temperatures between 700 and 1600 °C (Allen and Maxwell, 1952).

and by Heinrich Rubens who was appointed to a professorship of physics at the Technische Hochschule of Berlin in 1896.

This group made enormous advances in the determination of the properties of black-body radiation. In particular, they emphasised the importance of developing sources of radiation which were as uniform as possible. The preferred solution was to bring a cavity to as uniform a temperature as possible and then allow the radiation to pass outwards through an opening (Fig. 2.4*a*). In addition, Rubens pioneered the techniques of precise photometric measurement in the infrared regions of the spectrum which were to be crucial in the coming years.

Wien (1896) followed up his studies of the spectrum of thermal radiation by attempting to derive the radiation law from theory. We need not go into his ideas, but he derived an expression for the radiation law which was consistent with his displacement law and which provided an excellent fit to all the data available in 1896. Wien's theory suggested that, consistent with his displacement law, the spectrum should have the form

$$u(\nu) = \frac{8\pi\alpha}{c^3} \nu^3 e^{-\beta\nu/T} \,. \tag{2.29}$$

This is *Wien's law*, written in our notation. There are two unknown constants in the formula, α and β; the constant $8\pi/c^3$ has been included on the right-hand side for reasons which will become apparent in a moment. Rayleigh's comment on Wien's paper was terse:

> 'Viewed from the theoretical side, the result appears to me to be little more than a conjecture.' (Rayleigh, 1900)

The importance of the formula was that it gave an excellent account of all the experimental data available at the time and therefore could be used in theoretical studies. Typical black-body spectra are shown in Fig. 2.4b. The wavelength regions close to the maxima were best defined by the experimental data – the uncertainties in the wings of the energy distribution were much greater.

Returning to Planck's paper of March 1900, the next step was the definition of the entropy S of an oscillator,

$$S = -\frac{E}{\beta v} \ln \frac{E}{\alpha v e}, \tag{2.30}$$

where E is its energy, α and β are the constants in Wien's law and e is the base of natural logarithms. In fact, Planck worked backwards from Wien's law to determine this relation.[8]

The next step was crucial for Planck. The second derivative of (2.30) with respect to E is needed in order to apply (2.28) to the black-body spectrum:

$$\frac{d^2 S}{dE^2} = -\frac{1}{\beta v}\frac{1}{E}. \tag{2.31}$$

β, v and E are all necessarily positive quantities and so $\partial^2 S/\partial E^2$ is necessarily negative. Therefore, according to (2.28), Wien's law is entirely consistent with the second law of thermodynamics. The simplicity of the expression for the second derivative of the entropy with respect to energy profoundly impressed Planck who remarked:

> 'I have repeatedly tried to change or generalise the equation for the electromagnetic entropy of an oscillator in such a way that it satisfies all theoretically sound electromagnetic and thermodynamic laws but I was unsuccessful in this endeavour.' (Planck, 1900a)

In his paper presented to the Prussian Academy of Sciences in May 1899 he stated:

> 'I believe that this must lead me to conclude that the definition of radiation entropy and therefore Wien's energy distribution law necessarily result from the application of the principle of the increase of entropy to the electromagnetic radiation theory and therefore the limits of validity of this law, in so far as they exist at all, coincide with those of the second law of thermodynamics.' (Planck, 1899)

Planck had gone too far – in fact, many negative functions of energy would have satisfied Planck's requirement so far as the second law of thermodynamics was concerned.

These calculations were presented to the Prussian Academy of Sciences in June 1900. By October 1900, Rubens and Kurlbaum had shown beyond any doubt that Wien's law was inadequate to explain the spectrum of black-body radiation at low frequencies and high temperatures. They showed that, at low frequencies and high temperatures, the intensity of radiation was proportional to temperature. This is clearly inconsistent with Wien's law because, if $u(v) \propto v^3 e^{-\beta v/T}$, then for $\beta v/T \ll 1$, $u(v) \propto v^3$ and is independent of temperature. Therefore, the functional dependence must depart from (2.30) and (2.31) for small values of v/T.

Rubens and Kurlbaum showed Planck their results before they were presented in October 1900 and he was given the opportunity to make some remarks about their implications. The result was his paper entitled *On an improvement of Wien's spectral distribution* (Planck,

1900b). Here is what Planck did. At low frequencies $\nu/T \rightarrow 0$, the experiments of Rubens and Kurlbaum showed that $u(\nu) \propto T$. The thermodynamic relations between S, E and T are

$$E \propto T, \qquad \frac{dS}{dE} = \frac{1}{T}.$$

Hence

$$\frac{dS}{dE} \propto \frac{1}{E}, \qquad \frac{d^2S}{dE^2} \propto \frac{1}{E^2}. \tag{2.32}$$

Therefore, d^2S/dE^2 must change its functional dependence upon E between large and small values of ν/T. Wien's law remains good for large values of ν/T and leads to

$$\frac{d^2S}{dE^2} \propto \frac{1}{E}. \tag{2.33}$$

The standard technique for combining functions of the form (2.32) and (2.33) is to try an expression of the form

$$\frac{d^2S}{dE^2} = -\frac{a}{E(b+E)}, \tag{2.34}$$

which has exactly the required properties for large and small values of E, $E \gg b$ and $E \ll b$ respectively. The rest of the analysis is straightforward.[9] Integrating with respect to E,

$$\frac{dS}{dE} = -\int \frac{a}{E(b+E)}\, dE = -\frac{a}{b}[\ln E - \ln(b+E)] = \frac{1}{T}, \tag{2.35}$$

and hence

$$E = \frac{b}{e^{b/aT} - 1}. \tag{2.36}$$

Then,

$$u(\nu) = \frac{8\pi\nu^2}{c^3} E = \frac{8\pi\nu^2}{c^3}\frac{b}{e^{b/aT} - 1}. \tag{2.37}$$

From the high frequency, low temperature limit, we can find the constants from Wien's law (2.29):

$$u(\nu) = \frac{8\pi\nu^2}{c^3}\frac{b}{e^{b/aT}} = \frac{8\pi\alpha}{c^3}\frac{\nu^3}{e^{\beta\nu/T}}.$$

Thus b must be proportional to the frequency ν. We can now write Planck's formula in its primitive form:

$$u(\nu) = \frac{A\nu^3}{(e^{\beta\nu/T} - 1)}. \tag{2.38}$$

Of equal importance for the next part of the story is that Planck was also able to find an expression for the entropy of an oscillator by integrating dS/dE. From (2.35),

$$S = -a\left[\frac{E}{b}\ln\frac{E}{b} - \left(1 + \frac{E}{b}\right)\ln\left(1 + \frac{E}{b}\right)\right], \tag{2.39}$$

with $b \propto \nu$. Planck's new formula could now be confronted with the experimental evidence.

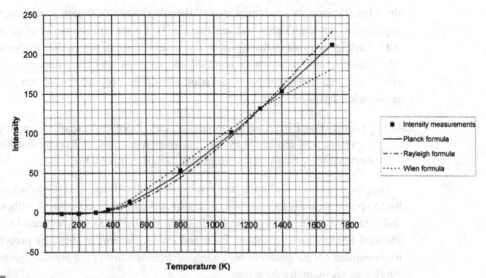

Fig. 2.5 Comparison of the radiation formulae of Planck (2.38, solid line), Wien (2.29, dotted line) and Rayleigh (2.40, dot-dashed line) with the intensity of black-body radiation at 8.85 μm as a function of temperature as measured by Rubens and Kurlbaum (filled boxes). Similar experiments were carried out at longer wavelengths, (24 + 31.6) μm and 51.2 μm, at which there was little difference between the predictions of the Planck and Rayleigh functions. (Replotted from the data presented by Rubens and Kurlbaum (1901) in the same format as their original presentation.)

2.5 Comparison of the laws for black-body radiation with experiment

In their presentation of 25 October 1900, Rubens and Kurlbaum compared their new precise measurements of the spectrum of black-body radiation with five different formulae. Two of these were empirical relations proposed by Thiesen and by Lummer and Jahnke. The others were:

- Wien's relation (2.29);
- Planck's formula (2.38);
- Rayleigh's result (2.27), but modified to avoid the divergence of the Rayleigh–Jeans spectrum at high frequencies, the problem known as the *ultraviolet catastrophe*. Rayleigh was well aware of 'the difficulties which attend the Boltzmann–Maxwell doctrine of the partition of energy' and that (2.27) must break down at high frequencies because the spectrum of black-body radiation does not increase as ν^2 to infinite frequency. In the fifth paragraph of his short paper, he stated, 'If we introduce the exponential factor, the complete expression is', in our notation,

$$u(\nu) = \frac{8\pi \nu^2}{c^3} kT \, e^{-\beta \nu/T}. \tag{2.40}$$

The data presented by Rubens and Kurlbaum (1901) are replotted in Fig. 2.5 showing only the formulae proposed by Wien, Planck and Rayleigh. Rubens and Kurlbaum concluded that Planck's formula was superior to all the others and that it gave precise agreement with experiment. Rayleigh's proposal was found to be a poor representation of the experimental data. Rayleigh was somewhat upset by the tone of voice in which they discussed his result. When his scientific papers were republished two years later he remarked on his important conclusion that the intensity of radiation should be proportional to temperature at low frequencies. He pointed out,

> 'This is what I intended to emphasise. Very shortly afterwards the anticipation above expressed was confirmed by the important researches of Rubens and Kurlbaum who operated with exceptionally long waves.' (Rayleigh, 1902)

Despite this success, Planck had not explained anything. The formula he derived was based upon essentially thermodynamic arguments guided by experiment without theoretical understanding at the microscopic level. The formula (2.38) was presented to the German Physical Society on 19 October 1900 and on 14 December 1900, he presented another paper entitled *On the theory of the energy distribution law in the normal spectrum* (Planck, 1900c). In his memoirs, he wrote

> 'After a few weeks of the most strenuous work of my life, the darkness lifted and an unexpected vista began to appear.' (Planck, 1925)

2.6 Planck's theory of black-body radiation

In his scientific biography, Planck wrote:

> 'On the very day when I formulated this law, I began to devote myself to the task of investing it with a true physical meaning. This quest automatically led me to study the interrelation of entropy and probability – in other words, to pursue the line of thought inaugurated by Boltzmann.' (Planck, 1950)

Planck recognised that the way forward involved adopting a point of view which he had rejected in essentially all his previous work. He was not a specialist in statistical physics and we will find that his analysis did not follow the precepts of classical statistical mechanics. Despite the basic flaws in his argument, he discovered the essential role which *quantisation* plays in accounting for the spectrum of black-body radiation.

Planck's analysis began by following Boltzmann's procedure. There is a fixed total energy E to be divided among the N oscillators and energy elements ε are introduced. Therefore, there are $r = E/\varepsilon$ energy elements to be shared among the oscillators. Rather than following Boltzmann's procedure in statistical physics, however, Planck simply worked out the *total number of ways* in which r energy elements can be distributed over the N oscillators. The answer is[10]

$$\frac{(N + r - 1)!}{r!(r - 1)!}. \tag{2.41}$$

Now Planck made the crucial step in his argument. He *defined* (2.41) to be the probability p to be used in Boltzmann's expression for the entropy

$$S = C \ln p .$$

Let us see where this leads. N and r are very large indeed and so we can use Stirling's approximation for $n!$:

$$n! = (2\pi n)^{1/2} \left(\frac{n}{e}\right)^n \left(1 + \frac{1}{12n} + \cdots \right) . \qquad (2.42)$$

We need to take the logarithm of (2.41) and so we can use an even simpler approximation, $n! \approx n^n$, and so

$$p = \frac{(N+r-1)!}{r!(N-1)!} \approx \frac{(N+r)!}{r!N!} \approx \frac{(N+r)^{N+r}}{r^r N^N} . \qquad (2.43)$$

Therefore,

$$\begin{cases} S = C[(N+r)\ln(N+r) - r\ln r - N\ln N , \\ r = \dfrac{E}{\varepsilon} = \dfrac{N\overline{E}}{\varepsilon} , \end{cases} \qquad (2.44)$$

where $\overline{E}$ is the average energy of the oscillators. Therefore,

$$S = C\left\{ N\left(1 + \frac{\overline{E}}{\varepsilon}\right) \ln\left[N\left(1 + \frac{\overline{E}}{\varepsilon}\right)\right] - \frac{N\overline{E}}{\varepsilon} \ln \frac{N\overline{E}}{\varepsilon} - N\ln N \right\} . \qquad (2.45)$$

The average entropy per oscillator $\overline{S}$ is therefore

$$\overline{S} = \frac{S_N}{N} = C\left[\left(1 + \frac{\overline{E}}{\varepsilon}\right) \ln\left(1 + \frac{\overline{E}}{\varepsilon}\right) - \frac{\overline{E}}{\varepsilon} \ln \frac{\overline{E}}{\varepsilon} \right] . \qquad (2.46)$$

But this looks rather familiar. The relation (2.46) is exactly the expression (2.39) for the entropy of an oscillator which Planck had derived to account for the spectrum of black-body radiation, namely:

$$S = a\left[\left(1 + \frac{\overline{E}}{b}\right) \ln\left(1 + \frac{\overline{E}}{b}\right) - \frac{\overline{E}}{b} \ln \frac{\overline{E}}{b} \right] ,$$

with the requirement $b \propto \nu$. Thus, the energy elements ε must be proportional to frequency and Planck wrote this requirement in the familiar form

$$\varepsilon = h\nu , \qquad (2.47)$$

where h is rightly known as *Planck's constant*. This is the origin of the concept of *quantisation*. According to the procedures of classical statistical mechanics, we ought now to allow $\varepsilon \to 0$, but evidently we cannot obtain the expression for the entropy of an oscillator unless the energy elements do not disappear, but have finite magnitude $\varepsilon = h\nu$. Therefore, the expression for the energy density of black-body radiation is

$$u(\nu) = \frac{8\pi h\nu^3}{c^3} \frac{1}{e^{h\nu/CT} - 1} . \qquad (2.48)$$

Finally, what about C? Planck pointed out that C is a universal constant relating the entropy to the probability. Therefore, any suitable law such as the perfect gas law which determines its value determines C for all processes. For example, we could use the results of the Joule expansion of a perfect gas, treated classically and statistically. Then, the ratio $C = k = R/N_A$, where R is the gas constant and N_A is Avogadro's number, the number of molecules per gram molecule.[11] Hence, the Planck distribution in its final form is:

$$u(v) = \frac{8\pi h v^3}{c^3} \frac{1}{e^{hv/kT} - 1} . \tag{2.49}$$

Integrating, the total energy density of radiation u in the black-body spectrum is,

$$u = \int_0^\infty u(v)\, dv = \frac{8\pi h}{c^3} \int_0^\infty \frac{v^3\, dv}{e^{hv/kT} - 1} = \left(\frac{8\pi^5 k^4}{15 c^3 h^3}\right) T^4 = a T^4 \tag{2.50}$$

where

$$a = \frac{8\pi^5 k^4}{15 c^3 h^3} = 7.566 \times 10^{-16}\ \mathrm{J\ m^{-3}\ K^{-4}} .$$

We have recovered the Stefan–Boltzmann law for the energy density of radiation u. We can relate this energy density to the energy emitted per second from the surface of a black-body maintained at temperature T. The rate at which the energy is reradiated[12] is $\frac{1}{4}uc$ and so

$$I = \tfrac{1}{4}uc = \frac{ac}{4} T^4 = \sigma T^4 = \left(\frac{2\pi^5 k^4}{15 c^2 h^3}\right) T^4 = 5.67 \times 10^{-8} T^4\ \mathrm{W\ m^{-2}} . \tag{2.51}$$

This provides the determination of the value of the Stefan–Boltzmann constant σ in terms of fundamental constants.

What is one to make of Planck's argument? There are two fundamental concerns:

(1) Planck certainly does not follow Boltzmann's procedure for finding the equilibrium energy distribution of the oscillators. What he defines as a probability is not really a probability of anything drawn from any parent population. Planck had no illusions about this. In his own words:

> 'In my opinion, this stipulation basically amounts to a definition of the probability W; for we have absolutely no point of departure, in the assumptions which underlie the electromagnetic theory of radiation, for talking about such a probability with a definite meaning.'[13]

Einstein repeatedly pointed out this weak point in Planck's argument:

> 'The manner in which Mr Planck uses Boltzmann's equation is rather strange to me in that a probability of a state W is introduced without a physical definition of this quantity. If one accepts this, then Boltzmann's equation simply has no physical meaning.' (Einstein, 1912)

(2) A second problem concerns a logical inconsistency in Planck's analysis. On the one hand, the oscillators can only take energies $E = r\varepsilon$ and yet a classical result (2.1) has been used to work out the rate of radiation of the oscillator. Implicit in that analysis

is the assumption that the energies of the oscillators can vary continuously rather than take only discrete values.

These were major stumbling blocks and nobody quite understood the significance of what Planck had done. The theory did not in any sense gain immediate acceptance, there being no further papers by Planck or others until Einstein's papers of 1905 and 1906. Nonetheless, the concept of quantisation had been introduced through the energy elements ε without which it is not possible to reproduce the Planck distribution. In fact, in 1906 Einstein showed that, if Planck had followed strictly Boltzmann's procedure, he would have obtained the same answer whilst maintaining the essential concept of energy quantisation (Sect. 3.3).

Why did Planck find the correct expression for the radiation spectrum, despite the fact that the statistical procedures he used were more than a little suspect? It seems quite likely that Planck worked backwards. It was suggested by Rosenfeld, and endorsed by Klein on the basis of an article by Planck of 1943, that he started with the expression of the entropy of an oscillator (2.39) and worked backwards to find W from $\exp(S/k)$. This results in the permutation formula on the right-hand side of (2.43), which is more or less exactly the same as (2.41) for large values of N and r. The expression (2.41) was a well-known formula in permutation theory and appears early in Boltzmann's exposition of the fundamentals of statistical physics. Planck then regarded (2.46) as the definition of entropy according to statistical physics. If this is indeed what happened, it in no sense diminishes Planck's achievement.

2.7 Planck and 'natural units'

Planck appreciated that there are two fundamental constants in his theory of black-body radiation, Boltzmann's constant k and the new constant h, which Planck was to refer to as the *quantum of action*. The fundamental nature of k was apparent from its appearance in the kinetic theory as the gas constant per molecule and as the constant C in Boltzmann's relation $S = k \ln W$. Both constants could be determined with considerable precision from the experimentally measured form of the spectrum of black-body radiation (2.49), and from the value of the Stefan–Boltzmann constant (2.50) or (2.51). Combining the known value of the gas constant R with his new determination of k, Planck found a value of Avogadro's number, $N_A = 6.175 \times 10^{23}$ molecules per mole, by far the best estimate known at that time. The present adopted value is $N_A = 6.022 \times 10^{23}$ molecules per mole.

The electric charge carried by a gram equivalent of monovalent ions was known from electrolytic theory and is known as Faraday's constant. Knowing N_A precisely, Planck was able to derive the elementary unit of charge and found it to be $e = 4.69 \times 10^{-10}$ esu, corresponding to 1.56×10^{-19} C. Again, Planck's value was by far the best available at that time, contemporary experimental values ranging between 1.3 and 6.5×10^{-10} esu. The present standard value is 1.602×10^{-19} C.

Table 2.1 Planck's system of natural units		
Time	$t_{\mathrm{Pl}} = (Gh/c^5)^{1/2}$	10^{-43} s
Length	$l_{\mathrm{Pl}} = (Gh/c^3)^{1/2}$	4×10^{-35} m
Mass	$m_{\mathrm{Pl}} = (hc/G)^{1/2}$	5.4×10^{-8} kg $\equiv 3 \times 10^{19}$ GeV

Equally compelling for Planck was the fact that h, in conjunction with the constant of gravitation G and the velocity of light c, enabled a set of 'natural' units to be defined in terms of fundamental constants (Table 2.1). Often these *Planck units* are written in terms of $\hbar = h/2\pi$, in which case their values are 2.5 times smaller than those quoted in the table. It is evident that the natural units of time and length are very small indeed, while the mass is much greater than that of any known elementary particle. A century later, these quantities were to play a central role in the physics of the very early Universe.

2.8 Planck and the physical significance of h

It was a number of years before the truly revolutionary nature of what Planck had achieved in these crucial last months of 1900 was appreciated. Perhaps surprisingly, he wrote no papers on the subject of quantisation over the next five years. The next publication which casts some light on his understanding was his text *Lectures on the Theory of Thermal Radiation* of 1906 (Planck, 1906). Thomas Kuhn gives a detailed analysis of Planck's thoughts on quantisation through the period 1900–1906 (Kuhn, 1978). What is clear from Kuhn's analysis is that Planck undoubtedly believed that the classical laws of electromagnetism were applicable to the processes of the emission and absorption of radiation, despite the introduction of the finite energy elements in his theory of quantisation. Planck describes two versions of Boltzmann's procedure in statistical physics, the first being the version described in Sect. 2.6, in which the energies of the oscillators take values 0, ε, 2ε, 3ε, and so on. There is a second version in which the molecules were considered to lie within the energy ranges 0 to ε, ε to 2ε, 2ε to 3ε, and so on. This procedure leads to exactly the same statistical probabilities as the first version. In a subsequent passage, in which the motions of the oscillators are traced in phase space, he again refers to the energies of the trajectories corresponding to certain energy ranges U to $U + \Delta U$, the ΔU eventually being identified with $h\nu$. Thus, Planck regarded quantisation as referring to the average properties of the oscillators.

Planck had little to say about the nature of the quantum of action h, but he was well aware of its fundamental importance. In his words,

'The thermodynamics of radiation will therefore not be brought to an entirely satisfactory state until the full and universal significance of the constant h is understood.' (Kuhn, 1978)

Planck spent many years trying to reconcile his theory with classical physics, but he failed to find any physical significance for h beyond its appearance in the radiation formula. In his words:

'My futile attempts to fit the elementary quantum of action somehow into the classical theory continued for a number of years and they cost me a great deal of effort. Many of my colleagues saw in this something bordering on tragedy. But I feel differently about it. For the thorough enlightenment I thus received was all the more valuable. I now knew for a fact that the elementary quantum of action played a far more significant part in physics than I had originally been inclined to suspect and this recognition made me see clearly the need for the introduction of totally new methods of analysis and reasoning in the treatment of atomic problems.' (Planck, 1950)

It was not until after about 1908 that Planck fully appreciated the quite fundamental nature of quantisation, which has no counterpart in classical physics. His original view was that the introduction of energy elements was

'a purely formal assumption and I really did not give it much thought except that no matter what the cost, I must bring about a positive result.' (Planck, 1931a)

Later in the same letter, he writes:

'It was clear to me that classical physics could offer no solution to this problem and would have meant that all energy would eventually transfer from matter into radiation. In order to prevent this, a new constant is required to assure that energy does not disintegrate. But the only way to recognise how this can be done is to start from a definite point of view. This approach was opened to me by maintaining the two laws of thermodynamics. The two laws, it seems to me, must be upheld under all circumstances. For the rest, I was ready to sacrifice every one of my previous convictions about physical laws.'

3 Einstein and quanta 1900–1911

3.1 Einstein in 1905

The next great steps were taken by Albert Einstein and it is no exaggeration to state that he was the first person to appreciate the full significance of quantisation and the reality of quanta. He showed that these are fundamental aspects of all physical phenomena, rather than just a 'formal device' for accounting for the Planck distribution. From 1905 onwards, he never deviated from his belief in the reality of quanta – it was some considerable time before the great figures of the day conceded that Einstein was indeed correct.

Einstein completed what we would now call his undergraduate studies in August 1900. Between 1902 and 1904, he wrote three papers on the foundations of Boltzmann's statistical mechanics. In 1905, Einstein was 26 and employed as 'technical expert, third class' at the Swiss patent office in Bern. In that year, he completed his doctoral dissertation on *A new determination of molecular dimensions*, which he presented to the University of Zurich on 20 July 1905. In the same year, he published three papers which are among the greatest classics in the literature of physics.[1] Any one of them would have ensured that his name remained a permanent fixture in the scientific literature. These papers are:

(1) *On a heuristic point of view concerning the production and transformation of light* (Einstein, 1905a);
(2) *On the motion of small particles suspended in stationary liquids required by the molecular-kinetic theory of heat* (Einstein, 1905b);
(3) *On the electrodynamics of moving bodies* (Einstein, 1905c).

The third paper is Einstein's paper on the special theory of relativity which was described briefly in Sect. 1.5. The second paper confirmed the correctness of the kinetic theory while the first introduces the concept of light quanta to describe the spectrum of black-body radiation. Einstein confessed that he was no mathematician and the mathematics needed to understand his papers of 1905 is no more than is taught in the first two years of an undergraduate physics course. His genius lay in his extraordinary physical intuition, which enabled him to see deeper into physical problems than his contemporaries. The three great papers were not the result of a sudden burst of creativity, but the product of deep pondering about the fundamental problems of physics over almost a decade. His deliberations on all three topics suddenly came to fruition almost simultaneously in 1905. Despite their obvious differences, the three papers have a striking commonality of approach. In each case, Einstein stands back from the specific problem at hand and studies the underlying physical principles.

3.2 Einstein on Brownian motion

The second paper is more familiarly known by the title of a subsequent paper published in 1906 entitled *On the theory of Brownian motion* (Einstein, 1906a) and is a reworking of some of the results of his doctoral dissertation. Brownian motion is the irregular motion of microscopic particles in fluids and had been studied in detail in 1828 by the botanist Robert Brown, who had noted the ubiquity of the phenomenon. The motion results from the statistical effect of very large numbers of collisions between molecules of the fluid and the microscopic particles. Although each impact is very small, the net result of a very large number of them randomly colliding with the particle is a 'drunken man's walk'. Einstein was not certain about the applicability of his analysis to Brownian motion, writing in the introduction to his paper:

> 'It is possible that the movements to be discussed here are identical with the so-called 'Brownian molecular motion'; however, the information available to me regarding the latter is so lacking in precision that I can form no judgment in the matter.'

In his autobiographical notes, he stated that he wrote the paper 'without knowing that observations concerning Brownian motion were already long familiar' (Einstein, 1979).

Einstein begins with Stokes' formula for the force acting on a sphere of radius a moving at speed v through a medium of kinematic viscosity ν, $F = 6\pi\,\nu\,a\,v$, where a is the radius of the sphere and ν the coefficient of kinematic viscosity of the fluid. By considering the one-dimensional diffusion of the particles in the steady state, he found the diffusion coefficient for the particles in the medium, $D = kT/6\pi\nu a$, and from this the one-dimensional distance the particle diffuses $\langle\lambda_x^2\rangle = 2Dt$ in time t. The result is his famous formula for the mean squared distance travelled by the particle in time t in one dimension,

$$\langle\lambda_x^2\rangle = \frac{kTt}{3\pi\nu a},\tag{3.1}$$

where T is the temperature and k Boltzmann's constant. Crucially, Einstein had discovered the relation between the molecular properties of fluids and the observed diffusion of macroscopic particles. In his estimates of the magnitude of the effect for particles 1 μm in diameter, he needed a value for Avogadro's number N_A and he used the values which Planck and he had found from their studies of the spectrum of black-body radiation (see Sect. 3.3 below). He predicted that such particles would diffuse about 6 μm in one minute. In the last paragraph of the paper, Einstein states:

> 'Let us hope that a researcher will soon succeed in solving the problem presented here, which is so important for the theory of heat!'

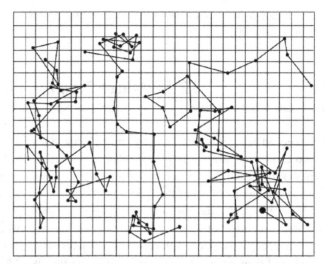

Fig. 3.1 Three tracings of the motion of colloidal particles of radius 0.53 μm, as seen under the microscope. Successive positions every 30 seconds are joined by straight line segments, the spacings of the grid lines being 3.2 μm. This drawing is based upon diagrams in Perrin's paper of 1909 (Perrin, 1909).

Precise observations of Brownian motion were difficult at that time, but in 1908, Jean Perrin (1909) carried out a meticulous series of brilliant experiments which confirmed in detail all Einstein's predictions (Fig. 3.1). This work convinced everyone, even the sceptics, of the reality of molecules. In Perrin's words,

> 'I think it is impossible that a mind free from all preconception can reflect upon the extreme diversity of the phenomena which thus converge to the same result without experiencing a strong impression, and I think it will henceforth be difficult to defend by rational arguments a hostile attitude to molecular hypotheses.' (Perrin, 1910)

Einstein was well aware of the importance of this calculation for the theory of heat – the agitational motion of the particles observed in Brownian motion *is* heat, the macroscopic particles reflecting the motion of the molecules on the microscopic scale.

3.3 On a heuristic viewpoint concerning the production and transformation of light (Einstein 1905a)

The first paper listed in Sect. 3.1 is often referred to as Einstein's paper on the photoelectric effect but that scarcely does justice to its profundity. In a letter to his friend Conrad Habicht in May 1905, Einstein wrote:

> 'I promise you four papers . . . the first of which I could send you soon, since I will soon receive the free reprints. The paper deals with radiation and the energetic properties of light and is very revolutionary, . . . ' (Einstein, 1993)

Here are the opening paragraphs of Einstein's great paper. They demand attention, like the opening of a great symphony.

> 'There is a profound formal difference between the theoretical ideas which physicists have formed concerning gases and other ponderable bodies and Maxwell's theory of electromagnetic processes in so-called empty space. Thus, while we consider the state of a body to be completely defined by the positions and velocities of a very large but finite number of atoms and electrons, we use continuous three-dimensional functions to determine the electromagnetic state existing within some region, so that a finite number of dimensions is not sufficient to determine the electromagnetic state of the region completely . . .
>
> The undulatory theory of light, which operates with continuous three-dimensional functions, applies extremely well to the explanation of purely optical phenomena and will probably never be replaced by any other theory. However, it should be kept in mind that optical observations refer to values averaged over time and not to instantaneous values. Despite the complete experimental verification of the theory of diffraction, reflection, refraction, dispersion and so on, it is conceivable that a theory of light operating with continuous three-dimensional functions will lead to conflicts with experience if it is applied to the phenomena of light generation and conversion.' (Einstein, 1905b)

In other words, there may well be circumstances under which Maxwell's theory of the electromagnetic field cannot explain all electromagnetic phenomena and Einstein specifically gives as examples the spectrum of black-body radiation, photoluminescence and the photoelectric effect. His proposal is that, for some purposes, it may be more appropriate to consider light to be

> 'discontinuously distributed in space. According to the assumption considered here, in the propagation of a light ray emitted from a point source, the energy is not distributed continuously over ever-increasing volumes of space, but consists of a finite number of energy quanta localised at points in space that move without dividing, and can be absorbed and generated only as complete units.'

He ends by hoping that

> 'the approach to be presented will prove of use to some researchers in their investigations.'

Notice how Einstein's proposal differs from Planck's approach. Planck found that the 'energy elements' $\varepsilon = h\nu$ associated with the oscillators in thermal equilibrium at temperature T must not vanish. These oscillators are the source of the electromagnetic radiation in the black-body spectrum, but Planck had absolutely nothing to say about the radiation emitted by them. He firmly believed that the waves emitted by the oscillators were the classical electromagnetic waves of Maxwell. In contrast, Einstein proposes that the radiation field itself should be quantised.

After the introduction, Einstein states Planck's formula (2.26) relating the average energy of an oscillator to the energy density of black-body radiation in thermodynamic equilibrium but does not hesitate to set the average energy of the oscillator $\overline{E} = kT$ according to the kinetic theory. He then writes the total energy in the black-body spectrum in the provocative form

$$\text{total energy density} = \int_0^\infty u(v)\,dv = \frac{8\pi kT}{c^3}\int_0^\infty v^2\,dv = \infty. \tag{3.2}$$

This is exactly the problem which had been pointed out by Rayleigh in 1900 and which led to his arbitrary introduction of the exponential factor to prevent the spectrum diverging at high frequencies (see Sect. 2.5). This phenomenon was later called the *ultraviolet catastrophe* by Paul Ehrenfest.

Einstein next goes on to show that, despite the high frequency divergence of the expression (3.2), it is a very good description of the black-body radiation spectrum at low frequencies and high temperatures, and hence the value of Boltzmann's constant k can be derived from that part of the spectrum alone. Einstein's value of k agreed precisely with Planck's estimate, which Einstein interpreted as meaning that Planck's estimate was independent of the details of the theory he had developed to account for the black-body spectrum.

We have already emphasised the central role which entropy plays in the thermodynamics of radiation. Einstein now derives a suitable form for the entropy of the spectrum of black-body radiation using only thermodynamics and the observed form of the radiation spectrum. Entropies are additive, and since, in thermal equilibrium, we may consider the radiation of different wavelengths to be independent, we can write the entropy of the radiation enclosed in volume V as

$$S = V \int_0^\infty \phi[u(v), v]\,dv. \tag{3.3}$$

The function ϕ is the entropy of the radiation per unit frequency interval per unit volume. The aim of the calculation is to find an expression for the function ϕ in terms of the spectral energy density $u(v)$ and frequency v. No other quantities besides the temperature T can be involved in the expression for the equilibrium spectrum, as was shown by Kirchhoff (see Sect. 1.6). The problem had already been solved by Wien but Einstein gives an elegant proof of the result,[2]

$$\frac{\partial \phi}{\partial u} = \frac{1}{T}, \tag{3.4}$$

resulting in a pleasant symmetry between the relations:

$$\begin{cases} S = \int_0^\infty \phi\,dv & E = \int_0^\infty u(v)\,dv, \\ \dfrac{dS}{dE} = \dfrac{1}{T} & \dfrac{\partial \phi}{\partial u} = \dfrac{1}{T}. \end{cases} \tag{3.5}$$

Einstein now uses (3.5) to work out the entropy of black-body radiation. Rather than use Planck's formula, he uses Wien's formula since, although it is not correct for low frequencies and high temperatures, it is the correct law in the region where the classical theory breaks down and so the analysis of this part of the spectrum is likely to give insight into where the classical calculation goes wrong.

First, Einstein writes down the form of Wien's law as derived from experiment. In the notation of (2.29),

$$u(v) = \frac{8\pi\alpha}{c^3} \frac{v^3}{e^{\beta v/T}} \,.$$ (3.6)

Taking logarithms, we can therefore find an expression for $1/T$,

$$\frac{1}{T} = \frac{1}{\beta v} \ln \frac{8\pi\alpha v^3}{c^3 u(v)} = \frac{\partial \phi}{\partial u} \,.$$ (3.7)

The expression for ϕ is found by integration:

$$\frac{\partial \phi}{\partial u} = -\frac{1}{\beta v} \left(\ln u + \ln \frac{c^3}{8\pi\alpha v^3} \right) ,$$

$$\phi = -\frac{u}{\beta v} \left(\ln \frac{uc^3}{8\pi\alpha v^3} - 1 \right) .$$ (3.8)

Now, consider the energy density of radiation in the spectral range v to $v + \Delta v$ which has energy $\varepsilon = Vu\,\Delta v$, where V is the volume. The entropy associated with this radiation is

$$S = V\phi\,\Delta v = -\frac{\varepsilon}{\beta v} \left(\ln \frac{\varepsilon c^3}{8\pi\alpha v^3 V \,\Delta v} - 1 \right) .$$ (3.9)

Suppose the volume changes from V_0 to V, while the total energy remains constant. Then, the entropy change is

$$S - S_0 = \frac{\varepsilon}{\beta v} \ln(V/V_0).$$ (3.10)

But this formula looks familiar. Einstein shows that this entropy change is exactly the same as that found in the Joule expansion of a perfect gas according to elementary statistical mechanics. Boltzmann's relation $S = k \ln W$ can be used to work out the entropy difference $S - S_0$ between the initial and final states, $S - S_0 = k \ln W/W_0$, where the Ws are the probabilities of these states. In the initial state, the system has volume V_0 and the particles move randomly throughout this volume. The probability that a single particle occupies a smaller volume V is V/V_0 and hence the probability that all N end up in this volume V is $(V/V_0)^N$. Therefore, the entropy difference for a gas of N particles is

$$S - S_0 = kN \ln(V/V_0).$$ (3.11)

Einstein notes that (3.10) and (3.11) are formally identical. He immediately concludes that the radiation behaves thermodynamically as if it consisted of discrete particles, their number N being equal to $\varepsilon/k\beta v$. In Einstein's own words,

> 'Monochromatic radiation of low density (within the limits of validity of Wien's radiation formula) behaves thermodynamically as though it consisted of a number of independent energy quanta of magnitude $k\beta v$.'

Rewriting this result in Planck's notation, since $\beta = h/k$, the energy of each quantum is hv.

Einstein then works out the average energy of the quanta according to Wien's formula for black-body radiation. The energy in the frequency interval ν to $\nu + d\nu$ is ε and the number of quanta is $\varepsilon/k\beta\nu$. Therefore, the average energy is

$$\overline{E} = \frac{\int_0^\infty (8\pi\alpha/c^3)\nu^3\,e^{-\beta\nu/T}\,d\nu}{\int_0^\infty (8\pi\alpha/c^3)(\nu^3/k\beta\nu)\,e^{-\beta\nu/T}\,d\nu} = k\beta\frac{\int_0^\infty \nu^3\,e^{-\beta\nu/T}\,d\nu}{\int_0^\infty \nu^2\,e^{-\beta\nu/T}\,d\nu} = k\beta \times \frac{3T}{\beta} = 3kT\,.$$

(3.12)

The average energy of the quanta is closely related to the mean kinetic energy per particle in the black-body enclosure, $\frac{3}{2}kT$.

So far, Einstein has stated that the radiation 'behaved as though' it consisted of a number of independent particles. Is this just another 'formal device'? The last sentence of Sect. 6 of his paper leaves the reader in no doubt:

> 'the next obvious step is to investigate whether the laws of emission and transformation of light are also of such a nature that they can be interpreted or explained by considering light to consist of such energy quanta.'

Einstein considers three phenomena which cannot be explained by classical electromagnetic theory.

1. *Stokes' rule* is the observation that the frequency of photoluminescent emission is less than the frequency of the incident light. This is explained as a consequence of the conservation of energy. If the incoming quanta each have energy $h\nu_1$, the re-emitted quanta can at most have this energy. If some of the energy of the quanta is absorbed by the material before re-emission, the emitted quanta of energy $h\nu_2$ must have $h\nu_2 \leq h\nu_1$.
2. *The photoelectric effect.* This is the most famous result of the paper because Einstein made a definite quantitative prediction on the basis of the theory expounded above. Ironically, the photoelectric effect had been discovered by Hertz in 1887 in the same experiments which fully validated Maxwell's equations. Perhaps the most remarkable feature of the effect was Lénárd's discovery that the energies of the electrons emitted from the metal surface are independent of the intensity of the incident radiation (Lénárd, 1902).

   Einstein's proposal provided an immediate solution to this problem. Radiation of a given frequency consists of quanta of the same energy $h\nu$. If one of these is absorbed by the material, the electron may receive sufficient energy to remove it from the surface against the forces which bind it to the material. If the intensity of the light is increased, more electrons are ejected, but their energies remain unchanged. Einstein wrote this result in the following form. The maximum kinetic energy which the ejected electron can have, E_k, is

$$E_k = h\nu - W\,,$$

(3.13)

   where W is the amount of work necessary to remove the electron from the surface of the material, the *work function* of the material. Experiments to estimate the magnitude of the work function involve placing the photocathode in an opposing potential so that, when the potential reaches some value V, the ejected electrons can no longer reach the anode and the photoelectric current falls to zero. This occurs at the potential at which

$E_k = eV$. Therefore,

$$V = \frac{h}{e}v - \frac{W}{e}.$$ (3.14)

In Einstein's words,

> 'If the formula derived is correct, then V must be a straight line function of the frequency of the incident light, when plotted in Cartesian coordinates, whose slope is independent of the nature of the substance investigated.'

Thus, the quantity h/e, the ratio of Planck's constant to the electronic charge, can be found directly from the slope of this relation. Nothing was known about the dependence of the photoelectric effect upon the frequency of the incident radiation at that time. It was only in 1916 that Millikan's meticulous experiments confirmed precisely Einstein's prediction.

3. *Photoionisation of gases.* The third piece of experimental evidence was the fact that the energy of each photon has to be greater than the ionisation potential of the gas if photoionisation is to take place. Einstein showed that the smallest energy quanta for the ionisation of air were approximately equal to the ionisation potential determined independently by Stark. Once again, the quantum hypothesis is in agreement with experiment.

This is the work described in Einstein's Nobel Prize citation of 1921.

3.4 The quantum theory of solids

In 1905, Einstein was not at all clear that Planck and he were actually describing the same phenomenon. In 1906, however, he showed that the two approaches were, in fact, the same (Einstein, 1906b). Then, later in the same year, he came to the same conclusion by a different argument, and went on to extend the idea of quantisation to solids (Einstein, 1906c).

In the first of these papers, Einstein asserts that he and Planck are actually describing the same phenomena of quantisation.

> 'At that time [1905] it seemed to me as though Planck's theory of radiation formed a contrast to my work in a certain respect. New considerations which are given in the first section of this paper demonstrate to me, however, that the theoretical foundation on which Planck's radiation theory rests differs from the foundation that would result from Maxwell's theory and the electron theory and indeed differs exactly in that Planck's theory implicitly makes use of the hypothesis of light quanta just mentioned.' (Einstein, 1906b)

These arguments are developed further in the second paper of 1906. Einstein demonstrated that, if Planck had followed Boltzmann's procedures, he would have ended up with the Boltzmann expression for the probability that a state of energy $E = r\varepsilon$ is occupied, even although the limit $\varepsilon \to 0$ is not taken:

$$p(E) \propto e^{-E/kT}.$$

It is assumed that the energy of the oscillator is quantised in units of ε. Thus, if there are N_0 oscillators in the ground state, the number in the $r = 1$ state is $N_0\,e^{-\varepsilon/kT}$, in the $r = 2$ state $N_0 e^{-2\varepsilon/kT}$, and so on. Therefore, the average energy of the oscillator is

$$\overline{E} = \frac{N_0 \times 0 + \varepsilon N_0\,e^{-\varepsilon/kT} + 2\varepsilon N_0\,e^{-2\varepsilon/kT} + \cdots}{N_0 + N_0\,e^{-\varepsilon/kT} + N_0\,e^{-2\varepsilon/kT} + \cdots}$$

$$= \frac{N_0\varepsilon\,e^{-\varepsilon/kT}\left[1 + 2(N_0\,e^{-\varepsilon/kT}) + 3(N_0\,e^{-\varepsilon/kT})^2 + \cdots\right]}{N_0\left[1 + e^{-\varepsilon/kT} + (e^{-\varepsilon/kT})^2 + \cdots\right]}. \tag{3.15}$$

We recall the following series:

$$\frac{1}{1-x} = 1 + x + x^2 + x^3 + \cdots\,; \qquad \frac{1}{(1-x)^2} = 1 + 2x^2 + 3x^3 + \cdots, \tag{3.16}$$

and hence the mean energy of the oscillator is

$$\overline{E} = \frac{\varepsilon\,e^{-\varepsilon/kT}}{1 - e^{-\varepsilon/kT}} = \frac{\varepsilon}{e^{\varepsilon/kT} - 1}. \tag{3.17}$$

Thus, using the proper Boltzmann procedure, Planck's relation for the mean energy of the oscillator is recovered, provided the energy element ε does not vanish. Einstein's approach indicates clearly the origin of the departure from the classical result. The mean energy $\overline{E} = kT$ is recovered in the classical limit $\varepsilon \to 0$ from (3.17). Notice that, by allowing $\varepsilon \to 0$, the averaging takes place over a continuum of energies which the oscillator might take. In that limit, equal volumes of phase space are given equal weights in the averaging process, and this is the origin of the classical equipartition theorem. Einstein shows that Planck's formula requires this assumption to be wrong. Rather, only those volumes of phase space with energies $0, \varepsilon, 2\varepsilon, 3\varepsilon \ldots$ should have non-zero weights and these should all be equal.

Einstein then relates this result directly to his previous paper on light quanta:

> 'we must assume that for ions which can vibrate at a definite frequency and which make possible the exchange of energy between radiation and matter, the manifold of possible states must be narrower than it is for the bodies in our direct experience. We must in fact assume that the mechanism of energy transfer is such that the energy can assume only the values $0, \varepsilon, 2\varepsilon, 3\varepsilon \ldots$' (Einstein, 1906c)

But this is only the beginning of the paper. Much more is to follow – Einstein puts it beautifully.

> 'I now believe that we should not be satisfied with the result. For the following question forces itself upon us. If the elementary oscillators that are used in the theory of the energy exchange between radiation and matter cannot be interpreted in the sense of the present kinetic molecular theory, must we not also modify the theory for the other oscillators that are used in the molecular theory of heat? There is no doubt about the answer, in my opinion. If Planck's theory of radiation strikes to the heart of the matter, then we must also expect to find contradictions between the present kinetic molecular theory and experiment in other areas of the theory of heat, contradictions that can be resolved by the route just traced. In my opinion, this is actually the case, as I try to show in what follows.' (Einstein, 1906c)

The problem discussed by Einstein concerns the heat capacities of solids. According to the *Dulong and Petit law*, the heat capacity per mole of a solid is $3R$. This result can be derived simply from the equipartition theorem. The model of the solid consists of N_A atoms per mole and it is supposed that they can all vibrate in the three independent directions, x, y, z. According to the equipartition theorem, the internal energy per mole of the solid should therefore be $3N_A kT$, since each independent mode of vibration is awarded an energy kT. The heat capacity per mole follows directly by differentiation: $C = \partial U / \partial T = 3N_A k = 3R$.

It was known that some materials do not obey the Dulong and Petit law in that they have significantly smaller heat capacities than $3R$ – this was particularly true for light elements such as carbon, boron and silicon. In addition, by 1900, it was known that the heat capacities of some elements change rapidly with temperature and only attain the value $3R$ at high temperatures.

The problem is readily solved if Einstein's quantum hypothesis is adopted. For oscillators, the classical formula for the average energy of the oscillator kT should be replaced by the quantum formula

$$\overline{E} = \frac{h\nu}{e^{h\nu/kT} - 1}.$$

Now atoms are complicated systems, but let us suppose for simplicity that, for a particular material, they all vibrate at the same frequency, the *Einstein frequency* ν_E, and that these vibrations are independent. Since each atom has three independent modes of vibration, the internal energy is

$$U = 3N_A \frac{h\nu_E}{e^{h\nu_E/kT} - 1}, \tag{3.18}$$

and the heat capacity is

$$\frac{dU}{dT} = 3N_A h\nu_E (e^{h\nu_E/kT} - 1)^{-2} e^{h\nu_E/kT} \frac{h\nu_E}{kT^2},$$

$$= 3R \left(\frac{h\nu_E}{kT}\right)^2 \frac{e^{h\nu_E/kT}}{(e^{h\nu_E/kT} - 1)^2}. \tag{3.19}$$

Einstein compared the experimentally determined variation of the heat capacity of diamond with his formula with the results shown in Fig. 3.2. The decrease in the heat capacity at low temperatures is apparent, although the experimental points lie slightly above the predicted relation at low temperatures.

We can now understand why light elements have smaller heat capacities than the heavier elements. Presumably, the lighter elements have higher vibrational frequencies than heavier elements and hence, at a given temperature, ν_E / T is larger and the heat capacity is smaller. To account for the experimental data shown in Fig. 3.2, the frequency ν_E must lie in the infrared waveband. As a result, all vibrations at higher frequencies make only a vanishingly small contribution to the heat capacity. As expected, there is strong absorption at infrared wavelengths corresponding to frequencies $\nu \approx \nu_E$. Einstein compared his estimates of ν_E with the strong absorption features observed in a number of materials and found remarkable agreement, granted the simplicity of the model.

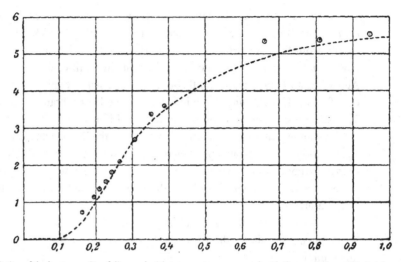

Fig. 3.2 The variation of the heat capacity of diamond with temperature compared with the prediction of Einstein's quantum theory. The abscissa is T/θ_E, where $k\theta_E = h\nu_E$, and the ordinate the molar heat capacity in calories mole^{-1}. This diagram appears in Einstein's paper of 1906 and uses the results of Heinrich Weber which were listed in the tables of Landolt and Börnstein (Einstein, 1906c).

The most important prediction of the theory was that the heat capacities of all solids should decrease to zero at low temperatures, as indicated in Fig. 3.2. This was of key importance from the point of view of furthering the acceptance of Einstein's ideas. At about this time, Walther Nernst began a series of experiments to measure the heat capacities of solids at low temperatures. His motivation was to test his *heat theorem*, or the *third law of thermodynamics*, which he had developed theoretically in order to understand the nature of chemical equilibria. The heat theorem enabled the calculation of chemical equilibria to be carried out precisely and also led to the prediction that the heat capacities of all materials should tend to zero at low temperatures. As recounted by Frank Blatt,

> '... shortly after Einstein had assumed a junior faculty position at the University of Zurich [in 1909], Nernst paid the young theorist a visit so that they could discuss problems of common interest. The chemist George Hevesy ... recalls that among his colleagues it was this visit by Nernst that raised Einstein's reputation. He had come as an unknown man to Zurich. Then, Nernst came, and the people at Zurich said, "This Einstein must be a clever fellow, if the great Nernst comes so far from Berlin to Zurich to talk to him."' (Blatt, 1992)

3.5 Debye's theory of specific heats

Einstein took little interest in the heat capacities of solids after 1907 but his ideas on quantisation were taken significantly further by Pieter Debye in an important paper of 1912

(Debye, 1912). Einstein was well aware of the fact that the assumption that the atoms of a solid vibrate independently was a crude approximation. Debye took the opposite approach of returning to a continuum picture, almost identical to that developed by Rayleigh in his treatment of the spectrum of black-body radiation (Sect. 2.3.4). Debye realised that the collective modes of vibration of the solid could be represented by the complete set of normal modes which follows from fitting waves into a box, as described by Rayleigh. Each independent mode of oscillation of a solid as a whole should be awarded an energy

$$\overline{E} = \frac{\hbar\omega}{\exp(\hbar\omega/kT) - 1} \,, \tag{3.20}$$

according to Einstein's prescription, where ω is the angular frequency of vibration of the mode. The number of modes $\mathcal{N}$ had been evaluated by Rayleigh according to the procedure described in Sect. 2.3.4 and was shown to be

$$d\mathcal{N} = \frac{L^3 \omega^2}{2\pi^2 c_s^3} d\omega \,, \tag{3.21}$$

where c_s is the speed of propagation of the waves in the material. Just as in the case of electromagnetic radiation, we need to determine the number of independent polarisation states for the wave modes. In this case, there are two transverse modes and one longitudinal mode, corresponding to the independent directions in which the material can be stressed by the wave, and so in total there are $3\,d\mathcal{N}$ modes, each of which is awarded the energy (3.20). Debye makes the assumption that these modes have the same speed of propagation and that it is independent of the frequency of the modes. Therefore, the total internal energy of the material is

$$\begin{aligned} U &= \int_0^{\omega_{\max}} \frac{\hbar\omega}{\exp(\hbar\omega/kT) - 1} \, 3 \, d\mathcal{N} \\ &= \frac{3}{2\pi^2} \left(\frac{kTL}{\hbar c_s} \right)^3 \int_0^{x_{\max}} \frac{x^3}{e^x - 1} \, dx \,, \end{aligned} \tag{3.22}$$

where $x = \hbar\omega/kT$.

The problem is now to determine the value of $x_{\max}$. Debye introduced the idea that there must be a limit to the total number of modes in which energy could be stored. In the high temperature limit, he argued that the total energy should not exceed that given by the classical equipartition theorem, namely $3NkT$. Since each mode of oscillation has energy kT in this limit, there should be a maximum of $3N$ modes in which energy is stored. Therefore, recalling that there are $3\mathcal{N}$ modes, Debye's condition can be found by integrating (3.21)

$$3N = 3 \int_0^{\omega_{\max}} d\mathcal{N} = 3 \int_0^{\omega_{\max}} \frac{L^3 \omega^2}{2\pi^2 c_s^3} \, d\omega \,,$$

$$\omega_{\max}^3 = \frac{6\pi^2 N}{L^3} c_s^3 \,. \tag{3.23}$$

It is conventional to write $x_{\max} = \hbar\omega_{\max}/kT = \theta_D/T$, where θ_D is known as the *Debye temperature*. Therefore, the expression for the total internal energy of the material (3.22) can be rewritten

$$U = 9RT \left(\frac{T}{\theta_D} \right)^3 \int_0^{\theta_D/T} \frac{x^3}{e^x - 1} \, dx , \tag{3.24}$$

for one mole of the material. This is the famous expression derived by Debye for the internal energy per mole of the solid.

To find the heat capacity, it is simplest to consider an infinitesimal increment dU associated with a single frequency ω and then integrate over x as before,

$$C = \frac{dU}{dT} = 9R \left(\frac{T}{\theta_D} \right)^3 \int_0^{\theta_D/T} \frac{x^4}{(e^x - 1)^2} \, dx . \tag{3.25}$$

This integral cannot be written in closed form, but it provides a better fit to the data on the heat capacity of solids than Einstein's expression (3.19). This is particularly true for the data at low temperatures. If $T \ll \theta_D$, the upper limit to the integral in (3.25) can be set to infinity, and then the integral has the value $4\pi^4/15$. Therefore, at low temperatures $T \ll \theta_D$, the heat capacity depends upon temperature as

$$C = \frac{dU}{dT} = \frac{12\pi^4}{5} R \left(\frac{T}{\theta_D} \right)^3 . \tag{3.26}$$

Rather than decreasing exponentially at low temperatures, the heat capacity varies as T^3.

There is a simple interpretation of ω_{max} in terms of wave propagation in the solid. From (3.23), the maximum frequency is

$$\nu_{max} = \left(\frac{3}{4\pi} \right)^{1/3} \left(\frac{N_A}{L^3} \right)^{1/3} c_s , \tag{3.27}$$

for one mole of the solid, where N_A is Avogadro's number. But $L/N_A^{1/3}$ is just the typical interatomic spacing, a. Therefore, (3.27) states that $\nu_{max} \approx c_s/a$, that is, the minimum wavelength of the waves $\lambda_{min} = c_s/\nu_{max} \approx a$. This makes a great deal of sense physically. On scales less than the interatomic spacing a, the concept of collective vibration of the atoms of the material ceases to have any meaning.

3.6 Fluctuations of particles and waves – Einstein (1909)

Einstein's startling new ideas on light quanta did not gain immediate acceptance by the scientific community at large. Most of the major figures in physics rejected the idea that light could be considered to be made up of discrete quanta. In a letter to Einstein of 1907, Planck wrote:

> 'I look for the significance of the elementary quantum of action (light quantum) not in vacuo but rather at points of absorption and emission and assume that processes in vacuo are accurately described by Maxwell's equations. At least, I do not yet find a compelling reason for giving up this assumption which for the time being seems to be the simplest.' (Planck, 1907)

Planck continued to reject the light quantum hypothesis as late as 1913. In 1909, Lorentz, who was generally regarded as the leading theoretical physicist in Europe, and whom Einstein held in the highest esteem, wrote:

> 'While I no longer doubt that the correct radiation formula can only be reached by way of Planck's hypothesis of energy elements, I consider it highly unlikely that these energy elements should be considered as light quanta which maintain their identity during propagation.' (Lorentz, 1909)

Einstein never deviated from his conviction concerning the reality of quanta and continued to find other ways in which the experimental features of black-body radiation lead inevitably to the conclusion that light consists of quanta. In one of his most impressive papers written in 1909, he showed how fluctuations in the intensity of the black-body radiation spectrum provide further evidence for the quantum nature of light (Einstein, 1909). Notice how the theme of fluctuations and stochastic processes keeps reappearing in Einstein's physical understanding. I have given a detailed treatment of the theory of fluctuations in the number densities of particles and waves in Sects. 15.2.1 and 15.2.2 of *TCP2*. Here the results of these calculations are summarised.

3.6.1 Particles in a box

A box is divided into N equal cells and a large number of particles n is distributed randomly among them. If n is very large, the mean number of particles in each cell is roughly the same, but there is a real scatter about the mean value because of statistical fluctuations. Suppose p is the probability of a single cell being occupied and q the probability that it is not occupied so that $p + q = 1$. The probability distribution can be worked out exactly using permutation theory and then converted to a continuous distribution. The result is a normal, or Gaussian, distribution $p(x)\,\mathrm{d}x$, which can be written

$$p(x)\,\mathrm{d}x = \frac{1}{(2\pi\sigma^2)^{1/2}}\exp\left(-\frac{x^2}{2\sigma^2}\right)\mathrm{d}x \qquad (3.28)$$

where $\sigma^2 = npq$ is the variance; x is measured with respect to the mean value np. If the box is divided into N cells, the probability of a particle being in a single cell in one experiment is $p = 1/N$, $q = (1 - 1/N)$. The total number of particles is n. Therefore, the average number of particles per sub-box is n/N and the variance about this mean value, that is, the mean squared statistical fluctuation about the mean, is

$$\sigma^2 = \frac{n}{N}\left(1 - \frac{1}{N}\right). \qquad (3.29)$$

If N is large, $\sigma^2 = n/N$ and is the average number of particles in each cell, that is, $\sigma = (n/N)^{1/2}$. This is the well-known result that, for large values of N, the mean is equal to the variance. This is the origin of the useful rule that the fractional fluctuation about the average value is $1/M^{1/2}$ where M is the number of discrete objects counted.

3.6.2 Fluctuations of randomly superposed waves

The random superposition of waves is different in important ways. Suppose the electric field $\boldsymbol{E}$ at some point in space is the random superposition of the electric fields from N sources, where N is very large. For simplicity, we consider only propagation in the z-direction and only one of the two linear polarisations of the waves, E_x or E_y. We also assume that the frequencies ν and the amplitudes ξ of all the waves are the same, the only difference being their random phases. Then, the quantity $E_x^* E_x = |\boldsymbol{E}|^2$ is proportional to the Poynting vector flux density in the z-direction for the E_x component and so is proportional to the energy density of the radiation, where E_x^* is the complex conjugate of E_x. Since the phases of the waves are random,

$$\langle E_x^* E_x \rangle = N \xi^2 \propto u_x . \tag{3.30}$$

This is a familiar result. For incoherent radiation, meaning for waves with random phases, the total energy density is equal to the sum of the energies in all the waves.

A similar calculation can be carried out for the fluctuations in the average energy density of the waves. We work out the quantity $\langle (E_x^* E_x)^2 \rangle$ with respect to the mean value (3.30). Recalling that $\langle \Delta n^2 \rangle = \langle n^2 \rangle - \langle \overline{n} \rangle^2$,

$$\Delta u_x^2 \propto \langle (E_x^* E_x)^2 \rangle - \langle E_x^* E_x \rangle^2 . \tag{3.31}$$

As shown in Sect. 15.2.2 of *TCP2*,

$$\Delta u_x^2 = u_x^2 , \tag{3.32}$$

that is, *the fluctuations in the energy density are of the same magnitude as the energy density of the radiation field itself.* Despite the fact that the radiation measured by a detector is a superposition of a large number of waves with random phases, the fluctuations in the fields are as large as the magnitude of the total intensity. The physical meaning of this calculation is clear. Every pair of waves of frequency ν interferes to produce fluctuations in intensity of the radiation $\Delta u \approx u$. Notice that this analysis refers to waves of random phase ϕ and of a particular angular frequency ω, that is, what we would refer to as waves corresponding to a single mode.

3.6.3 Fluctuations in black-body radiation

Einstein begins his paper of 1909 by reversing Boltzmann's relation between entropy and probability:

$$W = e^{S/k} . \tag{3.33}$$

Consider the radiation in the frequency interval ν to $\nu + d\nu$. As before, we write $\varepsilon = V u(\nu) \, d\nu$. Now divide the volume into a large number of cells and suppose that $\Delta \varepsilon_i$ is the fluctuation in the ith cell. Then the entropy of this cell is

$$S_i = S_i(0) + \left(\frac{\partial S}{\partial U} \right) \Delta \varepsilon_i + \tfrac{1}{2} \left(\frac{\partial^2 S}{\partial U^2} \right) (\Delta \varepsilon_i)^2 + \cdots . \tag{3.34}$$

But, averaging over all cells, we know that there is no net fluctuation, $\sum_i \Delta \varepsilon_i = 0$, and therefore

$$S = \sum S_i = S(0) + \frac{1}{2} \left(\frac{\partial^2 S}{\partial U^2} \right) \sum (\Delta \varepsilon_i)^2. \tag{3.35}$$

Therefore, using (3.33), the probability distribution of the fluctuations is

$$W \propto \exp \left[\frac{1}{2} \left(\frac{\partial^2 S}{\partial U^2} \right) \frac{\sum (\Delta \varepsilon_i)^2}{k} \right]. \tag{3.36}$$

This is the sum of a set of normal distributions which, for any individual cell, can be written

$$W_i \propto \exp \left[-\frac{1}{2} \frac{(\Delta \varepsilon_i)^2}{\sigma^2} \right] \quad \text{where} \quad \sigma^2 = -\frac{k}{(\partial^2 S/\partial U^2)}. \tag{3.37}$$

Notice that we have obtained a physical interpretation for the second derivative of the entropy with respect to energy, which had played a prominent part in Planck's original analysis (see equation (2.28)).

Let us now find σ^2 for a black-body spectrum:

$$u(\nu) = \frac{8\pi h \nu^3}{c^3} \frac{1}{e^{h\nu/kT} - 1}. \tag{3.38}$$

Inverting (3.38),

$$\frac{1}{T} = \frac{k}{h\nu} \ln \left(\frac{8\pi h \nu^3}{c^3 u} + 1 \right). \tag{3.39}$$

We now express this result in terms of the total energy in the cavity in the frequency interval ν to $\nu + d\nu$, $\varepsilon = V u \, d\nu$. As before, $dS/dU = 1/T$ and we may identify ε with U. Therefore,

$$\frac{\partial S}{\partial \varepsilon} = \frac{k}{h\nu} \ln \left(\frac{8\pi h \nu^3}{c^3 u} + 1 \right) = \frac{k}{h\nu} \ln \left(\frac{8\pi h \nu^3 V \, d\nu}{c^3 \varepsilon} + 1 \right),$$

$$\frac{\partial^2 S}{\partial \varepsilon^2} = -\frac{k}{h\nu} \frac{1}{\left(\dfrac{8\pi h \nu^3 V \, d\nu}{c^3 \varepsilon} + 1 \right)} \times \frac{8\pi h \nu^3 V \, d\nu}{c^3 \varepsilon^2},$$

$$\frac{k}{\partial^2 S/\partial \varepsilon^2} = - \left(h\nu\varepsilon + \frac{c^3}{8\pi \nu^2 V \, d\nu} \varepsilon^2 \right) = -\sigma^2. \tag{3.40}$$

In terms of fractional fluctuations,

$$\frac{\sigma^2}{\varepsilon^2} = \left(\frac{h\nu}{\varepsilon} + \frac{c^3}{8\pi \nu^2 V \, d\nu} \right). \tag{3.41}$$

Einstein noted that the two terms on the right-hand side have quite specific meanings. The first term originates from the Wien part of the spectrum and, if we suppose the radiation consists of photons, each of energy $h\nu$, it corresponds to the statement that the fractional fluctuation in the intensity is just $1/N^{1/2}$ where N is the number of photons, that is,

$$\Delta N/N = 1/N^{1/2}. \tag{3.42}$$

According to the considerations of Sect. 3.6.1, this is exactly the result expected if light consists of discrete particles.

Let us now look more closely at the second term. It originates from the Rayleigh–Jeans part of the spectrum. We ask, 'How many independent modes are there in the box in the frequency range ν to $\nu + d\nu$?' We have already shown in Sect. 2.3.4 that there are $8\pi \nu^2 V \, d\nu / c^3$ modes (see (2.24) *et seq.*). We have also shown in Sect. 3.6.2 that the fluctuations associated with each wave mode have magnitude $\Delta \varepsilon^2 = \varepsilon^2$. When we add together randomly all the independent modes in the frequency interval ν to $\nu + d\nu$, we add their variances and hence

$$\frac{\langle \delta E^2 \rangle}{E^2} = \frac{1}{N_{\text{mode}}} = \frac{c^3}{8\pi \nu^2 V \, d\nu} ,$$

which is exactly the same as the second term on the right-hand side of (3.41).

Thus, the two parts of the fluctuation spectrum correspond to particle and wave statistics, the former corresponding to the Wien part of the spectrum and the latter to the Rayleigh–Jeans part. The amazing aspect of this formula for the fluctuations is that we recall that we add together the variances due to independent causes and the equation

$$\frac{\sigma^2}{\varepsilon^2} = \left(\frac{h\nu}{\varepsilon} + \frac{c^3}{8\pi \nu^2 V \, d\nu} \right) \tag{3.43}$$

states that *we should add independently the variances of the 'wave' and 'particle' fluctuations of the radiation field to find the total magnitude of the fluctuations.* This remarkable expression was to have long-term resonances in the struggles to interpret quantum mechanics once the theory reached its definitive form in the late 1920s.

3.7 The First Solvay Conference

Among those who were persuaded of the importance of quanta was Walther Nernst, who at that time was measuring the low temperature heat capacities of various materials. As recounted in Sect. 3.4, Nernst visited Einstein in Zurich in March 1910 and they compared Einstein's theory with his recent experiments. These experiments showed that Einstein's predictions of the low temperature variation of the specific heat with temperature (3.19) gave a good description of the experimental results. As Einstein wrote to his friend Jakob Laub after the visit,

> 'I consider the quantum theory certain. My predictions with respect to specific heats seem to be strikingly confirmed. Nernst, who has just been here, and Rubens are eagerly occupied with experimental tests, so that people will soon be informed about this matter.' (Einstein, 1910)

By 1911, Nernst was convinced, not only of the importance of Einstein's results, but also of the theory underlying them. The outcome of the meeting with Einstein was dramatic. The number of papers on quanta began to increase rapidly, as Nernst popularised the results of the quantum theory of solids.

Fig. 3.3 The participants in the First Solvay Conference on physics, Brussels 1911 (Langevin and De Broglie, 1912). From left to right at the table: Nernst, Brillouin, Solvay, Lorentz, Warburg, Perrin, Wien, Skłodowska-Curie, Poincaré. From left to right standing: Goldschmidt, Planck, Rubens, Sommerfeld, Lindemann, de Broglie, Knudsen, Hasenöhrl, Hostelet, Herzen, Jeans, Rutherford, Onnes, Einstein, Langevin.

Nernst was a friend of the wealthy Belgian industrialist Ernest Solvay and he persuaded him to sponsor a meeting of a select group of physicists to discuss the issues of quanta and radiation. The idea was first mooted in 1910, but Planck urged that the meeting should be postponed for a year. As he wrote,

> 'My experience leads me to the opinion that scarcely half of those you envisage as participants have a sufficiently lively conviction of the pressing need for reform to be motivated to attend the conference ... Of the entire list you name, I believe that besides ourselves [only] Einstein, Lorentz, W. Wien and Larmor are deeply interested in the topic.' (Planck, 1910)

By the following year, matters were very different. Debye, Haas, Hasenöhrl, Schidlof, Weiss and Wilson had published papers on the quantum hypothesis.

The eighteen official participants met on 29 October 1911 in the Hotel Metropole in Brussels and the meeting took place between the 30th of that month and 3 November (Fig. 3.3). By this time, the majority of the participants took the quantum hypothesis seriously. Two of them were against quanta – Jeans and Poincaré. Five were initially neutral – Rutherford, Brillouin, Skłodowska-Curie, Perrin, Knudsen. The eleven others were basically pro-quanta – Lorentz (chairman), Nernst, Planck, Rubens, Sommerfeld, Wien, Warburg, Langevin, Einstein, Hasenöhrl and Onnes. The secretaries were Goldschmidt, de Broglie and Lindemann; Solvay, who hosted the conference, was there, as well as his

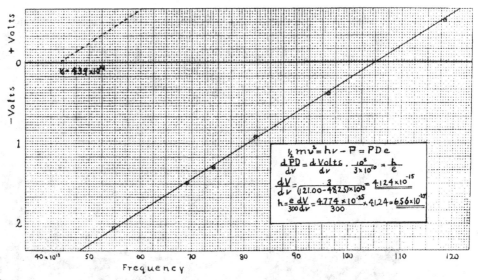

Fig. 3.4   Millikan's results on the photoelectric effect compared with the predictions of Einstein's quantum theory (Millikan, 1916).

collaborators Herzen and Hostelet. The physicists who took a neutral position had done so because they were unfamiliar with the arguments.

The conference had a profound effect in that it provided a forum at which all the arguments could be presented. In addition, all the participants wrote their lectures for publication beforehand and these were then discussed in detail. These discussions were recorded and the full proceedings published within a year of the event in the important volume *La Théorie du rayonnement et les quanta: Rapports et discussions de la réunion tenue à Bruxelles, du 30 octobre au 3 novembre 1911* (Langevin and De Broglie, 1912) Thus, all the important issues were made available to the scientific community in one volume. As a result, the next generation of students became fully familiar with the arguments and many of them set to work immediately to tackle the problems of quanta. Furthermore, these problems began to be appreciated beyond the central European German-speaking scientific community.

A particularly significant convert was Poincaré who immediately tackled the issue of whether or not the introduction of what he called 'discontinuities' were essential in order to understand the spectrum of black-body radiation. In his detailed analysis of the problem, he came to the conclusion that if $w(\varepsilon)$ is the probability density of Planck's resonators, the measured spectrum in the Wien region could only be accounted for if the function was a discontinuous function of the energy ε (Poincaré, 1912). It had to be zero for all values of ε except $\varepsilon = 0$, $h\nu$, $2h\nu$, $3h\nu$.... Poincaré's paper was so compelling that even Jeans conceded that he had to accept the quantum hypothesis in its entirety.

It would be wrong, however, to believe that everyone was suddenly convinced of the existence of quanta. In his paper of 1916 on his famous series of experiments in which he verified the dependence of the photoelectric effect upon frequency (Fig. 3.4), Millikan stated:

'We are confronted however by the astonishing situation that these facts were correctly and exactly predicted nine years ago by a form of quantum theory which has now been generally abandoned.' (Millikan, 1916)

Millikan refers to Einstein's 'bold, not to say reckless, hypothesis of an electromagnetic light corpuscle of energy $h\nu$ which flies in the face of the thoroughly established facts of interference.' (Millikan, 1916).

3.8 The end of the beginning

The 1911 Solvay Conference is a convenient point at which to conclude the introduction to the history of quantum mechanics. The essential role of quanta and the fundamental significance of Planck's constant h could not be ignored. At the same time, something had gone spectacularly wrong with classical physics, but it was far from clear what was to replace it. The remarkable analyses of Planck and Einstein had shown that oscillators and radiation are quantised, but how did it all fit together?

The next phase from 1911 to 1924 I have designated the era of the *old quantum theory* during which many of the pieces began to fall into place and which were eventually to be subsumed into the full theory of quantum mechanics, but the old theory was doomed to failure. Nonetheless, these endeavours uncovered many of the essential features of quantum processes which were only to be rationalised within the context of quantum mechanics during the dramatic years from 1925 to 1930.

PART II

THE OLD QUANTUM THEORY

The Bohr model of the hydrogen atom

Following the success of the 1911 Solvay Conference and the rapid dissemination of the proceedings, the emphasis of research shifted towards the understanding of the spectra of atoms and molecules. With the availability of precision spectroscopic techniques, the bewildering variety of spectral features of atoms and molecules became apparent. The efforts described in Sect. 1.6 indicate how regularities were found in the patterns of spectral lines, the culmination of these investigations being the discovery of the formula for the Balmer series and the various formulae to account for the principal, diffuse and sharp series of the lines in the spectra of sodium, potassium, magnesium, calcium and zinc. As Planck remarked in 1902,

'If the question concerning the nature of white light may thus be regarded as being solved, the answer to the closely related but no less important question – the question concerning the nature of light of the spectral lines – seems to belong among the most difficult and complicated problems, which have ever been posed in optics or electrodynamics.' (Planck, 1902)

4.1 The Zeeman effect: Lorentz and Larmor's interpretations

In 1862, Faraday attempted to measure the change in wavelength of spectral lines when the source of the lines was placed in a strong magnetic field, but failed to observe any positive effect (Jones, 1870). Inspired by this negative result, Pieter Zeeman repeated the experiment and discovered the broadening of the D lines of sodium when a sodium flame was placed between the poles of a strong electromagnet (Zeeman, 1896a). Zeeman used a high quality Rowland grating with a radius of 10 feet and 14 938 lines per inch but the 10 kG produced by the magnet was insufficient to resolve the broadened lines (Fig. 4.1). By the end of October 1896, Zeeman was convinced that the broadening of the spectral lines was a real effect, the broadening being proportional to the applied magnetic flux density, and his paper was presented on Saturday 31 October 1896 to the Science Section of the Dutch Academy of Sciences. Over that same weekend, Lorentz interpreted this result in terms of the splitting of spectral lines due to the motion of the 'ions' in the atoms in the magnetic field.

Lorentz had derived the correct expression for the force acting on a charge in combined electric and magnetic fields in 1892, the *Lorentz force*,

$$F = e(E + v \times B),$$
(4.1)

Fig. 4.1   The original electromagnet used by Zeeman in 1896 is in the Museum Boerhaave in Leiden. The experiment was reconstructed in 2002, on the occasion of the centennial anniversary of the Nobel Prize for Zeeman and Lorentz.

where E is the electric field strength and B is the magnetic flux density (Lorentz, 1892b). Lorentz carried out the following calculation.[1] It is supposed that the emission line is due to the vibration of oscillators within the atoms of the material, each characterised by a mass m and a spring constant k. In the presence of a uniform magnetic field in the z-direction, the equations of motion of the oscillator under the influence of the Lorentz force can be written,

$$m\frac{\mathrm{d}^2x}{\mathrm{d}t^2} = -kx + eB\frac{\mathrm{d}y}{\mathrm{d}t}, \tag{4.2}$$

$$m\frac{\mathrm{d}^2y}{\mathrm{d}t^2} = -ky - eB\frac{\mathrm{d}x}{\mathrm{d}t}, \tag{4.3}$$

$$m\frac{\mathrm{d}^2z}{\mathrm{d}t^2} = -kz. \tag{4.4}$$

The solution of (4.4) is $z = a\cos(\omega_0 t + p)$ where a and p are constants and ω_0 is the angular frequency of the oscillator, $\omega_0 = 2\pi\nu_0 = \sqrt{k/m}$. Lorentz found the following two solutions for motion in the x- and y-directions:

$$\begin{cases} x = a_1\cos(\omega_1 t + p_1); \\ y = -a_1\sin(\omega_1 t + p_1), \end{cases} \tag{4.5}$$

and

$$\begin{cases} x = a_2\cos(\omega_2 t + p_2); \\ y = a_2\sin(\omega_2 t + p_2), \end{cases} \tag{4.6}$$

where

$$\omega_1^2 - \frac{eB}{m}\omega_1 = \omega_0^2 \quad \text{and} \quad \omega_2^2 + \frac{eB}{m}\omega_2 = \omega_0^2 . \tag{4.7}$$

The angular frequencies ω_1 and ω_2 are just slightly displaced from ω_0 and so, for small departures $\Delta\omega_0$ from ω_0, we find

$$\Delta\omega_0 = \pm\frac{eB}{2m} \quad \text{or} \quad \Delta\nu_0 = \pm\frac{eB}{4\pi m} . \tag{4.8}$$

This analysis made definite predictions about the polarisation of the broadened lines. First of all, inspection of (4.5) and (4.6) shows that these motions correspond to circular motion in opposite senses about the magnetic field direction. Therefore, as viewed along the magnetic field direction, the radiation of the 'ions' should be circularly polarised in opposite senses on either side of the unperturbed frequency ν_0. Along this direction there should be no emission at ν_0 because the acceleration is along the line of sight.[2] On the other hand, when viewed perpendicular to the magnetic field direction, all three components should be observed. They are all linearly polarised, the central frequency component being polarised parallel to the field direction and the displaced components perpendicular to the field direction. Zeeman continued his careful measurements over the next two months and discovered that the polarisation properties of the broadened lines agreed with these expectations (Zeeman, 1896b). Several months later, the splitting of the blue cadmium line into separate lines was observed (Zeeman, 1897). For observations along the magnetic field direction two components were observed, while triplets were observed for measurements perpendicular to it.

Furthermore, according to (4.8), the splitting of the line depends upon the ratio e/m and so the broadening of the line enables this ratio to be found for the 'ions' responsible for the emission lines. Lorentz found a lower limit of 1000 for the value of e/m_e relative to that of the hydrogen ion. This came as a surprise to Lorentz and his colleagues since the ions responsible for electrolytic phenomena had values of e/m similar to that of the hydrogen ion. The Zeeman effect thus provided a means of studying the internal structure of atoms.

At about the same time, an alternative approach to the splitting of spectral lines was proposed by Joseph Larmor (Larmor, 1897). In the simplest case, consider a charged particle of mass m and electric charge e moving in a circular orbit. There is an electric current associated with this motion resulting in a magnetic moment μ which is related to the angular momentum of the particle L by the classical relation $\mu = (e/2m)L$. Suppose the axis of the magnetic dipole of the orbit is at an angle θ with respect to the magnetic field direction B. Then, there is a torque acting on the orbiting electron, the magnitude and direction of the torque being given by the vector relation $\Gamma = \mu \times B$. The action of the torque causes the angular momentum vector to precess about the direction of the magnetic field in the azimuthal ϕ direction since

$$\Gamma = \frac{dL}{dt} . \tag{4.9}$$

From the geometry of the precession, $|\mathrm{d}\boldsymbol{L}| = |\boldsymbol{L}|\sin\theta\,\mathrm{d}\phi$ and so the angular frequency of precession is

$$\omega_{\mathrm{p}} = \frac{\mathrm{d}\phi}{\mathrm{d}t} = \frac{|\mathrm{d}\boldsymbol{L}|}{\mathrm{d}t}\frac{1}{|\boldsymbol{L}|\sin\theta} = \frac{|\boldsymbol{\Gamma}|}{|\boldsymbol{L}|\sin\theta} = \frac{|\boldsymbol{\mu}||\boldsymbol{B}|\sin\theta}{|\boldsymbol{L}|\sin\theta} = \frac{eB}{2m}. \qquad (4.10)$$

This is exactly the same formula as derived by Lorentz. Thus, there were two 'ion' pictures involving the magnetic field – either the splitting was associated with the effect of the magnetic field upon a linear oscillator or with the precessional motion of an orbiting charge.

These discoveries gave significant insights into the physics of atoms, but the picture was soon clouded by the discoveries of Michelson (1897) and Preston (1898) that the spectral lines of atoms could be split into four, six or more components. These results were inconsistent with the Lorentz picture which became known as the *normal Zeeman effect*. The higher order splittings were referred to as the *anomalous Zeeman effect*, the explanation of which was at least 20 years in the future.

4.2 The problems of building models of atoms

By 1900, the existence of the electron was clearly established through the discovery of the Zeeman effect and the values of e/m determined from the discharge tube experiments of Thomson, Wiechert and Kaufmann described in Sect. 2.2.3. As Planck noted above, however, the problems of constructing models of atoms were formidable. Heilbron (1977) conveniently lists six basic questions which faced the model builders.

- The *nature of the positive charge* necessary to create electrically neutral atoms. Was charge neutrality provided by an equal number of positively charged electrons or by some other distribution of positive charge?
- The *number of electrons* in the atom was uncertain. From their charge-to-mass ratio, there could be thousands of electrons in the atom and that was perhaps not unreasonable in view of the large numbers of spectral lines observed in atomic spectra. Even the lightest elements have large numbers of spectral lines while several thousands of lines are observed in the spectrum of iron.
- There is nothing in classical physics which could establish a *natural length-scale for atoms*. A clue was at hand with the introduction of Planck's constant h in his epochal paper of 1900, but it was over a decade before Bohr showed how the concept of quantisation could be applied to determine the size of the hydrogen atom.
- The problem of the *collapse of the atom* due to the radiation of electromagnetic radiation by the orbiting electrons was a major stumbling block for atomic theorists. As we will see in Sect. 4.3.2, there are ways of minimising the problem, but these were to prove inadequate.
- Even if these problems could be overcome, there was still the problem of understanding the *origin of the various formulae for the spectral lines* discussed in Sect. 1.6.

- Finally, there remained the fundamental problem of understanding the *nature of the oscillators* responsible for the observation of spectral lines.

These issues were to be addressed by a combination of experiment and theory over the following decade, the pieces of the jigsaw gradually falling into place. Two of the problems were to find definitive solutions through the experiments of Thomson and Rutherford.

4.3 Thomson and Rutherford

4.3.1 Thomson and the numbers of electrons in atoms

In 1906, Thomson published his analysis of three different ways of estimating the number of electrons in atoms (Thomson, 1906). His model of the atom involved a swarm of electrons within a neutralising sphere of positive charge. The first approach involved working out the dispersion of light of different frequencies when it passes through a gas. He applied the method successfully to hydrogen and found that the number of electrons had to be approximately equal to the atomic mass number $A = 1$.

The second approach involved the scattering of X-rays by electrons. X-rays are scattered by the electrons in atoms by *Thomson scattering*, the theory of which was worked out by Thomson using the classical expression for the radiation of an accelerated electron (Thomson, 1907). It is straightforward to show that the cross-section for the scattering of a beam of incident radiation by an electron is

$$\sigma_T = \frac{e^4}{6\pi \epsilon_0^2 m_e^2 c^4} = \frac{8\pi r_e^2}{3} = 6.653 \times 10^{-29} \text{ m}^2, \tag{4.11}$$

the *Thomson cross-section*, where $r_e = e^2/4\pi \epsilon_0 m_e c^2$ is the classical electron radius.[3] In his model of the atom, it was assumed that the 'corpuscles' behave like free electrons – notice that the Thomson cross-section is independent of the frequency of the incident radiation. As Thomson states in his paper,

> 'Barkla has shown that in the case of gases the energy in the scattered radiation always bears, for the same gas, a constant ratio to the energy in the primary whatever be the nature of the rays, that is, whether they are hard or soft; and secondly, that the scattered energy is proportional to the mass of the gas. The first of these results is a confirmation of the theory, as the ratio of the energy scattered to that in the primary rays ... is independent of the nature of the rays; the second result shows that the number of corpuscles per cubic centimetre is proportional to the mass of the gas: from this it follows that the number of corpuscles in an atom is proportional to the mass of the atom, that is, to the atomic weight.'

For a beam of X-rays passing through air, Barkla had measured the scattered fraction of the X-ray intensity to be 2.4×10^{-4} cm^{-3} and consequently that there should be about 25 corpuscles per air molecule, roughly equal to the atomic mass number A of the air molecules.

The third approach involved the scattering of β-rays by matter. Thomson derived the formula for what was called 'multiple-scattering theory', the energy loss due to multiple electrostatic interactions between the fast electron and the electrons in atoms – this process is also known as *ionisation losses* in the context of the interactions of electrons with atoms.[4] Using estimates by Rutherford for the mean free path of β-rays, Thomson again found the result $n \sim A$. The conclusion of his paper was dramatic – most of the mass of atoms could not be due to the negatively charged electrons but must reside in the positive charge which held the atoms together. The X-ray experiments were continued by Barkla who showed that in fact, except for hydrogen, the number of electrons is roughly half the atomic weight $n \approx A/2$ (Barkla, 1911a).

4.3.2 The radiative and mechanical instability of atoms

The construction of atomic models was a major industry in the early years of the twentieth century, particularly in England, and has been splendidly surveyed by Heilbron (1977). A key question was, 'How are the electrons and the positive charge distributed inside atoms?' However they are distributed, they cannot be stationary because of *Earnshaw's theorem*, which states that any static distribution of electric charges is mechanically unstable, in that they either collapse or disperse to infinity under the action of electrostatic forces. The alternative is to place the electrons in orbits, what is often called the 'Saturnian' model of the atom, as advocated by Perrin (1901) and Nagaoka (1904a,b). Nagaoka was inspired by Maxwell's model for Saturn's rings and attempted to associate the spectral lines of atoms with small vibrational perturbations of the electrons about their equilibrium orbits.

A major problem with the Saturnian pictures was the radiative instability of the electron. Suppose the electron has a circular orbit of radius a. Then, equating the centripetal force to the electrostatic force of attraction between the electron and the nucleus of charge Ze,

$$\frac{Ze^2}{4\pi\epsilon_0 a^2} = \frac{m_e v^2}{a} = m_e |\ddot{r}|, \tag{4.12}$$

where $|\ddot{r}|$ is the centripetal acceleration. The rate at which the electron loses energy by radiation is given by (2.1). The kinetic energy of the electron is $E = \frac{1}{2}m_e v^2 = \frac{1}{2}m_e a|\ddot{r}|$. Therefore, the time it takes the electron to lose all its kinetic energy by radiation is

$$T = \frac{E}{|dE/dt|} = \frac{2\pi a^3}{\sigma_T c}. \tag{4.13}$$

Taking the radius of the atom to be $a = 10^{-10}$ m, the time it takes the electron to lose all its energy is about 3×10^{-10} s. Something is profoundly wrong. As the electron loses energy, it moves into an orbit of smaller radius, loses energy more rapidly and spirals into the nucleus.

The pioneer atom model builders were well aware of this problem. Fortunately, the wavelength of light λ is very much greater than the size of atoms a and so the solution was to place the electrons in orbits such that there would be no net acceleration when the acceleration vectors of all the electrons in the atom are added together. This requires, however, that the electrons are well ordered in their orbits about the nucleus. If, for example, there are two electrons in the atom, they can be placed in the same circular orbit on opposite sides of the nucleus and so, to first order, there is no net dipole moment as observed at infinity, and hence no dipole radiation. There is, however, a finite electric quadrupole moment and hence radiation at the level $(\lambda/a)^2$, relative to the intensity of dipole radiation, is expected. Since $\lambda/a \sim 10^{-3}$, the radiation problem can be significantly relieved. By adding more electrons to the orbit, the quadrupole moment can be cancelled out as well and so, by adding sufficient electrons to each orbit, the radiation problem can be reduced to manageable proportions. Thus, before Thomson's paper of 1906, the radiative instability could be overcome by assuming that a huge number of electrons were so disposed as to result in no net multipole moments of the electron distribution. Each orbit had to be densely populated with a well-ordered system of large numbers of electrons. This was the basis of Thomson's 'plum-pudding' model in which the well-ordered orbits were embedded in a sphere of positive charge. Thomson's result of 1906 that the numbers of electrons in atoms is of the same order as the atomic mass number meant that this problem could no longer be ignored. The radiative problem was particularly severe for hydrogen, which possesses only one electron.

The other problem with the Saturnian picture of Nagaoka was its mechanical instability. Nagaoka had been inspired by Maxwell's model of Saturn's rings in which stable oscillations were found when rings of particles were perturbed. He attempted to associate the spectral lines of atoms with these oscillations. In Maxwell's case the perturbations were stable under the attractive force of gravity between particles of the ring, but in the case of the repulsive electrostatic forces between electrons, the perturbations were unstable. This mechanical instability was inevitable, even if the radiative instability could be eliminated.

4.3.3 Rutherford, α-particles and the discovery of the atomic nucleus

The nature of β-rays as electrons was quickly assimilated into the armoury of the physicist, but what about the nature of the α-particles? In 1902, while Rutherford was at McGill University in Canada, he showed that the α-particles were deflected by electric and magnetic fields and that their value of e/m was roughly that of the charge-to-mass ratio of hydrogen ions (Rutherford, 1903). Rutherford took up the Langworthy Professorship of Physics at Manchester University in 1907 and, in the following year, demonstrated convincingly that α-particles are helium nuclei (Rutherford and Royds, 1909). A source of α-particles was inserted into a fine glass tube which could be inserted into an evacuated discharge tube. Before this 'needle' was inserted, when a high voltage was maintained across the discharge tube, no evidence for helium was observed. Once the needle was inserted into the tube, the α-particles passed through the thin walls of the glass tube, which were only 0.01 mm in thickness, and the characteristic lines of helium were observed in the discharge tube (Fig. 4.2). This was convincing evidence that α-particles are the nuclei of helium atoms.

The discovery of the nuclear structure of atoms resulted from a brilliant series of experiments carried out by Rutherford and his colleagues, Hans Geiger and Ernest Marsden, in the period 1909–1912. Rutherford had been impressed by the fact that α-particles could pass through thin films rather easily, suggesting that much of the volume of atoms is empty space, although there was clear evidence for small-angle scattering. Rutherford persuaded

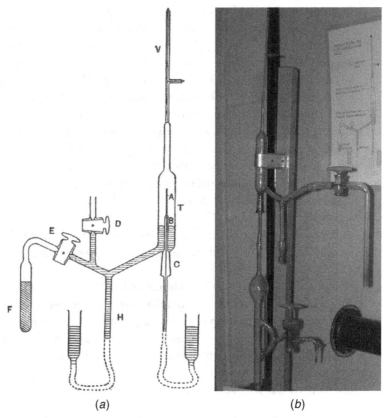

(a)   (b)

Fig. 4.2   (a) The apparatus with which Rutherford and Royds (1909) demonstrated that α-particles are the nuclei of helium atoms. The fine glass tube containing the source of α-particles, a sample of radium, is labelled A. (b) The original experiment in the Cavendish museum.

Marsden, who was still an undergraduate, to investigate whether or not α-particles were deflected through large angles on being fired at a thin gold foil target. To Rutherford's astonishment, a few particles were deflected by more than 90°, and a very small number almost returned along the direction of incidence. In Rutherford's words:

> 'It was quite the most incredible event that has ever happened to me in my life. It was almost as incredible as if you fired a 15-inch shell at a piece of tissue paper and it came back and hit you.' (Andrade, 1964)

Rutherford realised that it required a very considerable force to send the α-particle back along its track. In 1911 he hit upon the idea that, if all the positive charge were concentrated in a compact nucleus, the scattering could be attributed to the repulsive electrostatic force between the incoming α-particle and the positive nucleus. Rutherford was no theorist, but he used his knowledge of central orbits in inverse-square law fields of force to work out the properties of what became known as *Rutherford scattering*[5] (Rutherford, 1911). The orbit of the α-particle is a *hyperbola*, the angle of deflection ϕ being

$$\cot\frac{\phi}{2} = \left[\frac{4\pi\epsilon_0 m_\alpha}{2Ze^2}\right] p_0 v_0^2, \tag{4.14}$$

where p_0 is the collision parameter, v_0 is the initial velocity of the α-particle and Z the nuclear charge. It is straightforward to work out the probability that the α-particle is scattered through an angle ϕ. The result is

$$p(\phi) \propto \frac{1}{v_0^4} \operatorname{cosec}^4 \frac{\phi}{2}, \tag{4.15}$$

the famous $\operatorname{cosec}^4(\phi/2)$ law derived by Rutherford, which was found to explain precisely the observed distribution of scattering angles of the α-particles (Geiger and Marsden, 1913).

Rutherford had, however, achieved much more. The fact that the scattering law was obeyed so precisely, even for large angles of scattering, meant that the inverse-square law of electrostatic repulsion held good to very small distances indeed. They found that the nucleus had to have size less than about 10^{-14} m, very much less than the sizes of atoms, which are typically about 10^{-10} m.

Rutherford attended the First Solvay Conference in 1911, but made no mention of his remarkable experiments, which led directly to his nuclear model of the atom. Remarkably, this key result for understanding the nature of atoms made little impact upon the physics community at the time and it was not until 1914 that Rutherford was thoroughly convinced of the necessity of adopting his nuclear model of the atom. Before that time, however, someone else did – Niels Bohr, the first theorist to apply successfully quantum concepts to the structure of atoms.

4.4 Haas's and Nicholson's models of atoms

Bohr was not, however, the first physicist to attempt to introduce quantum concepts into the construction of atomic models. In 1910, a Viennese doctoral student, Arthur Erich Haas, realised that, if Thomson's sphere of positive charge were uniform, an electron would perform simple harmonic motion through the centre of the sphere, since the restoring force at radius r from the centre would be, according to Gauss's theorem in electrostatics,

$$f = m_e \ddot{r} = -\frac{eQ(\leq r)}{4\pi\epsilon_0 r^2} = -\left(\frac{eQ}{4\pi\epsilon_0 a^3}\right) r, \tag{4.16}$$

where a is the radius of the atom and Q the total positive charge. For a hydrogen atom, for which $Q = e$, the frequency of oscillation of the electron is

$$v = \frac{1}{2\pi}\left(\frac{e^2}{4\pi\epsilon_0 m_e a^3}\right)^{1/2} \tag{4.17}$$

Haas argued that the energy of oscillation of the electron, $E = e^2/4\pi\epsilon_0 a$, should be quantised and set equal to $h\nu$. Therefore,

$$h^2 = \frac{\pi m_e e^2 a}{\epsilon_0}\,. \tag{4.18}$$

Haas used (4.18) to show how Planck's constant could be related to the properties of atoms, taking for ν the short wavelength limit of the Balmer series, that is, allowing $m \rightarrow \infty$ in the Balmer formula (1.17) (Haas, 1910a,b,c). Haas's efforts were discussed by Lorentz at the 1911 Solvay Conference, but they did not attract much attention. According to Haas's approach, Planck's constant was simply a property of atoms as described by (4.18), whereas those already converted to quanta preferred to believe that h had much deeper significance.

The next clue was provided by the work of the Cambridge physicist John William Nicholson, who arrived at the concept of the quantisation of angular momentum. Nicholson (1911, 1912) had shown that, although the Saturnian model of the atom is unstable for perturbations in the plane of the orbit, perturbations perpendicular to the plane are stable for orbits containing up to five electrons – he assumed that the unstable modes in the plane of the orbit were suppressed by some unspecified mechanism. The frequencies of the stable oscillations were multiples of the orbital frequency and he compared these with the frequencies of the lines observed in the spectra of bright nebulae, particularly with the 'nebulium' and 'coronium' lines. Performing the same exercise for ionised atoms with one less orbiting electron, further matches to the astronomical spectra were obtained. The frequency of the orbiting electrons remained a free parameter, but when he worked out the angular momentum associated with them, Nicolson found that they turned out to be multiples of $h/2\pi$. When Bohr returned to Copenhagen from England in 1912, he was perplexed by the success of Nicholson's model, which seemed to provide a successful, quantitative model for the structure of atoms and which could account for the spectral lines observed in astronomical spectra.

4.5 The Bohr model of the hydrogen atom

Niels Bohr completed his doctorate on the electron theory of metals in 1911. Even at that stage, he had convinced himself that this theory was seriously incomplete and required further mechanical constraints on the motion of electrons at the microscopic level. He spent the following year in England, working for seven months with Thomson at the Cavendish Laboratory in Cambridge, and four months with Rutherford in Manchester. Bohr was immediately struck by the significance of Rutherford's model of the nuclear structure of the atom and began to devote all his energies to understanding atomic structure on that basis. He quickly appreciated the distinction between the chemical properties of atoms, which are associated with the orbiting electrons, and radioactive processes which are associated with activity in the nucleus. On this basis, he could understand the nature of the isotopes of a particular chemical species. Bohr also realised from the outset that the structure of atoms could not be understood on the basis of classical physics. The obvious way forward was to incorporate the quantum concepts of Planck and Einstein into the models of atoms. Einstein's statement, quoted in Sect. 3.4,

> '... for ions which can vibrate with a definite frequency, ... the manifold of possible states must be narrower than it is for bodies in our direct experience.' (Einstein, 1906c)

was precisely the type of constraint which Bohr was seeking. Such a mechanical constraint was essential to understand how atoms could survive the inevitable instabilities according to classical physics. How could these ideas be incorporated into models of atoms?

In the summer of 1912, Bohr wrote an unpublished memorandum for Rutherford, in which he made his first attempt at quantising the energy levels of the electrons in atoms (Bohr, 1912). He proposed relating the kinetic energy T of the electron to the frequency $\nu' = v/2\pi a$ of its orbit about the nucleus through the relation

$$T = \tfrac{1}{2} m_e v^2 = K\nu', \tag{4.19}$$

where K is a constant which he expected would be of the same order of magnitude as Planck's constant h. Bohr believed there must be some such non-classical constraint in order to guarantee the stability of atoms. Indeed, his criterion (4.19) absolutely fixed the kinetic energy of the electron about the nucleus. For a bound circular orbit,

$$\frac{mv^2}{a} = \frac{Ze^2}{4\pi\epsilon_0 a^2}, \tag{4.20}$$

where Z is the positive charge of the nucleus in units of the charge of the electron e. As is well known, the binding energy of the electron is

$$E = T + U = \tfrac{1}{2} m_e v^2 - \frac{Ze^2}{4\pi\epsilon_0 a} = -\frac{Ze^2}{8\pi\epsilon_0 a} = -T = \frac{U}{2}, \tag{4.21}$$

where U is the electrostatic potential energy. The quantisation condition (4.19) enables both v and a to be eliminated from the expression for the kinetic energy of the electron. A straightforward calculation shows that

$$T = \frac{m Z^2 e^4}{32\epsilon_0^2 K^2}, \tag{4.22}$$

which was to prove to be of central significance for Bohr. His memorandum containing these ideas was principally about issues such as the number of electrons in atoms, atomic volumes, radioactivity, the structure and binding of diatomic molecules and so on. There is no mention of spectroscopy, which he and Thomson considered too complex to provide useful information.

The breakthrough came in early 1913, when Hans Marius Hansen told Bohr about the Balmer formula for the wavelengths, or frequencies, of the spectral lines in the spectrum of hydrogen,

$$\frac{1}{\lambda} = \frac{\nu}{c} = R_\infty \left(\frac{1}{2^2} - \frac{1}{n^2} \right), \tag{4.23}$$

where $R_\infty = 1.097 \times 10^7 \text{ m}^{-1}$ is the *Rydberg constant* and $n = 3, 4, 5, \ldots$ As Bohr recalled much later,

'As soon as I saw Balmer's formula, the whole thing was clear to me.' (Bohr, 1963)

He realised immediately that this formula contained within it the crucial clue for the construction of a model of the hydrogen atom, which he took to consist of a single negatively charged electron orbiting a positively charged nucleus. He went back to his memorandum on the quantum theory of the atom, in particular, to his expression for the binding energy, or kinetic energy, of the electron (4.21). He realised that he could determine the value of his constant K from the expression for the Balmer series. The running term in $1/n^2$ can be associated with (4.23), if we write for hydrogen with $Z = 1$,

$$T = \frac{m_e e^4}{32\epsilon_0^2 n^2 K^2}. \tag{4.24}$$

Then, when the electron changes from an orbit with quantum number n to that with $n = 2$, the energy of the emitted radiation would be the difference in kinetic energies of the two states. Applying Einstein's quantum hypothesis, this energy should be equal to $h\nu$. Inserting the numerical values of the constants into (4.24), Bohr found that the constant K was exactly $h/2$. Therefore, the energy of the state with quantum number n is

$$E = -T = -\frac{m_e e^4}{8\epsilon_0^2 n^2 h^2}. \tag{4.25}$$

The angular momentum of the state could be found immediately by writing $T = \frac{1}{2}I\omega'^2 = 8\pi^4 m_e a^2 \nu'^2$, from which it follows that

$$J = I\omega' = \frac{nh}{2\pi}. \tag{4.26}$$

This is how Bohr arrived at the *quantisation of angular momentum* according to the old quantum theory. Perhaps most spectacularly, the theory enabled the value of the Rydberg constant to be expressed in terms of fundamental physical constants. From (4.25), it follows immediately that

$$R_\infty = \frac{m_e e^4}{8\epsilon_0^2 h^3 c} = 1.097 \times 10^7 \text{ m}^{-1}. \tag{4.27}$$

In the first paper of his famous trilogy (Bohr, 1913a,b,c), Bohr acknowledged that Nicholson had discovered the quantisation of angular momentum in his papers of 1912. These results were the inspiration for what became known as the *Bohr model of the atom*.

In addition to the Balmer series of hydrogen, Bohr's expression could account for the Paschen series of hydrogen which had been discovered in the near-infrared region of the spectrum by Friedrich Paschen in 1908 (Paschen, 1908). In this case, Bohr's formula became

$$\frac{1}{\lambda} = \frac{\nu}{c} = R_\infty \left(\frac{1}{m^2} - \frac{1}{n^2} \right), \tag{4.28}$$

where again $R_\infty = 1.097 \times 10^7 \text{ m}^{-1}$ but now $m = 3$ and $n = 4, 5, 6, \ldots$ The formula also predicted a series of lines with $m = 1$ and $n = 2, 3, 4, \ldots$, which was discovered by Theodore Lyman in 1914, the Lyman series of hydrogen (Lyman, 1914).

In the first paper of the trilogy of 1913, Bohr noted that a similar formula to (4.23) could account for the *Pickering series*, which had been discovered in 1896 by Edward Pickering in the spectra of stars (Pickering, 1896). In 1912, Alfred Fowler discovered similar series in laboratory experiments (Fowler, 1912). Bohr argued that singly ionised helium atoms would have exactly the same spectrum as hydrogen, but the wavelengths of the corresponding lines would be four times shorter, as observed in the Pickering series. Fowler objected, however, that the ratio of the Rydberg constants for singly ionised helium and hydrogen was not 4, but 4.00163 (Fowler, 1913a). Bohr realised that the problem arose from neglecting the contribution of the mass of the nucleus to the computation of the moments of inertia of the hydrogen atom and the helium ion. If the angular velocity of the electron and the nucleus about their centre of mass is ω, the condition for the quantisation of angular momentum is

$$\frac{nh}{2\pi} = \mu \omega R^2, \tag{4.29}$$

where $\mu = m_e m_N / (m_e + m_N)$ is the *reduced mass* of the atom, or ion, which takes account of the contributions of both the electron and the nucleus to the angular momentum; R is their separation. Therefore, the ratio of Rydberg constants for ionised helium and hydrogen should be

$$\frac{R_{He^+}}{R_H} = 4 \left(\frac{1 + \dfrac{m_e}{M}}{1 + \dfrac{m_e}{4M}} \right) = 4.00160, \tag{4.30}$$

where M is the mass of the hydrogen atom (Bohr, 1913d). Thus, precise agreement was found between the theoretical estimates and laboratory measurements of the ratio of Rydberg constants for hydrogen and ionised helium. In a further paper to *Nature*, Fowler acknowledged that Bohr's formula was indeed a better and more elegant explanation of the lines observed in the spectrum of singly ionised helium (Fowler, 1913b).

Bohr's theory of the hydrogen atom was a quite remarkable achievement and the first convincing application of quantum concepts to atoms. Bohr's dramatic results were persuasive evidence for many scientists that Einstein's quantum theory had to be taken really seriously for processes occurring on the atomic scale. In his biography of Bohr, Pais (1985) recounts the story of Hevesy's encounter with Einstein in September 1913. When Einstein heard of Bohr's analysis of the Balmer series of hydrogen, he remarked cautiously that Bohr's work was very interesting, and important if right. When Hevesy told him about the helium results, Einstein responded,

'This is an enormous achievement. The theory of Bohr must then be right.'

4.6 Moseley and the X-ray spectra of the chemical elements

Support for Bohr's model was not long in coming. Barkla continued his studies of the scattered X-ray emission of different elements and in 1908 he and Charles Sadler discovered

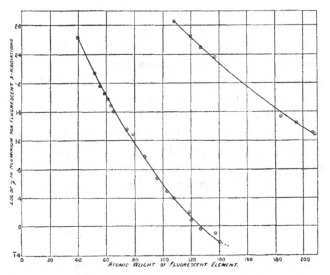

Fig. 4.3   Barkla's summary of his experiments on the *K* (left curve) and *L* components (right curve) of fluorescent X-rays from samples of different elements plotted against atomic weight (Barkla, 1911b). The ordinate is the logarithm of the quantity λ / ρ where λ is the absorption coefficient defined by $I = I_0 \, e^{-\lambda x}$.

that each element had a *characteristic X-ray signature* which was correlated with the atomic weight of the material (Barkla and Sadler, 1908). In these experiments, the absorption of the X-rays by thin aluminium sheets was used to measure the 'hardness' or 'softness' of the X-ray emission. For a number of elements the fluorescent emission consisted of two components, a 'soft' component which was readily absorbed and a 'hard' component which suffered very much less absorption. In 1911, he summarised the results of his numerous absorption experiments (Fig. 4.3), demonstrating that the materials had both hard and soft components, which he labelled *K* and *L* (Barkla, 1911b).

Henry Moseley was a member of Rutherford's team in Manchester, but rather than working on radioactivity, he studied the characteristic X-ray emission of the elements. Barkla's experiments provided a rough indication of the spectra of fluorescent X-rays but the picture changed dramatically with von Laue's discovery of the diffraction of X-rays by crystals (Sect. 2.2.1). Von Laue and his colleagues used the crystal materials as transmission gratings in which the diffracted X-rays passed through the crystal and the diffraction pattern was recorded on a photographic plate. In contrast, William and Lawrence Bragg realised that pure crystal samples such as rock salt could be used as a diffraction grating in which the spectrum of the X-rays could be found in reflection according to Bragg's law $n\lambda = 2d \sin\theta$, where λ is the wavelength of the radiation, d is the lattice spacing and θ is the angle between the crystal planes and the direction of incidence of the X-ray beam; n is an integer. The invention of the X-ray spectrometer enabled X-ray spectra to be recorded on photographic plates with high spectral resolution (Fig. 4.4).

In Moseley's experiments, different pure materials were inserted into an X-ray tube and the reflected X-rays were analysed spectroscopically by reflecting them from a carefully prepared sample of rock salt. He discovered that the reflected spectrum consisted of

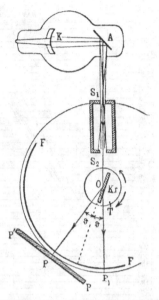

Fig. 4.4   A diagram illustrating the operation of William and Lawrence Bragg's rotating crystal X-ray spectrometer. This apparatus was used by Moseley in his experiments in which the nature of the *K* and *L* lines of the elements was elucidated (Sommerfeld, 1919).

continuum radiation superimposed upon which were strong X-ray lines. The continuum radiation was the *bremsstrahlung*, or braking radiation, of the energetic electrons decelerated in the material. The lines were responsible for the *K* and *L* components of the X-ray emission identified by Barkla. The *K* lines were split into two components which Moseley labelled K_α and K_β. Corresponding splittings were observed in the *L* lines at somewhat longer wavelengths. The K_α line was about five times stronger than the K_β line but the K_β lines had frequencies about 10% greater than those of the K_α lines.

By this date, it was well known that the number of electrons in atoms was roughly half the atomic weight and was correlated with the position of the element in the periodic table. It was also known that the atomic weights of many of the successive elements in the periodic table differed by about two mass units. In 1913, the Dutch lawyer Antonius Johannes van den Broek made the proposal that, starting with $Z = 1$ for hydrogen, each atom is characterised by the number of electrons which is exactly equal to the sequential order of the elements in the periodic table (Fig. 1.2) (van den Broek, 1913). Since there were gaps of more than two mass units in the periodic table, van den Broek proposed that these were filled by as yet undiscovered elements.

The most spectacular result of Moseley's experiments was the discovery of the correlation between the frequency of the X-ray lines and the atomic number Z, corresponding to the number of electrons in the neutral atom (Moseley, 1913, 1914). In his papers, he plotted these correlations separately for the K_α, K_β, L_α and L_β lines (Fig. 4.5). The remarkable linear correlation between the square root of the frequencies of the lines and the atomic number had a number of crucial consequences. Moseley wrote the correlation between the square root of the frequencies of the K_α lines and the atomic number in the somewhat

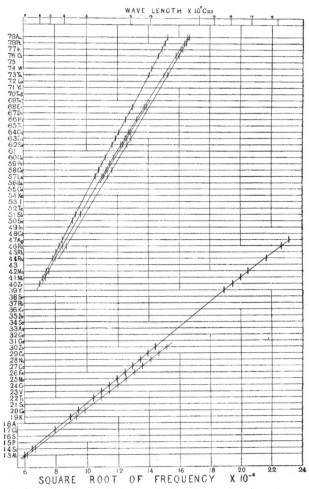

Fig. 4.5 Moseley's correlation diagram between frequency of the various X-ray lines and atomic number. The K_α, K_β lines (bottom half of diagram) and L_α, L_β lines (top half of diagram) are plotted separately and show perfect correlations with atomic number (Moseley, 1913, 1914).

provocative form

$$K_\alpha \text{ lines:} \quad \nu_\alpha = R_\infty (Z - 1)^2 \left(\frac{1}{1^2} - \frac{1}{2^2} \right), \tag{4.31}$$

where R_∞ is the same constant which appears in Rydberg's formula. For the L series, the correlations were described by

$$L_\alpha \text{ lines:} \quad \nu_\alpha = R_\infty (Z - 7.4)^2 \left(\frac{1}{2^2} - \frac{1}{3^2} \right). \tag{4.32}$$

Moseley wrote to Bohr in November 1913 that these results are 'extremely simple and largely what you would expect'. By this he meant that a number of features of these empirical formulae could be immediately explained in terms of Bohr's model of the atom. Rewriting Bohr's formula for the frequencies of the spectral lines for a nucleus of charge Ze, we can write

$$\frac{\nu}{c} = R_\infty Z^2 \left(\frac{1}{m^2} - \frac{1}{n^2} \right) , \qquad (4.33)$$

with m and n different values of the quantum number, generically to be called n, characterising the stationary states. The K_α line would correspond to transitions between the stationary states with $n = 1$ and $n = 2$, while the L_α lines would correspond to transitions from $n = 3$ to $n = 2$. For the K_α line, this would occur if an electron were removed from the $n = 1$ orbit and was replaced by an electron making a transition from the $n = 2$ to the $n = 1$ state. Similarly, the L_α lines would arise from an electron being removed from the $n = 2$ orbit and being replaced by one from the $n = 3$ orbital. It was a puzzle why the dependence upon atomic number should be $(Z - 1)^2$ and $(Z - 7.4)^2$ rather than Z^2. This was to remain a puzzle until the screening of the nucleus by the inner electrons was fully appreciated, but this realisation required a number of key additional features of atomic structure which were to be discovered over the following decade. Rutherford immediately appreciated the significance of Moseley's discovery. In his words,

'The original suggestion of van den Broek that the charge of the nucleus is equal to the atomic number and not to half the atomic weight seems to me very promising. The idea has already been used by Bohr in his theory of the constitution of atoms. The strongest and most convincing evidence in support of this hypothesis will be found in a paper by Moseley in the *Philosophical Magazine* of this month. He there shows that the frequency of the X-radiations from a number of elements can be simply explained if the number of unit charges on the nucleus is equal to the atomic number. It would appear that the charge of the nucleus is the fundamental constant which determines the physical and chemical properties of the atom, while the atomic weight, although it approximately follows the order of the nuclear charge, is probably a complicated function of the latter depending upon the detailed structure of the nucleus.' (Rutherford, 1913)

Conclusions (4) and (5) of Moseley's paper of 1914 read:

(4) The order of the atomic numbers is the same as that of the atomic weights, except where the latter disagrees with the order of the chemical properties.

(5) Known elements correspond with all the numbers between 13 and 79 except three. There are here three possible elements still undiscovered.

Conclusion 4 resulted in the reordering of the elements nickel ($Z = 28$) and cobalt ($Z = 27$), in accord with their chemical properties. Conclusion 5 resulted in the prediction of elements with atomic numbers 43, 61 and 75 (see Fig. 4.5) which were only discovered many years later as the elements technetium (Tc), promethium (Pm) and rhenium (Re) respectively. Tragically, Moseley was killed in action during the Gallipoli campaign in Turkey on 10 August 1915.

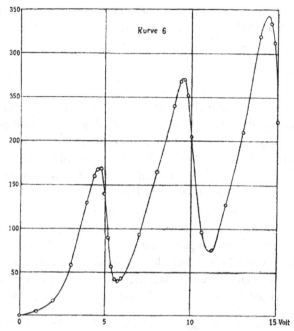

Fig. 4.6 Franck and Hertz' measurements to determine the ionisation potential of mercury vapour. In fact, the maxima in the graph correspond to the excitation of electrons from the ground state to the first excited state 4.9 eV above the ground level (Franck and Hertz, 1914).

4.7 The Franck–Hertz experiment

The reality of the stationary states within atoms was reinforced by the experiments of James Franck and Gustav Hertz (Franck and Hertz, 1914). Their objective was to measure the ionisation potentials of atoms by bombarding them with electrons which had been accelerated through a precisely known electrostatic potential. In their classic experiment of 1914, they aimed to measure the ionisation potential of mercury vapour. Their results are shown in Fig. 4.6 in which, at small voltages, the current plotted on the ordinate increases as the energy of the electrons measured in electron volts increases. At an accelerating voltage of 4.9 eV, however, there is a sudden decrease in the current. As the voltage is increased further, the current again increases and then suddenly decreases at 9.8 eV. In fact, these steep decreases were found at integral multiples of the voltage of the first decrease at 4.9 eV. In addition, they found evidence for an emission line at a wavelength of 253.6 nm, which would correspond to an energy of $h\nu = 4.9$ eV. They interpreted 4.9 eV as the ionisation potential of mercury vapour.

If this were correct it would have contradicted the basic postulates of the Bohr model of the atom since the mercury spectrum also displayed higher frequency Paschen lines with a series limit at 185.0 nm. Bohr interpreted the results of their experiments differently. He argued that the energy of 4.9 eV corresponded to the *excitation energy* of an electron from the ground state to the first excited state and that the line at 253.6 nm represented the emission associated with the transition of an electron from the first excited state to fill the vacancy in the ground state (Bohr, 1915). The steep decreases at multiples of the first excited state corresponded to the electron being accelerated through a sufficient voltage to remove two or more successive electrons from the ground state. In due course, Franck and Hertz agreed with Bohr's interpretation.

The significance of the Franck–Hertz experiment was that it provided compelling evidence for the existence of stationary states within atoms, quite independent of the spectroscopic data.

4.8 The reception of Bohr's theory of the atom

The experiments and their interpretation described in the last three sections indicated that, whatever reservations theorists might have had about the underlying physics of the Bohr model of the atom, the concepts of quantisation and stationary states within atoms had to be taken seriously. Despite the concerns about the stability of atoms according to Bohr's picture, many results were beginning to fall into place and the theorists could not neglect them. The succeeding years would witness a major increase in activity in experimental and theoretical physics dedicated to the study of quanta. The Bohr picture contained only a single quantum number n which labelled the energies of the stationary states. This was the first step along a chain of events which would lead to the appreciation that more quantum numbers and selection rules would have to be introduced to understand the plethora of quantum phenomena observed in atoms. In these endeavours, Bohr was fortunate in involving Sommerfeld in tackling these problems.

5 Sommerfeld and Ehrenfest – generalising the Bohr model

5.1 Introduction

Bohr's success in accounting for the frequencies observed in the spectral series of hydrogen was rightly regarded as a triumph, despite the fact that it violated the classical laws of mechanics and electromagnetism. It could not account, however, for the spectra of helium and heavier elements. The Bohr model was the simplest possible model for the dynamics of a single electron in the electrostatic potential of a positively charged point nucleus, in that it involved only quantised circular orbits defined by a *single quantum number n*, what became known as the *principal quantum number*. At the 1911 Solvay Conference, before Bohr's announcement of his model for the hydrogen atom, Poincaré had raised the issue of how the quantisation conditions could be extended to systems of more than one degree of freedom. The problem was attacked by both Planck and Sommerfeld. Their approaches ended up being essentially the same, although expressed in somewhat different language. We will follow Sommerfeld's approach.

In 1891 Michelson had shown that the Hα and Hβ lines of the Balmer series displayed very narrow splittings (Michelson, 1891, 1892). Although incompatible with Bohr's theory, the problem was set aside in the face of the other remarkable successes of the theory. Sommerfeld suspected that the explanation lay in the fact that Bohr's quantisation condition involved only a single degree of freedom. In his papers of 1915 and 1916, he extended the quantisation of the orbits of the electron to more than one degree of freedom and accounted for the splitting of the lines of the Balmer series once a special relativistic treatment of the model was adopted (Sommerfeld, 1915a,b, 1916a). These advances are beautifully described in his influential book *Atombau und Spektrallinien* (*Atomic Structure and Spectral Lines*) (Sommerfeld, 1919) which was based upon his lecture courses on atomic spectra and their interpretation. Van der Waerden (1967) remarks that,

> 'it was mainly from this book that the young physicists who created Quantum Mechanics in 1925–26 learned Quantum Theory.'

Let us demonstrate exactly what Sommerfeld achieved.

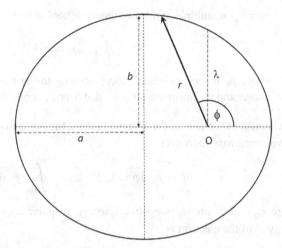

Illustrating the geometry of an ellipse with eccentricity $\epsilon = 0.5$ in the (r, ϕ) coordinate system used in Sect. 5.2.

5.2 Sommerfeld's extension of the Bohr model to elliptical orbits

The obvious extension of the Bohr model was to elliptical rather than circular orbits. The equation for an ellipse in pedal, or (r, ϕ), coordinates is

$$\frac{\lambda}{r} = 1 + \epsilon \cos\phi \, , \tag{5.1}$$

where $\lambda = a(1 - \epsilon^2)$ is the semi-latus rectum, a is the semi-major axis of the ellipse and ϵ is its eccentricity. r is the radial distance from a focus to a point on the ellipse and ϕ is the angle between the major axis and the radius vector r. The semi-minor axis of the ellipse is $b = a(1 - \epsilon^2)^{1/2}$. This geometry is illustrated in Fig. 5.1.

It is apparent that there are now two independent parameters which define the orbit of the electron, r and ϕ. The Bohr model involves the quantisation of the angular coordinate ϕ through the quantisation of angular momentum $m_e v r = n_\phi h/2\pi = n_\phi \hbar$, where r and v are constants.[1] Bohr and Sommerfeld realised that this quantisation condition could also be written in the form

$$\int_0^{2\pi} p_\phi \, \mathrm{d}\phi = n_\phi h \, , \tag{5.2}$$

where p_ϕ is the angular momentum and ϕ the azimuthal angle. Sommerfeld, widely recognised as one of the leading mathematical physicists in Europe, appreciated that p_ϕ and ϕ are canonical coordinates in Hamiltonian mechanics and so (5.2) provides the prescription for the quantisation condition of such pairs of coordinates. As expressed by Jammer (1989),

'Sommerfeld postulated that the stationary states of a periodic system with f degrees of freedom are determined by the condition that the 'phase integral for every coordinate is an integral multiple of the quantum of action' or that for $k = 1, 2, \ldots, f$,

$$\oint p_k \, dq_k = n_k h \,, \tag{5.3}$$

where p_k is the momentum corresponding to the coordinate q_k, n_k is a non-negative integer and the integration is extended over a period of q_k.'

Specifically, in the case of the elliptical orbits, the generalisation of the quantum conditions to two coordinates becomes

$$\oint_0^{2\pi} p_\phi \, d\phi = n_\phi h \quad \text{and} \quad \oint_{\text{orbit}} p_r \, dr = n_r h \,, \tag{5.4}$$

where n_ϕ and n_r are non-negative integers. In polar coordinates $q_k \equiv (r, \phi)$, the kinetic energy T of the electron is

$$T = \frac{m_e}{2} (\dot{r}^2 + r^2 \dot{\phi}^2) \,. \tag{5.5}$$

Then, according to the prescription developed in Sect. 5.4.3 (equations 5.52, 5.53 and 5.61), the corresponding momenta $p_k = \partial T / \partial \dot{q}_k$ are

$$p_\phi = m_e r^2 \dot{\phi} \quad \text{and} \quad p_r = m_e \dot{r} \,, \tag{5.6}$$

which are the azimuthal angular momentum and radial momentum of the electron respectively.

The quantum condition for $(\dot{\phi}, \phi)$ yields exactly the same result given by the circular Bohr model, namely,

$$\int_0^{2\pi} p_\phi \, d\phi = n_\phi h \,; \quad 2\pi p_\phi = n_\phi h \quad \text{or} \quad p_\phi = n_\phi \hbar \,. \tag{5.7}$$

p_ϕ is therefore a constant of the motion. To derive the quantum condition for the radial component of the momentum, we first rewrite (5.1) as follows:

$$\frac{1}{r} = \frac{1}{a} \frac{1 + \epsilon \cos \phi}{1 - \epsilon^2} \,. \tag{5.8}$$

Taking the derivative of (5.8) with respect to ϕ and then dividing by (5.8),

$$\frac{1}{r} \frac{dr}{d\phi} = \frac{\epsilon \sin \phi}{1 + \epsilon \cos \phi} \,. \tag{5.9}$$

We now need to work out $p_r \, dr$. This is achieved by the following relations:

$$p_r = m_e \dot{r} = m_e \frac{dr}{d\phi} \dot{\phi} = \frac{p_\phi}{r^2} \frac{dr}{d\phi} \,; \quad dr = \frac{dr}{d\phi} d\phi \,. \tag{5.10}$$

Therefore, using (5.9),

$$p_r \, dr = p_\phi \left(\frac{1}{r} \frac{dr}{d\phi} \right)^2 d\phi = p_\phi \epsilon^2 \frac{\sin^2 \phi \, d\phi}{(1 + \epsilon \cos \phi)^2} \,. \tag{5.11}$$

Unlike p_ϕ, p_r is *not* a constant of the motion. The quantum condition for the radial component of the momentum therefore becomes

$$\oint p_r \, \mathrm{d}r = p_\phi \epsilon^2 \int_0^{2\pi} \frac{\sin^2 \phi}{(1 + \epsilon \cos \phi)^2} \, \mathrm{d}\phi = n_r h \, . \tag{5.12}$$

Dividing by the quantum condition (5.7), we find

$$\frac{n_r}{n_\phi} = \frac{\epsilon^2}{2\pi} \int_0^{2\pi} \frac{\sin^2 \phi}{(1 + \epsilon \cos \phi)^2} \, \mathrm{d}\phi \, . \tag{5.13}$$

This is the result we have been seeking. The right-hand side of (5.13) depends only upon the eccentricity ϵ and, because n_ϕ and n_r are integers, the ellipticity is also quantised. Sommerfeld (1919) evaluated the integral (5.13) in Section 6 of the *Mathematical Notes and Addenda* of his book and found the result

$$1 - \epsilon^2 = \frac{n_\phi^2}{(n_\phi + n_r)^2} \, , \tag{5.14}$$

indicating that, of all possible ellipticities, only those involving the integers n_ϕ and n_r are allowed.

Next, the energies of the orbits are evaluated. The kinetic energy of the electron is

$$T = \frac{m_e}{2}(\dot{r}^2 + r^2 \dot{\phi}^2) = \frac{1}{2m_e}\left(p_r^2 + \frac{p_\phi^2}{r^2}\right) \, . \tag{5.15}$$

Using (5.10), this can be rewritten

$$T = \frac{p_\phi^2}{2m_e r^2}\left[\left(\frac{1}{r}\frac{\mathrm{d}r}{\mathrm{d}\phi}\right)^2 + 1\right] \, . \tag{5.16}$$

Next, we use (5.8) and (5.9) to eliminate r from this expression,

$$T = \frac{p_\phi^2}{m_e a^2 (1 - \epsilon^2)^2}\left[\frac{1 + \epsilon^2}{2} + \epsilon \cos \phi\right] \, . \tag{5.17}$$

The potential energy of the electron can also be written in a form independent of r using (5.8):

$$U = -\frac{Ze^2}{4\pi \epsilon_0 r} = -\frac{Ze^2}{4\pi \epsilon_0 a}\frac{1 + \epsilon \cos \phi}{1 - \epsilon^2} \, , \tag{5.18}$$

where the charge of the nucleus is Ze, in other words, Z is the atomic number of the nucleus. Therefore, the total energy of the electron in its elliptical orbit is

$$E_{\text{tot}} = T + U = \frac{p_\phi^2}{m_e a^2 (1 - \epsilon^2)^2}\left[\frac{1 + \epsilon^2}{2} + \epsilon \cos \phi\right] - \frac{Ze^2}{4\pi \epsilon_0 a}\frac{1 + \epsilon \cos \phi}{1 - \epsilon^2} \, . \tag{5.19}$$

The total energy is time-independent and so must be independent of ϕ which changes by 2π radians per orbit. The terms in $\cos \phi$ in (5.19) must therefore sum to zero and so we can find the value of a

$$a = \frac{4\pi \epsilon_0 \, p_\phi^2}{Ze^2 m_e (1 - \epsilon^2)} \, . \tag{5.20}$$

With the terms in $\cos\phi$ set to zero, the total energy of the electron is

$$E_{\text{tot}} = T + U = \frac{p_\phi^2}{m_e a^2 (1 - \epsilon^2)^2} \left[\frac{1 + \epsilon^2}{2} \right] - \frac{Ze^2}{4\pi \epsilon_0 a} \frac{1}{1 - \epsilon^2} . \tag{5.21}$$

Using the result (5.20) for a, we find the simple expression,

$$E_{\text{tot}} = -\frac{Ze^2}{8\pi \epsilon_0 a} , \tag{5.22}$$

recalling that a is the semi-major axis of the ellipse. The final step is to work out the quantised values of a using (5.7) and (5.14),

$$E = -\frac{Ze^2}{8\pi \epsilon_0 a} = -\frac{Z^2 e^4 m_e (1 - \epsilon^2)}{32\pi^2 \epsilon_0^2 p_\phi^2} = -\frac{Z^2 e^4 m_e}{8\epsilon_0^2 h^2 (n_\phi + n_r)^2} . \tag{5.23}$$

This is the remarkable result obtained by Sommerfeld in 1916. Even in his book of 1919, he can scarcely contain his excitement. He writes,

> 'This result is of the greatest consequence and is superlatively simple: *we have found for the energy of elliptical orbits the same value as . . . for circular orbits, with the one difference that the* **quantum number** *n in the latter case* **is replaced by the quantum sum**, $n_\phi + n_r$. *Each of the quantised ellipses of our family has an amount of energy equivalent to that of a definite Bohr circle.*'[2]

This result is identical to (4.25) for the hydrogen atom for which $Z = 1$ and $n_r = 0$. The range of potential quantum transitions has been greatly increased. Considering transitions between the upper (u) and lower (l) quantum states characterised by (n_ϕ^u, n_r^u) and (n_ϕ^l, n_r^l), evidently $(n_\phi^u + n_r^u)$ must be greater than $(n_\phi^l + n_r^l)$ and then the energy of the photon emitted in the transition is

$$\nu = \frac{Z^2 e^4 m_e}{8\epsilon_0^2 h^3} \left[\frac{1}{(n_\phi^l + n_r^l)^2} - \frac{1}{(n_\phi^u + n_r^u)^2} \right] , \tag{5.24}$$

which for hydrogen becomes

$$\frac{1}{\lambda} = \frac{\nu}{c} = R_\infty \left[\frac{1}{(n_\phi^l + n_r^l)^2} - \frac{1}{(n_\phi^u + n_r^u)^2} \right] . \tag{5.25}$$

Thus, the spectral series is again identical with the Balmer and other series of hydrogen, but the number of ways in which the lines can be produced has been considerably increased. Sommerfeld's words, again printed in italics, state:

> '. . . *it has a deepened theoretical significance and its origin now has multiple roots. By the admission of elliptical orbits, the series has gained no extra lines and has lost none of its sharpness.*'

Sommerfeld developed his model of the atom using the following arguments. The case $n_\phi = 0$ corresponds to a degenerate ellipse, $\epsilon = 1$, which is a straight line joining the foci of the ellipse and so this trajectory would pass though the nucleus. Sommerfeld excluded this possibility so that $n_\phi = 1, n_r = 0$ is the lowest energy state. Secondly, when $n_r = 0$,

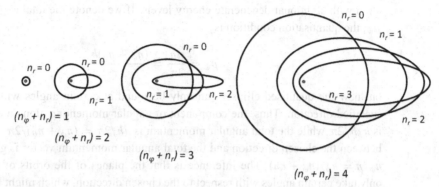

Fig. 5.2 The quantised circular and elliptical orbits parameterised by the quantum numbers n_ϕ and n_r according to the Sommerfeld model. Orbits with the same values of $(n_\phi + n_r)$ have the same energies.

the orbits become circles $\epsilon = 0$ as can be seen from (5.14) for the ellipticity of the electron's orbit. The resulting families of orbits for $(n_\phi + n_r) = 1, 2, 3,$ and 4 and $n_r = 1, 2,$ and 3 are shown in Fig. 5.2 in which the orbits have been drawn to scale with respect to the foci of the ellipses.

But, much more has been achieved. Although the unperturbed hydrogen atom now has numerous ways of creating sharp Balmer lines, when the atom is placed in an electric or magnetic field, the lines associated with different initial and final states are split because of the different kinematics of the electron in the perturbed elliptical orbits. Specifically, the Stark and Zeeman effects should display plentiful fine structure (see Sects. 7.2 and 7.3). Consider for example the case of the transition $(n_\phi^u + n_r^u) = 4$ to $(n_\phi^l + n_r^l) = 3$. We can combine any of the orbits with $(n_\phi^u + n_r^u) = 4$ in Fig. 5.2 with any of those with $(n_\phi^l + n_r^l) = 3$, in other words, there are $(n_\phi^u + n_r^u) \times (n_\phi^l + n_r^l) = 4 \times 3 = 12$ ways in which the line can be produced. Likewise, for the Balmer series for which $(n_\phi^u + n_r^u) = 2$ and $(n_\phi^l + n_r^l) = 3, 4, 5, \ldots$ the degeneracies would be $3 \times 2 = 6$ for Hα, $4 \times 2 = 8$ for Hβ, $5 \times 2 = 10$ for Hγ, $6 \times 2 = 12$ for Hδ and so on. Sommerfeld appreciated that not all of these would be realised in nature and so there had to be additional *principles of selection*, or *selection rules*, which determine which transitions are allowed. Initially, he provided a set of empirical rules to avoid excessive numbers of lines.

Sommerfeld proceeded to the next generalisation of the model by considering quantisation in three dimensions. Spherical polar coordinates are the natural extension of the polar coordinates considered so far, the three independent coordinates now being (r, θ, ϕ), where θ is the polar angle measured from the axis of the (r, ϕ) coordinate system used above. The quantisation conditions follow exactly the same prescription as (5.4), namely,

$$\oint_{\text{orbit}} p_r \, \mathrm{d}r = n_r h \, ; \qquad \oint_0^{2\pi} p_\phi \, \mathrm{d}\phi = n_\phi h \, ; \qquad \oint_0^\pi p_\theta \, \mathrm{d}\theta = n_\theta h \, , \qquad (5.26)$$

where the new quantum number n_θ is associated with the conjugate pair (p_θ, θ). Carrying out a similar analysis to the above two-dimensional calculation, Sommerfeld found that the energy levels had to satisfy the relation $n = n_\phi + n_\theta$, where n, n_ϕ, n_θ are non-negative integers. Once again, the energy levels corresponded precisely to those of the hydrogen atom with additional degenerate energy levels. If we denote the total angular momentum p_ψ, the quantisation condition is

$$p_\psi = \frac{nh}{2\pi} = \frac{(n_\phi + n_\theta)h}{2\pi} \, , \tag{5.27}$$

but now the quantised ellipses can only be found at certain angles with respect to any specified direction. Thus, the component of angular momentum along a chosen direction is $n_\phi h/2\pi$ while the total angular momentum is $nh/2\pi = (n_\phi + n_\theta)/2\pi$ and so the angle between the chosen direction and the total angular momentum vector is given by $\cos\alpha = n_\phi/n = n_\phi/(n_\phi + n_\theta)$. The inference is that the planes of the orbits of the ellipses can only take certain angles with respect to the chosen direction, which might be defined by an imposed electric or magnetic field. This was the discovery of *space quantisation*.

5.3 Sommerfeld and the fine-structure constant

Perhaps the most remarkable results of Sommerfeld's paper concerned the extension of the Bohr model to include the effects of special relativity. The quantisation conditions remain the same as in (5.4),

$$\oint_0^{2\pi} p_\phi \, d\phi = n_\phi h \quad \text{and} \quad \oint_{\text{orbit}} p_r \, dr = n_r h \, . \tag{5.28}$$

In Sommerfeld's picture, the effect of relativity is to relieve the degeneracy of the orbits because the total energies of the orbits are now slightly different. A simple interpretation of the effect is that the elliptical orbits have different energies because the electrons acquire high speeds in close encounters with the nucleus and so the relativistic corrections differ from orbit to orbit.

Sommerfeld's analysis in his book *Atombau und Spektrallinien* (Sommerfeld, 1919) is a brilliant and clear exposition of the energy changes of the orbits of the electrons when the effects of special relativity are taken into account. The analysis is somewhat lengthy and so we summarise the key results of the analysis. The first realisation is that the ellipses are no longer stationary in space, but the perihelion of the electron's orbit, the point of closest approach to the nucleus, precesses about the nucleus. This is illustrated by the diagram from Sommerfeld's book (Fig. 5.3). In polar coordinates, Sommerfeld writes the expression for the precessing elliptical orbit as

$$\frac{1}{r} = C_1 + C_2 \cos \gamma\phi \, , \tag{5.29}$$

which differs from the case of the stationary ellipse (5.1) by the inclusion of the factor γ in the cosine term. The quantity γ is defined to be

$$\gamma^2 = 1 - \frac{p_0^2}{p^2} \, , \tag{5.30}$$

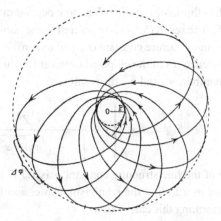

Fig. 5.3 Illustrating the precession of the elliptical orbits when the effects of special relativity are taken into account. This diagram appears as Fig. 110 in Sommerfeld's book *Atombau und Spektrallinien* (Sommerfeld, 1919).

where p is now the relativistic three-momentum and is quantised according to the rule $\oint p\,\mathrm{d}\phi = n_\phi h$. p_0 is defined to be the quantity $p_0 = Ze^2/4\pi\epsilon_0 c$. In the case of circular Bohr orbits, it is straightforward to show that $\gamma = (1 - v^2/c^2)^{1/2}$, in other words, the inverse of my normal convention for the Lorentz factor $\gamma = (1 - v^2/c^2)^{-1/2}$. Just for this section, we will use γ in Sommerfeld's sense. Inspection of (5.29) shows that, since $\gamma \leq 1$, the orbit does not close up after $\phi = 2\pi$ radians, but after $\gamma\phi = 2\pi$ radians. This slight change per orbit is illustrated by the angle $\Delta\phi$ in Fig. 5.3. The orbits return to the standard elliptical form however if we introduce the coordinate $\psi = \gamma\phi$. Sommerfeld shows that the equation for the ellipse now becomes

$$\frac{1}{r} = \frac{1}{a}\frac{1 + \epsilon\cos\gamma\phi}{1 - \epsilon^2}, \tag{5.31}$$

while the momenta corresponding to r, ϕ are

$$p_\phi = mr^2\dot{\phi}, \quad p_r = m\dot{r}, \tag{5.32}$$

where the momenta are relativistic three-momenta, that is, in (5.32) $m = m_e(1 - v^2/c^2)^{-1/2}$. The one important difference is that the quantisation condition in the radial direction now corresponds to the integration over a single ellipse in the ψ coordinate, that is,

$$\int_{\phi=0}^{2\pi} p_\phi\,\mathrm{d}\phi = n_\phi h \quad \text{and} \quad \int_{\psi=0}^{\psi=2\pi} p_r\,\mathrm{d}r = n_r h. \tag{5.33}$$

Carrying out the same procedure as in the non-relativistic case, Sommerfeld found that the ellipticities of the orbits are given by

$$1 - \epsilon^2 = \frac{n_\phi^2 - \alpha^2 Z^2}{\left[n_r + \sqrt{n_\phi^2 - \alpha^2 Z^2}\right]^2}, \tag{5.34}$$

where $\alpha = e^2/2\epsilon_0 hc$ is the *fine-structure constant*, for reasons which will be apparent in a moment. It is already clear that the degeneracy of the energy levels has been relieved because the ellipticities of the orbits now depend differently upon n_ϕ and n_r as compared with (5.14). The result (5.14) is recovered in the non-relativistic limit, which is obtained by setting the fine-structure constant α equal to zero.

Next, the energies of the elliptical orbits are evaluated. In the two-dimensional case with quantum numbers n_ϕ and n_r, the result is

$$E_{\text{tot}} = m_e c^2 \left\{ 1 + \frac{\alpha^2 Z^2}{\left[n_r + \sqrt{n_\phi^2 - \alpha^2 Z^2} \right]^2} \right\}^{-1/2} - m_e c^2 . \tag{5.35}$$

The value of the fine-structure constant α is $1/137.036$ and so (5.35) can be expanded to fourth order in α to find the relativistic expression for the energy levels of hydrogen-like atoms. Performing this calculation,

$$E_{\text{tot}} = \frac{Z^2 e^4 m_e}{8 \epsilon_0 h^2} \left\{ \frac{1}{(n_\phi + n_r)^2} + \frac{\alpha^2 Z^2}{(n_\phi + n_r)^4} \left[\frac{n_r}{n_\phi} + \frac{1}{4} \right] \right\} . \tag{5.36}$$

This expression has a number of remarkable features. First, in the non-relativistic limit $c \to \infty$, $\alpha \to 0$, the second term in curly brackets disappears and we recover (5.23) for the energies of the elliptical orbits. Secondly, the second term in curly brackets is entirely associated with the effects of special relativity. The term associated with the factor $1/4$ in this term corresponds to a shift in energy of the energy level labelled $(n_\phi + n_r)$ as a whole and does not result in any splitting of the energy level. In contrast, the term associated with the factor n_r/n_ϕ results in splitting of the energy levels and Sommerfeld identified this as the cause of the splitting of the lines of the Balmer series. This is why the quantity $\alpha = e^2/2\epsilon_0 hc = 1/137.036$ is referred to as the *fine-structure constant*.

For hydrogen, the effect is most pronounced in the optical waveband for the Hα line, in particular, with the splitting of the lower energy state with $(n_\phi + n_r) = 2$. The splitting of this level corresponds to the difference between the energies of the orbits with $(n_\phi = 2, n_r = 0)$ and $(n_\phi = 1, n_r = 1)$. Inserting these values into (5.36), the difference in energy corresponds to a frequency shift

$$\Delta \nu_{\text{H}} = \frac{\Delta E}{h} = \frac{Z^2 e^4 m_e}{8 \epsilon_0 h^3} \left[\frac{\alpha^2}{16} \right] = R_\infty \frac{\alpha^2}{16} \equiv 0.365 \, \text{cm}^{-1} . \tag{5.37}$$

This was found to be in good agreement with Michelson's measurements of fine structure in the Hα and Hβ lines of hydrogen. Even better agreement was found from Paschen's measurement of the splitting of the Balmer line of ionised helium He$^+$ which has the same form of fine structure but the effect is 16 times greater because of the Z^4 factor in the expression for the fine-structure displacement of the energy levels. Referring Paschen's result to the expected splitting for the hydrogen Balmer lines, the measured value was $\Delta \nu_{\text{H}} = 0.3645 \pm 0.0045 \, \text{cm}^{-1}$.

Sommerfeld next applied the theory of fine-structure splitting to the K and L X-ray fluorescence lines into (K_α, K_β) and (L_α, L_β) discussed in Sect. 4.6 and seen in Fig. 4.5. In the case of the K lines, Moseley's formula (4.31) showed that the effective nuclear charge was $(Z - 1)$ while (4.32) showed it was $(Z - b)$ for L lines. Sommerfeld carried out his own analysis of the L lines and found $b = 3.5$. Using the 'shielded' charges in the expressions for the splitting of the L X-ray energy levels, the splittings expected for the elements shown in Fig. 4.5 were expected to be vastly greater than that of hydrogen because of the roughly Z^4 dependence upon the effective nuclear charge. Extrapolating from the L_α and L_β X-ray lines of the heavy elements back to the case of hydrogen, Sommerfeld again found $\Delta \nu_H = 0.365$ cm^{-1}.

Sommerfeld was rightly triumphant. He fully appreciated the fact that the success of the theory not only provided evidence for the elliptical orbits of electrons in atoms, but also provided a demonstration of the correctness of the special relativistic formula for the momentum of the electron. In his words:

> 'Thus, the observation of the fine-structure discloses the whole mechanism of the intra-atomic motions as far as the motion of the perihelion of the elliptic orbits. The complex of facts contained in the fine-structures has just the same importance for the special theory of relativity and for the atomic structure, as the motion of Mercury's perihelion for the general theory of relativity.'

These results provided the basic elements for the old quantum theory which was to be greatly elaborated by Bohr and his associates. Sommerfeld took the generalised quantum conditions

$$\oint p_k \, dq_k = n_k h \qquad (5.38)$$

to be the ultimate foundations of quantum theory, statements which were 'unproved and perhaps incapable of being proved'. He also realised that much more powerful mathematical tools were already available for carrying out calculations in the quantum theory. In fact, these considerations were to lead to the natural mathematical language of quantum mechanics.

5.4 A mathematical interlude – from Newton to Hamilton–Jacobi

It is worthwhile reviewing how the methods of higher mechanics provided the route to the analysis of problems in quantum theory. As Sommerfeld remarked,

> 'It is truly a royal route for quantum problems.'

The history of the development of more and more powerful ways of expressing the content of Newton's laws of motion can be appreciated from the historical sequence of approaches:

- Newton's laws of motion,
- D'Alembert's principle,
- Hamilton's principle,
- The principle of least action,
- Generalised coordinates and Lagrange's equations,
- The canonical equations of Hamilton,
- Transformation theory of mechanics and the Hamilton–Jacobi equations,
- Action–angle variables.

Very often, various more advanced approaches provide much more straightforward routes to the solutions to problems and provide a deeper appreciation of the basic features of dynamical systems, including *conservation laws* and *normal modes of oscillation* of a mechanical or dynamical system.

5.4.1 Newton's laws of motion and principles of 'least action'

Some of the most powerful approaches involve finding that function which minimises the value of some quantity subject to well-defined boundary conditions. In the formal development of mechanics, the procedures are stated axiomatically. Let us take as an example what Feynman refers to as the *principle of minimum action*. Consider the case of the dynamics of a particle in a conservative field of force, that is, one which is derived as the gradient of a scalar potential, $F = -\text{grad } V$.

We introduce a set of axioms which enables us to work out the path of the particle subject to this force field. First, we define the quantity

$$\mathcal{L} = T - V = \tfrac{1}{2}mv^2 - V \,, \tag{5.39}$$

the difference between the kinetic energy T and the potential energy V of the particle in the field. To derive the trajectory of the particle between two fixed endpoints in the field in a fixed time interval t_1 to t_2, we find that path which minimises the function

$$S = \int_{t_1}^{t_2} \left(\tfrac{1}{2}mv^2 - V \right) \mathrm{d}t = \int_{t_1}^{t_2} \left[\tfrac{1}{2}m \left(\frac{\mathrm{d}r}{\mathrm{d}t} \right)^2 - V \right] \mathrm{d}t \,. \tag{5.40}$$

These statements are to be regarded as equivalent to Newton's laws of motion or, rather, what Newton accurately called his 'axioms'. $\mathcal{L}$ is called the *Lagrangian*.

These axioms are consistent with Newton's first law of motion. If there are no forces present, $V = $ constant and hence we minimise $S = \int_{t_1}^{t_2} v^2 \, \mathrm{d}t$. The minimum value of S must correspond to a constant velocity v between t_1 and t_2. If the particle travelled between the endpoints by accelerating and decelerating in such a way that t_1 and t_2 are the same, the integral must be greater than that for constant velocity between t_1 and t_2 because v^2 appears in the integral and a basic rule of analysis tells us that $\langle v^2 \rangle \geq \langle \overline{v} \rangle^2$. Thus, in the absence of forces, $v = $ constant, Newton's first law of motion.

To proceed further, we need the techniques of the *calculus of variations*. Suppose $f(x)$ is a function of a single variable x. The minimum value of the function corresponds to the value of x at which $\mathrm{d}f(x)/\mathrm{d}x$ is zero. Approximating the variation of the function about the minimum at $x = 0$ by a power series,

$$f(x) = a_0 + a_1 x + a_2 x^2 + a_3 x^3 + \cdots \,, \tag{5.41}$$

the function can only have $\mathrm{d}f/\mathrm{d}x = 0$ at $x = 0$ if $a_1 = 0$. At this point, the function is approximately a parabola since the first non-zero coefficient in the expansion of $f(x)$ about the minimum is the term in x^2 – for small displacements x from the minimum, the change in the function $f(x)$ is second order in x.

The same principle is used to find the path of the particle when S is minimised. If the true path of the particle is $x_0(t)$, another path between t_1 and t_2 is given by

$$x(t) = x_0(t) + \eta(t) \tag{5.42}$$

where $\eta(t)$ describes the deviation of $x(t)$ from the minimum path $x_0(t)$. Just as we can define the minimum of a function as the point at which there is no first-order dependence of the function upon x, so we can define the minimum of the function $x(t)$ as that function for which the dependence on $\eta(t)$ is at least *second order*, that is, there should be no linear term in $\eta(t)$.

Substituting (5.42) into (5.40),

$$S = \int_{t_1}^{t_2} \left[\frac{m}{2} \left(\frac{dx_0}{dt} + \frac{d\eta}{dt} \right)^2 - V(x_0 + \eta) \right] dt$$

$$= \int_{t_1}^{t_2} \left\{ \frac{m}{2} \left[\left(\frac{dx_0}{dt} \right)^2 + 2\frac{dx_0}{dt} \cdot \frac{d\eta}{dt} + \left(\frac{d\eta}{dt} \right)^2 \right] - V(x_0 + \eta) \right\} dt \,. \tag{5.43}$$

We now seek to eliminate first-order quantities in $d\eta$ and hence we can drop the term $(d\eta/dt)^2$. We expand $V(x_0 + \eta)$ to first order in η by a Taylor expansion,

$$V(x_0 + \eta) = V(x_0) + \nabla V \cdot \eta \,. \tag{5.44}$$

Substituting (5.44) into (5.43) and preserving only quantities to first order in η, we find

$$S = \int_{t_1}^{t_2} \left[\frac{m}{2} \left(\frac{dx_0}{dt} \right)^2 - V(x_0) + m\frac{dx_0}{dt} \cdot \frac{d\eta}{dt} - \nabla V \cdot \eta \right] dt \,. \tag{5.45}$$

Now the first two terms inside the integral are the minimum path and so are a constant. We therefore need to ensure that the last two terms have no dependence upon η, the condition for a minimum. We therefore need consider only the last two terms,

$$S = \int_{t_1}^{t_2} \left(m\frac{dx_0}{dt} \cdot \frac{d\eta}{dt} - \eta \cdot \nabla V \right) dt \,. \tag{5.46}$$

We integrate the first term by parts so that η alone appears in the integrand, that is,

$$S = m \left[\eta \cdot \frac{dx_0}{dt} \right]_{t_1}^{t_2} - \int_{t_1}^{t_2} \left[\frac{d}{dt} \left(m\frac{dx_0}{dt} \right) \cdot \eta + \eta \cdot \nabla V \right] dt \,. \tag{5.47}$$

The function η must be zero at t_1 and t_2 because that is where the path begins and ends and the endpoints are fixed. Therefore, the first term on the right-hand side of (5.47) is zero and we can write

$$S = -\int_{t_1}^{t_2} \left\{ \eta \cdot \left[\frac{d}{dt} \left(m\frac{dx_0}{dt} \right) + \nabla V \right] \right\} dt = 0 \,. \tag{5.48}$$

This must be true for arbitrary perturbations about $x_0(t)$ and hence the term in square brackets must be zero, that is,

$$\frac{d}{dt} \left(m\frac{dx_0}{dt} \right) = -\nabla V \,. \tag{5.49}$$

We have recovered Newton's second law of motion since $F = -\nabla V$, that is,

$$F = \frac{d}{dt}\left(m\frac{dx_0}{dt}\right) = \frac{dp}{dt} . \tag{5.50}$$

Thus, our alternative formulation of the laws of motion in terms of an action principle is exactly equivalent to Newton's statement of the laws of motion.

These procedures can be generalised to take account of conservative *and* non-conservative forces such as those which depend upon velocity, for example, friction and the force on a charged particle in a magnetic field (Goldstein, 1950). The key point is that we have a prescription which involves writing down the kinetic and potential energies (T and V respectively) of the system and then forming the Lagrangian $\mathcal{L}$ and finding the minimum value of the function S. The advantage of this procedure is that it is often a straightforward matter to write down these energies in some suitable set of coordinates. We therefore need rules which tell us how to find the minimum value of S in any set of coordinates which is convenient for the problem in hand. These are the *Euler–Lagrange equations*.

5.4.2 The Euler–Lagrange equations

Let us consider a system of N particles interacting through a scalar potential function V. The positions of the N particles are given by the vectors $[r_1, r_2, r_3, \ldots, r_N]$ in Cartesian coordinates. Since three numbers are needed to describe the position of each particle, for example, x_i, y_i, z_i, the vector describing the positions of all the particles has $3N$ coordinates. For greater generality, we wish to transform these coordinates into a different set of coordinates which we write as $[q_1, q_2, q_3, \ldots, q_{3N}]$. The set of relations between these coordinates can be written

$$q_i = q_i(r_1, r_2, r_3, \ldots, r_N), \quad \text{and hence} \quad r_i = r_i(q_1, q_2, q_3, \ldots, q_{3N}) . \tag{5.51}$$

This amounts to no more than a change of variables.

The aim of the procedure is to write down the equations for the dynamics of the particles, that is, an equation for each independent coordinate, in terms of the coordinates q_i rather than r_i. We are guided by the analysis of the previous subsection on action principles to form the quantities T and V, the kinetic and potential energies respectively, in terms of the new set of coordinates and then to find the stationary value of S, that is

$$\mathcal{L} = T - V , \quad \delta S = \delta \int_{t_1}^{t_2} (T - V)\,dt = 0 , \tag{5.52}$$

where δ means 'take the variation about a particular value of the coordinates' as discussed in the previous subsection. This formulation is called *Hamilton's principle* and as before $\mathcal{L}$ is the *Lagrangian* of the system. Notice that Hamilton's principle makes no reference to the coordinate system to be used in the calculation.

The kinetic energy of the system is

$$T = \sum_i m_i \dot{r}_i^2 . \tag{5.53}$$

In terms of our new coordinate system, we can write without loss of generality

$$r_i = r_i(q_1, q_2, \ldots, q_{3N}, t) \quad \text{and} \quad \dot{r}_i = \dot{r}_i(\dot{q}_1, \dot{q}_2, \dot{q}_3, \ldots, \dot{q}_{3N}, q_1, q_2, \ldots, q_{3N}, t).$$
(5.54)

Notice that we have now included explicitly the time dependence of r_i and $\dot{r}_i$. Therefore, we can write the kinetic energy as a function of the coordinates $\dot{q}_i$, q_i and t, that is, $T(\dot{q}_i, q_i, t)$, where we understand that all the values of i from 1 to $3N$ are included. Similarly, we can write the expression for the potential energy entirely in terms of the coordinates q_i and t, that is, $V(q_i, t)$. We therefore need to find the stationary values of

$$S = \int_{t_1}^{t_2} [T(\dot{q}_i, q_i, t) - V(q_i, t)] \, dt = \int_{t_1}^{t_2} \mathcal{L}(\dot{q}_i, q_i, t) \, dt.$$
(5.55)

We repeat the analysis of Sect. 5.4.1 in which we found the condition for S to be independent of first-order perturbations about the minimum path. As before, we let $q_0(t)$ be the minimum solution and write the expression for another function $q(t)$ in the form

$$q(t) = q_0(t) + \eta(t).$$
(5.56)

Now we insert the trial solution (5.56) into (5.55),

$$S = \int_{t_1}^{t_2} \mathcal{L}[\dot{q}_0(t) + \dot{\eta}(t), q_0(t) + \eta(t), t] \, dt.$$

Performing a Taylor expansion to first order in $\dot{\eta}(t)$ and $\eta(t)$,

$$S = \int_{t_1}^{t_2} \mathcal{L}[\dot{q}_0(t), q_0(t), t] \, dt + \int_{t_1}^{t_2} \left[\frac{\partial \mathcal{L}}{\partial \dot{q}_i} \dot{\eta}(t) + \frac{\partial \mathcal{L}}{\partial q_i} \eta(t) \right] \, dt.$$
(5.57)

Setting the first integral equal to S_0 and integrating the term in $\dot{\eta}(t)$ by parts,

$$S = S_0 + \left[\frac{\partial \mathcal{L}}{\partial \dot{q}_i} \eta(t) \right]_{t_1}^{t_2} - \int_{t_1}^{t_2} \left[\frac{d}{dt} \left(\frac{\partial \mathcal{L}}{\partial \dot{q}_i} \right) \eta(t) - \frac{\partial \mathcal{L}}{\partial q_i} \eta(t) \right] \, dt.$$
(5.58)

Again because $\eta(t)$ must always be zero at the endpoints, the first term in square brackets disappears and the result can be written

$$S = S_0 - \int_{t_1}^{t_2} \eta(t) \left[\frac{d}{dt} \left(\frac{\partial \mathcal{L}}{\partial \dot{q}_i} \right) - \frac{\partial \mathcal{L}}{\partial q_i} \right] \, dt.$$

We require this integral to be zero for all first-order perturbations about the minimum solution. Therefore, the condition is

$$\frac{\partial \mathcal{L}}{\partial q_i} - \frac{d}{dt} \left(\frac{\partial \mathcal{L}}{\partial \dot{q}_i} \right) = 0.$$
(5.59)

The equations (5.59) represents $3N$ second-order differential equations for the time evolution of the $3N$ coordinates and are known as the *Euler–Lagrange equations*. They are no more than Newton's laws of motion written in the q_i coordinate system which can be chosen to be the most convenient for the particular problem at hand.

5.4.3 Hamilton's equations

Another way of developing the equations of mechanics and dynamics is to convert (5.59) into a set of $6N$ first-order differential equations by introducing the generalised momenta p_i. There was no mention of the components of the momenta of the particles p_i in the last subsection but these can be introduced by taking the derivative of the Lagrangian (5.52) in Cartesian coordinates with respect to the component of velocity $\dot{x}_i$, and we find

$$\frac{\partial \mathcal{L}}{\partial \dot{x}_i} = \frac{\partial T}{\partial \dot{x}_i} - \frac{\partial V}{\partial \dot{x}_i} = \frac{\partial T}{\partial \dot{x}_i} = \frac{\partial}{\partial \dot{x}_i} \sum_j \frac{1}{2} m_j \left(\dot{x}_j^2 + \dot{y}_j^2 + \dot{z}_j^2 \right) = m_i \dot{x}_i = p_i \ . \tag{5.60}$$

This expression indicates how the concept of momentum can be generalised within the framework of the Lagrangian approach to mechanics. The quantity $\partial \mathcal{L}/\partial \dot{q}_i$ is defined to be p_i, the *canonical momentum*, conjugate to the coordinate q_i, by

$$p_i = \frac{\partial \mathcal{L}}{\partial \dot{q}_i} \ . \tag{5.61}$$

The p_i defined by (5.61) do not necessarily have the dimensions of linear velocity times mass if q_i is not in Cartesian coordinates. Furthermore, if the potential is velocity dependent, even in Cartesian coordinates, the generalised momentum is not identical with the usual definition of mechanical momentum. An example of this is the motion of a charged particle in an electromagnetic field in Cartesian coordinates for which the Lagrangian is

$$\mathcal{L} = \sum_i \frac{1}{2} m_i \dot{r}_i^2 - \sum_i e_i \phi(x_i) + \sum_i e_i A(x_i) \cdot \dot{r}_i \ , \tag{5.62}$$

where e_i is the charge of the particle i, $\phi(x_i)$ is the electrostatic potential at the particle i and $A(x_i)$ the vector potential at the same position. Then, the x-component of the canonical momentum of the ith particle is

$$p_{ix} = \frac{\partial \mathcal{L}}{\partial \dot{x}_i} = m_i \dot{x}_i + e_i A_x \ , \tag{5.63}$$

and similarly for the y- and z-coordinates.

With these definitions, a number of conservation laws can be derived.[3] For example, if the Lagrangian does not depend upon the coordinate q_i, (5.59) immediately shows that the generalised momentum is conserved, $\mathrm{d}\dot{p}_i/\mathrm{d}t = 0$, $p_i = $ constant. Another standard result is that the total energy of the system in a conservative field of force with no time-dependent constraints is given by the *Hamiltonian H* which can be written

$$H = \sum_i p_i \dot{q}_i - \mathcal{L}(q, \dot{q}) \ . \tag{5.64}$$

It looks as though H depends upon p_i, $\dot{q}_i$ and q_i but, in fact, we can rearrange the equation to show that H is a function of only p_i and q_i. Let us take the total differential of H in the usual way, assuming $\mathcal{L}$ is time independent. Then

$$\mathrm{d}H = \sum_i p_i \, \mathrm{d}\dot{q}_i + \sum_i \dot{q}_i \, \mathrm{d}p_i - \sum_i \frac{\partial \mathcal{L}}{\partial \dot{q}_i} \mathrm{d}\dot{q}_i - \sum_i \frac{\partial \mathcal{L}}{\partial q_i} \, \mathrm{d}q_i \ . \tag{5.65}$$

Since $p_i = \partial\mathcal{L}/\partial\dot{q}_i$, the first and third terms on the right-hand side cancel. Therefore,

$$dH = \sum_i \dot{q}_i \, dp_i - \sum_i \frac{\partial\mathcal{L}}{\partial q_i} \, dq_i \, . \tag{5.66}$$

This differential depends only on the increments dp_i and dq_i and hence we can compare dH with its formal expansion in terms of p_i and q_i:

$$dH = \sum_i \frac{\partial H}{\partial p_i} dp_i + \sum_i \frac{\partial H}{\partial q_i} dq_i \, .$$

It follows immediately that

$$\frac{\partial H}{\partial q_i} = -\frac{\partial\mathcal{L}}{\partial q_i} \, ; \quad \frac{\partial H}{\partial p_i} = \dot{q}_i \, . \tag{5.67}$$

Since

$$\frac{\partial\mathcal{L}}{\partial q_i} = \frac{d}{dt}\left(\frac{\partial\mathcal{L}}{\partial\dot{q}_i}\right) , \tag{5.68}$$

we find from the Euler–Lagrange equation,

$$\frac{\partial H}{\partial q_i} = -\dot{p}_i \, .$$

We thus reduce the equations of motion to the pair of relations

$$\dot{q}_i = \frac{\partial H}{\partial p_i} \, , \quad \dot{p}_i = -\frac{\partial H}{\partial q_i} \tag{5.69}$$

This pair of equations is known as *Hamilton's equations*. They are first-order differential equations for each of the $3N$ coordinates. We are now treating the p_is and the q_is on the same footing. If V is independent of $\dot{q}$, H is just the total energy $T + V$ expressed in terms of the coordinates p_i and q_i.

5.4.4 The Hamilton–Jacobi equations and action–angle variables

Why do we have to delve deeper into the principles of classical mechanics? Hamilton's equations (5.69) are often difficult to solve but they can be simplified to tackle specific problems. It turned out that the appropriate tools were available for application to the quantum theory through Carl Jacobi's extension of Hamilton's equations. In his classic textbook *Classical Mechanics*, Goldstein (1950) remarks:

> 'For a long time action–angle variables remained an esoteric technique of classical mechanics used only by astronomers. The situation changed rapidly with the advent of Bohr's quantum theory of the atom, for it was found that the quantum conditions could be stated most simply in terms of action variables. In classical mechanics the action variables possess a continuous range of values, but this is no longer the case in quantum mechanics. The quantum conditions of Sommerfeld and Wilson required that the motion be limited to such orbits for which the "proper" action variables had discrete values which were integral multiples of h, the quantum of action.'

These 'esoteric' procedures had been used with considerable success in astronomical problems, for example, in Delauney's *Théorie du mouvement de la lune* (1860, 1867) and in Charlier's *Die Mechanik des Himmels* (1902). It is no surprise that Karl Schwarzschild, theoretical astrophysicist and the Director of the Potsdam Observatory, was one of the principal contributors to the introduction of Hamilton–Jacobi theory into the solution of problems in quantum theory.

The mathematical procedures which result in the Hamilton–Jacobi equations and their application to astronomical and quantum problems are elegantly expounded by Goldstein (1950). We present here a simplified version of that story aimed at deriving the useful solutions as directly as possible. The essence of the approach is to transform from one set of coordinates to another, while preserving the Hamiltonian form of the equations of motion and ensuring that the new variables are also canonical coordinates. This approach is sometimes called the *transformation theory of mechanics*. Thus, suppose we wish to transform between the set of coordinates q_i to a new set Q_i. We need to ensure that the independent momentum coordinates simultaneously transform into canonical coordinates in the new system, that is, there should also be a set of coordinates P_i corresponding to p_i. Thus, the problem is to define the new coordinates (Q_i, P_i) in terms of the old coordinates (q_i, p_i):

$$Q_i = Q_i(q, p, t), \quad P_i = P_i(q, p, t) \,. \tag{5.70}$$

We need to find the transformations which result in the (Q_i, P_i) being canonical coordinates such that

$$\dot{Q}_i = \frac{\partial K}{\partial P_i}, \quad \dot{P}_i = -\frac{\partial K}{\partial Q_i} \tag{5.71}$$

where K is now the Hamiltonian in the (Q_i, P_i) coordinate system. The transformations which result in (5.71) are known as *canonical transformations*. They are also known as *contact transformations*. In the old system, the coordinates satisfied Hamilton's variational principle,

$$\delta \int_{t_1}^{t_2} \left[\sum p_i \dot{q}_i - H(q, p, t) \right] dt = 0 \,, \tag{5.72}$$

and so the same rule must hold true in the new set of coordinates,

$$\delta \int_{t_1}^{t_2} \left[\sum P_i \dot{Q}_i - K(Q, P, t) \right] dt = 0 \,. \tag{5.73}$$

The equations (5.72) and (5.73) must be valid simultaneously. The clever trick is to note that the integrands inside the integrals in (5.72) and (5.73) need not be equal, but can differ by the total time derivative of some arbitrary function S. This works because the integral of S between the fixed endpoints is a constant and so disappears when the variations in (5.72) and (5.73) are carried out,

$$\delta \int_{t_1}^{t_2} \frac{dS}{dt} dt = \delta(S_2 - S_1) = 0 \,. \tag{5.74}$$

The function S is known as the *generating function* because once it is specified, the transformation equations (5.70) are completely determined. It looks as though S is a function of the four coordinates (q_i, p_i, Q_i, P_i), but in fact they are related by the functions (5.70) and so S can be defined by any two of (q_i, p_i, Q_i, P_i). Goldstein (1950) and Lindsay and Margenau (1957) show how the various transformations between the coordinate systems can be written. For example, if we choose to write the transformations in terms of p_i, P_i, the transformation equations are

$$p_i = \frac{\partial S}{\partial q_i}, \quad P_i = -\frac{\partial S}{\partial Q_i} \quad \text{and} \quad K = H + \frac{\partial S}{\partial t} . \tag{5.75}$$

Alternatively, if we choose q_i, P_i, the transformation equations become

$$p_i = \frac{\partial S}{\partial q_i}, \quad Q_i = \frac{\partial S}{\partial P_i} \quad \text{with} \quad K = H + \frac{\partial S}{\partial t} . \tag{5.76}$$

The other two sets of transformations are included in the endnote.[4]

Example Let us first work out the motion of a one-dimensional *harmonic oscillator* in this new formalism. The mass of the oscillator is m and the spring constant k. From these, we can define the angular frequency ω by $\omega^2 = k/m$. The Hamiltonian is therefore

$$H = \frac{p^2}{2m} + \frac{kq^2}{2} . \tag{5.77}$$

Now introduce a generating function defined by

$$S = \frac{m\omega}{2} q^2 \cot Q , \tag{5.78}$$

which depends only upon q and Q. We can therefore use (5.75) to find the transformations between coordinate systems. There is no explicit dependence of S upon time t and so, from the third equation of (5.75), $H = K$. We can now find the coordinates p and q in terms of the new coordinates P and Q. From the first and second equations of (5.75)

$$p = \frac{\partial S}{\partial q} = m\omega q \cot Q , \quad P = -\frac{\partial S}{\partial Q} = \frac{m\omega}{2} q^2 \operatorname{cosec}^2 Q . \tag{5.79}$$

These can be reorganised to give expressions for p and q in terms of P and Q:

$$q = \sqrt{\frac{2}{m\omega}} \sqrt{P} \sin Q , \quad p = \sqrt{2m\omega} \sqrt{P} \cos Q . \tag{5.80}$$

Substituting these expressions for p and q into the Hamiltonian (5.77), we find

$$H = K = \frac{p^2}{2m} + \frac{kq^2}{2} = \omega P . \tag{5.81}$$

We now find the solutions for P and Q from Hamilton's equations

$$\dot{P} = -\frac{\partial K}{\partial Q} , \quad \dot{Q} = \frac{\partial K}{\partial P} , \tag{5.82}$$

and so, from (5.81), it follows that

$$\dot{P} = 0 \,, \dot{Q} = \omega \quad \text{and so} \quad P = \alpha \,, Q = \omega t + \beta \,, \tag{5.83}$$

where α and β are constants. These solutions can now be substituted into (5.80) and we find

$$p = \sqrt{2m\omega\alpha} \, \cos(\omega t + \beta) \,, \quad q = \sqrt{\frac{2\alpha}{m\omega}} \, \sin(\omega t + \beta) \,. \tag{5.84}$$

Notice how economically we can determine the constants of the motion once we have found an appropriate generating function S.

The question is then, 'Can we find the appropriate generating functions for the problem at hand?' For the cases in which we are interested in which there is no explicit dependence upon the time t and in which the variables in the partial differential equations are separable, the answer is 'Yes'. We begin with the relation between the Hamiltonian H and the total energy of the system E. For the cases we are interested in,

$$H(q_i, p_i) = E \,. \tag{5.85}$$

We can now replace p_i by its definition in terms of the partial differentials of the generating function S, $p_i = \partial S / \partial q_i$ so that

$$H\left(q_i, \frac{\partial S}{\partial q_i}\right) = E \,. \tag{5.86}$$

Writing out the partial derivatives, this is a set of n first-order partial differential equations for the generating function S,

$$H\left(q_1, q_2, \ldots, q_n, \frac{\partial S}{\partial q_1}, \frac{\partial S}{\partial q_2}, \ldots, \frac{\partial S}{\partial q_n}\right) = E \,. \tag{5.87}$$

Equation (5.87) is known as the *Hamilton–Jacobi equation* in time-independent form. S is also referred to as *Hamilton's principal function*.

We are interested in periodic solutions of the equations of motion for conservative motion under a central field of force. There is no explicit dependence of the coordinate transformations upon time and so the transformation equations (5.76) become

$$p_i = \frac{\partial S}{\partial q_i} \,, \quad Q_i = \frac{\partial S}{\partial P_i} \quad \text{with} \quad K = H \,. \tag{5.88}$$

We now define a new coordinate

$$J_i = \oint p_i \, dq_i \,, \tag{5.89}$$

which is known as a *phase integral*, the integration of the momentum variable p_i being taken over a complete cycle of the values of q_i. In the case of Cartesian coordinates, it can be seen that J_i has the dimensions of angular momentum and is known as an *action variable*, by analogy with Hamilton's definition in his *principle of least action*. During one cycle the values of q_i vary continuously between finite values from q_{min} to q_{max} and back again to q_{min}. As a result, from the definition (5.89), the action J_i is a constant. If q_i does not appear in the Hamiltonian, as occurs in the cases of interest to us, the variable is said to be *cyclic*.

Accompanying J_i there is a conjugate quantity w_i,

$$w_i = \frac{\partial S}{\partial J_i} . \tag{5.90}$$

The corresponding canonical equations are

$$\dot{w}_i = \frac{\partial K}{\partial J_i} , \quad \dot{J}_i = -\frac{\partial K}{\partial w_i} , \tag{5.91}$$

where K is the transformed Hamiltonian. w_i is referred to as an *angle variable* and so the motion of the particle is now described in terms of (J_i, w_i) or *action–angle variables*. If the transformed Hamiltonian K is independent of w_i, it follows that

$$\dot{w}_i = \text{constant} = \nu_i , \quad \dot{J}_i = 0 , \tag{5.92}$$

where the ν_is are constants. Therefore,

$$w_i = \nu_i t + \gamma_i , \quad J_i = \text{constant} . \tag{5.93}$$

The constants ν_i are associated with the frequency of motion. If we take the integral of w_i round a complete cycle of q_i, we find

$$\Delta w_i = \oint \frac{\partial w_i}{\partial q_k} \, dq_k = \oint \frac{\partial}{\partial q_k} \left(\frac{\partial S}{\partial J_i} \right) dq_k . \tag{5.94}$$

Taking the differentiation outside the integral,

$$\Delta w_i = \frac{\partial}{\partial J_i} \oint \left(\frac{\partial S}{\partial q_k} \right) dq_k = \frac{\partial J_k}{\partial J_i} , \tag{5.95}$$

the last equality resulting from the definition $p_k = \partial S/\partial q_k$. Thus, $\Delta w_i = 1$, if $i = k$ and $\Delta w_i = 0$, if $i \neq k$. In other words, each independent variable w_i changes from 0 to 1 when the corresponding q_i goes through a complete cycle.

It is apparent from these considerations that action–angle variables are the ideal tools for studying the orbits of electrons in the old quantum theory, Sommerfeld's 'royal route for quantum problems'. The introduction of action–angle variables into quantum theory was pioneered by Schwarzschild and Epstein in the contexts to be described in Sect. 7.2 (Schwarzschild, 1916; Epstein, 1916a). As remarked by Jammer (1989),

'... it almost seemed as if Hamilton's method had expressly been created for treating quantum mechanical problems.'

Let us give the example of Sommerfeld's treatment of the Bohr model in three dimensions.

5.5 Sommerfeld's model of the atom in three dimensions

The best way of illustrating how these procedures provide a natural set of mathematical tools for quantum problems is to repeat Sommerfeld's analysis of the quantisation of angular momentum in three dimensions. In spherical polar coordinates, the kinetic energy is

$$T = \frac{m}{2} \left(\dot{r}^2 + r^2 \dot{\theta}^2 + r^2 \sin^2 \theta \, \dot{\phi}^2 \right) . \tag{5.96}$$

From (5.61), the canonical momenta in (r, θ, ϕ) coordinates are $p_r = m\dot{r}$, $p_\theta = mr^2 \dot{\theta}$ and $p_\phi = mr^2 \sin^2 \theta \, \dot{\phi}$. Therefore, the Hamiltonian is

$$H = \frac{1}{2m} \left(p_r^2 + \frac{p_\theta^2}{r^2} + \frac{p_\phi^2}{r^2 \sin^2 \theta} \right) - \frac{k}{r} , \tag{5.97}$$

where $k = Ze^2/4\pi\epsilon_0$. We can therefore immediately convert this equation into a Hamilton–Jacobi equation:

$$H = \frac{1}{2m} \left[\left(\frac{\partial W}{\partial r} \right)^2 + \frac{1}{r^2} \left(\frac{\partial W}{\partial \theta} \right)^2 + \frac{1}{r^2 \sin^2 \theta} \left(\frac{\partial W}{\partial \phi} \right)^2 \right] - \frac{k}{r} = E , \tag{5.98}$$

where we have written Hamilton's function as W, rather than S.[5] The total energy associated with the Hamiltonian H is E and is a constant of the motion. We now write the function W in separable form,

$$W = W_r(r) + W_\theta(\theta) + W_\phi(\phi) . \tag{5.99}$$

First, we consider the partial differential term in ϕ in (5.98). This relation must be true for all ϕ and so

$$\frac{\partial W}{\partial \phi} = \alpha_\phi = \text{constant} . \tag{5.100}$$

Therefore (5.98) becomes

$$\frac{1}{2m} \left[\left(\frac{\partial W}{\partial r} \right)^2 + \frac{1}{r^2} \left\{ \left(\frac{\partial W}{\partial \theta} \right)^2 + \frac{\alpha_\phi^2}{\sin^2 \theta} \right\} \right] - \frac{k}{r} = E . \tag{5.101}$$

Now the term in curly brackets in (5.101) involves only θ and so must also be a constant,

$$\left(\frac{\partial W}{\partial \theta} \right)^2 + \frac{\alpha_\phi^2}{\sin^2 \theta} = \alpha_\theta^2 . \tag{5.102}$$

Replacing the term in curly brackets by α_θ^2, the Hamilton–Jacobi equation becomes

$$\left(\frac{\partial W}{\partial r} \right)^2 + \frac{\alpha_\theta^2}{r^2} = 2m \left(E + \frac{k}{r} \right) . \tag{5.103}$$

The three equations (5.100), (5.102) and (5.103) are conservation equations for motion in the three independent coordinates (r, θ, ϕ). The first (5.100) is simply the conservation of angular momentum about the fixed polar axis. There is also, however, angular momentum about the θ-axis. Conservation of the total angular momentum p is given by (5.102) as may be seen by comparing the expression for conservation of angular momentum for motion in a plane

$$H = \frac{1}{2m} \left(p_r^2 + \frac{p}{r^2} \right) - \frac{k}{r} = E , \tag{5.104}$$

with (5.97). The third equation (5.103) corresponds to the conservation of total energy. Now we convert these results to action–angle variables. These are defined by

$$J_1 \equiv J_\phi = \oint p_\phi \, d\phi = \oint \frac{\partial W_\phi}{\partial \phi} \, d\phi \, , \tag{5.105}$$

$$J_2 \equiv J_\theta = \oint p_\theta \, d\theta = \oint \frac{\partial W_\theta}{\partial \theta} \, d\theta \, , \tag{5.106}$$

$$J_3 \equiv J_r = \oint p_r \, dr = \oint \frac{\partial W_r}{\partial r} \, dr \, . \tag{5.107}$$

Using (5.100), (5.102) and (5.103), these integrals can be written

$$J_\phi = \oint \alpha_\phi \, d\phi \, , \tag{5.108}$$

$$J_\theta = \oint \sqrt{\alpha_\theta^2 - \frac{\alpha_\phi^2}{\sin^2 \theta}} \, d\theta \, , \tag{5.109}$$

$$J_r = \oint \sqrt{2mE + \frac{2mk}{r} - \frac{\alpha_\theta^2}{r^2}} \, dr \, . \tag{5.110}$$

The first integral is trivial,

$$J_\phi = 2\pi \alpha_\phi \, . \tag{5.111}$$

The second integral is most simply evaluated by comparing the expressions for the kinetic energy in spherical polar and plane polar coordinates, as was done in the comparison of (5.104) with (5.97),

$$p_r \dot{r} + p_\theta \dot{\theta} + p_\phi \dot{\phi} = p_r \dot{r} + p \dot{\psi} \, , \tag{5.112}$$

where p is the total angular momentum and ψ is the azimuthal angle in the plane of the orbit. Hence, replacing $p_\theta \dot{\theta}$ by $p \dot{\psi} - p_\phi \dot{\phi}$, the action integral for J_θ can be written

$$J_\theta = \oint p \, d\psi - \oint p_\phi \, d\phi \, . \tag{5.113}$$

Taking both integrals round 2π, we find

$$J_\theta = 2\pi (p - p_\phi) = 2\pi (\alpha_\theta - \alpha_\phi) \, . \tag{5.114}$$

The third integral for the radial action therefore becomes

$$J_r = \oint \sqrt{2mE + \frac{2mk}{r} - \frac{(J_\theta + J_\phi)}{4\pi^2 r^2}} \, dr \, . \tag{5.115}$$

On performing this integral, we find a relation between the energy of the system and the action integrals J_ϕ, J_θ and J_r. This calculation is carried out by Goldstein by contour integration with the final remarkable result

$$H \equiv E = -\frac{2\pi^2 mk^2}{(J_r + J_\theta + J_\phi)^2} \, . \tag{5.116}$$

This calculation has been entirely classical and the values of the action variables are continuous. Sommerfeld now applied his quantum conditions to the three action integrals (5.105), (5.106) and (5.107) with the result

$$J_\phi = \oint p_\phi \, \mathrm{d}\phi = n_\phi h \, , \tag{5.117}$$

$$J_\theta = \oint p_\theta \, \mathrm{d}\theta = n_\theta h \, , \tag{5.118}$$

$$J_r = \oint p_r \, \mathrm{d}r = n_r h \, , \tag{5.119}$$

where n_ϕ, n_θ and n_r are integers. Now only certain orientations of the orbits of the electrons in space are allowed and the energy levels are again degenerate in the non-relativistic limit. Inserting these values into (5.115) and setting $k = Ze^2/4\pi\epsilon_0$, the energies of the three-dimensional elliptical orbits are

$$E = -\frac{Z^2 e^4 m}{8\epsilon_0^2 h^2 (n_r + n_\theta + n_\phi)^2} \, , \tag{5.120}$$

exactly the same form of result as (5.23).

Let us now work out the frequencies ν_i associated with each independent coordinate. From (5.91) and (5.92),

$$\nu_i = \dot{w}_i = \frac{\partial H}{\partial J_i} \, , \tag{5.121}$$

and so from (5.116), we find

$$\nu = \frac{\partial H}{\partial J_r} = \frac{\partial H}{\partial J_\theta} = \frac{\partial H}{\partial J_\phi} = \frac{4\pi^2 m k^2}{(J_r + J_\theta + J_\phi)^3} \, . \tag{5.122}$$

Thus, the frequencies are the same for all three independent coordinates, r, θ and ϕ.

These results are the basis of quantisation in the old quantum theory. The rules are as follows:

- Quantisation in the ϕ-coordinate corresponds to quantising the projection of the angular momentum onto some chosen axis, which we can choose to be the z-direction of a Cartesian coordinate system. The quantisation rule is $J_z = n_\phi h/2\pi = mh/2\pi = m\hbar$, where m came to be known as the *magnetic quantum number* for reasons to be discussed later.
- The combined quantisation in the θ- and ϕ-coordinates is given by $J = J_\theta + J_\phi = (n_\theta + m)\hbar$. This can be written $J = k\hbar$ where k is known as the *azimuthal quantum number* and defines the quantised total angular momentum.
- The quantisation in the radial r direction is given by (5.116)–(5.119) and corresponds to $J_r = (n_r + n_\theta + m)\hbar$. The quantum number $(n_r + n_\theta + m)$ defines the total energy of the orbit given by (5.120) and is known as the *principal quantum number*.[6]

We have concentrated upon the non-relativistic case of hydrogen-like atoms in which the electron moves in a strictly inverse-square law electrostatic field of force. For other atoms, the field experienced by the electron is generally not an inverse-square law and

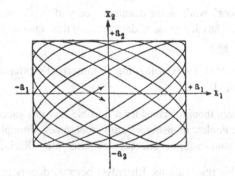

Fig. 5.4   Illustrating the motion of an electron in the x–y plane when the frequencies of oscillation are different in the x- and y-directions (Sommerfeld, 1919).

then the frequencies of oscillation of the independent coordinates need not be the same. In particular, they need not be related by the ratios of integers. In these cases, the overall motion of the electron is not simply periodic, although each of the coordinates undergoes simple periodic motion. When the associated frequencies are not all rational fractions of each other, the motion is referred to as *conditionally periodic*. In the two-dimensional case in Cartesian coordinates, conditionally periodic motion can be represented by Lissajou figures as illustrated in Fig. 5.4. The motion of the electron is represented by oscillations at slightly different frequencies in the x- and y-directions. If the ratio of frequencies is a ratio of integers, the loci will eventually repeat, but in the more general case in which the ratio can be irrational, the loci never join up and the loci will fill the complete x–y plane.

These were very considerable triumphs and many of these features will reappear in a very different guise in the full quantum theory of Heisenberg and Schrödinger. In 1916, the tools of action–angle variables were clearly the way ahead and these had already been developed to a high degree of sophistication by the dynamical astronomers. These techniques flourished and became the preferred methods of the theoretical physicist. While the model of the hydrogen atom was a triumph, the classical Bohr–Sommerfeld model ran into almost insuperable difficulties with multi-electron systems. Fortunately, the techniques for dealing with small perturbations to planetary orbits, which would result in slow changes of conserved quantities as the system evolved, had been pioneered by the astronomers and so these tools were available. They were to be used to great effect in understanding the influence of electric and magnetic fields upon the Bohr–Sommerfeld model of the atom.

5.6 Ehrenfest and the adiabatic principle

Ehrenfest was no admirer of quantum theory. As he remarked in a letter to Lorentz of 25 August 1913,

'Bohr's work on the quantum theory of the Balmer formula (in the *Philosophical Maga-zine*) has driven me to despair. . . . If this is the way to reach the goal, I must give up doing physics.'

Again in May 1916, following Sommerfeld's extension of the Bohr model, he wrote to Sommerfeld

'Even though I consider it horrible that this success will help the preliminary, but still incomplete monstrous Bohr model to new triumphs, I nevertheless heartily wish physics at Munich further success along this path.' (Klein, 1970)

Despite his reservations, Ehrenfest became deeply involved in issues in quantum physics and, in particular, introduced the concept of *adiabatic invariance* into the formalism of the old quantum theory.

The seeds of this concept were already present in the discussions at the 1911 Solvay Conference. Lorentz raised the issue of whether or not a quantised pendulum whose string is being shortened remained in a quantised state. Without hesitation, Einstein replied,

'If the length of the pendulum is changed infinitely slowly, its energy remains $h\nu$ if it was originally $h\nu$.' (Einstein, 1912)

This is an example of adiabatic invariance. Sommerfeld carried out this calculation in his book *Atomic Spectra and Spectral Lines* (1919). Let us repeat it here and clarify exactly what Einstein meant. The pendulum has length l and the bob mass m. The instantaneous angle from the vertical direction is ϕ, the maximum amplitude ϕ_0 being assumed to be small. The angular frequency of oscillation of the pendulum is then $\omega_0 = 2\pi\nu_0 = \sqrt{g/l}$. The tension S in the string is the sum of the gravitational and centripetal forces acting on the bob and so

$$S = mg\cos\phi + ml\dot{\phi}^2 . \tag{5.123}$$

Now the length of the string is shortened very slowly so that there are many swings of the pendulum as its length decreases by dl. The average work done in shortening the string by dl is

$$dW = S\,|dl| = -mg\,\overline{\cos\phi}\,dl - ml\,\overline{\dot{\phi}^2}\,dl , \tag{5.124}$$

where the bars indicate averages taken over a period of the pendulum swing. The signs are negative since the increment of length dl is negative. Taking the motion of the pendulum to be

$$\phi = \phi_0 \sin(\omega_0 t + \gamma) , \tag{5.125}$$

we find

$$\overline{\cos\phi} = 1 - \frac{1}{2}\overline{\phi^2} = 1 - \frac{1}{4}\phi_0^2 , \quad \overline{\dot{\phi}^2} = \frac{\phi_0^2\omega_0^2}{2} = \frac{g\phi_0^2}{2l} . \tag{5.126}$$

Therefore,

$$dW = -\left[mg\left(1 - \frac{1}{4}\phi_0^2\right) + \frac{mg\phi_0^2}{2}\right]dl = -mg\left(1 + \frac{\phi_0^2}{4}\right)dl . \tag{5.127}$$

The first term in dW, $-mg dl$, is the work needed to raise the mean position of the bob by dl while the second represents the increase in the average kinetic energy of motion of the pendulum. The total energy is, as usual, twice the mean kinetic energy, $E = 2\overline{E_{\text{kin}}}$, where

$$\overline{E_{\text{kin}}} = \frac{m}{2} l^2 \overline{\dot{\phi}^2} = mgl \frac{\phi_0^2}{4} . \tag{5.128}$$

Differentiating this expression, the change in total energy is

$$d\overline{E} = mg \frac{\phi_0^2}{2} \, dl + mgl\phi_0 \, d\phi_0 . \tag{5.129}$$

Equating (5.129) to the work done in increasing the motion of the pendulum, the term $-mg\phi_0^2 dl/4$ from (5.127), we find

$$-\frac{3}{4}\phi_0 \, dl = l \, d\phi_0 , \tag{5.130}$$

and so, integrating,

$$\frac{3}{4}\log l = -\log \phi_0 + \text{constant} , \qquad l^{3/4}\phi_0 = \text{constant} . \tag{5.131}$$

Thus, as the string of the pendulum shortens, its angular amplitude increases, although the physical amplitude of the swing decreases, $\Delta x = \phi_0 l \propto l^{1/4}$. Since g is a constant, the frequency of oscillation increases as $\omega \propto l^{-1/2}$ We can now work out the relation between the total energy of oscillation of the pendulum and its frequency from (5.128). We obtain the key result

$$\frac{\overline{E}}{\nu} = \text{constant} . \tag{5.132}$$

This is what Einstein meant in his response to Lorentz.

The quantity $\overline{E}/\nu$ is an example of an *adiabatic invariant*. Adiabatic invariance had been in the literature in various guises since the understanding of the first and second laws of thermodynamics in the mid-1850s. In classical thermodynamics, adiabatic processes play a central role in defining the underlying structure of the theory. To emphasise this point, let me repeat the description of what is involved in reversible adiabatic processes from my treatment in *Theoretical Concepts in Physics* (Longair, 2003).

'A reversible process is one which is carried out infinitely slowly so that, in passing from the state A to state B, the system passes through an infinite number of equilibrium states. Since the process takes place infinitely slowly, there is no friction or turbulence and no sound waves are generated. At no stage are there unbalanced forces. At each stage, we make only an infinitesimal change to the system. The implication is that, by reversing the process precisely, we can get back to the point from which we started and nothing will have changed in either the system or its surroundings. . . .

Let us emphasise this point by considering in detail how we could carry out a reversible isothermal expansion. Suppose we have a large heat reservoir at temperature T and a cylinder with gas in thermal contact with it also at temperature T. No heat flows if the two are at the same temperature. But if we make an infinitesimally small movement of the piston outwards, the gas in the cylinder cools infinitesimally and so an infinitesimal amount of heat flows into the gas by virtue of the temperature difference. This small amount of energy brings the gas back to T. The system is reversible because, if we compress the gas at T slightly, it heats up and heat flows from the gas into the reservoir. Thus, provided we consider only infinitesimal changes, the heat flow process occurs reversibly.

Clearly, this is not possible if the reservoir and the piston are at different temperatures. In this case, we cannot reverse the direction of heat flow by making an infinitesimal change in the temperature of the cooler object. This makes the important point that, in reversible processes, the system must be able to evolve from one state to another by passing through an infinite set of equilibrium states which we join together by infinitesimal increments of work and energy flow.

To reinforce this point, let us repeat the argument for an adiabatic expansion. The cylinder is completely thermally isolated from the rest of the Universe. Again, we perform each step infinitesimally slowly. There is no flow of heat in or out of the system and there is no friction. Therefore, since each infinitesimal step is reversible, we can perform the whole expansion by adding lots of them together.'

In the case of a reversible adiabatic expansion, the pressure and volume are related by the expression $pV^\gamma = \text{constant}$, where γ is the ratio of the specific heat capacities at constant pressure and constant volume respectively, $\gamma = C_p/C_V$. In terms of pressure and temperature, $pT^{\gamma/(\gamma-1)} = \text{constant}$. At all stages in the expansion, the system is in thermal equilibrium at temperature T and so the total energy of the system is $E = V\varepsilon$, where ε is the energy density of the gas. The relation between pressure and internal energy density is $p = (\gamma - 1)\varepsilon$ and so, since $pV = RT$, the relation between the energy E and the temperature T for one mole of gas is

$$E = \frac{RT}{\gamma - 1}, \quad \text{that is,} \quad \frac{E}{T} = \text{constant}. \tag{5.133}$$

This is a further example of an *adiabatic invariant*. But, we have seen this before. The derivation of Wien's displacement law described in Sect. 1.7.2 involved the use of two adiabatic invariants. From (1.31) and (1.32), we find that the total energy $E = V\varepsilon$, the frequency ν and the temperature T are related by the invariants

$$E/\nu = \text{constant}, \quad \nu/T = \text{constant} \quad \text{and so} \quad E/T = \text{constant}. \tag{5.134}$$

Ehrenfest's interest in adiabatic invariance resulted from his perplexity about Planck's and Einstein's derivation of the form of the Planck spectrum in which classical and quantum concepts were mixed together. Ehrenfest's concern was that the Stefan–Boltzmann law and Wien's displacement law are derived by purely classical arguments as we demonstrated in Sects. 1.7.1 and 1.7.2 and yet the highly non-classical procedure of quantisation had to be included to derive the correct form of the black-body spectrum. By 1913, Ehrenfest had resolved the conundrum when he realised that the quantities which were quantised were in fact adiabatic invariants (Ehrenfest, 1913). As shown by (5.134), E/ν is an invariant in the adiabatic expansion of a gas of radiation. If the rule of quantisation is applied to this adiabatic invariant, we immediately find

$$E/\nu = nh \quad \text{where} \quad n = 0, 1, 2, \ldots \tag{5.135}$$

Thus, the rule of quantisation is preserved during the adiabatic expansion from T_2 to T_1. Besides resolving Ehrenfest's concern, it provided a new approach to the issue of which quantities were to be subject to the rules of quantisation. At the time, little attention was paid to Ehrenfest's ideas, but Bohr and Einstein recognised its importance. Einstein called this approach to quantisation Ehrenfest's *adiabatic hypothesis*, namely

> 'If a system be affected in a reversible adiabatic way, allowed motions are transformed into allowed motions.'

In his paper of 1916, Ehrenfest stated,

> '... the hypothesis gives restrictions to the arbitrariness which exists otherwise in the introduction of the quanta.' (Ehrenfest, 1916)

This paper of 1916 presented a much more general and abstract definition of the concept of adiabatic invariance. In Ehrenfest's words,

> 'Let the coordinates of the system be denoted by $q_1, \ldots, q_n$. The potential energy Φ may contain besides the coordinates q certain 'parameters' $a_1, a_2 \ldots$, the values of which can be altered infinitely slowly. The kinetic energy T may be a homogeneous quadratic function of the velocities $\dot{q}_1, \ldots, \dot{q}_n$, the coefficients of which are functions of q and may be of $a_1, a_2, \ldots$. By changing the parameters from the values $a_1, a_2, \ldots$ to $a_1', a_2', \ldots$ in an infinitely slow way, a given motion $\beta(a)$ is transformed into another motion $\beta(a')$. This special type of influencing upon the system may be called *"a reversible adiabatic affection"*, the motions $\beta(a)$ and $\beta(a')$, *"adiabatically related to each other."*'

This last statement means that the process must be a reversible adiabatic change.

This hypothesis immediately accounted for one of the remarkable results of Sommerfeld's development of the Bohr model of the atom, namely that the period of the quantised circular and elliptical orbits of the same principal quantum number had the same energies and frequencies. It could be shown that the model with circular orbits is adiabatically related, in Ehrenfest's sense, to elliptical orbits of the same frequency. Consequently, since E/ν is an adiabatic invariant and equal to nh, it follows that the orbits must have the same energies.

The final step in the argument concerned the appropriate sets of coordinates which should be employed in applying the quantum conditions. The answer came from the studies of Schwarzschild and Epstein whose papers were concerned with the Stark effect, but which contained within them the correct answer (Schwarzschild, 1916; Epstein, 1916a). The answer was that, for systems of dynamical equations which were separable, the quantum conditions should be applied in the action–angle coordinates discussed in Sects. 5.4.4 and 5.5. Specifically, they showed that the action variables J are the quantities to be identified with adiabatic invariants and that these lead directly to Sommerfeld's quantum conditions (5.117)–(5.119). It is now clear why Planck referred to h as the *quantum of action*.

These results led to a great deal of mathematical study of the conditions under which the dynamical equations would be separable and also to the issue of the degeneracy of the orbits in the non-relativistic Bohr–Sommerfeld model. Burgers, a student and then collaborator of Ehrenfest's at Leiden, went on to study·the conditions under which the quantisation rules should be applied to systems in which the Hamiltonian is time-variable, that is, the time-dependent version of the Hamilton–Jacobi equation (5.87), which is

$$H\left(q_1, q_2, \ldots, q_n, \frac{\partial S}{\partial q_1}, \frac{\partial S}{\partial q_2}, \ldots, \frac{\partial S}{\partial q_n}, t\right) + \frac{\partial S}{\partial t} = 0 \,. \tag{5.136}$$

The conditions are not dissimilar from those for the time-independent equation and in addition he showed that the action variables J are adiabatic invariants (Burgers, 1916). These analyses provided the framework for the more detailed study of quantum phenomena, in particular the Zeeman and Stark effects which are discussed in Chap. 7.

5.7 The developing infrastructure of quantum theory

After only a few years, Bohr's dramatic insight into the nature of quantum phenomena in atoms and molecules had been placed on a much more secure theoretical foundation. The apparent arbitrariness of the quantum conditions had been replaced by the understanding that the quantities which had to be quantised were the adiabatic invariants of classical physics. Furthermore, the appropriate mathematical tools for treating quantised systems had been established – the action–angle variables of the transformation theory of mechanics, which were derived from Hamiltonian mechanics and dynamics. This mathematical technique, which had been developed for applications in celestial mechanics, suddenly became the tool of choice of theorists for the study of quantum problems. This brought with it the full apparatus of higher mechanics which was to become the bread and butter of the theoretical physicist. These developments had a long-range impact upon quantum theory since many of the mathematical techniques were to find application in somewhat different guises when Heisenberg's and Schrödinger's versions of the theory of quantum processes were enunciated.

The exploitation of this newly won understanding was the immediate goal of the theorists who still had a long way to go before the theory could account for the details of quantum phenomena.

Einstein coefficients, Bohr's correspondence principle and the first selection rules

6.1 The problem of transitions between stationary states

The understanding of adiabatic invariants as the quantities which are subject to the Bohr–Sommerfeld quantisation conditions and their mathematical description through Hamilton–Jacobi theory and action-angle variables provided the tools for determining the energies of stationary states of quantum systems and hence the *frequencies* of the radiation emitted in transitions between them. The theory had, however, nothing to say about the physical processes by which the transitions take place nor about the intensities and polarisations of the resulting radiation. Bohr proposed to address these issues through his *correspondence principle*. In simple terms, the principle states that in the limit of large quantum numbers, when $\Delta n \ll n$, the radiation processes between stationary states should approach the classical results derived from Maxwell's theory of electromagnetic radiation and so provide information about the intensity and polarisation properties of the radiation.

This correspondence principle had already been foreshadowed in Planck's theory of the spectrum of black-body radiation of 1900. The expression for the equilibrium energy density of radiation in an enclosure at temperature T, derived in Sect. 2.6, is

$$u(\nu) = \frac{8\pi h \nu^3}{c^3} \frac{1}{e^{h\nu/kT} - 1}. \tag{6.1}$$

In Planck's derivation of this result, it was assumed that the energy levels of the oscillators are quantised for the emission of radiation throughout the electromagnetic spectrum and yet, at frequencies $h\nu \ll kT$, (6.1) reduces to the classical formula, the Rayleigh–Jeans law,

$$u(\nu) = \frac{8\pi h \nu^3}{c^3} \frac{1}{e^{h\nu/kT} - 1} \rightarrow \frac{8\pi \nu^2}{c^3} kT. \tag{6.2}$$

This expression was derived by Rayleigh using purely classical arguments, as shown in Sect. 2.3.4. This result was to be elaborated by Bohr into his correspondence principle, but before that Einstein made a crucial advance in tackling the issue of quantum transitions from a probabilistic perspective.

6.2 *On the quantum theory of radiation* (Einstein 1916)

During the years 1911–1916, Einstein was fully preoccupied with the formulation of general relativity, one of the most remarkable intellectual achievements in the history of physics. Following the 1911 Solvay Conference, he wrote relatively little about quantum physics until he returned to the origin of the spectrum of black-body radiation in 1916. As recounted in Chaps. 4 and 5, by that time, the general opinion had shifted in favour of the view that quanta and quantum theory had to be taken seriously in understanding the physics of atoms.

It is important to recall that Einstein's proposal of 1905 that light is composed of discrete quanta was significantly more revolutionary than Planck's introduction of quantisation which only applied to the sources of the radiation. The quotations from the writings of Planck in 1907 and Lorentz in 1909 in Sect. 3.6 indicate that they preferred to think of the emission, absorption and propagation of radiation strictly in terms of Maxwell's classical electromagnetic theory. In contrast, Einstein never deviated from his belief in the reality of light quanta. His paper on fluctuations in the intensity of black-body radiation described in Sect. 3.6 was an example of this continuing pursuit. The paper of 1916 was a further contribution to his crusade to convince his colleagues of the reality of light quanta. The paper is best remembered today for its introduction of what are now known as *Einstein's A and B coefficients* but it also introduces concepts which were to be important in the development of the theory of quanta, particularly the introduction of transition probabilities.

Following Bohr's pioneering efforts of 1913, the emphasis of quantum physics shifted to the understanding of the details of atomic spectra as discussed in Chaps. 4 and 5. Let us look at Einstein's paper of 1916 in a little detail (Einstein, 1916). He begins by noting the formal similarity between the Maxwell–Boltzmann distribution for the velocity distribution of the molecules in a gas and Planck's formula for the black-body spectrum. Einstein shows how these distributions can be reconciled through his new derivation of the Planck spectrum, which gives insight into what he refers to as the 'still unclear processes of emission and absorption of radiation by matter.' The paper begins with a description of a quantum system consisting of a large number of molecules which can occupy a discrete set of states Z_1, Z_2, Z_3, ... with corresponding energies ε_1, ε_2, ε_3, ... According to classical statistical mechanics, the relative probabilities W_n of these states being occupied in thermodynamic equilibrium at temperature T are given by Boltzmann's relation

$$W_n = g_n \exp\left(-\frac{\varepsilon_n}{kT}\right) , \qquad (6.3)$$

where the g_n are the *statistical weights*, or *degeneracies*, of the states Z_n, meaning the number of states with exactly the same energy ε_n. As Einstein remarks in his paper

> '[(6.3)] expresses the farthest-reaching generalisation of Maxwell's velocity distribution law.'

Consider two quantum states of the gas molecules, Z_m and Z_n with energies ε_m and ε_n respectively, such that $\varepsilon_m > \varepsilon_n$. Bohr's model associates the frequency of the radiation emitted in such a transition as $h\nu = \varepsilon_m - \varepsilon_n$, but the mechanism by which this takes place is not specified. According to Einstein's picture, the transition is associated with the emission of a quantum of radiation of energy $h\nu$, which for convenience we will refer to as a *photon*.[1] Similarly, when a photon of energy $h\nu$ is absorbed, the molecule changes from the state Z_n to Z_m. Einstein's aim was to develop a purely quantum approach to the emission and absorption of radiation.

The quantum description of these processes follows by analogy with the classical processes of the emission and absorption of radiation – Jammer (1989) remarks that this is an early manifestation of what was to be embodied in Bohr's *correspondence principle*.

- *Induced emission and absorption* By analogy with the classical case, if an oscillator is excited by waves of the same frequency ν as the oscillator, it either gains or loses energy, depending upon the phase of the wave relative to that of the oscillator, that is, the work done on the oscillator can be either positive or negative. The magnitude of the positive or negative work done is proportional to the energy density u of the incident waves with frequency ν. The quantum mechanical equivalents of these processes are those of *induced absorption*, in which the molecule absorbs the photon and is consequently excited from the state Z_n to Z_m, and *induced emission*, in which the molecule emits a photon under the influence of the incident radiation field. The probabilities of these processes are written

$$\text{\textit{Induced absorption}} \qquad \mathrm{d}W = B_n^m u \, \mathrm{d}t \, ,$$

$$\text{\textit{Induced emission}} \qquad \mathrm{d}W = B_m^n u \, \mathrm{d}t \, .$$

The lower indices refer to the initial state and the upper indices to the final state. B_n^m and B_m^n are constants for a particular pair of energy states, and are referred to as coefficients associated with 'changes of state by induced absorption and emission'.

- *Spontaneous emission* Einstein noted that an electric dipole oscillator emits radiation 'spontaneously' in the absence of excitation by an external field. The corresponding process at the quantum level is called *spontaneous emission*, the probability of the emission of a photon taking place in the time interval $\mathrm{d}t$ without external causes being

$$\mathrm{d}W = A_m^n \, \mathrm{d}t \, . \tag{6.4}$$

Here Einstein used the analogy with the radioactive decay law, $N = N_0 \exp(-\alpha t)$, in which the nucleus decays spontaneously. This can be interpreted in terms of a probability of a decay occurring in the time interval $\mathrm{d}t$ by the rule $p(t)\,\mathrm{d}t = \alpha\,\mathrm{d}t$. He remarked,

'One can hardly think of it in any other way except as a radioactive reaction.'

We now seek the spectrum of the energy density of radiation $u(\nu)$ in thermal equilibrium. The relative numbers of molecules with energies ε_m and ε_n in thermal equilibrium are given by the Boltzmann relation (6.3) and so, in order to leave the equilibrium distribution unchanged under the processes of spontaneous and induced emission and induced absorption of radiation, the probabilities must balance, that is,

$$\underbrace{g_n \mathrm{e}^{-\varepsilon_n/kT} B_n^m u}_{\text{absorption}} = \underbrace{g_m \mathrm{e}^{-\varepsilon_m/kT} \left(B_m^n u + A_m^n \right)}_{\text{emission}} \, . \tag{6.5}$$

In the limit $T \to \infty$, the radiation energy density $u \to \infty$, and the induced processes dominate the equilibrium. Allowing $T \to \infty$, $A_m^n \ll B_m^n u$ and so (6.5) becomes

$$g_n B_n^m = g_m B_m^n .$$
(6.6)

Reorganising (6.5), the equilibrium radiation spectrum u can be written

$$u = \frac{A_m^n / B_m^n}{\exp\left(\dfrac{\varepsilon_m - \varepsilon_n}{kT}\right) - 1} .$$
(6.7)

But, this is Planck's radiation law. Einstein had already demonstrated in 1905 that, in the Wien limit $h\nu \gg kT$, light can be considered to consist of a gas of photons. In that limit,

$$u = \frac{A_m^n}{B_m^n} \exp\left(-\frac{\varepsilon_m - \varepsilon_n}{kT}\right) \propto \nu^3 \exp\left(-\frac{h\nu}{kT}\right) .$$
(6.8)

Therefore, we find the following relations

$$\frac{A_m^n}{B_m^n} \propto \nu^3, \quad \varepsilon_m - \varepsilon_n = h\nu .$$
(6.9)

The value of the constant in (6.8) can be found from the Rayleigh–Jeans limit of the black-body spectrum, $\varepsilon_m - \varepsilon_n / kT \ll 1$. From (6.2), it follows that

$$u(\nu) = \frac{8\pi \nu^2}{c^3} kT = \frac{A_m^n}{B_m^n} \frac{kT}{h\nu} \quad \text{and so} \quad \frac{A_m^n}{B_m^n} = \frac{8\pi h\nu^3}{c^3} .$$
(6.10)

The A_m^n and B_m^n coefficients are associated with atomic processes at the microscopic level. Once A_m^n or B_m^n or B_n^m is known, the other coefficients can be found from (6.6) and (6.10) immediately. Einstein wrote exuberantly to his friend Michele Besso on 11 August 1916,

> 'A splendid flash came to me concerning the absorption and emission of radiation ... A surprisingly simple derivation of Planck's formula, I would say *the* derivation. Everything completely quantum.'

This important analysis occupies only the first three sections of Einstein's paper. The remainder of the paper concerns the transfer of momentum as well as energy between matter and radiation. We will not go through the details of the argument, but summarise its physical content. According to standard kinetic theory, when molecules collide in a gas in thermal equilibrium, there are fluctuations in the momentum transfer between molecules which amounts to

$$\frac{\overline{\Delta^2}}{\tau} = 2RkT ,$$
(6.11)

where Δ is the momentum transfer to a molecule during a short time interval τ and R is a constant related to the 'frictional' force acting upon the moving molecules. Note the striking similarity with Einstein's famous formula (3.1) for the diffusion of microscopic particles undergoing Brownian motion. Now suppose the density of atoms is reduced to an extremely low value so that the momentum transfer is dominated by collisions between photons and the few remaining atoms, which can be considered collisionless. According to Einstein's quantum hypothesis, the photons have energy $h\nu$ and momentum $h\nu/c$. Therefore, the particles should be brought into thermal equilibrium at temperature T entirely through collisions between photons and particles. This was the reason that Einstein needed his equations for spontaneous emission and induced absorption and emission of radiation since these determine the transfer of energy and, more important in the present context, momentum between the particles and the radiation. In the rest of the paper of 1916, Einstein showed that, assuming the momentum transfers occur randomly in directional collisions between photons and electrons, the variance of the fluctuations in the momentum transfer is exactly the same expression as (6.11). This could not happen if the energy re-radiated by the particles was isotropic because then there would be no random component in the momentum transfer process. The key result was that, when a molecule emits or absorbs a quantum $h\nu$, there must be a positive or negative change in the momentum of the molecule of magnitude $|h\nu/c|$, even in the case of spontaneous emission. In Einstein's words,

> 'Outgoing radiation in the form of spherical waves does not exist. During the elementary process of radiative loss, the molecule suffers a recoil of magnitude $h\nu/c$ in a direction which is determined only by "chance", according to the present state of the theory.'

His view of the importance of the calculation is summarised in the version of the paper published in 1917.

> 'The most important thing seems to me to be the momenta transferred to the molecule [atom] in the processes of absorption and emission. If any of our assumptions concerning the transferred momenta were changed [(6.11)] would be violated. It hardly seems possible to reach agreement with this relation, which is demanded by the [kinetic] theory of heat, in any other way than on the basis of our assumption.'

Direct experimental evidence for the correctness of Einstein's conclusion was only obtained in 1923 from Compton's X-ray scattering experiments (Compton, 1923). These showed that photons undergo collisions in which they behave like particles, the *Compton effect* or *Compton scattering*. It is a standard result of special relativity that the increase in wavelength of a photon in collision with a stationary electron is

$$\lambda' - \lambda = \frac{hc}{m_e c^2}(1 - \cos\theta), \tag{6.12}$$

where θ is the angle through which the photon is scattered.[2] Implicit in this calculation is the conservation of relativistic three-momentum in which the momentum of the photon is $h\nu/c$.

6.3 Bohr's correspondence principle

Einstein's introduction of the concept of spontaneous and induced emission processes made a profound impression upon Bohr who saw a way of relating the spontaneous transition probabilities A_m^n to classical electrodynamics through application of what became known as the *correspondence principle*. Allusion has already been made in Sect. 6.1 to the result that the Planck spectrum reduces to the classical Rayleigh–Jeans formula in the limit of very low frequencies. In addition, Bohr had shown that the same equivalence applied to transitions between states of very large quantum numbers n for which the emitted frequencies are also very low. Let us first demonstrate this result for the Bohr model of the hydrogen atom with circular orbits.

We showed in Sect. 4.5 that the kinetic energy of an electron in a state with principal quantum number n is

$$T = \frac{1}{2}m_e v^2 = \frac{m_e e^4}{8\epsilon_0^2 n^2 h^2} \,, \tag{6.13}$$

from which it follows that the velocity of the electron in its circular orbit is given by

$$v^2 = \frac{e^4}{4\epsilon_0^2 n^2 h^2} \,. \tag{6.14}$$

The frequency of rotation of the electron about the nucleus ν_e is

$$\nu_e = \frac{v}{2\pi r} = \frac{v^2}{2\pi vr} = \frac{m_e v^2}{2\pi J} = \frac{T}{\pi J} \,, \tag{6.15}$$

where $J = nh/2\pi$ is the quantised angular momentum of the electron. Therefore,

$$\nu_e = \frac{m_e e^4}{4\epsilon_0^2 h^3 n^3} \,. \tag{6.16}$$

Let us now work out the frequency of emission of radiative transitions between stationary states with large values of n, the principal quantum number according to the Bohr picture. From (4.28), we find

$$\nu = \frac{m_e e^4}{8\epsilon_0^2 h^3} \left(\frac{1}{n^2} - \frac{1}{(n + \Delta n)^2} \right) \,, \tag{6.17}$$

where $\Delta n \ll n$ is the difference in principal quantum number between the upper and lower states. We can therefore carry out a Taylor expansion of the term in $(n + \Delta n)^{-2}$ and find that

$$\nu = \frac{m_e e^4}{4\epsilon_0^2 h^2 n^3} \Delta n \,. \tag{6.18}$$

Now, $\Delta n = 1, 2, \ldots$ and so we find that the frequency of rotation (6.16) is exactly equal to the emission frequency of the transition for $\Delta n = 1$, $\nu = \nu_e$. Thus, in this simplest case, we see that there is an exact correspondence between the frequency of rotation of the electron in its orbit about the nucleus, which classically gives rise to electromagnetic radiation at the orbital frequency of the electron, and the frequency associated with the difference in energies of the quantised stationary states of the electron. This is the feature which Bohr exploited in his correspondence principle. His proposal was to use the theory of the electromagnetic radiation of the orbiting electron according to Maxwell's electromagnetic theory to determine the intensity and polarisation properties of the emission lines according to quantum theory.

Notice another important feature of (6.18). The frequencies of lines with $\Delta n = 2, 3, 4, \ldots$ are exact harmonics of the orbital frequency ν_e, in other words, $\nu = \tau \nu_e$, where $\tau = \Delta n$ takes the values $\tau = 2, 3, 4, \ldots$. This will prove to be an important result in the transition from classical to quantum mechanics (Sect. 10.3).

The above argument can be readily extended to more general orbits using the action–angle variables introduced in Sects. 5.4 and 5.5. It is immediately apparent from (5.122) that the quantum frequency ν for non-relativistic elliptical orbits for hydrogen-like atoms follows exactly the same quantum rules as for circular orbits, where the principal quantum number is $(n_r + n_\theta + n_\phi)$ in the notation of Sect. 5.5. In his major paper of 1918, Bohr established this equivalence in the following way (Bohr, 1918a). According to the formalism of action–angle variables, (5.121) states that the frequency associated with the k coordinate of a periodic, or more generally conditionally periodic, system is

$$\nu_k = \frac{\partial H}{\partial J_k} = \frac{\partial E}{\partial J_k} , \tag{6.19}$$

where J_k is the action variable associated with the k coordinate and H the Hamiltonian, which is just the total energy E. This can be compared with the quantum expressions

$$h\nu_q = \Delta E , \quad \Delta J_k = (n'_k - n_k)h = \tau_k h , \tag{6.20}$$

where n'_k and n_k are the principal quantum numbers describing the states before and after the transition and τ_k takes the values $1, 2, \ldots$. In the limit of large quantum numbers, in which the energy differences between neighbouring states are small, we can replace the differences in E and J_k between states by their partial differentials, that is,

$$\frac{\Delta E}{\Delta J_k} \approx \frac{\partial E}{\partial J_k} . \tag{6.21}$$

This relation encapsulates Bohr's correspondence principle, the left-hand side being quantum and the right-hand side classical. Therefore,

$$h\nu_q = \Delta E = \frac{\Delta E}{\Delta J_k} \Delta J_k \approx \left(\frac{\partial E}{\partial J_k} \right) \Delta J_k = \tau_k \nu_k h . \tag{6.22}$$

In other words, $\nu_q = \tau_k \nu_k$. For the case $\tau = 1$, we recover the exact equivalence between the classical and the quantum results.

There is also an equivalence for higher harmonics of ν_k in the classical theory as well. The values of $\tau_k > 1$ correspond to harmonics of the fundamental frequency ν_k and this has a counterpart in the classical theory of the emission of electromagnetic radiation of an electron in orbit about the nucleus at frequency ν. The necessary tools have already been described in Sect. 2.3.1. In particular, from (2.3), the average rate of radiation of a dipole of dipole moment $p_0 = ex_0$ oscillating at angular frequency ω_0 is

$$-\left(\frac{dE}{dt} \right)_{\text{average}} = \frac{\omega_0^4 e^2 x_0^2}{12\pi \epsilon_0 c^3} = \frac{\omega_0^4 p_0^2}{12\pi \epsilon_0 c^3} . \tag{6.23}$$

Therefore, the procedure is to decompose the periodic motion of the electron in its orbit into a complete set of orthogonal dipoles and this is achieved by describing the motion in terms of Fourier series.

Consider first a periodic system of one degree of freedom. Then, the Bohr–Sommerfeld quantisation condition for the system is

$$\oint p \, \mathrm{d}q = nh \, . \tag{6.24}$$

The displacement x of the charged particle in its periodic, but not necessarily harmonic, motion can be expressed as a Fourier series,

$$x = \sum_{\tau} C_{\tau} \cos 2\pi(\tau\omega t + c_{\tau}) \, , \tag{6.25}$$

where C_{τ} and c_{τ} are constants and the sum extends over all integral values of $\tau = 1, 2, \ldots$. ω is the frequency associated with the periodic motion of the electron about the nucleus.[3] In this one-dimensional example, the quantities C_{τ} are the amplitudes of oscillating dipoles with frequencies ω, 2ω, 3ω, $\ldots$ Thus, according to classical electrodynamics, the particle emits a series of lines with frequencies ω, 2ω, $\ldots$ and the intensity of each line is given by (6.23) with the dipole moment p_0 being given by $D_{\tau} = eC_{\tau}$. In other words, the intensities of the lines are determined by the squares of the absolute values of the Fourier components $|D_{\tau}|^2$. Bohr concluded

> 'We must therefore expect that for large values of n, these coefficients will on the quantum theory determine the *probability of spontaneous transition* from a given stationary state for which $n = n'$ to a neighbouring state for which $n = n'' = n' - \tau$.' (Bohr, 1918a)

Bohr had already noted that this result would be obtained in the case of an electron moving in an elliptical orbit about the nucleus:

> 'The possibility of an emission of a radiation of such a frequency may also be interpreted from analogy with ordinary electrodynamics, as an electron rotating around a nucleus in an elliptical orbit will emit a radiation which according to Fourier's theorem can be resolved into homogeneous components, the frequencies of which are $\tau\omega$, if ω is the frequency of revolution of the electron.' (Bohr, 1913b)

This was only the beginning of the story. In general, the orbits of the electrons will not be simple ellipses because the electric field is not a purely $1/r$ potential when other electrons are present and so the motion is conditionally periodic rather than periodic. In two dimensions, the motion can be visualised as shown in Fig. 5.4. Therefore, we need the general k-dimensional Fourier series for general conditionally periodic motion. In Bohr's notation, the coordinates of the conditionally periodic system are $q_1, q_2, \ldots, q_s$ and so the displacement ξ of the particles in any direction can be expressed as a function of time by what Bohr calls an 's-double infinite Fourier series' of the form

$$\xi = \sum C_{\tau_1, \tau_2, \ldots, \tau_s} \cos 2\pi[(\tau_1\omega_1 + \cdots + \tau_s\omega_s)t + c_{\tau_1, \tau_2, \ldots, \tau_s}] \, , \tag{6.26}$$

where the sum extends over all positive and negative values of τ_s and each ω is the mean frequency of oscillation of each independent coordinate q_k. The values of $C_{\tau_1, \tau_2, \ldots, \tau_s}$ depend upon the constants of motion α_i derived from the Hamilton–Jacobi equations, such as those described by (5.100)–(5.103). Notice that, unlike the degenerate elliptical orbits, in general, $\omega_1 \neq \omega_2 \neq \cdots \neq \omega_s$, and so the ωs are not simply related to each other by ratios of integers. Notice also that the Fourier series contains not only harmonics of each fundamental frequency ω_k, but also 'cross-terms' $(\tau_k \omega_k + \tau_l \omega_l)$ which correspond to coupling between independent modes of oscillation.

Bohr's insight was that the spontaneous transition probability A_m^n should be identified with the dipole moments of the corresponding Fourier components of (6.26). Thus, if we write the dipole moment associated with the transition $n \to m$ as $D_{n \to m} = eC_{n \to m}$, we equate the classical radiation formula (6.23) to the spontaneous emission luminosity of a single electron transition,

$$A_m^n h \nu = \frac{\omega_{n \to m}^4 |D_{n \to m}|^2}{12 \pi \epsilon_0 c^3} . \tag{6.27}$$

Therefore,

$$A_m^n = \frac{\omega_{n \to m}^3}{6 \epsilon_0 c^3 h} |D_{n \to m}|^2 . \tag{6.28}$$

Strictly speaking, these correspondences should only apply for transitions Δn between stationary states with large values of n such that $\Delta n \gg n$. Bohr, however, had little hesitation in applying the correspondence principle to small values of n as well. In addition to interpreting the intensity and polarisation properties of the emissions lines, the principle also led to the concept of *selection rules*.

6.4 The first selection rules

The expression (6.26) indicates that there are very large numbers of stationary states and correspondingly an even greater number of possible transitions, but not all of these are realised in nature. In his paper of 1918, Bohr enunciated the first selection rules which would limit the number of permitted transitions between stationary states. Clearly, if the Fourier component $D_{n \to m}$ in (6.28) is zero, the probability of the transition is zero and the transition is *forbidden* for such electric dipole transitions.

Let us consider first the case of a one-dimensional harmonic oscillator. Because the system is periodic, we can simplify the Fourier series (6.26) by writing

$$p = \frac{1}{2} \sum_{\tau = -\infty}^{+\infty} D_\tau \exp(i \tau \omega_0 t) , \tag{6.29}$$

where the factor $1/2$ is included since the summation spans the range $-\infty < \tau < +\infty$. The time variation of the dipole moment of the oscillator is harmonic and so

$$p = e x_0 \cos \omega_0 t = \frac{e x_0}{2} \left(e^{i \omega_0 t} + e^{-i \omega_0 t} \right) . \tag{6.30}$$

To find the Fourier components we multiply both sides by $\exp(-i \tau' \omega_0 t)$ and integrate over one cycle of the oscillation. The only Fourier components which are non-zero are those corresponding to $\tau = 1$ and $\tau = -1$,

$$D_1 = D_{-1} = \frac{1}{2}ex_0 \quad \text{for} \quad \tau = \pm 1, \quad \text{and} \quad D_\tau = 0 \quad \text{if} \quad \tau \neq \pm 1. \tag{6.31}$$

Therefore, the expression for the spontaneous transition probability is

$$A^n_m = \frac{\omega_0^3}{6\epsilon_0 c^3 h} e^2 x_0^2, \tag{6.32}$$

with the selection rule $\Delta n = \pm 1$. Since the quantised energy levels of the harmonic oscillator are

$$E_n = nh\nu_0 = \frac{1}{2}m\omega_0^2 x_0^2, \tag{6.33}$$

using the relation $\omega_0 = 2\pi\nu_0$, the transition probability can also be written

$$A^n_m = \frac{\omega_0^2 e^2}{6\pi\epsilon_0 mc^3} n. \tag{6.34}$$

These are rather remarkable properties of the quantised harmonic oscillator. It can be seen that, because of the equal spacings between the energy levels and the selection rule $\Delta n = \pm 1$, for all transitions, the frequency of the emitted radiation is exactly equal to that of the oscillator for all permitted transitions. Notice also that although Bohr's correspondence principle was intended only to provide an estimate of the radiation rate for high energy levels, the results hold true for transitions between *all energy levels* of the harmonic oscillator. As remarked by Jammer (1989), these remarkable features of the quantised harmonic oscillator are fortunate features of the early conceptual development of quantum theory. In his pioneering investigations of 1900, Planck had assumed that the quantisation of the harmonic oscillator resulted in equal energy differences between the energy levels and that these differences corresponded to the frequency of the oscillator through the relations $\Delta E = \varepsilon = h\nu$ and $\nu = \nu_0$. Notice also that this equally spaced quantum 'ladder' is found because the oscillation is harmonic and so derived from a harmonic potential.

The same procedure was adopted by Bohr for transitions between the stationary states of atoms. The transition probabilities are associated with the amplitudes of the components of the Fourier series, but now the transitions in each of the r, θ and ϕ coordinates had to be considered. Let us consider the simplest case of hydrogen-like atoms in which the orbits of the electrons are ellipses, the periods of electrons with the same principal quantum number n being fixed. The solution of this problem was discussed in Sect. 5.5, the energy levels being given by (5.120) and the frequencies associated with the three orthogonal coordinates r, θ and ϕ given by (5.122) being all the same, namely,

$$E = -\frac{Z^2 e^4 m}{8\epsilon_0^2 h^2 (n_r + n_\theta + n_\phi)^2}, \quad \nu_r = \nu_\theta = \nu_\phi = \frac{4\pi^2 mk^2}{(J_r + J_\theta + J_\phi)^3}. \tag{6.35}$$

Bohr pointed out in his paper of 1918 that in this case, the quantisation rules derived for the harmonic oscillator also apply since there is only a single frequency associated with the transition, just as in (6.30). Therefore similar selection rules apply, but now all three quantum numbers need to be considered.

The simplest approach is that presented by ter Haar (1967). The dipole moment can be resolved into components along the x-, y- and z-directions as follows:

$$P_x = er \cos\phi \sin\theta \,, \tag{6.36}$$

$$P_y = er \sin\phi \sin\theta \,, \tag{6.37}$$

$$P_z = er \cos\theta \,, \tag{6.38}$$

where (r, θ, ϕ) are standard spherical polar coordinates. The terms in θ and ϕ can be written in the form $e^{\pm i\theta}$ and $e^{\pm i\phi}$. For the case of hydrogen-like atoms, these factors as a function of time t can be written in the form

$$e^{i\phi(t)} = \sum_\tau A_\tau \, e^{\pm i\tau\omega_\phi t} \propto e^{\pm i\omega_\phi t} \,, \tag{6.39}$$

$$e^{i\theta(t)} = \sum_\tau B_\tau \, e^{\pm i\tau\omega_\theta t} \propto e^{\pm i\omega_\theta t} \,, \tag{6.40}$$

where A_τ and B_τ can be written in the same form of Fourier series as (6.29) and $\omega_\phi = 2\pi\nu_\phi = \omega_\theta = 2\pi\nu_\theta = 2\pi\nu_r$. Just as in the case of the harmonic oscillator, the only terms of the Fourier expansions which are non-zero are those with $\tau = \pm 1$ and so the selection rule for the quantum number in the θ-coordinate is

$$\Delta n_\theta = \pm 1 \,. \tag{6.41}$$

A similar result applies for the ϕ-component, but in addition, since there is no dependence of p_z on ϕ, $\nu_\phi = 0$ and so the selection rule $\Delta n_\phi = 0$ is also allowed. Therefore, for the ϕ-component, the quantisation conditions are:

$$\Delta n_\phi = 0, \, \pm 1 \,. \tag{6.42}$$

Bohr went on to show that these quantisation rules also apply to the case in which the frequencies ν_r, ν_θ and ν_ϕ are different, the case of conditionally periodic orbits. Thus, by using the correspondence principle, Bohr was able to reduce the number of possible transitions between stationary states to those which satisfied the above selection rules for electric dipole transitions.

6.5 The polarisation of quantised radiation and selection rules

Further insight into the origin of the selection rules came from considerations of the polarisation properties of the radiation emitted in quantum transitions. The basic premise of the Bohr–Sommerfeld model of atoms is that the angular momenta of stationary states are quantised in units of $h/2\pi$. In the simplest model with circular orbits, when a transition occurs, there is a change in the energy of the electron in the atom $h\nu$ and a corresponding change in angular momentum $h/2\pi$. The angular momentum is transferred to the emitted radiation. Bohr and Sommerfeld appreciated that since the radiation is quantised, there must be selection rules which determine which transitions between stationary states can result in electric dipole radiation.

The application of the correspondence principle to the conservation of angular momentum in the emission of radiation was worked out by Sommerfeld's assistant Adalbert Rubinowicz and published in an important paper of 1918 (Rubinowicz, 1918a). Sommerfeld's exposition in Chapter 5 of his book *Atomic Structure and Spectral Lines* (1919) provides a wonderful review of the problems of reconciling the classical and quantum pictures of the process of emission of radiation by atoms which confronted the pioneers of quantum theory. Rubinowicz's approach to the emission of an electron in a general elliptical orbit about the nucleus was adopted. They considered the radiation of a three-dimensional dipole P which has components

$$\boldsymbol{p}_x = a \, \exp{(i\alpha)}, \qquad (6.43)$$

$$\boldsymbol{P} = \boldsymbol{p} \, \exp{(i\omega t)}, \qquad \boldsymbol{p} = \boldsymbol{p}_x + \boldsymbol{p}_y + \boldsymbol{p}_z, \qquad \boldsymbol{p}_y = b \, \exp{(i\beta)}, \qquad (6.44)$$

$$\boldsymbol{p}_z = c \, \exp{(i\gamma)}. \qquad (6.45)$$

The amplitudes of the dipole in the x-, y- and z-directions are a, b and c and the phases of the oscillators are determined by the quantities α, β and γ. If $\alpha = \beta = \gamma = 0$, the dipole is a linear oscillator with axis in the direction of P. A dipole rotating about the z-axis in the x–y plane could be described by $a = b$, $c = 0$, $\alpha = 0$, $\beta = \pm\pi/2$.

The properties of the radiation of an oscillating dipole were described in Sect. 2.3.1. More specifically, in the far field limit, the instantaneous electric field of the radiation is given by

$$E_\theta = \frac{|\ddot{\boldsymbol{p}}| \sin\theta}{4\pi\varepsilon_0 c^2 r}. \qquad (6.46)$$

The rate of energy flow per unit area per second at distance r is given by the magnitude of the Poynting vector $\boldsymbol{S} = \boldsymbol{E} \times \boldsymbol{H} = E_\theta^2/Z_0 \, \mathbf{i}_r$, where $Z_0 = (\mu_0/\varepsilon_0)^{1/2}$ is the impedance of free space. The rate of energy flow through the area $r^2 \, d\Omega$ subtended by solid angle $d\Omega$ at angle θ and at distance r from the charge is therefore

$$S r^2 \, d\Omega = -\left(\frac{dE}{dt}\right) d\Omega = \frac{|\ddot{\boldsymbol{p}}|^2 \sin^2\theta}{16\pi^2 Z_0 \varepsilon_0^2 c^4 r^2} r^2 \, d\Omega = \frac{|\ddot{\boldsymbol{p}}|^2 \sin^2\theta}{16\pi^2 \varepsilon_0 c^3} \, d\Omega. \qquad (6.47)$$

To find the total radiation rate $-dE/dt$, we integrate over solid angle. Because of the symmetry of the emitted intensity with respect to the acceleration vector, we can integrate over the solid angle defined by the circular strip between the angles θ and $\theta + d\theta$, $d\Omega = 2\pi \sin\theta \, d\theta$:

$$-\left(\frac{dE}{dt}\right) = \int_0^\pi \frac{|\ddot{\boldsymbol{p}}|^2 \sin^2\theta}{16\pi^2 \varepsilon_0 c^3} 2\pi \sin\theta \, d\theta. \qquad (6.48)$$

We find the result

$$-\left(\frac{dE}{dt}\right) = \frac{|\ddot{\boldsymbol{p}}|^2}{6\pi\varepsilon_0 c^3} = \frac{q^2 |\boldsymbol{a}|^2}{6\pi\varepsilon_0 c^3}. \qquad (6.49)$$

This result is sometimes referred to as *Larmor's formula*, the result given in (2.1).

These results illustrate the well-known features of the radiation of an accelerated charge that the radiation is emitted symmetrically with respect to the acceleration vector and the energy loss rate is given by the Poynting vector flux which is radial. This energy also transfers the momentum of the waves to infinity, the momentum flow rate per unit area per second being S/c. However, because of the symmetry of the emitted radiation, there is no overall momentum loss by the particle. These results remain true for a linear oscillator and for the rotating dipole.

There is, however, a difference in the case of the angular momentum, or the moment of the momentum, of the radiation field. The momentum per unit volume at distance r from the dipole is S/c^2 and so the moment of the momentum of the emitted radiation is

$$M = r \times \left(\frac{S}{c^2}\right). \tag{6.50}$$

This quantity has to be integrated over all space. This calculation is carried out by Sommerfeld in Appendix 9 of his book. In the case in which the dipole rotates about the z-axis in the x–y plane, the angular momentum of the radiation transferred per unit time is

$$N_z = \frac{W}{\omega} \frac{2ab \sin \gamma}{a^2 + b^2}, \tag{6.51}$$

where W is the total energy radiated by the orbiting electron and $\gamma = \beta - \alpha$ is the phase difference between the oscillations in the x- and y-directions. Let us check that this result makes sense. In the cases in which only one component of the oscillator is present, say $b = 0$, there is evidently no net angular momentum in the radiation, as would be expected from the symmetry of the radiated emission. Likewise, if $\gamma = 0$ so that the oscillations in the x- and y-directions are in phase, there is no angular momentum since the particle is a linear oscillator at some fixed angle in the x–y plane. If, however, the phase difference is $\pm\pi/2$ and $a = b$, the quantity $2ab \sin \gamma /(a^2 + b^2)$ takes maximum and minimum values of ± 1. These values correspond to *circular polarisation in opposite senses* when viewed along the z-direction. Notice that these results are no more than the law of conservation of angular momentum for the emission of radiation. An electron in a circular orbit, for example, loses angular momentum as it radiates away energy and this is transferred to the radiation field which takes away the angular momentum as circularly polarised radiation.

Sommerfeld now translates these classical results to the quantum emission of radiation. The energy removed in the quantum emission of radiation is $W = h\nu$ and so, since $\omega = 2\pi\nu$,

$$N_z = \frac{h}{2\pi} \frac{2ab \sin \gamma}{a^2 + b^2}. \tag{6.52}$$

Now, according to the rules for the quantisation of angular momentum $J_\phi = n_\phi h/2\pi$ and so, when a quantum transition Δn_ϕ is made,

$$\Delta J_\phi = \frac{h}{2\pi} \Delta n_\phi. \tag{6.53}$$

Equating (6.52) and (6.53), we find

$$\Delta n_\phi = \frac{2ab \sin \gamma}{a^2 + b^2}. \tag{6.54}$$

Now, we have just shown that the extremal values of the right-hand side of (6.53) are ± 1 and so the only possible integral values of Δn_ϕ are

$$\Delta n_\phi = +1 , \qquad \Delta n_\phi = -1 , \qquad \Delta n_\phi = 0 . \qquad (6.55)$$

Sommerfeld's analysis resulted in two key results for the quantum emission of atoms:

- **The rule of selection** states: *the azimuthal quantum number can at most alter by one unit at a time in changes of configuration of the atom.*
- **The rule of polarisation** demands that *if the azimuthal quantum number alters by ± 1, the light is circularly polarised; if the quantum number remains constant, the light is linearly polarised.*

As we will discuss in the next chapter, Rubinowicz was to apply these concepts to the understanding of the Stark and Zeeman effects in hydrogen and other atoms with some success (Rubinowicz, 1918b).

Thus, Bohr's and Rubinowicz's somewhat different approaches resulted in the same result that there are necessarily *selection rules* which determine which transitions in atoms are to be permitted. Note that there is an important difference between their approaches. Bohr worked exclusively in terms of the correspondence principle in which the transitions between states were subject to the quantum rules and the selection rules were found by analogy. In his picture, the radiation of the atoms themselves was determined by Maxwell's equations. In contrast, Rubinowicz considered both the atoms and the radiation to be part of a single quantum system, as he described in his paper of 1921 (Rubinowicz, 1921). His analysis endowed the emitted photons with energy and angular momentum.

6.6 The Rydberg series and the quantum defect

The Bohr model of the atom could successfully account in considerable detail for the spectral properties of the hydrogen atom, but atoms with more than one electron were a much greater challenge. Nonetheless, the fact that the Rydberg series of atoms such as sodium, potassium, magnesium, calcium and zinc had similar forms to that of the hydrogen atom, as illustrated by the Rydberg formulae (1.19)–(1.21), suggested that the same general principles were involved. It is striking, for example, that the spectrum of sodium is similar to that of hydrogen. What was needed was a suitable form for the electrostatic potential experienced by an electron in the combined field of the nucleus and the other orbiting electrons.

Sommerfeld tackled this problem in his book *Atomic Structure and Spectral Lines* (1919) by adopting a more general form for the electrostatic potential experienced by the electron. He wrote

$$U = -\frac{eE}{4\pi\epsilon_0 r} + V , \quad \text{where} \quad V = \frac{e^2}{4\pi\epsilon_0 r}\left[c_1 \left(\frac{a}{r}\right)^2 + c_2 \left(\frac{a}{r}\right)^3 + \cdots \right] , \qquad (6.56)$$

where the scale a is chosen to be the radius of the first circular Bohr orbit, $a = h^2 \epsilon_0 / \pi m_e e^2$. Assuming the electron moves in a planar orbit defined by the coordinates r, ϕ in this central electrostatic potential, the analysis is exactly the same as in Sect. 5.5, but now the potential term k/r in (5.104) and similar formulae is replaced by (6.56). Thus, the quantisation condition for ϕ takes the usual form $p_\phi = n_\phi h / 2\pi$. In the radial coordinate, Sommerfeld showed that, including only the first additional term in c_1 and carrying out the same type of integration which resulted in (5.116), the expression for the energies of the stationary states of the electron arc of the form

$$E = -\frac{R_\infty}{(n_r + n_\phi + d)} , \tag{6.57}$$

where $d = Z c_1 / n^3$. Thus, the additional term in d depends upon the nature of the additional term in the potential. Similar results were obtained when the next term in c_2 was included in the expression for the electrostatic potential. Evidently, the inclusion of these additional terms leads to energy levels of the type needed to account for the Rydberg formulae (1.19)–(1.21).

The origin of this modification to the standard Bohr formula can be understood from the argument presented by ter Haar (1967). Considering for illustration the example of the sodium atom, the Bohr–Sommerfeld picture would involve a nucleus with charge $11e$, surrounded by 11 orbiting electrons. If a single electron is responsible for the spectral line, we can consider its motion in the combined field of the nucleus and the other 10 electrons. The potential due to the 10 other electrons can be represented by a spherically symmetric central field of force, an assumption which was to find some justification in the shell model of the electron distribution in atoms. Therefore, when the electron is far from the nucleus, it experiences a potential $-e^2 / 4\pi \epsilon_0 r$ as the nucleus is shielded from all but $1/11$ of the nuclear charge. On the other hand, when close to the nucleus, the electron experiences the full nuclear electrostatic potential, $-11 e^2 / 4\pi \epsilon_0 r$.

The specific form of the potential is not known, but we can derive the general features of the solution from the following line of reasoning. To find the energy of the stationary states, we need to find the integral equivalent to (5.110) which was evaluated for a pure inverse-square law electrostatic potential. Replacing the term $2mk/r$ by the general potential term $2mU(r)$, the action integral becomes,

$$J_r = \oint \sqrt{2mE + 2mU(r) - \frac{\alpha_\theta^2}{r^2}} \, dr , \tag{6.58}$$

since, according to classical mechanics, the radial component of the momentum of the electron is given by the expression

$$p_r^2 = 2mU(r) - \frac{\alpha_\theta^2}{r^2} + 2mE . \tag{6.59}$$

For the case of a pure inverse-square law of electrostatic attraction, $U(r) = k/r$ and then the motion of the electron can be represented by a diagram in which the square of the radial momentum is plotted against distance from the nucleus (Fig. 6.1a). The total energy of the bound orbit is E and is a negative quantity. At the points r_{min} and r_{max}, the radial component

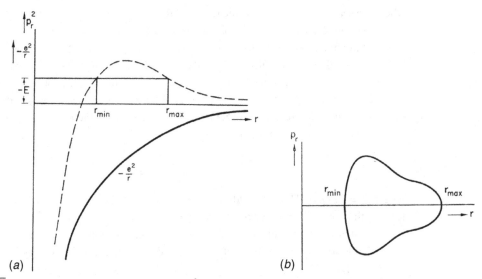

Fig. 6.1 (*a*) Illustrating the motion of an electron in a p_r^2–r diagram for an inverse-square law force of electrostatic attraction. (*b*) The same motion plotted in a p_r–r diagram (ter Haar, 1967).

of the momentum goes to zero since all the momentum is then in the azimuthal component and the integral (6.58) is taken between these limits. Correspondingly, the radial motion of the electron can be represented on a p_r–r diagram which is shown in Fig. 6.1*b*.

Now, the quantisation condition is

$$J_r = \oint p_r \, \mathrm{d}r = n_r h \,, \tag{6.60}$$

and so this integral corresponds to the area bounded by the locus in the p_r–r relation shown in Fig. 6.1*b*. In other words, whatever the shape of the locus in the p_r–r plane, the area is quantised in units of h. Let us now consider the case of the sodium atom. At large radii, the locus will have the same form as the hydrogen atom since it only feels a net single electronic charge. Close to the nucleus, the potential will be much deeper since the charge of the nucleus is $11e$. Therefore, we would expect the p_r^2–r relation to extend much closer to the nucleus, as illustrated in the comparison of the $U(r) \propto 1/r$ potential and the potential felt by the electron in the sodium atom in Fig. 6.2(i). The corresponding quantities in the p_r–r plane are shown in Fig. 6.2(ii). Now, the quantisation conditions for J_r for the sodium atom is that the area enclosed by the locus (b) in Fig. 6.2(ii) is equal to $n_r h$. This area is equal to the area enclosed by the locus for the hydrogen atom plus the shaded area which must be some fraction α of the total area. Therefore, we can write the quantisation condition

$$J_r = n_r h = \alpha h + \oint \sqrt{2mE + \frac{2mek}{r} - \frac{\alpha_\theta^2}{r^2}} \, \mathrm{d}r \,, \tag{6.61}$$

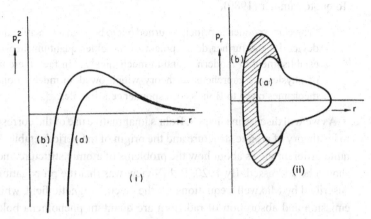

Fig. 6.2 (i) Illustrating the motion of an electron in a p_r^2–r diagram for the potential of a hydrogen-like atom (a) and a sodium atom (b). (ii) The same motions plotted in a p_r–r diagram (ter Haar, 1967).

and so

$$(n_r - \alpha)h = \oint \sqrt{2mE + \frac{2mk}{r} - \frac{\alpha_\theta^2}{r^2}}\, dr . \qquad (6.62)$$

If we now carry out exactly the same analysis which led to (5.120), we find

$$E = -\frac{me^4}{2\epsilon_0^2 h^2} \frac{1}{(n - \alpha)^2} , \qquad (6.63)$$

where $n = n_r + n_\theta + n_\phi$ is the principal quantum number. This is the Rydberg formula for the energy levels of atoms quoted in (1.16).[4]

6.7 Towards a more complete quantum theory of atoms

Over a matter of a few years, Bohr's theory of the hydrogen atom had been given a much more secure theoretical basis. The models of atoms were now fully three dimensional and the correspondence principle offered a route to understanding the intensity and polarisation properties of the emission of atoms. The selection rules limited the numbers of allowed transitions between stationary states. Despite the continuing concern about the incompatibilities between the quantum and classical pictures, there was optimism that these ideas provided the basis for the development of a more complete quantum theory of atoms.

The subtle use of the correspondence principle was behind many of the successes of the theory. Sommerfeld remarked that the correspondence principle was like,

'a magic wand that allowed the results of classical wave theory to be of use for the quantum theory.'

To quote Jammer (1989),

> 'The correspondence principle turned out to be a most versatile and productive conceptual device for the further development of the older quantum theory – and . . . even for the establishment of modern quantum mechanics. . . . In fact, there was rarely in the history of physics a comprehensive theory which owed so much to one principle as quantum mechanics owed to Bohr's correspondence principle.'

As we will discuss in Chap. 8, Bohr's imaginative use of the correspondence principle led to his theory of atomic structure and the origin of the periodic table. Bohr and Einstein held quite different views about how the problems of atomic structure and quantum phenomena should be addressed. By 1920, Bohr's view was that the propagation of light is accurately described by Maxwell's equations for the electromagnetic field, while the processes of the emission and absorption of radiation are quantum phenomena bolted onto a mechanical model of the atom. Einstein was critical of the insecure foundations of the old quantum theory, taking the view that the emission, absorption *and* propagation of radiation are all quantum processes. Despite these differing points of view, Einstein's admiration for Bohr's insight and achievements was unstinting. Many years later, Einstein wrote:

> 'That this insecure and contradictory foundation [of the quantum theory] was sufficient to enable a man of Bohr's instinct and tact to discover the major laws of the spectral lines and of the electron shells of atoms together with their significance for chemistry appeared to me like a miracle – and it appears to me as a miracle even today. This is the highest form of musicality in the sphere of thought.' (Einstein, 1949)

Understanding atomic spectra – additional quantum numbers

7.1 Optical spectroscopy, multiplets and the splitting of spectral lines

The achievements described in Chap. 6 represented a remarkable advance in the understanding of quantum phenomena, but there remained major challenges which were ultimately to undermine the successes of the old quantum theory. Continuing advances in spectroscopy enabled high resolution spectra to be obtained and, with the ability to place the sources of emission in strong electric and magnetic fields, the full complexity of atomic and molecular spectra became apparent. Atomic spectra display regularities, for example, the series spectra of elements such as sodium and calcium which could be described by the Rydberg formula (Sect. 1.6). Some of the most prominent spectral features consisted, however, of *multiplets*, meaning the splitting of a line into a number of separate lines with similar wavelengths. Examples of multiplets are illustrated in Fig. 7.1, derived from observations of the photosphere of the Sun from the Pic du Midi observatory.

The simplest lines are singlets, the example of the Hα line of the Balmer series of hydrogen being shown in Fig. 7.1a. In fact, the line is a very narrow doublet, which Sommerfeld attributed to the effects of special relativity upon the circular and elliptical orbits of electrons of the same principal quantum number (Sect. 5.3). The splittings we are interested in here are very much larger effects. The classic example of a doublet is the splitting of the sodium D line into two bright components labelled D_1 (589.592 nm) and D_2 (588.995 nm) (Fig. 7.1b). A number of lines are triplets, for example, the magnesium triplet at 516.7 nm, 517.3 nm and 518.4 nm shown in Fig. 7.1c.

In addition to multiplets, individual lines are split into a number of components in the presence of electric and magnetic fields, the Stark and Zeeman effects respectively. The discovery and interpretation of the Zeeman effect has already been discussed in Sect. 4.1. The simplest variant is the normal Zeeman effect, illustrated schematically in Fig. 7.2a. Generally, however, the splittings are more complex. An important example is the splitting of the sodium D_1 and D_2 lines, in which the longer wavelength D_1 line is split into four components with no central line, while the shorter wavelength D_2 line is split into six components, again with no central line (Fig. 7.2b). The Zeeman splitting of the triplet zinc line is illustrated in Fig. 7.2c, showing that one of the lines displays the normal Zeeman effect while the other two line splittings are similar to those of the sodium D lines. It was to prove to be a major challenge to understand the origins of the splitting of spectral lines by electric and magnetic fields.

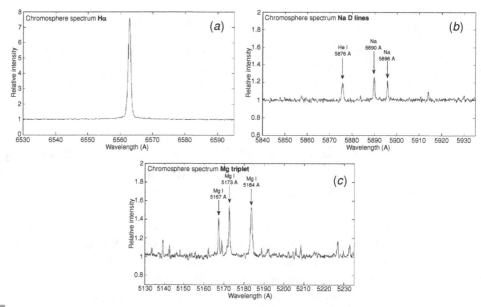

Fig. 7.1 Illustrating the multiplets of lines observed in optical spectra of the chromosphere of the Sun. (*a*) The singlet line of Hα. (*b*) The sodium D_1 (589.592 nm) and D_2 (588.995 nm) lines. The spectrum also includes the neutral helium line at 587.6 nm. (*c*) The magnesium triplet at 516.7 nm, 517.3 nm and 518.4 nm. (Courtesy of J-P. Rozelot, V. Desnoux and C. Buil. These observations were made at Pic Du Midi 'Lunette Jean Rösch' with the eShel spectrograph.)

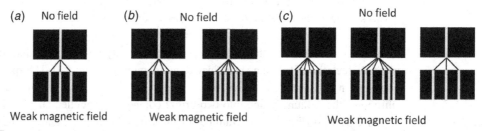

Fig. 7.2 Schematic diagrams illustrating the normal and anomalous Zeeman effects. (*a*) The Zeeman splitting of a singlet line. (*b*) The Zeeman splitting of the doublet D lines of sodium. (*c*) The Zeeman splitting of the triplet lines of zinc.

7.2 The Stark effect

Following the discovery of the splitting of emission lines in magnetic fields by Zeeman in 1896, Woldemar Voigt predicted in 1901 that there should be an analogous effect if atoms are subject to an electric field (Voigt, 1901). He estimated theoretically the magnitude of this effect for an elastically bound electron in an atom, but found the predicted splitting to be too small to be measured experimentally. Despite this discouraging prediction, Stark investigated the effect experimentally in 1913 using the emission of 'canal rays', the ions which passed through the perforated cathode of a discharge tube in the opposite direction

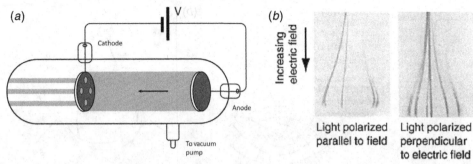

(*a*) A schematic diagram showing the experimental arrangement for the production of canal rays. Positively charged particles are accelerated from the anode to the cathode and pass through the holes in the cathode. The result is a set of 'canal rays' as illustrated in the diagram which strike the walls of the discharge tube. (*b*) The Stark effect showing the splitting of the 438.8 nm line of helium. The strength of the electric field increases down the diagram. In the left-hand panel, the light is observed parallel to the electric field while in the right-hand panel, the light is polarised perpendicular to the electric field (Foster, 1930).

to the electron beam (Fig. 7.3*a*) (Stark, 1913) – this apparatus was the precursor of the mass spectrograph which was to be perfected by Francis Aston. In his laboratory at the Technische Hochschule in Aachen, Stark discovered the splitting of the Balmer series of hydrogen and the lines of helium into a number of components when the canal rays were subjected to a strong electric field. Almost contemporaneously, Antonino Lo Surdo at Florence had detected a similar effect in his discharge tubes (Lo Surdo, 1913). Stark found that, when viewed perpendicular to the direction of the electric field, the $H\alpha$ and $H\beta$ lines were split into five components, the central components being polarised perpendicular to the direction of the field – the outer components were polarised parallel to the field. When viewed along the electric field direction, three unpolarised components were observed. The separation of the lines was symmetrical about the central line and, to a first approximation, proportional to the electric field strength. In subsequent experiments, Stark showed that there was no further splitting of the $H\alpha$ line, but $H\beta$ was split into 13 components, $H\gamma$ into 15 and $H\delta$ in 17 components in both the transverse and longitudinal directions (Stark, 1914). In the case of helium, six components were observed: three with polarisation parallel and three with polarisation perpendicular to the field direction. More complex splittings of the emission lines of helium and heavier elements were observed as greater electric field strengths became available (Fig. 7.3*b*).

In the same year, Warburg and Bohr realised that Bohr's model of the hydrogen atom offered an explanation of the Stark effect (Warburg, 1913; Bohr, 1914). Suppose the electric field acts in the z-direction. Then, in addition to the electrostatic force of the nucleus, $f = Ze^2/4\pi\epsilon_0 r^2$, there is a perturbing electrostatic force on the electron due to the field which does work $-eEz$, where z can be taken to be the distance from the nucleus in the z-direction. The nucleus is assumed to be infinitely heavy and so remains stationary. The change of energy of the stationary state is found by averaging z over the orbit, so that $\Delta E \sim eE\bar{z}$. The effect of the perturbing field clearly depends upon the orientation and ellipticity of the orbit of the electron, but it is apparent that the result will be to remove the

(a)   (b)

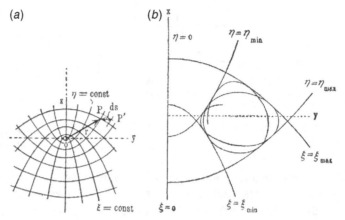

Fig. 7.4 (*a*) The system of confocal parabolae parameterised by coordinates ξ and η which form a set of parabolic coordinates used in the analysis of the Stark effect. (*b*) Illustrating the dynamics of the electron within the bounding quadrangle formed by the parabolae (ξ_{max}, ξ_{min}) and (η_{max}, η_{min}) (Sommerfeld, 1919).

degeneracy of the energy levels of hydrogen-like atoms because of the differing ellipticities of their orbits for a given principal quantum number (Fig. 5.2) – we recall that elliptical orbits, with the same semi-major axis and with the nucleus in one focus, all have the same total energy. Inspection of the orbits in Fig. 5.2 suggests that the actual splitting depends upon the quantum numbers n and n' and that $\bar{z}$ should be of the order $a = \epsilon_0 h^2 n^2 / \pi m_e e^2 Z$, the radius of a circular Bohr orbit with principal quantum number n. Hence, the change in energy ΔE_n of the nth energy level is expected to be of the order

$$\Delta E_n \sim \frac{\epsilon_0 h^2 E}{\pi m_e Z e} n^2 . \tag{7.1}$$

Bohr's estimates of the magnitude of ΔE_n provided a good account of the Stark effect in hydrogen, but he understood that a complete solution would require the determination of the perturbed orbit of the electron in the atom in the presence of a uniform electric field.

The solution was provided almost simultaneously by Paul Epstein[1] (1916a) and Karl Schwarzschild[2] (1916). They used the action–angle variables introduced in Sects. 5.4.4 and 5.5 and followed the procedures developed by Delaunay (1860) in his studies of the perturbing effects of distant planets upon the dynamics of the Earth–Moon system. Sommerfeld (1919) provides an elegant description of these calculations.

These authors began with the more general problem of the motion of an electron, or satellite, in the field of two centres of attraction. The astrophysicists had shown that the problem can be reduced to Hamilton–Jacobi form in which the coordinate system consists of a family of confocal ellipses and hyperbolae with the two planets as centres of attraction in each focus. If one of the centres of attraction is then taken to infinity while the force of attraction at the other is kept constant, the result is the same as a uniform field at the other centre of attraction. In this limiting case, the system of coordinates becomes a set of confocal parabolae parameterised by coordinates ξ and η as illustrated in Fig. 7.4*a*. To complete the three-dimensional coordinate system, the angle ψ is taken to be the polar angle with respect to an axis through the focus perpendicular to the $\xi-\eta$ plane. The transformations from (x, y) to (ξ, η) coordinates are:

$$\frac{y^2}{\xi^2} + 2x = \xi^2, \qquad \frac{y^2}{\eta^2} - 2x = \eta^2. \tag{7.2}$$

We now follow the procedures for finding the orbit of the electron in (ξ, η, ψ) coordinates according to the Hamilton–Jacobi prescription of Sect. 5.5. The total energy E_{tot} of the electron, or its Hamiltonian H, is found to be

$$E_{tot} = H = \frac{1}{2m_e(\xi^2 + \eta^2)}\left[p_\xi^2 + p_\eta^2 + \left(\frac{1}{\xi^2} + \frac{1}{\eta^2}\right)p_\psi^2 - \frac{Ze^2}{\pi\epsilon_0} + m_e eE(\xi^4 - \eta^4)\right]. \tag{7.3}$$

The equations for the velocities and momenta in (ξ, η, ψ) coordinates are given by Hamilton's equations (5.69), which become

$$\dot{\xi} = \frac{\partial H}{\partial p_\xi} = \frac{p_\xi}{m_e(\xi^2 + \eta^2)}, \quad \dot{\eta} = \frac{\partial H}{\partial p_\eta} = \frac{p_\eta}{m_e(\xi^2 + \eta^2)}, \quad \dot{\psi} = \frac{\partial H}{\partial p_\psi} = \frac{p_\psi}{m_e\xi^2\eta^2}. \tag{7.4}$$

$$\dot{p_\xi} = -\frac{\partial H}{\partial \xi}, \quad \dot{p_\eta} = -\frac{\partial H}{\partial \eta}, \quad \dot{p_\psi} = -\frac{\partial H}{\partial \psi} = 0. \tag{7.5}$$

The importance of working in parabolic coordinates is that the Hamiltonian (7.3) is separable and furthermore, from the last equation of (7.5), the p_ψ momentum is a constant, corresponding to the conservation of angular momentum. The details of the calculations to find the expressions for p_ξ and p_η are given by Sommerfeld (1919). The result is a system of conditionally periodic orbits similar to those illustrated in Fig. 5.4, but now in parabolic coordinates (Fig. 7.4b). Again the orbits just touch the bounding quadrangle formed by the parabolae ξ_{max}, ξ_{min} and η_{max} and η_{min}. The orbits would eventually pass through every point within the bounding quadrangle.

The next step is to apply the Bohr–Sommerfeld quantisation conditions (5.117)–(119) in the (ξ, η, ψ) coordinate system,

$$\oint p_\xi \, d\xi = n_1 h, \quad \oint p_\eta \, d\eta = n_2 h, \quad \int_0^{2\pi} p_\psi \, d\psi = n_3 h, \tag{7.6}$$

where n_1, n_2 and n_3 are positive integers. n_1 and n_2 are referred to as *parabolic quantum numbers* while n_3 is known as the *equatorial quantum number*. Carrying out the calculations for the energy levels of hydrogen-like atoms to first order in the electric field E, Epstein and Schwarzschild found the result

$$-W = \frac{m_e Z^2 e^4}{8\epsilon_0^2 h^2} \frac{1}{(n_1 + n_2 + n_3)^2} + \frac{3h^2\epsilon_0 E}{2\pi m_e Ze}(n_2 - n_1)(n_1 + n_2 + n_3). \tag{7.7}$$

We recognise immediately that the first term on the right-hand side of (7.7) is the same as the expression (5.120) for the energy levels of the Bohr–Sommerfeld atom. In addition, the quantisation conditions applied to these orbits in the presence of the electric field results in a splitting of each energy level, as given by the second term on the right-hand side of (7.7). The term is of the same form as that derived from our order of magnitude calculation (7.1), but now the dependence of the splitting of the energy levels on the quantum numbers n_1, n_2

and n_3 have been determined. The frequency splittings associated with transitions between an initial stationary state (m_1, m_2, m_3) and a final state (n_1, n_2, n_3) is

$$\Delta \nu = \frac{3h\epsilon_0 E}{2\pi m_e Z |e|}[(m_1 - m_2)(m_1 + m_2 + m_3) - (n_1 - n_2)(n_1 + n_2 + n_3)] . \quad (7.8)$$

Finally, Sommerfeld's selection rules for radiative transitions between stationary states could be applied to determine the predicted splittings of the lines of the Balmer series in hydrogen. Epstein found that these rules provided a complete description of Stark's data on the splittings of the lines of the Balmer series, as well as showing empirically that even-numbered differences in $m_3 - n_3$ give rise to polarisation in the direction of the external field, while odd-numbered differences give rise to polarisation perpendicular to the electric field. These calculations were of the greatest importance for promoting the case for the old quantum theory. In particular, as seen from (7.8), the splitting of the Balmer lines in the Stark effect depends upon Planck's constant h, in contrast to Lorentz' explanation of the normal Zeeman effect which does not depend upon h. The argument provides independent support for Bohr's postulate of the central importance of quantisation effects on the atomic scale. As Epstein (1916b) remarked

> 'We believe that the reported results prove the correctness of Bohr's atomic model with such striking evidence that even our conservative colleagues cannot deny its cogency. It seems that the potentialities of quantum theory as applied to this model are almost miraculous and far from being exhausted.'

Sommerfeld was equally effusive:

> 'All in all, we may regard the theory of the Stark effect as one of the most striking achievements of the quantum theory in atomic physics.' (Sommerfeld, 1919)

7.3 The Zeeman effect

In the mid-1890s, Lorentz and Larmor had accounted for Zeeman's discovery of the broadening and splitting of spectral lines in the presence of strong magnetic fields by the action of the magnetic field upon the oscillating or gyrating 'ions' in atoms (Sect. 4.1). The splitting of the line consisted of three components, a central line plus two lines, equally spaced on either side of it with displacements

$$\Delta \omega_0 = \pm \frac{eB}{2m_e} \quad \text{or} \quad \Delta \nu_0 = \pm \frac{eB}{4\pi m_e} . \quad (7.9)$$

This splitting was referred to as the *normal Zeeman effect*. Zeeman was fortunate in that he only observed broadening of the D lines of sodium, whereas he detected the characteristic triple splitting of the spectral lines of cadmium and zinc, which are singlet lines and display the normal Zeeman effect. When Preston carried out his experiments on the Zeeman effect in sodium using the powerful electromagnet at the Royal College of Science in Dublin, he found something quite different (Preston, 1898). In his words,

'It is interesting to notice that the two lines of sodium and the blue line 4800 (Å) of cadmium do not belong to the class which show as triplets. In fact, the blue cadmium line belongs to the weak-middled quartet class, while one of the D lines (D_2) shows as a sextet of fine bright lines. . . . the other D line (D_1) shows as a quartet . . . '

These splittings of the D-lines were confirmed by Cornu (1898), the D_1 line being a quadruplet and the D_2 line a sextuplet (Fig. 7.2b). These phenomena were referred to as the *anomalous Zeeman effect*.

Following the development of the Bohr–Sommerfeld quantum model of the atom, the challenge to theorists was to apply these concepts to the description of the Zeeman effect, in the hope that this would cast light upon the anomalous Zeeman effect. This was taken up by Debye (1916) and Sommerfeld (1916b) who used the formalism of action–angle variables which proved to be ideally matched to the requirements of quantum theory. The motion of the electron was determined by the combined influences of the electrostatic field of the nucleus and the Lorentz force $f = e(v \times B)$ associated with the magnetic field B. If the magnetic field is homogeneous and uniform along the z-direction, the equations of motion are (see also (4.2)–(4.4)):

$$m_e \ddot{x} = eB\dot{y} - \frac{\partial V}{\partial x} , \tag{7.10}$$

$$m_e \ddot{y} = -eB\dot{x} - \frac{\partial V}{\partial y} , \tag{7.11}$$

$$m_e \ddot{z} = -\frac{\partial V}{\partial z} , \tag{7.12}$$

where we have written $B = |B|$ and $V = -Ze^2/4\pi\epsilon_0 r$. To convert these equations into a form suitable for the application of the Hamilton–Jacobi equations, we need the Hamiltonian for the system. Because of the $v \times B$ term, no work is done by the magnetic field on the electron and so the total energy of the system is

$$E = \frac{1}{2}m_e\dot{x}^2 + \frac{1}{2}m_e\dot{y}^2 + \frac{1}{2}m_e\dot{z}^2 + V = \frac{1}{2m_e}\left(p_x^2 + p_y^2 + p_z^2\right) + V . \tag{7.13}$$

To convert this into a Hamiltonian, we need to replace the electron's three-momentum by its conjugate momenta which is $p_i = m_e\dot{x}_i + eA_i$, where A_i is one of the three components of the vector potential A, defined by $B = \text{curl } A$ and $i = 1, 2, 3$ (see Sect. 5.4.3). For a uniform magnetic field in the z-direction,

$$A = \frac{B}{2}\left(y\,i_x - x\,i_y\right) , \tag{7.14}$$

and so

$$p_1 = p_x + \frac{eyB}{2} , \quad p_2 = p_y - \frac{exB}{2} , \quad p_3 = p_z . \tag{7.15}$$

Therefore, the Hamiltonian for the electron in a uniform magnetic field is

$$H = E = \frac{1}{2m_e}\left[\left(p_1 - \frac{eyB}{2}\right)^2 + \left(p_2 + \frac{exB}{2}\right)^2 + p_3^2\right] + V . \tag{7.16}$$

The next step, which was carried out by Debye, is to transform the Hamiltonian from Cartesian (x, y, z) to spherical polar (r, θ, ϕ) coordinates. Carrying out this point transformation,

$$H = \frac{1}{2m_e}\left(p_r^2 + \frac{p_\theta^2}{r^2} + \frac{p_\phi^2}{r^2\sin^2\theta} - eBp_\phi\right) - \frac{Ze^2}{4\pi\epsilon_0 r} \, . \tag{7.17}$$

This Hamiltonian is separable and does not depend upon ϕ. ϕ is therefore a cyclic variable for which we can write $p_\phi = \text{constant} = \alpha_3$. Following the procedures described in Sect. 5.5, the Hamilton–Jacobi equation can be written

$$\left(\frac{\partial W}{\partial r}\right)^2 + \frac{1}{r^2}\left(\frac{\partial W}{\partial \theta}\right)^2 + \frac{\alpha_3^2}{r^2\sin^2\theta} - 2\omega m_e\alpha_3 - \frac{m_e e^2}{2\pi\epsilon_0 r} + 2m_e\alpha_1 = 0 \, , \tag{7.18}$$

where $\omega = eB/2m_e$ and α_1 is the negative total energy constant. The equation (7.18) is separable in the r and θ coordinates and the constants of the motion are given by

$$p_\theta = \frac{\partial W}{\partial \theta} = \left(\alpha_2^2 - \frac{\alpha_3^2}{\sin^2\theta}\right)^{1/2} \, , \tag{7.19}$$

$$p_r = \frac{\partial W}{\partial r} = \left(\omega m_e\alpha_3 + \frac{m_e e^2}{2\pi\epsilon_0 r} - 2m_e\alpha_1 - \frac{\alpha_2^2}{r^2}\right) \, , \tag{7.20}$$

where α_2 is the third constant of the motion. Finally, we apply the Bohr–Sommerfeld quantum conditions:

$$\oint p_\phi \, \mathrm{d}\phi = 2\pi\alpha_3 = m_3 h \, , \quad 2\int_{\theta_1}^{\theta_2} p_\theta \, \mathrm{d}\theta = m_2 h \, , \quad 2\int_{r_1}^{r_2} p_r \, \mathrm{d}r = m_1 h \, . \tag{7.21}$$

Carrying out the second and third integrals of (7.21), we find

$$2\pi(\alpha_2 - \alpha_3) = m_2 h \, , \quad \frac{e^2\sqrt{2m_e}}{4\epsilon_0\sqrt{\alpha_1 - \omega\alpha_3}} - 2\pi\alpha_2 = m_1 h \, . \tag{7.22}$$

The expression (7.22) can be inverted to express the quantised energy levels, parameterised by $E_n = \alpha_1$, as

$$E_n = \alpha_1 = \frac{m_e e^4}{8\epsilon_0^2 n^2 h^2} + \frac{m_3 h\omega}{2\pi} \, , \tag{7.23}$$

where $n = m_1 + m_2 + m_3$. From this expression, we find from Bohr's frequency condition,

$$\nu = \frac{m_e e^4}{8\epsilon_0^2 h^3}\left(\frac{1}{n'^2} - \frac{1}{n''^2}\right) + \frac{\omega}{2\pi}(m_3' - m_3'') \, , \tag{7.24}$$

where the single primed quantities characterise the final orbit and the double primed quantities the initial orbit. This is exactly the same as Bohr's original formula for the Balmer series of the hydrogen atom (4.28), but now it involves elliptical orbits parameterised by m_1, m_2 and m_3 and includes the effects of the magnetic field on the energy of the orbits. We recognise that m_3 is the azimuthal quantum number n_ϕ introduced in Sect. 6.4 and, following the considerations of Sect. 6.5, the selection rule for m_3 is the same as (6.55), namely

$$\Delta m_3 = +1 \,, \qquad \Delta m_3 = -1 \,, \qquad \Delta m_3 = 0 \,. \tag{7.25}$$

These splittings correspond exactly to the normal Zeeman effect and the polarisation properties are identical to those described in Sect. 6.5, namely the central line $\Delta m_3 = 0$ is linearly polarised and the $\Delta m_3 = \pm 1$ components are circularly polarised. It can also be observed that the Zeeman splitting term in (7.24) does not contain Planck's constant and so is an entirely classical term. Sommerfeld was well aware of the fact that, despite the more sophisticated tools used, the analysis provided no more than the classical results derived by Lorentz and Larmor. As expressed by Sommerfeld

> 'In the present state the quantum treatment of the Zeeman effect achieves just as much as Lorentz's theory, but not more. It can account for the normal triplet . . . but hitherto it has not been able to explain the complicated Zeeman types.'

Sommerfeld went on to consider the relativistic case, but this did not provide any further insights (Sommerfeld, 1916b).

Sommerfeld labelled m_3 the *magnetic quantum number* because of its appearance in (7.23) in the guise of $\omega = eB/2m_e$ where m_e is the mass of the electron – m_3 reappears in the full theory of quantum mechanics as the quantum number m. The quantum number m_3 corresponds to the projection of the total angular momentum vector, characterised by the quantum number k, onto the magnetic field direction and hence the energy term associated with the magnetic field can also be written

$$\frac{m_3 h \omega}{2\pi} = \left(\frac{kh}{2\pi} \right) \left(\frac{eB}{2m_e} \right) \cos\alpha \,, \tag{7.26}$$

where $|\boldsymbol{L}| = kh/2\pi$ is the magnitude of the total angular momentum and α is the angle between the magnetic field direction and the total angular momentum vector $\boldsymbol{L}$. Introducing the magnetic moment of the electron $\boldsymbol{\mu} = (e/2m_e) \boldsymbol{L}$ in its orbit about the nucleus, this energy can be written $\boldsymbol{\mu} \cdot \boldsymbol{B}$, the interaction energy between a magnetic dipole and a magnetic field. If we consider the magnetic moment of an electron in the lowest energy Bohr orbit, $k = 1$, we find that $|\boldsymbol{\mu}| = (e/2m_e)(h/2\pi) = eh/4\pi m_e$. Pauli introduced the term *Bohr magneton* to refer to this natural unit of magnetic moment which is written $\mu_B = eh/4\pi m_e = e\hbar/2m_e$.

7.4 The anomalous Zeeman effect

While the Bohr–Sommerfeld model of the atom could account for the normal Zeeman effect, it could not account for the much more complex splittings observed in the anomalous Zeeman effect. The origin of the multiplicities of spectral lines was clearly related to the anomalous Zeeman effect since singlet lines displayed the normal Zeeman effect while the higher multiplets all exhibited anomalous Zeeman effects. Although the Bohr–Sommerfeld model could account for the properties of hydrogen-like atoms, multiple electron systems proved to be intractable – models for helium with two electrons and a nucleus with two opposite electric charges could not be reconciled with the experimental results. Sommerfeld realised that, with the analyses of Zeeman splitting discussed in Sect. 7.3, he had gone as far as he could with the standard procedures of action–angle variables and a different approach would be required. His efforts were devoted to finding empirical relations which would give insight into the origin of the anomalous Zeeman effect.

The Rydberg series of elements such as sodium and calcium could be accounted for in terms of a modified variant of the Bohr–Sommerfeld model in which the shielding of the nuclear charge by the orbiting electrons was included in the expression for the electrostatic potential, as was demonstrated in Sect. 6.6. This modified model could also account for the multiplicities of spectral lines – just as in the Stark effect the degeneracy associated with the Balmer series was lifted by an applied electric field, so the internal electric field associated with the multiple electron systems could lead to the lifting of the degeneracy of the energy levels and change their energies. We recall that the degeneracy of the states of hydrogen-like atoms is associated with the assumption of the pure inverse-square law electrostatic potential of the nucleus – the presence of additional electrons results in a non-inverse-square law of attraction close to the nucleus.

These considerations led to a natural ordering of the various series in terms of the magnitudes of the quantum defects described by the quantities s, p and d in (1.19), (1.20) and (1.21). Sommerfeld had shown that the properties of the ellipses could be characterised by the two quantum numbers n, which determined the energy of the stationary state and k which determined the total angular momentum of the orbit (Sect. 5.5). The emission lines were assumed to be associated with the 'valance electrons', the outermost electrons of the atoms. Thus, the Rydberg terms with the smallest quantum defects could be associated with valance electrons which had circular orbits and so experienced the fully shielded nuclear charge whereas the 'penetrating' orbits with large eccentricities would be expected to have the greatest quantum defects. The quantum defects of the different series increase in the order $s, p, d, f, \ldots$. Therefore, it was argued that the series with the greatest quantum defects should have the most 'penetrating' orbits, corresponding to the maximum eccentricity for a given principal quantum number, that is, $k = 1$. On the other hand, those with the smallest quantum defects should be associated with circular orbits for which $n = k$. This led to the association of the quantum numbers $k = 1, 2, 3, 4, \ldots$ with the terms $s, p, d, f, \ldots$, corresponding to the designations associated with the sharp, principal, diffuse and fundamental series.

The issue arose whether or not the quantum numbers n and k, together with the selection rules for k, were sufficient to account for all the features observed in atomic spectra. The introduction of the magnetic quantum number showed that this feature had to be accommodated in the rules. Sommerfeld was also concerned by the fact that some transitions which would have been permitted according to Ritz's combination principle were not in fact observed (Sommerfeld, 1920a). This suggested that some other selection principle was at work, associated with an additional *inner quantum number*, which at Bohr's suggestion, he named j. Thus, each state would be characterised by three quantum number n, k and j. For example, for a state with $n = 5$, $k = 1$ and $j = 3$, Sommerfeld introduced the notation $^5s_3$, since $k = 1$ characterised the s states. From his analyses of the doublets and triplets of the diffuse series, Sommerfeld was able to assign numerical values of j such that the selection rule for j became $\Delta j = \pm 1$ or 0. Alfred Landé took the analysis further and showed that the transition $j = 0$ to $j = 0$ also had to be excluded. Sommerfeld was well aware of the fact that these were empirical rules without any underpinning theoretical justification.

To give more physical significance to the introduction of this additional quantum number, Sommerfeld and Landé developed their *magnetic core hypothesis* according to which the atomic core, consisting of the nucleus and the inner, non-valence, electrons, was attributed an angular momentum $sh/2\pi$, where s took values $1, 2, \ldots$ Associated with s was the corresponding magnetic moment $\boldsymbol{\mu} = (e/2m_e)\,\boldsymbol{L}$ which was to be aligned with the angular momentum vector of the core (Sommerfeld, 1923, 1924; Landé, 1921a,b, 1923a,c). This may be thought of as an 'inner Zeeman effect', the angular momentum being subject to the rules of quantisation and so the vector could only take specific quantised orientations with respect to any given axis. Such a hypothesis had already been introduced by Roschdest-wensky in connection with the doublet separation of the first line of the principal series of lithium (Roschdestwensky, 1920).

Let us return to some of the early results which stimulated Landé's interest in the anomalous Zeeman effect. Gradually the patterns of Zeeman splittings began to show significant regularities. In 1898, Preston formulated his law that

> 'All the lines of a given series of a substance exhibit the same pattern of components in a magnetic field; moreover, analogous spectral lines of the same series, even if belonging to different elements have the same Zeeman effect.' (Preston, 1898)

By 'series' is meant columns in the periodic table containing similar elements, such as the sequence of the alkaline metals sodium, potassium, rubidium (Group 1) or the alkaline earth metals magnesium, calcium, strontium, barium (Group 2). This was followed by the rule discovered by Runge and Paschen (1900) to the effect that,

> 'the frequency differences for lines of the same type (multiplicity) turned out to be the same.'

Of particular interest for interpreting the nature of the anomalous Zeeman effect was Runge's analysis of 1907 in which he provided a remarkably simple expression for the frequency, or wavelength, dependence of the splittings of the spectral lines (Runge, 1908). The normal Zeeman effect could be written

$$a = \Delta \nu_0 = \frac{\Delta \lambda_0}{\lambda_0^2} = \frac{eB}{4\pi m_e}. \tag{7.27}$$

In Runge's words,

> 'The hitherto observed complex splittings of spectral lines in a magnetic field exhibit the following peculiarity: the distances of the components from the centre are integral multiples of just a fraction of the normal separation a ... So far, the fractions $a/2$, $a/3$, $a/4$, $a/6$, $a/7$, $a/11$ and $a/12$ have been definitely observed. That is, ... $\delta \nu$ the frequency difference(s) were integral multiples of the quantity a/r, where ... the denominator r is an integer between 1 (for the normal Zeeman effect) and 12.'

The quantity r became known as the Runge denominator.

Exceptions to the rules concerning the appearance of the anomalous Zeeman effect were discovered by Voigt and Hansen in helium and lithium (Voigt and Hansen, 1912). Particularly in the case of lithium, it had been expected that the lines would display the same anomalous Zeeman effect as observed in sodium and potassium, but instead the normal Zeeman effect was observed. As stronger magnetic fields became available, Paschen and Back found that at large enough magnetic flux densities, the anomalous Zeeman effect was replaced by the normal effect (Paschen and Back, 1912). Thus, in the case of the oxygen triplet, for example, below 0.6 T, the three lines separated into the standard anomalous Zeeman effect. At 0.6 T, however, the effect of the strong magnetic field was that the different natural line components overlapped and, at greater magnetic flux densities, a transformation began to take place until at magnetic flux densities of 4 T, the pattern of line splittings reverted to the normal Zeeman effect with the standard three lines. This *Paschen–Back effect* could account for all the exceptions to Preston's law.

Sommerfeld appreciated that the interpretation of the anomalous Zeeman effect had the potential to unlock further insights into quantum physics. His concept was to give further physical meaning to the empirical rules associated with the anomalous Zeeman effect. In 1919, he introduced his *decomposition rule* starting from an analysis of the *Runge denominators* (Sommerfeld, 1920b). Sommerfeld began with Runge's rule written in the form $\Delta\nu_{an} = qa/r$, where q, the integral numerator, took values $0, \pm 1, \pm 2, \dots$ and the characteristic values of r were 1, 2 and 3 for singlet, doublet and triplet lines respectively. According to the precepts of quantum theory, the lines had to originate between stationary states of the atoms in the magnetic field which had energy displacements with respect to the unperturbed states ΔW_1 and ΔW_2, so that

$$\Delta\nu_{an} = \Delta\nu_1 - \Delta\nu_2 = \frac{\Delta W_1 - \Delta W_2}{h} . \tag{7.28}$$

To obtain Runge's rule, the frequency shifts $\Delta\nu_i$ also had to satisfy the rule and so

$$\Delta\nu_i = \frac{q_i a}{r_i} , \quad \text{where } i = 1, 2 , \tag{7.29}$$

where q_1 and r_i are the Runge numerators and denominators associated with the individual energy states. Hence, from (7.28) Runge's rule could be written in the following form

$$\Delta\nu_{an} = \frac{qa}{r} = \left(\frac{q_1 a}{r_1} - \frac{q_2 a}{r_2} \right) = a\frac{q_1 r_2 - q_2 r_1}{r_1 r_2} . \tag{7.30}$$

Now, Sommerfeld associated the Runge denominators with the properties of the energy levels themselves and not just the transitions. In his words,

> 'We denote the equation, $r = r_1 r_2$, as the magneto-optical decomposition rule and the Runge number observed in the anomalous Zeeman effect can be decomposed into Runge numbers of the first and second terms involved in the spectral lines.' (Sommerfeld, 1920b)

This was the beginning of a complex story of the development of empirical rules to describe the many features of the anomalous Zeeman effect.[3] The principal contributors to these developments were Landé, Sommerfeld and the young Werner Heisenberg. Landé rapidly became the authority on the anomalous Zeeman effect, particularly after he joined

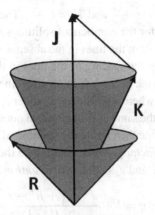

Fig. 7.5 Illustrating Landé's construction for the vector combination of angular momenta R and K to form the total angular momentum vector J. The separate angular momentum vectors R and K precess about the resultant vector J.

Paschen's group at Tübingen in 1922 as an extraordinary (associate) professor of theoretical physics. At that time, Tübingen was the leading centre for experimental studies of the anomalous Zeeman effect and Paschen needed Landé's theoretical expertise to make sense of the experimental results.

In 1919, Landé introduced the vector model of the atom to represent the addition of angular momenta by analogy with their vectorial representation in classical mechanics (Landé, 1919). He represented the angular momentum of the core by the vector R which was to be combined vectorially with the angular momentum K of the electron responsible for the optical emission to form a resultant vector J which represented the total angular momentum of the atom (Fig. 7.5). As in classical mechanics, the separate R and K vectors were assumed to precess about the resultant angular momentum vector J. In the presence of a magnetic field, J itself would precess about the magnetic field direction. R and K had to be combined according to the rules of quantisation and so would combine at different angles, resulting in a variety of different stationary states with different total energies. As a result of the magnetic interaction between the magnetic moments of the core and the outer electron, the energy levels were different and Landé attributed the different energies of the multiplets to this interaction.

Both Sommerfeld and Landé developed sets of empirical rules to describe the anomalous Zeeman effect. The vector model of the atom with integral quantum numbers ran into the difficulty that the singlet states would be associated with a core quantum number $s = 1$ which implied that the orbital quantum number $l = 0$, which was contrary to the assignment of $k = 1$ according to the standard Bohr–Sommerfeld model. A different assignment of quantum numbers was proposed by Landé (1921a; 1922) which was widely adopted. In it, half-integral quantum numbers were introduced with the following assignments: $R = 1/2, 2/2, 3/2, \ldots$ for singlet, doublet, triplet, $\ldots$ systems and $K = 1/2, 3/2, 5/2, \ldots$ for the $s, p, d, \ldots$ states with J half-integral for odd multiplicities and integral for even multiplicities as the quantum number associated with the sum of the R and K vectors. In this scheme, $R = s + \frac{1}{2}$ and $K = l + \frac{1}{2}$. The objective of these schemes was to account simultaneously for the numbers of splittings of the lines, their polarisation properties and their separation from the lines in the absence of a magnetic field.

To account for the spacing of the lines in the anomalous Zeeman effect, Landé worked by analogy with the normal Zeeman effect in which the energy levels were given by (7.23) $\Delta E_n = m_3 h \nu_0$, where ν_0 is the gyrofrequency. m_3 is the magnetic quantum number which we showed was the same as the azimuthal quantum number in the Bohr–Sommerfeld model of the atom. Landé then wrote the splitting of the lines in terms of the quantum number m with a *splitting factor g* which determined the actual splitting in terms of combinations of the vectors R, K and J so that $\Delta E = gmh\nu_0$. Empirically, he found the formula

$$g = 1 + \frac{J^2 - \frac{1}{4} + R^2 - K^2}{2(J^2 - \frac{1}{4})} = 1 + \frac{\overline{J}^2 + \overline{R}^2 - \overline{K}^2}{2\overline{J}^2}, \tag{7.31}$$

where

$$\overline{J} = \sqrt{(J + \tfrac{1}{2})(J - \tfrac{1}{2})}, \quad \overline{R} = \sqrt{(R + \tfrac{1}{2})(R - \tfrac{1}{2})} \quad \text{and} \quad \overline{K} = \sqrt{(K + \tfrac{1}{2})(K - \tfrac{1}{2})}. \tag{7.32}$$

This formula closely resembles the modern expression for the *Landé g-factor*.

In order to give more physical content to the theory, Landé next reverted to the vector model of the atom, but now using $\overline{J}$, $\overline{R}$ and $\overline{K}$ in place of J, R and K. From the geometry of Fig. 7.5, the cosine rule can be used to find the angle between the vectors $\boldsymbol{J}$ and $\boldsymbol{R}$ and then

$$g = 1 + \frac{\overline{R}\,\cos(J, R)}{\overline{J}}. \tag{7.33}$$

If $\boldsymbol{H}$ is the direction of the magnetic field, then the quantisation of the angular momentum vector associated with J is given by the magnetic quantum number m such that $m = \overline{J}\cos(J, H)$ and so

$$mg = \overline{J}\,\cos(J, H) + \overline{R}\,\cos(J, R)\cos(J, H). \tag{7.34}$$

Now the product of the two cosines in the last term of (7.34) is just the average value of $\cos(R, H)$ and so the value of mg can be written

$$mg = \overline{K}\,\cos(K, H) + 2\overline{R}\,\cos(R, H). \tag{7.35}$$

This is a rather dramatic result since it might have been expected that, if the angular momentum vectors had been added together, then their projections on the magnetic field direction would have resulted in

$$mg = \overline{K}\,\cos(K, H) + \overline{R}\,\cos(R, H), \tag{7.36}$$

in other words, the second term associated with the magnetic core contributed twice what might have been expected according to Larmor's formula for the precession of the vector about the magnetic field direction. Landé made the suggestion that there might be an anomalous magnetic moment associated with the core such that the ratio of the angular momentum to magnetic moment was only half of the Larmor value, that is, an *anomalous gyromagnetic ratio*. Remarkably, there was independent experimental evidence for the anomalous gyromagnetic ratio from the experiments of Einstein and de Haas and confirmed by the careful experiments by Beck, as discussed in the next section.

Landé's achievement was quite remarkable, but there remained a lot to be explained, in particular, what was the significance of the introduction of $\overline{J} = [(J + \frac{1}{2})(J - \frac{1}{2})]^{1/2}$? Sommerfeld used a slightly different notation for J such that $J = j + \frac{1}{2}$ and so the quantity $\overline{J}$ is

$$\overline{J} = \sqrt{j(j+1)} \,. \tag{7.37}$$

To the modern reader, this has a familiar ring but in 1924 its significance was obscure. Landé's achievement was to infer these new features of quantum physics from analysis of the anomalous Zeeman effect, *before* the invention of quantum mechanics.

7.5 The Barnett, Einstein–de Haas and Stern–Gerlach experiments

7.5.1 The anomalous gyromagnetic ratio

The value of the gyromagnetic ratio for atoms had been the subject of various experiments which originated in attempts to understand the phenomenon of the magnetisation of materials in terms of the magnetic moment per unit volume associated with what were termed 'Ampèrian currents'. To recapitulate the argument, an electron in a circular orbit corresponds to a circular current $I = ev$, where v is the frequency of rotation of the electron about the nucleus. If the area of the orbit is $A = \pi r^2$, the magnetic moment of the electron is $\mu = IA = evA$. The angular momentum of the electron is $L = m_e vr$, and so, since $v = v/2\pi r$, the gyromagnetic ratio is defined to be

$$\frac{L}{\mu} = \frac{2m_e}{e} \,. \tag{7.38}$$

The same result is obtained for the bulk properties of a material if the orbits are aligned. Therefore, there should be a relation between the magnetisation of a material and its angular momentum. In macroscopic terms,

$$\frac{|L|}{|\mu|} = \frac{2m_e}{e} = 1.13 \times 10^{-10} \quad \text{in SI units} \,. \tag{7.39}$$

Samuel Jackson Barnett realised that it would be possible to measure such an effect using a cylinder of iron which initially is at rest with zero magnetisation. Then, if the cylinder acquires angular momentum about its long axis, it should become magnetised. After a long series of careful experiments carried out at the Physical Laboratory at Ohio State University, he eventually found a positive effect and measured its magnitude, what became known as the *Barnett effect* (Barnett, 1915).

A similar experiment was proposed by Einstein and Wander Johannes de Haas who looked for an 'AC' effect by switching the direction of magnetisation of the cylinder at the frequency of the torsion filament which was used to suspend the cylinder (Einstein and de Haas, 1915). They found the expected value given by (7.39) with an accuracy of about 10%, but this was about a factor of 2 greater than that found by Barnett. Over the next few years, these tricky experiments were repeated by Emil Beck, Gustav Arvidsson and the Barnetts with the result that the gyromagnetic ratio converged on a value half of that given by (7.39) (Beck, 1919a,b; Arvidsson, 1920; Barnett and Barnett, 1922). This phenomenon became known as the *anomalous gyromagnetic ratio*.

7.5.2 Space quantisation and the Stern–Gerlach experiment

The possibility of demonstrating experimentally spatial quantisation as predicted by Sommerfeld's analysis of the Stark effect was taken up by Otto Stern at the University of Frankfurt, despite the scepticism of the theorists who viewed quantisation as simply a formal set of procedures. In 1920, Stern used the method of atomic beams to measure the mean velocity of neutral silver atoms and found indeed that the experimental value agreed with the expectations of Maxwell's kinetic theory (Stern, 1920). He also planned to measure the velocity dispersion of the atoms, but this would have required many refinements of the experimental set-up. Instead, he turned to the issue of measuring space quantisation experimentally. He was joined at Frankfurt by Walther Gerlach, a gifted experimenter, to attempt the much more difficult task of demonstrating the splitting of spectral lines in a strong magnetic field gradient. It is a straightforward calculation to show that, although there is no net force on a magnetic dipole in a uniform magnetic field, there is a net force if the magnetic field is inhomogeneous. The net force on the dipole is

$$F = (m \cdot \nabla)B = |m| \cos\theta \, \frac{\partial B_z}{\partial z} \,, \tag{7.40}$$

where in the last expression it is assumed that the magnetic field gradient is in the z-direction and θ is the angle between the axis of the dipole and the magnetic field direction, in this case the z-direction. Classically, the magnetic moment could take any angle with respect to the magnetic field direction and so it would be expected that there would be a random distribution of deflection angles. If, however, space quantisation is a real physical phenomenon, the deflections would only take place at specific deflection angles θ. According to the Bohr–Sommerfeld model, an electron in orbit within an atom would have azimuthal angular momentum $p_\phi = h/2\pi$ and would assume only three orientations with respect to the magnetic field direction given by

$$\cos\theta = \frac{n_1}{n_\phi} = 0, \pm 1 \,. \tag{7.41}$$

Bohr had argued in 1918 that the case $n_1 = 0$ would be unstable because then the plane of the orbit of the electron would lie in the direction of the magnetic field and this he argued was an unstable configuration (Bohr, 1918b). Therefore, it was expected that the beam would only be split into two components corresponding to $\cos\theta = \pm 1$. Equally, the magnetic-core theory of fine-structure splitting by Sommerfeld and Landé predicted a splitting of the s-states of atoms into only two fine-structure lines with no central component, corresponding to $s = \pm\frac{1}{2}$.

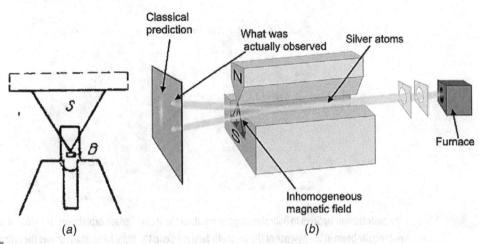

Classical prediction

What was actually observed

Silver atoms

N

S

B

Inhomogeneous magnetic field

Furnace

(a)   (b)

Fig. 7.6 (a) A cross-section through the magnet used by Stern and Gerlach to create an inhomogeneous magnetic field distribution in the vertical z-direction (Gerlach and Stern, 1922a). (b) Illustrating the deflection of beams of silver atoms through the inhomogeneous magnetic field of the Stern–Gerlach experiment.

The experiment proved to be a considerable challenge (Fig. 7.6). The deflection of the beam was expected to amount to only

$$s = 1.12 \times 10^{-5} \left(\frac{\partial B_z}{\partial z} \right) \frac{l^2}{T} \text{ cm} , \tag{7.42}$$

where the field gradient is measured in gauss per centimetre, the temperature in kelvin and l in cm. For a gradient of 10^4 G cm^{-1} and a length through the magnetic field of 3 cm, the deflection of silver atoms at a temperature of 1000 °C (1273 K) only amounted to about 10^{-2} mm. By obtaining the support of various companies for a powerful electromagnet, for cryogens and vacuum pumps, they worked ceaselessly from the summer of 1921 to March 1922 on the experiment. Stern moved to the University of Rostock at the beginning of January 1922 and Gerlach carried on the experiments himself. In their original set-up, the narrow beam of silver atoms, which was produced by heating a platinum strip coated with silver to a temperature just above the boiling point of silver at 960 °C, was collimated by small circular holes (Fig. 7.6b), but Gerlach hit upon the idea that it would be better to replace the second hole with a narrow slit and so increase the intensity of the beam and allow a 'differential' means of observing the splitting. By February 1922, Gerlach observed the splitting of the beam of silver atoms into two separate beams. Following further experiments, their paper was published in March 1922 (Gerlach and Stern, 1922a). In it, they wrote

> 'The atomic beam splits up into two distinct beams in a magnetic field. No undeflected atoms can be detected . . . *In these results we see direct experimental proof of the directional [space-] quantisation in a magnetic field.*'

The famous postcard showing the result of the experiment from Stern to Bohr is shown in Fig. 7.7. They also measured precisely the gradient of the magnetic flux density ($\partial B_z / \partial z$)

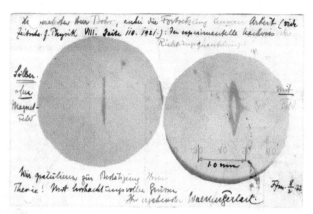

Fig. 7.7 The postcard sent by Stern to Bohr showing the results of the Stern–Gerlach experiment. The left image shows the rectangular beam in the absence of the magnetic field gradient; the right-hand image shows the splitting of the beam into two components. The greatest field gradient occurs in the centre of the beam.

through which the beam of silver atoms passed and, during the Easter vacation, they went on to show that the magnetic moment of the silver atoms in the ground state was one Bohr magneton per gram-atom (Gerlach and Stern, 1922b). The importance of the result was not lost on the physics community. As Paschen wrote to Gerlach,

> 'Your experiment proves for the first time the reality of Bohr's stationary states.' (Gerlach, 1969)

The Stern–Gerlach experiment is undoubtedly one of the great achievements of experimental physics, but it raised many issues about the fundamentals of quantum physics. Stern pointed out that, if space quantisation was a real physical effect, it should give rise to birefringence, or double refraction, in materials since there would be a difference in the refractive indices along and perpendicular to the magnetic field direction (Stern, 1921). Such an effect had never been observed. In addition, there were problems in understanding how the alignment of the orbits could come about according to classical physics. Einstein and Ehrenfest (1922) pointed out that the alignment of the magnetic moments of an initially random angular distribution would not take place instantaneously. Classically, the alignment would involve the exchange of angular momentum between atoms and their estimate of the time-scale for this process was 10^{11} seconds, far longer than the 10^{-4} seconds that it took the beam of atoms to pass through the magnetic field distribution of the magnets. It was inferred that there was something seriously wrong with the mechanical and dynamical laws at the atomic level. This was only the beginning of the difficulties with the old quantum theory – much worse was about to follow.

Bohr's model of the periodic table and the origin of spin

8.1 Bohr's first model of the periodic table

We now need to retrace our steps and follow Bohr's activities from his great Trilogy of 1913 to his model for the periodic table of 1922. In 1914, Bohr petitioned the Danish government to create a professorship for him in theoretical physics and this was granted two years later. In the meantime, he returned to Manchester as Schuster Reader in physics before taking up his appointment as Professor of Theoretical Physics in Copenhagen in 1916. In 1917, Bohr successfully petitioned the physics faculty of Copenhagen University to found an Institute for Theoretical Physics with Bohr as its founding director. The Institute was officially opened in 1921, but the strain of setting up the new institute combined with his continuing, almost obsessive, research programme into the fundamentals of quantum theory, took a heavy toll and in 1921 he suffered a serious bout of ill-health. Despite this, he remained the driving force behind the attack on the problems of quantum theory on a very broad front. Thanks to his tireless efforts and inspiration, Copenhagen became one of the two major centres for the development of quantum theory, the other being in Göttingen, through the 1920s when the foundations of the old theory were to be cut away and replaced by completely new concepts. The Institute for Theoretical Physics, commonly referred to as the 'Bohr Institute', became formally the Niels Bohr Institute in 1965, three years after his death.[1]

Already in the second paper of the Trilogy of 1913, Bohr aspired to account not just for the spectrum of hydrogen, but for the structure of all the atoms in the periodic table and their chemical properties (Bohr, 1913b). He had convinced himself that radioactive decay was associated with the nucleus of the Rutherford atom while the chemical properties were associated with the system of electrons orbiting the nucleus. His first efforts to define the electronic structure of atoms in the periodic table are engagingly told by Heilbron (1977). Heilbron summarises the four assumptions upon which Bohr made his first tentative steps into atomic structure: (i) all the electrons lie in circular orbits in the same plane through the nucleus; (ii) the populations of the innermost quantised rings increase with atomic number; (iii) each electron, no matter how far from the nucleus, has angular momentum $h/2\pi$ in its ground state; (iv) the ground state is characterised by that which results in the lowest energy for the given total angular momentum. Bearing in mind the need to account for the periodic structure of the periodic table, he came up with an assignment of the numbers of electrons to different shells in atoms given in Table 8.1.

Unlike the more theoretically minded physicists, Bohr lay considerable store in understanding the chemical properties of the atoms, for example, their valences and the similarity of the properties of the elements in the different columns of the periodic table. The stability

Table 8.1 Bohr's electron configurations for the first 24 elements of the periodic table (Bohr, 1913b). The numbers in brackets show the number of electrons in the $n = 1, 2, 3, \ldots$ shells.

1	H	(1)	7	N	(4,3)	13	Al	(8,2,3)	19	K	(8,8,2,1)
2	He	(2)	8	O	(4,2,2)	14	Si	(8,2,4)	20	Ca	(8,8,2,2)
3	Li	(2,1)	9	F	(4,4,1)	15	P	(8,4,3)	21	Sc	(8,8,2,3)
4	Be	(2,2)	10	Ne	(8,2)	16	S	(8,4,2,2)	22	Ti	(8,8,2,4)
5	B	(2,3)	11	Na	(8,2,1)	17	Cl	(8,4,4,1)	23	Y	(8,8,4,3)
6	C	(2,4)	12	Mg	(8,2,2)	18	A	(8,8,2)	24	Cr	(8,8,4,2,2)

of the circular orbits with increasing numbers of electrons also played a role in his thinking, following the hints provided by Nicholson's analysis of the stability of rings of electrons to oscillations perpendicular to the orbital planes of the electrons (see Sect. 4.4). Since the electrons were all assumed to have circular orbits in the same plane, the outermost 'valence' ring was shielded from the nuclear charge by the inner completed shells. As a result, the valence shells of similar elements such as the alkali metals, lithium, sodium and potassium were given the same outer electron structure, as did the alkaline earth metals, beryllium, magnesium, calcium. This model had little lasting appeal, but it is indicative of Bohr's adventurous spirit in attempting to encompass a wide range of quantum and atomic phenomena in a single scheme in the context of his great innovations of 1913. The concept of shells of electrons was introduced, although Bohr was well aware of the provisional nature of his assignments of quantum numbers to the different elements. This was the beginning of the attempts by many theorists to create models of atoms including the quantum concepts introduced by Bohr. Notice that the electron orbits are entirely determined by a single quantum number, the principal quantum number n. As a result, this model is often referred to as a *one-quantum structure*.

Following this foray into atomic structure, Bohr lay aside this aspect of his researches. As he wrote to Moseley in 1913, 'For the present I have stopped speculating on atoms.' Others including Ladenburg, Vegard, Langmuir and Bury proposed alternative schemes, but none of these gained general acceptance by the community of scientists. Bohr was profoundly impressed by Sommerfeld's extension of the Bohr atom to elliptical orbits and the explanation of the fine-structure splitting of spectral lines as the effects of special relativity upon the orbits, which were described in Chaps. 5 and 6. As Bohr wrote to Sommerfeld in March 1916,

> 'I thank you so much for your paper, which is so beautiful and interesting. I do not think that I have ever read anything which has given me so much pleasure.'[2]

At the same time, the analyses of the Stark and Zeeman effects by Schwarzschild, Epstein, Sommerfeld and Debye greatly advanced understanding of the nature of quantum effects within atoms. Bohr had planned to write a comprehensive review of quantum phenomena in 1916 in an attempt to present the theory in a logically consistent manner, but he delayed publication in the light of these advances. In fact, he did not publish the results of his deliberations until 1918 when his important papers under the title *On the quantum theory of line spectra* surveyed all that was known about atomic spectra and also advocated

Ehrenfest's adiabatic hypothesis (Sect. 5.6) as a guiding principle for the formulation of quantum theory (Bohr, 1918a,b). These papers contained the first formulation of Bohr's correspondence principle as well as the selection rules for permitted quantum transitions (see Chap. 6).

For the next few years, Bohr was largely preoccupied with the construction of the Institute of Theoretical Physics in Copenhagen and he published relatively few papers, but in 1921 he was spurred into action by a letter to *Nature* by Norman R. Campbell which proposed that the static models of atoms and molecules developed by Lewis and Langmuir were 'not really inconsistent' with the Bohr–Sommerfeld model (Campbell, 1920). Bohr disagreed profoundly with this statement and went much further in advocating a model of atomic structure based strictly upon the quantum principles he had enunciated in his papers of 1918 and the correspondence principle. In his words,

'Thus, by means of a closer examination of the progress of the binding process this principle offers a simple argument for concluding that these electrons are arranged in groups in a way which reflects the periods exhibited by the chemical properties of the elements within a sequence of increasing numbers. In fact, if we consider the binding of a large number of electrons by a nucleus of higher positive charge, this argument suggests that after the first two electrons are bound in one-quantum orbits, the next eight electrons will be bound in two-quantum orbits, the next eighteen in three-quantum orbits, the next thirty two in four-quantum orbits.' (Bohr, 1921a)

This announcement generated a great deal of interest in the physics and chemistry community which was anxious to hear about Bohr's apparent solution of the problem of understanding atomic structure and the nature of the periodic table of the elements. Bohr refined his proposals in a further letter to *Nature* later that year (Bohr, 1921b). The opportunity to give a complete picture of his thinking about quantum physics and the theory of the periodic table came in 1922 with the invitation to deliver the Wolfskehl lectures at Göttingen, which were to have a lasting impact upon the development of quantum theory.

8.2 The Wolfskehl lectures and Bohr's second theory of the periodic table

The mathematician Paul Wolfskehl died in 1906 and bequeathed 100,000 Marks to the *Königliche Gesellschaft der Wissenschaften* of Göttingen as a prize for the first person to provide a complete proof of Fermat's *Last Theorem*,[3] the statement that there are no integers x, y, z and n which satisfy the equation $x^n + y^n = z^n$ for $x, y, z \neq 0$ and $n > 2$. Until the prize was awarded, the interest on the bequest should be used by the Göttingen Academy to advance the mathematical sciences. Under Hilbert's direction, an annual series of lectures was set up to attract distinguished scientists in the areas of mathematics and physics to deliver a series of lectures on frontier topics in these disciplines. The first set of lectures was delivered by Poincaré in 1909, subsequent lecturers including Lorentz, Sommerfeld, Einstein, von Smoluchowski, Mie and Planck. The invitation to Bohr was sent in November

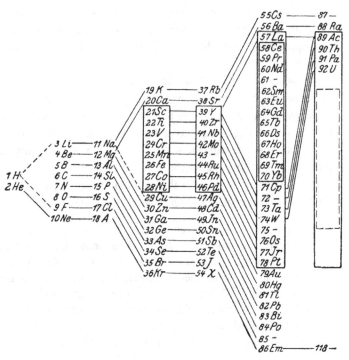

Fig. 8.1   The periodic table as presented by Thomsen which was adopted by Bohr in his analysis of the electronic configuration of atoms (Thomsen, 1895a,b).

1920 and, once his health had recovered, he delivered his seven lectures during the period 12–22 June 1922. The lectures attracted a galaxy of those working on the key problems of quantum physics from all over Germany, as well as Ehrenfest from Leiden and Klein and Oseen from Bohr's Institute in Copenhagen. Sommerfeld brought with him from Munich the 20-year old Werner Heisenberg, while 22-year old Wolfgang Pauli came from Hamburg. The audience totalled about 100 physicists and mathematicians. Bohr took the occasion very seriously and used it to present his personal survey of the entire field of quantum and atomic physics. Soon, the event became known as the *Bohr Festspiele*, or 'Bohr Festival' and was to have a major influence on the future directions which those present took to the study of quantum physics.

The contents of the lectures are described in the fourth volume of Bohr's collected works (Bohr, 1977). The first three lectures on the 12, 13 and 14 June 1922 concerned: Lecture 1: Review of the history of quanta and quantisation and the Bohr model of the atom of 1913; Lecture 2: The Bohr–Sommerfeld model of the atom, Ehrenfest's adiabatic hypothesis, multiply periodic systems and the relativistic theory of electron orbits; Lecture 3: The interpretation of spectra, the Zeeman and Stark effects, the correspondence principle, space quantisation and the Stern–Gerlach experiment. These topics have been discussed in earlier chapters. Lectures 4, 5, 6 and 7 were devoted to Bohr's interpretation of the periodic table in atomic terms. In this analysis, he used the form of the periodic table presented by his compatriot, Julius Thomsen, shown in Fig. 8.1, in which the lines join elements with similar chemical properties (Thomsen, 1895a,b). The important innovation of Thomsen's scheme,

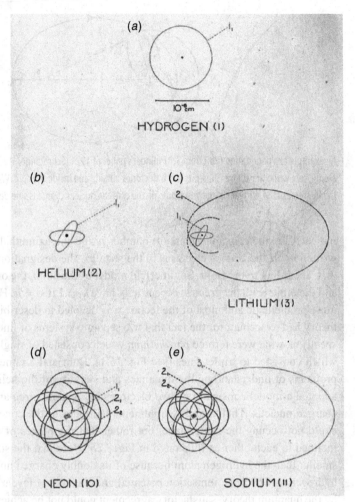

Fig. 8.2 Bohr's models for the orbits of electrons in (*a*) hydrogen, (*b*) helium, (*c*) lithium, (*d*) neon and (*e*) sodium atoms. In the original colour versions of this diagram, Fig. 8.4 and Fig. 8.5, odd principal quantum numbers $n = 1, 3, 5, \ldots$ are coloured red and even values $n = 2, 4, 6, \ldots$ black (Kramers and Holst, 1923).

as compared with that of Mendeleyev (Fig. 1.2), was the identification of the noble gases, helium (2), neon (10), argon (18), krypton (36) and xenon (54) as elements of zero valence which completed the various periods of the periodic table. As expressed by Thorpe in his *History of Chemistry, Vol. 2. From 1850 to 1910*,

> 'The valency of such an element would be zero, and therefore in this respect also it would represent a transitional stage in the passage from the univalent electronegative elements of the seventh to the univalent electropositive elements of the first group.' (Thorpe, 1910)

Lecture 4 was devoted to the first period of the table, the atoms hydrogen and helium. The hydrogen atom was in excellent agreement with the Bohr–Sommerfeld theory and was illustrated in the first of a beautiful set of plates prepared specially for the occasion by Bohr (Fig. 8.2*a*). These diagrams use a notation in which the state of the electron is

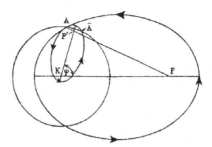

An example of a penetrating orbit from Schrödinger's paper of 1921 (Schrödinger, 1921). The circle represents the atomic core while the ellipses illustrate the trajectories outside and inside the core. Within the core, the electron feels the full force of the nuclear charge while outside the core the nuclear charge is shielded by the inner electrons.

indicated by the principal quantum number n and the azimuthal quantum number k and written n_k. If there are x electrons in the state n_k, the designation $(n_k)_x$ is used. We recall that, according to the Bohr–Sommerfeld model of the atom, the orbits are circular if $k = n$ and the orbits with the greatest eccentricity have $k = 1$ if $k \neq n$. Helium proved to be much more problematic and most of the lecture was devoted to describing these difficulties. The theory had to account for the fact that two separate systems of lines appeared in the helium spectrum, what were termed *parahelium*, which consisted of singlet lines, and *orthohelium* which consisted to triplet lines (see Fig. 16.1). Bohr and Kramers had wrestled with the problems of understanding the dynamics and stability of the helium atom and eventually adopted a model consisting of two electrons in separate circular orbits about the doubly charged nucleus. Their stability arguments led them to the conclusion that the electrons could not occupy the same plane, but rather that the planes of the two orbits had to be inclined to each other as illustrated in Fig. 8.2*b*. Note that the size of the helium atom is smaller than the hydrogen atom because of its doubly charged nucleus. Unlike the case of hydrogen in which the ionisation potential and the energy levels were precisely predicted by the quantum theory, satisfactory agreement could not be achieved for the helium atom.

Undaunted, Bohr proceeded to the rest of the periodic table in Lecture 5 and 6. There were two important influences on his thinking. The first was what he called his *Aufbauprinzip*, or 'building-up principle', in which the atomic structure of successive elements is obtained by adding an electron to the pre-existing structure of the previous element in the periodic table, noting at the same time that the charge on the nucleus increased, providing stronger binding to the nucleus. The second concept was the use of *penetrating orbits* to account for the valence electrons. The elliptical orbits of the Bohr–Sommerfeld model now came into their own, particularly thanks to a foray into atomic structure by Arnold Schrödinger. In his paper of 1921, he solved the problem of the dynamics of an electron in a quasi-elliptical orbit which penetrates within the atomic core of the inner electrons (Fig. 8.3) (Schrödinger, 1921). In the outer region, the electron experiences the electrostatic force of a single electronic charge, while inside the circular inner orbit, it feels the full force of the nucleus. The small ellipse embedded within the circular orbit shows the resulting kinematics. In addition, Schrödinger was able to account for the Rydberg formula of the alkali metals, the reasoning being similar to the arguments described in Sect. 6.6.

Bohr argued that the third electron in lithium could not be in a 1_1 orbit because it would be too tightly bound to the nucleus. The solution was to place it in the next shell as a 2_1 elliptical orbit about the helium nucleus, as illustrated in Fig. 8.2c. Notice the distortions of the ellipse close to the 'helium core' and the consequent precession of the elliptical orbit about the nucleus. The 2_1 orbit of the electron is evidently less tightly bound to the nucleus and becomes the valence electron of lithium. Bohr placed the next electron in a 2_1 orbit to create the beryllium atom with double valence. Proceeding beyond boron, Bohr found himself in uncertain waters. The one fixed point was that, by the time he reached the next noble gas neon, the structure should be stable with what we would now call a 'closed shell'. He achieved this by filling up the $n = 2$ state with 2_1 and 2_2 orbits, placing four electrons in each and so accounting for the atomic number $A = 10$. The structure he had in mind is shown in Fig. 8.2d. The diagram is to be interpreted as four 2_1 orbits and four 2_2 orbits in three dimensions as explained by Kramers and Helge Holst in their popular book (Kramers and Holst, 1923). Bohr's dilemma is indicated in the published form of his electron assignments in the periodic table of Table 8.2 in which boron is tentatively given a 2_2 electron and the filling of the remaining electrons to the shell left uncertain.

With the $n = 2$ shell filled, Bohr immediately assigned the eleventh electron to a 3_1 orbit which gave it a similar valence role to that of lithium (Fig. 8.2e) while the twelfth electron was also assigned a 3_1 orbit corresponding to the magnesium atom. In filling up the rest of the $n = 3$ shell, there were now elliptical 3_2 orbits available and Bohr proposed that the completion of the $n = 3$ shell at argon should consist of four electrons with 3_1 orbits and four with 3_2 orbits (Fig. 8.4a). The next step was to continue to the $n = 4$ shell and it was straightforward to assign one 4_1 electron to potassium and a second 4_1 electron to calcium. However, now he had to incorporate the elements from scandium to nickel. These elements have similar chemical properties and this could be explained if the inner $n = 3$ shell was completed, including the circular 3_3 orbits, rather than electrons being added to the $n = 4$ level. Bohr advanced arguments that, in filling up the $n = 3$ levels rather than proceeding with the $n = 4$ levels, these atoms would be in a lower state of total energy. The building up of the elements proceeded in this way, with closed shells at krypton (Fig. 8.4c) and xenon (Fig. 8.4d). The rare earth elements from cerium (58) to ytterbium (70) were associated with the completion of inner shells with the result that their chemical properties would be very similar. In his Nobel lecture of 1922, Bohr went so far as to state that

> 'Indeed, it is scarcely an exaggeration to say that if the existence of the rare earths had not been established by direct experimental investigation, the occurrence of a family of elements of this character within the sixth period of the natural system of the elements might have been theoretically predicted.' (Bohr, 1922)

Bohr proceeded to assign electrons to periods up to $n = 7$, his sketch of the radium atom, drawn at twice the scale of the earlier diagrams, showing clearly the 'shell' structure of the electron distributions (Fig. 8.5). He even suggested that the unknown element with $A = 118$ should be stable.

Bohr implied that his assignments of the electrons to different shells were based upon detailed calculations, including application of the correspondence principle to atomic structure. In fact, it seems that the assignments were primarily based upon symmetry, the need

Table 8.2 Bohr's model of 1922 for the distribution of electrons in atoms of the periodic table.

Z	El	1_1	2_1	2_2	3_1	3_2	3_3	4_1	4_2	4_3	4_4	5_1	5_2	5_3	5_4	5_5	6_1	6_2	6_3	6_4	6_5	6_6	7_1	7_2
1	H	1																						
2	He	2																						
3	Li	2	1																					
4	Be	2	2																					
5	B	2	2	(1)																				
10	Ne	2	4	4																				
11	Na	2	4	4	1																			
12	Mg	2	4	4	2																			
13	Al	2	4	4	2	1																		
18	A	2	4	4	4	4																		
19	K	2	4	4	4	4		1																
20	Ca	2	4	4	4	4		2																
21	Sc	2	4	4	4	4	1	(2)																
22	Ti	2	4	4	4	4	2	(2)																
29	Cu	2	4	4	6	6	6	1																
30	Zn	2	4	4	6	6	6	2																
31	Ga	2	4	4	6	6	6	2	1															
36	Kr	2	4	4	6	6	6	4	4															
37	Rb	2	4	4	6	6	6	4	4			1												
38	Sr	2	4	4	6	6	6	4	4			2												
39	Y	2	4	4	6	6	6	4	4	1		(2)												
40	Zr	2	4	4	6	6	6	4	4	2		(2)												
47	Ag	2	4	4	6	6	6	6	6	6		1												
48	Cd	2	4	4	6	6	6	6	6	6		2												
49	In	2	4	4	6	6	6	6	6	6		2	1											
54	X	2	4	4	6	6	6	6	6	6		4	4											
55	Cs	2	4	4	6	6	6	6	6	6		4	4				1							
56	Ba	2	4	4	6	6	6	6	6	6		4	4				2							
57	La	2	4	4	6	6	6	6	6	6		4	4	1			(2)							
58	Ce	2	4	4	6	6	6	6	6	6	1	4	4	1			(2)							
59	Pr	2	4	4	6	6	6	6	6	6	2	4	4	1			(2)							
71	Cp	2	4	4	6	6	6	8	8	8	8	4	4	1			(2)							
72	–	2	4	4	6	6	6	8	8	8	8	4	4	2			(2)							
79	Au	2	4	4	6	6	6	8	8	8	8	6	6	6			1							
80	Hg	2	4	4	6	6	6	8	8	8	8	6	6	6			2							
81	Tl	2	4	4	6	6	6	8	8	8	8	6	6	6			2	1						
86	Em	2	4	4	6	6	6	8	8	8	8	6	6	6			4	4						
87	–	2	4	4	6	6	6	8	8	8	8	6	6	6			4	4					1	
88	Ra	2	4	4	6	6	6	8	8	8	8	6	6	6			4	4	1				1	
89	Ac	2	4	4	6	6	6	8	8	8	8	6	6	6			4	4	1				(2)	
90	Th	2	4	4	6	6	6	8	8	8	8	6	6	6			4	4	2				(2)	
118	?	2	4	4	6	6	6	8	8	8	8	8	8	8	8		6	6	6				4	4

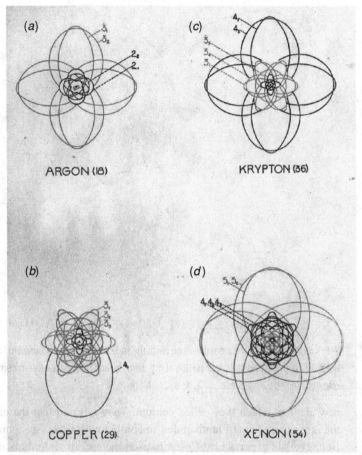

Fig. 8.4 Bohr's models for the orbits of electrons in (*a*) argon, (*b*) copper, (*c*) krypton and (*d*) xenon atoms. In the original colour version of this diagram, odd principal quantum numbers $n = 1, 3, 5, \ldots$ are coloured red and even values $n = 2, 4, 6, \ldots$ black (Kramers and Holst, 1923).

to account for the similarities of the chemical elements, his deep pondering about how the quantum concepts associated with the Bohr–Sommerfeld model could be extended to the complete periodic table, and intuition. Bohr's closest collaborator, Hans Kramers, probably had a better understanding of Bohr's meaning of the correspondence principle and its applications than anyone. In his popular book with Holst, he emphasised the provisional nature of the assignments of the electron distributions and of the arguments which led to these (Kramers and Holst, 1923). As remarked by Kragh, no evidence has been found that Bohr's conclusions were based upon detailed mathematical calculations (Kragh, 1985).

Bohr's assignments of electronic structures to the elements had a major success in predicting the properties of the unknown element with atomic number $A = 72$. According to his scheme, it should have properties similar to zirconium with $A = 40$ (see Table 8.2). A spanner was thrown in the works with the claim that an element with $A = 72$ had been discovered by the French scientists in a sample of rare earths which they interpreted as a

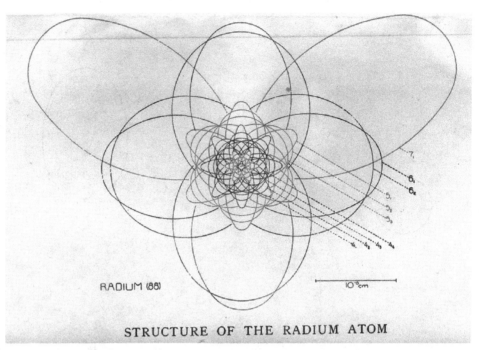

RADIUM (88)

STRUCTURE OF THE RADIUM ATOM

Fig. 8.5   Bohr's model of the radium atom drawn on twice the scale of the diagrams shown in Figs. 8.2 and 8.4 (Kramers and Holst, 1923). In the original colour version of this diagram, odd principal quantum numbers, $n = 1, 3, 5 \ldots$ are coloured red and even values $n = 2, 4, 6, \ldots$ black.

new element which they called celtium. They inferred that the element should belong to the rare earth group of lanthanides, in conflict with Bohr's assignment shown in Table 8.2. Bohr eventually persuaded Coster to examine ores of zirconium and found that indeed the element with $A = 72$ was present with high abundance in all such samples. The element was named hafnium, after the Latin name for Copenhagen, Hafnia. Bohr announced this discovery in his Nobel Prize lecture of 1922.[4]

Bohr's lectures had a major impact, particularly upon the younger participants. Although provisional, Bohr's model for the periodic table was an attempt using plausible quantum physics to put order into many different aspects of the periodic table of the chemical elements. As remarked later by Pauli,

> 'It made a strong impression upon me that Bohr at that time and in later discussions was looking for a *general* explanation that should hold for the closing of *every* electron shell and in which the number 2 was considered as essential as 8.' (Pauli, 1964)

The Bohr model provided a framework for many different attacks upon the problems of the quantum physics of atoms and molecules.

8.3 X-ray levels and Stoner's revised periodic table

It is striking that Bohr had been able to achieve so much employing only two quantum numbers, the principal quantum number n and the azimuthal quantum number k. There is no mention of the inner quantum number, j or J, introduced by Sommerfeld and Landé to account for the anomalous Zeeman effect. Most of Bohr's attention had concentrated upon the interpretation of optical data and the valencies of the chemical elements which are associated with the outer electronic structures of atoms. In contrast, X-ray emission lines and X-ray absorption edges provided direct information about the innermost shells of atoms. Bohr discussed the importance of the X-ray measurements in his seventh and final Wolfskehl lecture, but no mention is made of the inner quantum number.

In Moseley's pioneering X-ray experiments of 1913 and 1914, the K and L X-ray lines were used to order the chemical elements according to their atomic numbers, as discussed in Sect. 4.6 (Moseley, 1913, 1914). The interpretation of Moseley's data in terms of the Bohr model was taken up by Kossel (1914, 1916). He reasoned that, when an electron is ejected from the innermost $n = 1$ stationary state of a Bohr atom, this vacancy is refilled by an electron jumping from a higher energy state, in the process emitting a photon with energy equal to the difference in energies of the initial and final states. It was thus expected that there would be a series of lines associated with transitions originating from the $n = 2$, $n = 3, \ldots$ states. This model accounted for the fact that the characteristic X-ray emission lines were always observed in emission and not in absorption. In Kossel's interpretation, the K and L shells corresponded to final states with principal quantum numbers $n = 1$ and 2. Although not stated explicitly, Kossel assumed that there is a maximum number of electrons which could occupy each shell which becomes particularly stable when it is filled with the maximum number of electrons.

X-ray spectroscopy techniques developed dramatically over the succeeding years,[5] the spectral resolution of the measurements increasing by a factor of at least 100. A review of these early developments is presented by Manne Siegbahn (1962), one of the pioneers of precision X-ray spectroscopy, in which he pointed out that the X-ray spectroscopists had access to wavelengths in the range 0.1–20 Å, far exceeding the range available to optical spectroscopists. The high spectral resolution enabled details of the various X-ray multiplets to be studied in detail. The clustering of X-ray lines into K and L shells is illustrated by the X-ray spectrum of silver shown in Fig. 8.6a. Although there is a great deal of fine structure, the separation into K and L shells is clear. In addition to these emission spectra, the X-rays are absorbed by atoms, but rather than emission lines, absorption edges were observed, as illustrated by the variation of the absorption coefficient of X-rays by silver atoms (Fig. 8.6b). The wavelengths of the absorption edges are shorter than those of the emission lines of the same series. This was attributed to the fact that absorption of X-radiation from, say, the K level requires the photons to have energy $\varepsilon \geq E_K$, the ionisation energy from the K shell.

It can be seen from Fig. 8.6b that the K series is a singlet, in that there is only a single absorption edge, whereas the L series has three absorption edges. Proceeding to higher levels for elements with greater atomic numbers, it was found that the M series had five edges and N seven. These results found a natural interpretation as multiplets of the different shells of the atoms. Whilst a given level had the same principal quantum number n, different sublevels would be associated with the different allowed values of k. This interpretation was, however, in conflict with Bohr's assignments of the number of sublevels in the L shells which were only defined by the two quantum numbers $n = 2$ and $k = 2$ and 1. There are only two possible combinations of these quantum numbers ($n = 2, k = 2$) and ($n = 2, k = 1$)

Table 8.3 Landé's inferred distribution of the quantum numbers among the K, L and M X-ray levels (Landé, 1923b), including the equivalent optical terms derived from optical spectroscopy (Stoner, 1924). Note that, for consistency of notation, J is used for the inner quantum number rather than Stoner's j.

Level	K	L			M				
Sublevel		L_I	L_{II}	L_{III}	M_I	M_{II}	M_{III}	M_{IV}	M_V
n	1	2	2	2	3	3	3	3	3
k	1	1	2	2	1	2	2	3	3
J	1	1	1	2	1	1	2	2	3
Optical term	1σ	2σ	$2\pi_2$	$2\pi_1$	3σ	$3\pi_2$	$3\pi_1$	$3\delta_2$	$3\delta_1$

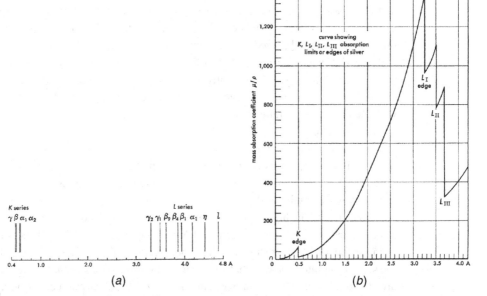

Fig. 8.6 (*a*) A schematic diagram showing the X-ray emission line spectrum of silver, clearly separating the K from the L series of lines. (*b*) The absorption coefficient of silver as a function of wavelength, showing the X-ray edges associated with absorption from the K and L levels (Semat, 1962).

which would have different energies since the former is a circular orbit and the latter an ellipse. This conflicted with the observation that the L state is a triplet and correspondingly the M series is a quintuplet rather than a triplet. Evidently, an additional quantum number was needed and this was provided by the inner quantum number J. Landé proposed the distribution of electrons among the K, L and M levels shown in Table 8.3. Accompanying these assignments was a set of selection rules in which k would change by 1 and J by 1 or 0.

In 1924 Table 8.3 was used by Edmund Stoner, then in the final year of his postgraduate studies with Rutherford and Fowler at Cambridge, to make the next major advance in

Element	Atomic number	Level (n)	Sublevel (k, j)						
			I	II	III	IV	V	VI	VII
			1, 1	2, 1	2, 2	3, 2	3, 3	4, 3	4, 4
He	2	K (1)	2						
Ne	10	L (2)	2	2	4				
A	18	M (3)	2	2	4	(4	6)		
Kr	36	N (4)	2	2	4	(4	6)	(6	8)
Xe	54	O (5)	2	2	4	(4	6)		
Nt	86	P (6)	2	2	4				

Fig. 8.7   Stoner's modification of Bohr's distribution of electrons in the shells of the inert gases (Stoner, 1924). The numbers of electrons in a given atom is indicated by *all* the numbers above and to the left of the heavy line drawn under each atom's symbol (Heilbron, 1977).

defining the electronic structures of atoms. He was well aware of the difficulties of the Bohr scheme and favoured Landé's scheme. He noted in addition that there is a strong analogy with the inferred structure of the atom based upon Landé's interpretation of the anomalous Zeeman effect. The X-ray levels could be related to the optical terms inferred from the analysis of optical spectra. The last line of Table 8.3 shows the corresponding optical terms for the doublet terms of the alkali metals in Stoner's notation. In his paper, Stoner (1924) wrote:

'The number of electrons in each completed level is equal to double the sum of the inner quantum numbers as assigned, there being in the K, L, M, N levels, when completed, 2, $8 (= 2 + 2 + 4)$, $18 (= 2 + 2 + 4 + 4 + 6)$, ... electrons. It is suggested that the number of electrons associated with each sub-level separately is also equal to double the inner quantum number.'

Thus, according to Stoner's arithmetic and referring to Table 8.3,

The K level contains $2 \times 1 = 2$ electrons;
The L_I sublevel contains $2 \times 1 = 2$ electrons;
The L_II sublevel contains $2 \times 1 = 2$ electrons;
The L_III sublevel contains $2 \times 2 = 4$ electrons.

In this way, Stoner built up the periodic table, Fig. 8.7 showing the distribution of electrons for the closed shells of the inert gases. The sublevels of a given shell, indicated by roman capital numerals, are characterised by the azimuthal quantum number k and the inner quantum number J which takes values k and $k - 1$. Therefore, in Stoner's scheme, the maximum occupancy of a k shell is $2 \times [k + (k - 1)] = 4k - 2$. For a given value of n, this resulted in the filling of the sublevels with 2 $(k = 1)$, 6 $(k = 2)$, 10 $(k = 3)$, 14 $(k = 4)$,

18 ($k = 5$), ... The number of electrons in a completely filled shell with principal quantum number n is 2, 8, 18, 32, 50 ..., in other words $2n^2$. The total numbers of electrons in a given atom is indicated by *all* the numbers above and to the left of the heavy line drawn under each atom's symbol. As compared with Bohr's scheme, there was a greater concentration of electrons in the outer subgroups and the closure of the inner subgroups at earlier stages. Stoner's scheme was immediately adopted by Sommerfeld who remarked:

> 'Being based on the incontestable experience as to the number and order of X-ray levels, and on the association of quantum numbers with these, Stoner's scheme is much more trustworthy than Bohr's. It has an arithmetic rather than geometric-mechanical character; without assuming any symmetry of orbits it exploits not some, but all available data of X-ray spectroscopy.' (Sommerfeld, 1925)

In his memoirs, Stoner wrote:

> 'Probably no other single paper of mine has attracted so much attention. . . . It is of interest to note, however, that an explicit statement is effectively made of what later became known as the Pauli exclusion principle, though it is presented more as having been arrived at inductively from experimental findings rather than as a basic axiom for a deductive treatment of electron distribution as in Pauli's paper. . . . ' (Bates, 1969)

8.4 Pauli's exclusion principle

Stoner's insights impressed Wolfgang Pauli, in particular his statement:

> 'Twice the inner quantum number does give the observed term multiplicity as revealed by the spectra in a weak magnetic field . . . In other words, the number of possible states of the (core plus electron) system is equal to twice the inner quantum number, these $2J$ states being always possible and equally probable, but only manifesting themselves separately in the presence of the external field.' (Stoner, 1924)

Pauli inferred that a natural explanation of the shell structure of the atom was to postulate that the state of electrons in atoms is determined by four quantum numbers, n the principal quantum number, k the azimuthal quantum number, J the inner quantum number and m, the component of angular momentum in the direction of an applied field, where $-J \leq m \leq J$, *provided only one electron is allowed to occupy each of these states*. These rules enabled Pauli to recover Stoner's results described in the last section. But the implications were much deeper and were to have profound implications for the future development of quantum mechanics. Pauli had enunciated the principle of *exclusion*, according to which within an atom not more than one electron can occupy a stationary state with a single set of quantum numbers, the *Pauli exclusion principle*. In his words,

> 'There never exist two or more equivalent electrons in an atom which, in strong magnetic fields, agree in all quantum numbers n, k, J and m. If there exists in the atom an electron for which these quantum numbers (in the external field) have definite values, this state is "occupied".' (Pauli, 1925)

Pauli was impressed by the fact that the exclusion principle could account for the absence of certain states in, for example, the alkaline earth metals such as calcium, which would involve two electrons with the same quantum numbers. The physical significance of the internal quantum number J and the origin of the exclusion principle itself were, however, obscure. Pauli associated the quantum number J with the valance electron itself rather than with the core and, in the limit of strong magnetic fields when the Paschen–Back effect dominates, he showed that J behaves like orbital angular momentum. He referred to the property of the quantum number J of the valance electron as 'classically not-describable two-valuedness'. He was very close to discovering the spin of the electron and its magnetic moment, but drew back from that final step.[6]

8.5 The spin of the electron

The discovery of the spin and magnetic moment of the electron by Samuel Goudsmit and George Uhlenbeck has an amusing history. The concept of the spin of the electron was first explored by Ralph Kronig in 1925 while he held a travelling fellowship from Columbia University. During his visit to Tübingen, then the centre of spectroscopic studies of atoms and molecules, Landé showed him a letter from Pauli describing his work on the exclusion principle and the need for four quantum numbers. Kronig hit upon the idea that the J quantum number was associated with the intrinsic spin of the electron. If the orbits of the electrons resembled a planetary system, he reasoned that the electron itself would be spinning, just like the planets. He then associated a magnetic moment of one Bohr magneton with the spinning electron. This concept immediately suggested an origin for the splitting of the D lines of sodium into D_1 and D_2 components. The electron experiences a magnetic field B because of its motion through the electric field of the nucleus of magnitude $B = (1/c^2)(v \times E)$. Consequently, there is an interaction energy associated with the magnetic moment of the electron and the induced magnetic field of magnitude $\mu \cdot B$. Because the J quantum number took up two orientations with respect to the magnetic field direction, their interaction energies were different and gave excellent agreement with the observed splitting of the sodium D lines. Furthermore, this model was also in full agreement with Landé's semi-empirical 'relativistic splitting rule'.

There were however concerns. Kronig went on to apply the magnetic coupling argument to hydrogen atoms and found an answer similar to that found by Sommerfeld in his relativistic treatment of the hydrogen atom. The worry was that Sommerfeld's model accounted very well for the observed fine-structure splitting of hydrogen and so there would have to be some remarkable fine tuning if the combination of effects were to result in the observed splitting. In fact, Kronig was always a factor of 2 out in obtaining the required compensation. There was another concern which was that the velocity of the surface of a classical electron would be rotating at a speed far exceeding the speed of light. This can be demonstrated by a simple classical calculation. The Stern–Gerlach experiment was interpreted as evidence that the angular momentum of the valence electron is $\hbar/2$. Therefore, the speed of rotation of the surface of the electron would be given by the relation $mvr = \hbar/2$. Using the expression for the classical electron radius found by equating the rest mass energy of the electron to its electrostatic potential energy $r = e^2/4\pi\epsilon_0 m_e c^2$, $v = c/2\alpha$, where the fine-structure constant $\alpha = e^2/2\epsilon_0 h \approx 1/137$. Thus, the speed of rotation of the surface of the electron would far exceed the speed of light. Kronig discussed these ideas with Pauli, Kramers and Heisenberg who were strongly critical of the proposal, Pauli being particularly dismissive. Kronig let the matter rest.

After the publication of Pauli's paper on the exclusion principle, the idea was revived by Uhlenbeck and Goudsmit (1925a). They were impressed by two considerations. The first, pointed out to them by Ehrenfest, was the calculation by Abraham that if the charge of a rotating sphere is distributed over its surface, the magnetic moment is twice that if the charge is uniformly distributed through the sphere (Abraham, 1903). The second was the fact that Pauli's paper required four quantum numbers to describe the motion of the electron. Since they equated the number of quantum numbers with the number of degrees of freedom, they could associate three of these with the kinematics of the motion of the electron, but what about the fourth? Their guess was that it was associated with the intrinsic angular momentum of the electron, rather than with that of the core. They were well aware that a consequence would be that the rotation speed of the surface of the electron would far exceed the speed of light. Pauli's 'classically not-describable two-valuedness' was to be associated with two allowed orientations of the spin axis of the electron with respect to any chosen axis. Finally, they proposed that the ratio of the magnetic moment of the electron to its angular momentum was twice that associated with the electron's orbital motion. Ehrenfest encouraged them to write a short paper for *Die Naturwissenschaften* and suggested that they should send it to Lorentz for comments. The following week they received a long response from Lorentz on the properties of rotating electrons. It revealed further flaws with the picture, including the fact that the magnetic energy would far exceed the rest mass energy of the electron and in fact would be greater than that of the proton. Uhlenbeck and Goudsmit decided against publication, but Ehrenfest had already sent the paper off to *Die Naturwissenschaften* in which it was published on 20 November 1925.

Reaction to the paper was mixed, Bohr welcoming the concept referring to it as 'a very welcome supplement to our ideas of atomic structure' as an addendum to the version of Uhlenbeck and Goudsmit's paper which appeared in *Nature* (Uhlenbeck and Goudsmit, 1925b). Kronig reiterated his concerns about the concept of electron spin and introduced a new problem, the effect of the spins of electrons which were assumed to be present in the nucleus (Kronig, 1925). In 1926, however, the consensus of opinion changed when the factor of 2 problem for the splitting of the fine structure of hydrogen was solved by Llewellyn Thomas (1926), who showed that a second-order relativistic term had been neglected in working out the magnitude of the spin–orbit coupling of the spinning electron about the nucleus, the effect referred to as *Thomas precession*.[7] Specifically, the energy shift associated with the interaction between the magnetic moment of the electron μ and the magnetic flux density B which it experiences in its instantaneous rest frame is

$$\Delta E = -\boldsymbol{\mu} \cdot \boldsymbol{B} = \frac{2\mu_B}{\hbar m_e e c^2} \frac{1}{r} \frac{\partial U(r)}{\partial r} \boldsymbol{L} \cdot \boldsymbol{S}, \qquad (8.1)$$

where $U(r)$ is the potential energy of the electron in the field of the nucleus and the vectors L and S are the orbital and spin angular momentum vectors respectively. The corresponding Thomas precession term, which takes account of time dilation between the frames of reference of the orbiting electron and the external frame, amounts to

$$\Delta E = -\frac{\mu_B}{\hbar m_e e c^2}\frac{1}{r}\frac{\partial U(r)}{\partial r}\, L \cdot S,\qquad(8.2)$$

that is, exactly half the magnetic spin–orbit interaction and with the opposite sign. This calculation persuaded Pauli that electron spin had to be taken really seriously and was to lead to his introduction of Pauli spin matrices in the following year.

The problems of understanding electron spin uncovered by the pioneers of quantum mechanics were well-founded. It turned out that the concept was very effective in understanding the features of atomic spectra and could be readily incorporated into Landé's vector model of the atom. It was emphasised by Pauli, however, that electron spin has the property of 'classically not-describable two-valuedness' – spin is an *intrinsic property of the electron itself*. It is not an angular momentum in the sense of classical mechanics, but rather, as Pauli expressed it, 'an essentially quantum-mechanical property'. This can be appreciated from the fact that it vanishes if $h \rightarrow 0$. We will return to these features of spin in Sect. 16.6.

The wave–particle duality

Despite Einstein's advocacy and Millikan's dramatic verification of Einstein's expression for the relation between the frequency of the incident radiation and the stopping voltage of the photoelectrons in the photoelectric effect (Sect. 3.7), the concept of light quanta was not taken seriously by the majority of physicists. As expressed by Mehra and Rechenberg,[1]

> '. . . the large majority of physicists working on quantum theory . . . subscribed to what Planck, Nernst, Rubens and Warburg wrote in 1913 about Einstein: "that he may have sometimes missed the target in his speculations as, for example, in his theory of light-quanta." And during the following decade hardly anyone took light-quanta seriously until there appeared a paper to settle this question completely: this was the paper presented by Arthur Holly Compton in December, 1922.' (Compton, 1923)

Compton demonstrated that the law of scattering of energetic X-rays by electrons in atoms follows precisely from the assumption that the quanta of light have energy $\varepsilon = h\nu$ and momentum $\boldsymbol{p} = (h\nu/c)\,\boldsymbol{i}_k$. This was irrefutable evidence for the *wave–particle duality* which was to play a central role in the development of quantum theory.

9.1 The Compton effect

The story begins with Thomson's classical analysis of the scattering of radiation by free electrons which was described in Sect. 4.3.1. The cross-section for Thomson scattering is

$$\sigma_{\mathrm{T}} = \frac{e^4}{6\pi \epsilon_0^2 m_{\mathrm{e}}^2 c^4} = \frac{8\pi r_{\mathrm{e}}^2}{3} = 6.653 \times 10^{-29}\ \mathrm{m}^2\,, \tag{9.1}$$

where $r_{\mathrm{e}} = e^2/4\pi \epsilon_0 m_{\mathrm{e}} c^2$ is the classical electron radius. The angular distribution of the radiation is given by the expression for the differential cross-section for Thomson scattering

$$\mathrm{d}\sigma_{\mathrm{T}} = \frac{r_{\mathrm{e}}^2}{2}(1 + \cos^2 \alpha)\,\mathrm{d}\Omega\,, \tag{9.2}$$

where α is the angle between the incident beam and the direction of the scattered radiation. Integrating (9.2) over solid angle gives (9.1).[2] Barkla and Ayres showed that this formula was a good description of the scattering of soft X-rays by electrons (Barkla and Ayres, 1911). For more energetic X-rays and γ-rays, however, the agreement was not so good and in particular, the scattered X- and γ-rays were of lower energy than the incident radiation (Gray, 1913). Gray repeated these experiment in 1920, again estimating the energies of the scattered X-rays by their absorption properties. He concluded that:

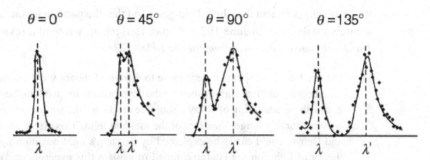

$\theta = 0°$ $\quad$ $\theta = 45°$ $\quad$ $\theta = 90°$ $\quad$ $\theta = 135°$

λ $\qquad$ $\lambda\ \lambda'$ $\qquad$ $\lambda\ \ \lambda'$ $\qquad$ $\lambda\ \ \lambda'$

Fig. 9.1 The results of Compton's X-ray scattering experiments showing the increase in wavelength of the K_α line of molybdenum as the deflection angle θ increases (Compton, 1923). The unscattered wavelength λ of the K_α line is 0.7107 Å (0.07107 nm).

'the results we have obtained would be explained if we could always look on a beam of X-rays or gamma-rays as a mixture of waves of definite frequencies, and if rays of a definite frequency were altered in wavelength during the process of scattering, the wavelength increasing with the angle of scattering.' (Gray, 1920)

With the invention of the X-ray spectrograph by the Braggs, the tools were available for more precise studies of the scattering of high energy X-rays (Bragg and Bragg, 1913a,b).

In 1921, the challenge was taken up by Compton who was the first investigator to use a Bragg spectrometer in combination with a 'recording device' to analyse the wavelengths of the scattered radiation. He soon confirmed the result that the scattered wavelength was greater than the incident wavelength, stating,

'in addition to scattered radiation, there appeared in the secondary rays a type of fluorescent radiation, whose wavelength was nearly independent of the substance used as the radiator, depending only upon the wavelength of the incident rays and the angle at which the secondary rays are examined.' (Compton, 1922)

From the beginning the challenge was to understand the mechanism by which the wavelength of the radiation increased on scattering. Compton examined what would be expected if 'each quantum of X-ray energy were concentrated in a single particle and would act as a unit on a single electron'. The result of his calculations was his famous formula for the change of wavelength of the X-rays as a function of scattering angle θ,

$$\Delta\lambda = \left(\frac{h}{m_e c}\right)(1 - \cos\theta). \tag{9.3}$$

In his classic experiment, he used the K_α line of molybdenum as the primary radiation and graphite as the scatterer. The results were reported to the American Physical Society in April 1923 and published a month later in the *Physical Review* (Compton, 1923). In fact, the same relation had been worked out by Debye and submitted to the *Physikalische Zeitschrift* only a month before Compton's presentation (Debye, 1923). The change in wavelength and the intensity of the radiation as a function of scattering angle θ were in excellent agreement with the expectation of Einstein's light quantum hypothesis (Fig. 9.1). Initially, not all the experimenters could reproduce Compton's results, but by the end of 1924, there was general agreement that these held good and that the particle nature of light quanta was securely established. In June 1929, Werner Heisenberg wrote in a review article, entitled *The development of the quantum theory 1918–1928*,

> 'At this time [1923] experiment came to the aid of theory with a discovery which would later become of great significance for the development of quantum theory. Compton found that with the scattering of X-rays from free electrons, the wavelength of the scattered rays was measurably longer than that of the incident light. This effect, according to Compton and Debye, could easily be explained by Einstein's light quantum hypothesis; the wave theory of light, on the contrary, failed to explain this experiment. With that result, the problems of radiation theory which had hardly advanced since Einstein's works of 1906, 1909 and 1917 were opened up.' (Heisenberg, 1929)

In his reminiscences in the year before he died, Compton stated:

> 'These experiments were the first to give, at least to physicists in the United States, a conviction of the fundamental validity of the quantum theory.' (Compton, 1961)

9.2 Bose–Einstein statistics

One of the intriguing questions about Planck's derivation of the black-body energy distribution is why he obtained the correct answer using the 'wrong' statistical procedures. One answer is that he may well have worked backwards from his definition of the entropy of a set of oscillators in thermal equilibrium, as discussed in Sect. 2.6. The deeper answer is that Planck had stumbled by accident upon the correct method of evaluating the statistics of indistinguishable particles. These procedures were first demonstrated by the Indian mathematical physicist Satyendra Nath Bose in a manuscript entitled *Planck's law and the hypothesis of light quanta*, which he sent to Einstein in 1924 (Bose, 1924). Einstein immediately appreciated its deep significance, translated it into German himself and arranged for it to be published in the journal *Zeitschrift für Physik*. Bose's paper and his collaboration with Einstein led to the establishment of the method of counting indistinguishable particles known as *Bose–Einstein statistics*, which differ radically from classical Boltzmann statistics.

Bose was not really aware of the profound implications of his derivation of the Planck spectrum. To paraphrase Pais's account of his paper, Bose introduced three new features into statistical physics:

(i) Photon number is not conserved.
(ii) Bose divides phase space into coarse-grained cells and works in terms of the numbers of particles per cell. The counting explicitly requires that, because the photons are taken to be identical, each possible distribution of states should be counted only once. Thus, Boltzmann's axiom of the distinguishability of particles is gone.
(iii) Because of this way of counting, the statistical independence of particles has gone.

These are profound differences as compared with classical Boltzmann statistics.[3] As Pais remarks,

'The astonishing fact is that Bose was correct on all three counts. (In his paper, he commented on none of them.) I believe there had been no such successful shot in the dark since Planck introduced the quantum in 1900.' (Pais, 1985)

The argument starts by dividing the volume of phase space into elementary cells.[4] Consider one of these cells, which we label k and which has energy ε_k and degeneracy g_k, the latter meaning the number of available states with the same energy ε_k within that cell. Now suppose there are n_k particles to be distributed over these g_k states and that the particles are identical. Then, the number of different ways the n_k particles can be distributed over these states is

$$\frac{(n_k + g_k - 1)!}{n_k!(g_k - 1)!} \, , \tag{9.4}$$

using standard procedures in statistical physics. This is the key step in the argument and differs markedly from the corresponding Boltzmann result. In the standard Boltzmann procedure, all possible ways of distributing the particles over the energy states are included in the statistics, whereas in (9.4) duplications of the same distribution are eliminated because of the factorials in the denominator.[5] Notice that this is the point at which the statistical independence of the particles is abandonned. The particles cannot be placed randomly in all the cells since duplication of configurations is not allowed.

The result (9.4) refers only to a single cell in phase space and we need to extend it to all the cells which make up the phase space. The total number of possible ways of distributing the particles over all the cells is the product of all numbers such as (9.4), that is,

$$W = \prod_k \frac{(n_k + g_k - 1)!}{n_k!(g_k - 1)!} \, . \tag{9.5}$$

We have not specified yet how the $N = \sum_k n_k$ particles are to be distributed among the k cells. To do so, we ask as before, 'What is the arrangement of n_k over the states which results in the maximum value of W?' At this point, we return to the recommended Boltzmann procedure. First, Stirling's theorem is used to simplify $\ln W$:

$$\ln W = \ln \prod_k \frac{(n_k + g_k - 1)!}{n_k!(g_k - 1)!} \approx \sum_k \ln \frac{(n_k + g_k)^{n_k + g_k}}{(n_k)^{n_k}(g_k)^{g_k}} \, . \tag{9.6}$$

Now we maximise W subject to the constraints $\sum_k n_k = N$ and $\sum n_k \varepsilon_k = E$. Using the method of undetermined multipliers as before,

$$\delta(\ln W) = 0 = \sum_k \delta n_k \{[\ln(g_k + n_k) - \ln n_k] - \alpha - \beta \varepsilon_k\} \, ,$$

so that

$$[\ln(g_k + n_k) - \ln n_k] - \alpha - \beta \varepsilon_k = 0 \, ,$$

and finally

$$n_k = \frac{g_k}{e^{\alpha + \beta \varepsilon_k} - 1} .$$

(9.7)

This is known as the *Bose–Einstein distribution* and is the correct statistics for counting indistinguishable particles.

In the case of black-body radiation, we do not need to specify the number of photons present. We can see this from the fact that the distribution is determined solely by one parameter – the total energy, or the temperature of the system. Therefore, in the method of undetermined multipliers, we can drop the restriction on the total number of particles. The distribution automatically readjusts to the total amount of energy present, and so $\alpha = 0$. Therefore,

$$n_k = \frac{g_k}{e^{\beta \varepsilon_k} - 1} .$$

(9.8)

By inspecting the low frequency behaviour of the Planck spectrum, we find that $\beta = 1/kT$, as in the classical case.

Finally, the degeneracy of the cells in phase space g_k for radiation in the frequency interval v to $v + dv$ has already been worked out in our discussion of Rayleigh's approach to the origin of the black-body spectrum (Sect. 2.3.4). One of the reasons for Einstein's enthusiasm for Bose's paper was that Bose had derived this factor entirely by considering the phase space available to the photons, rather than appealing to Planck's or Rayleigh's approaches, which relied upon results from classical electromagnetism. Planck's analysis was entirely electromagnetic and Rayleigh's argument proceeded by fitting electromagnetic waves within a box with perfectly conducting walls. Bose considered the photons to have momenta $p = hv/c$ and so the volume of momentum, or phase, space for photons in the energy range hv to $h(v + dv)$ is, using the standard procedure,

$$dV_p = V \, dp_x \, dp_y \, dp_z = V 4\pi p^2 \, dp = \frac{4\pi h^3 v^2 \, dv}{c^3} V ,$$

(9.9)

where V is the volume of real space. Now Bose considered this volume of phase space to be divided into elementary cells of volume h^3, following ideas first stated by Planck in his 1906 lectures, and so the numbers of cells in phase space was

$$dN_v = \frac{4\pi v^2 \, dv}{c^3} V .$$

(9.10)

He needed to take account of the two polarisation states of the photon and so Rayleigh's result was recovered,

$$dN = \frac{8\pi v^2}{c^3} \, dv \quad \text{with} \quad \varepsilon_k = hv .$$

(9.11)

We immediately find the expression for the spectral energy density of the radiation

$$u(v) \, dv = \frac{8\pi h v^3}{c^3} \frac{1}{e^{hv/kT} - 1} \, dv .$$

(9.12)

This is Planck's expression for the black-body radiation spectrum and it has been derived using Bose–Einstein statistics for indistinguishable particles.

Einstein did not stop there, but went on to apply these new procedures to the statistical mechanics of an ideal gas (Einstein, 1924, 1925). As he stated, the application of these statistics to a monatomic gas leads to a 'far-reaching formal relationship between radiation and gas'. In particular, he realised that the expression he had derived for the fluctuations in black-body radiation in his remarkable paper of 1909 (Sect. 3.6) must apply equally for the statistics of monatomic gases as well. It will be recalled that his expression (3.43) for the fluctuations of black-body radiation consists of two parts:

$$\frac{\sigma^2}{\varepsilon^2} = \left(\frac{h\nu}{\varepsilon} + \frac{c^3}{8\pi\nu^2 V \, d\nu}\right). \tag{9.13}$$

This is a dramatic result. Simply from the statistics of indistinguishable particles, the expression for the fluctuations consists of one term associated with the statistics of non-interacting particles according to the Maxwell–Boltzmann prescription and a second term associated with interference phenomena due to the wave properties of the particles. For these reasons, Einstein was particularly intrigued by the work of Louis de Broglie, which was reported by Langevin at the Fourth Solvay Conference in April 1924. De Broglie had made another 'shot in the dark', by ascribing wave properties to the electron. This conjecture was to have profound implications for the development of quantum mechanics.

9.3 De Broglie waves

The experimental and theoretical development of quantum physics was largely concentrated in the major centres in Germany, Göttingen, Munich and Berlin, as well as at Bohr's Institute in Copenhagen. Following the end of the First World War, communications were broken between France and Germany and the rapid exchange of results and ideas was not possible until the 1920s. Paris was somewhat off the main track of the development of quantum theory, but in many ways this was an advantage for the remarkable insights which resulted from Louis de Broglie's research. Unlike the majority of German physicists, de Broglie had no hesitation in adopting Einstein's light-quantum hypothesis. From the very beginning of his doctoral studies, his objective was to find techniques for reconciling the wave and particle descriptions of the behaviour of light. But he went much further and single-handedly introduced the concept of matter waves which completed the wave–particle duality of matter as well as light. Let us follow his reasoning.

De Broglie's three important papers were published in *Comptes Rendus (Paris)* in 1923 and were brought together in a single paper in English in the *Philosophical Magazine* in 1924 (de Broglie, 1923a,b,c, 1924b). He noted that there are two different ways of associating a frequency with a moving electron. He began by associating a frequency with the rest mass energy of the particle through Planck's relation $h\nu_0 = m_0 c^2$. If the particle moves at velocity v, the frequency associated with the particle in the external frame of reference will be greater since then $h\nu = \gamma m_0 c^2$, where $\gamma = (1 - v^2/c^2)^{-1/2}$ is the Lorentz factor. On the other hand, because of time dilation between the frame of reference of the moving particle and the external observer, time intervals are increased and so the 'internal frequency' is observed with a lower frequency $\nu_1 = m_0 c^2/\gamma$. The relation between these frequencies is

$$\nu_1 = \nu/\gamma^2 \,. \tag{9.14}$$

How are these frequencies to be interpreted? He identified ν with a 'fictitious wave' associated with the motion of the particle which propagated at a 'superluminal' velocity c^2/υ. Because the wave moved at a speed greater than the speed of light, it could not transport energy. At time $t = 0$, the location of the particle and the wave coincided. At time t, the particle had moved to the position $x = \upsilon t$ and the amplitude of its internal periodic motion would be observed to be $\sin(2\pi \nu_1 x/\upsilon)$. At the same time, the amplitude of the 'fictitious wave' would be $\sin[2\pi \nu(t - x\upsilon/c^2)]$ since its propagation speed is c^2/υ. But, because of (9.14), it follows that the two wave motions are always in phase. Inverting the argument, if the two waves are always to remain in phase, it follows that the 'fictitious wave' must propagate at speed c^2/υ.

Now de Broglie identifies the speed of the fictitious wave, c^2/υ, with the phase velocity υ_{ph} of the wave. Then, the packet of energy associated with the wave will be propagated at the group velocity υ_{g} which in modern notation we write

$$\upsilon_{\text{g}} = \frac{d\omega}{dk} \,. \tag{9.15}$$

We can write $k = 2\pi/\lambda = 2\pi\nu/\upsilon_{\text{ph}}$ and so

$$\upsilon_{\text{g}} = \frac{d\nu}{d\left(\nu/\upsilon_{\text{ph}}\right)} \quad \text{with} \quad \nu = \frac{1}{h}\frac{m_0 c^2}{(1 - \beta^2)^{1/2}} \quad \text{and} \quad \upsilon_{\text{ph}} = c^2/\upsilon = c/\beta \,, \tag{9.16}$$

where we have written the speed of the electron as $\beta = \upsilon/c$. Evaluating the differential to obtain υ_{g}, we find $\upsilon_{\text{g}} = \upsilon$, that is, *the group velocity* of the fictitious waves is exactly equal to the velocity of the electron. As de Broglie writes in his paper (de Broglie, 1923a),

'The velocity of the moving body is the energy velocity of a group of waves having frequencies $\nu = \dfrac{1}{h}\dfrac{m_0 c^2}{(1 - \beta^2)^{1/2}}$ and velocities $\dfrac{c}{\beta}$ corresponding to very slightly different values of β.'

This represents a key insight into how to reconcile the wave and particle properties of light quanta. But much more is to follow.

Next, de Broglie notes a further suggestive result by comparing the path of a light ray according to *Fermat's principle of least time* in optics with that of a particle according to the *principle of least action*. Fermat's principle is that, in a medium of variable refractive index, the path of a light ray is that which minimises the travel time between any two fixed points, $\delta \int ds/\upsilon = 0$, where ds is the element of path length. Since the frequency is a constant and $\upsilon = \lambda \nu$, this can equally be written

$$\delta \int \frac{ds}{\lambda} = 0 \,, \tag{9.17}$$

where λ the wavelength. Now, $\lambda = \upsilon_{\text{ph}}/\nu$ and since

$$\nu = \frac{1}{h}\frac{m_0 c^2}{(1 - \beta^2)^{1/2}} \quad \text{and} \quad \upsilon_{\text{ph}} = c^2/\upsilon = c/\beta \,, \tag{9.18}$$

we find

$$\delta \int \frac{ds}{\lambda} = \delta \int \frac{v\,ds}{v_{ph}} = \delta \int \frac{m_0 \beta c}{h\sqrt{1-\beta^2}}\,ds = 0 \,. \tag{9.19}$$

Now, de Broglie treats the motion of a particle according to the *principle of least action*, which can be written in the form

$$\delta \int \sum_{r=1}^{n} \dot{q}_r \frac{\partial \mathcal{L}}{\partial \dot{q}_r}\,dt = 0 \,, \tag{9.20}$$

where $\mathcal{L}$ is the Lagrangian of the particle.[6] For a relativistic particle of mass m_0 moving in a stationary electric potential $\phi(r)$, the Lagrangian takes the form

$$\mathcal{L} = -\frac{m_0 c^2}{\gamma} - q\phi(r) = -m_0 c^2 \left(1 - \frac{v^2}{c^2}\right)^{1/2} - q\phi(r)$$

$$= -m_0 c^2 \left(\frac{\dot{x}^2 + \dot{y}^2 + \dot{z}^2}{c^2}\right)^{1/2} - q\phi(r) \,, \tag{9.21}$$

where $\gamma = (1 - v^2/c^2)^{-1/2}$ is the Lorentz factor.[7] Taking $q_1 = x$, $q_2 = y$, $q_3 = z$, (9.20) becomes

$$\delta \int \frac{m_0 v^2}{(1 - \beta^2)^{1/2}}\,dt = \delta \int \frac{m_0 \beta c}{(1 - \beta^2)^{1/2}}\,ds = 0 \,. \tag{9.22}$$

This is exactly the same expression as (9.19). De Broglie immediately concluded:

'The rays of the phase wave are identical with the paths which are dynamically possible.'

Now de Broglie makes the key assumption.

'It seems quite necessary that the phase wave shall find the electron in phase with itself. That is to say "The motion can only be stable if the phase wave is tuned with the length of the path." '

To achieve this, there must be an integral number of wavelengths around the orbit. Then, (9.19) can be written

$$\int \frac{ds}{\lambda} = \int_0^{T_r} \frac{m_0 \beta^2 c^2}{h\sqrt{1-\beta^2}}\,dt = n \,, \tag{9.23}$$

where n is an integer and T_r the period of revolution. We can now rewrite the last equality in (9.23) in terms of the momentum of the electron.

$$\int_0^{T_r} \frac{m_0 \beta^2 c^2}{h\sqrt{1-\beta^2}}\,dt = \oint \frac{\gamma m_0 v}{h}\,ds = n \,, \tag{9.24}$$

and so

$$\oint p\,ds = \oint p_\phi\,d\phi = nh \,, \tag{9.25}$$

exactly the Bohr–Sommerfeld quantisation condition.

There is another way of looking at this result. Consider the circular motion of the electron and its associated wave according to the Bohr model of the hydrogen atom. If T_r is the period of the orbit of the electron, the 'fictitious wave' and the electron would meet on their circular orbit after a time τ given by equating the distance the fictitious wave travels round the orbit, $(c^2/v)\tau$, to the distance moved by the electron plus the circumference of the orbit, $v\tau + vT_r$. Therefore, writing $ds = vdt$, we find

$$\tau = \frac{v^2/c^2}{1 - v^2/c^2} \, T_r \, . \tag{9.26}$$

Therefore, the internal phase of the electron is changed by

$$2\pi \nu_1 \tau = 2\pi \nu_0 (1 - \beta^2)^{1/2} \times \frac{\beta^2}{1 - \beta^2} \, T_r = 2\pi \frac{m_0 c^2}{h} \frac{T_r \beta^2}{(1 - \beta^2)^{1/2}} \, . \tag{9.27}$$

This should be set equal to $2\pi n$ so that the electron and fictitious wave remain in phase. For circular orbits, we therefore recover again the relation

$$\int_0^{T_r} \frac{m_0 \beta^2 c^2}{h\sqrt{1 - \beta^2}} \, dt = \oint \frac{\gamma m_0 v}{h} \, ds = n \, . \tag{9.28}$$

This is de Broglie's famous formula for the quantisation of the 'fictitious waves' which travel at the phase velocity c^2/v round the electron's orbit. The associated group velocity at which the energy of the wave-packet travels round the orbit is just the velocity of the particle v. As he remarked later in his paper, 'I think that these ideas may be considered as a kind of synthesis of optics and dynamics'. The second paper also contained the prediction that the conclusions of his papers applied to electrons as well as to quanta of radiation and that

'a stream of electrons passing through a sufficiently narrow hole should also exhibit diffraction phenomena.' (de Broglie, 1923b)

The closing paragraph of his great paper shows due caution and yet optimism that these considerations represented a significant advance towards a proper theory of quantum mechanics.

'Many of these ideas may be criticised and perhaps reformed, but it seems that now little doubt should remain of the existence of light quanta. Moreover, if our opinions are received, as they are grounded on the relativity of time, all the enormous experimental evidence of the "quantum" will turn in favour of Einstein's conceptions.'

De Broglie went on to submit his doctoral thesis *Researches on the theory of quanta* in November 1923 to the Faculty of Science at the University of Paris (de Broglie, 1924a). The examining committee, consisting of Perrin, Cartan, Mauguin and Langevin, praised the striking originality of de Broglie's researches but was sceptical of the physical reality of the waves associated with the electrons. When asked about experimental tests of the hypothesis, de Broglie proposed the diffraction of beams of electron by crystals. It was not realised that experimental evidence for diffraction effects had already been discovered, as we recount in the next section.

A second important outcome was that Langevin discussed de Broglie's researches at the Fourth Solvay Conference held in April 1924, in particular with Einstein. At that time, Einstein was working on the implications of Bose's paper described in the last section and realised the significance of de Broglie's hypothesis for his theory of ideal gases, in particular, for the energy fluctuation formula and the fluctuation component due to the wave aspects of the particles. Einstein received a copy of de Broglie's dissertation in December 1924. He realised the significance of these ideas stating:

> 'I shall discuss the interpretation in greater detail because I believe that it involves more than merely an analogy.' (Einstein, 1925)

9.4 Electron diffraction

Einstein discussed de Broglie's thesis with Born who in turn brought it to the attention of James Franck, head of the department of experimental physics at Göttingen, and Born's student Walter Elsasser. When Elsasser suggested that electron diffraction experiments might be attempted, Franck commented that this:

> 'would not be necessary since Davisson's experiments had already established the expected effect.'

What Franck was referring to were the experiments of Davisson and Kunsman which had shown considerable structure in the angular distribution of electrons scattered from nickel surfaces (Davisson and Kunsman, 1921). They had interpreted these results in terms of the scattering of electrons which penetrate one or more of the outer electron shells of the nickel atoms at the surface of the crystal. Elsasser and Franck interpreted the structure in the angular distribution of the scattered electrons differently as the peaks associated with the scattering of the electrons by the nickel crystal. The characteristics of the scattered electrons were similar to the well-known phenomena of the scattering of X-rays by crystal surfaces and agreed with the expectations of de Broglie's theory if the waves associated with the electrons have wavelength $\lambda = h/mv$. Elsasser also showed that interference effects could also explain the *Ramsauer–Townsend effect,* the fact that the cross-section for scattering of slow electrons by rare gases, particularly that of argon, decreases to a very low value below electron energies of 25 eV (Ramsauer, 1921; Townsend and Bailey, 1922). Elsasser's note on the interpretation of these phenomena was published in *Die Naturwissenschaften* in August 1925 (Elsasser, 1925).

Davisson was not, however, convinced but continued his experiments on the scattering of electrons by nickel crystals. The remarkable series of events which led to his famous discovery is recounted in the opening paragraphs of his paper with Germer of 1927 (Davisson and Germer, 1927).

> 'The investigation reported in this paper was begun as the result of an accident which occurred in this laboratory in April 1925. At that time we were continuing an investigation, first reported in 1921 (Davisson and Kunsman, 1921), of the distribution-in-angle of electrons scattered by a target of ordinary (poly-crystalline) nickel. During the course of this work a liquid-air bottle exploded at a time when the target was at a high temperature; the experimental tube was broken, and the target heavily oxidized by the inrushing air.

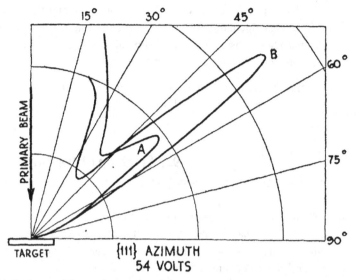

Fig. 9.2 The results of the Davisson and Germer electron scattering experiments. The curves labelled A and B show the diffraction maximum before (A) and after (B) the heating and recrystallisation of the nickel crystals for the '54 volt' electron beam (Davisson and Germer, 1927).

> The oxide eventually reduced and a layer of the target removed by vaporization, but only after prolonged heating at various temperatures in hydrogen and in vacuum.
>   When the experiments were continued it was found that the distribution-in-angle of the scattered electrons had been completely changed.'

What had happened was that the original nickel sample consisted of a number of separate crystals. The effect of the processing subsequent to the accident was that the separate crystals were fused into a single nickel crystal with much improved diffraction capabilities. The improvement is illustrated in Fig. 9.2 which shows the famous diffraction maximum at an angle of 50° using the original crystals and the larger single crystal. This maximum disappeared if the accelerating voltage was appreciably changed. The separation a between the planes of atoms in the nickel crystal, the grating constant, was 2.15 Å and so Bragg's formula for the diffraction of X-rays, $n\lambda = 2\pi a \sin\theta$, can be used to estimate the wavelength λ of the incident beam, where n is an integer and θ is the angle between the incident ray and the scattering planes. This formula resulted in a wavelength of $\lambda = 1.65$ Å, in perfect agreement with the expectation of de Broglie's formula $\lambda = h/mv = (150/V)^{1/2}$ Å for an accelerating voltage $V = 54$ V.

In fact these results were not published until 1927, although they were well-known through discussions with Born, Hartree and others at the 1926 meeting of the British Association for the Advancement of Science held in Oxford. The clinching experiments were carried out by George P. Thomson, the son of J. J. Thomson, and Andrew Reid at the University of Aberdeen. They passed a collimated beam of electrons through a thin film of celluloid and the resulting diffraction pattern was similar to that observed in the diffraction of X-rays of the same energy as the electrons (Thomson and Reid, 1927). In

Fig. 4.—Gold.   Fig. 5.—Celluloid.   Fig. 6.—Film X.

Fig. 9.3 Examples of photographs of the diffraction rings observed in electron diffraction experiments in which a beam of electrons is incident upon thin films of different materials (Thomson, 1928).

a subsequent paper, Thomson (1928) demonstrated diffraction rings associated with the passage of electrons through thin films of gold, celluloid and other substances (Fig. 9.3). There followed numerous experiments demonstrating the wave nature of the diffraction of beams of electrons. The 1937 Nobel Prize for physics was awarded to Davisson and Thomson. As famously remarked by Jammer (1989),

'. . . Thomson, the father, was awarded the Nobel prize for having shown that the electron is a particle, and Thomson, the son, for having shown that the electron is a wave.'

These electron diffraction experiments provided incontrovertible evidence for the correctness of de Broglie's deep insight into the wave–particle duality for light and matter. We have, however, now run far ahead of the chronological unfolding of the story. Let us review the status of the old quantum theory as it stood at the end of 1924.

9.5 What had been achieved by the end of 1924

The discussion of the last six chapters demonstrates the remarkable efforts undertaken by many outstanding physicists and mathematicians to come to terms with the challenges posed by the introduction of quantisation and quanta into the infrastructure of classical physics. It will be recognised that most of the distinctive features which a successful quantum theory had to encompass were already in place, but there was no coherent quantum theory which could accommodate all of them, let alone the deeper implications of what these discoveries implied for the understanding of the physical world. It is helpful to list what had been achieved by the end of 1924.

- The Bohr model of the atom and the concept of *stationary states* which electrons can occupy in orbits about the atomic nucleus (Chap. 4).
- The introduction of additional quantum numbers to account for multiplets in atomic spectra and the associated selection rules for permitted transitions (Chap. 5 and Sects. 6.4 and 6.5).

- Bohr's correspondence principle which enabled classical physical concepts to inform the rules of the old quantum theory (Sect. 6.3).
- The explanation of the Stark and Zeeman effects in terms of the effects of quantisation upon atoms in electric and magnetic fields (Chap. 7).
- Landé's vector model of the atom which could account for the splitting of spectral lines in terms of the addition of angular momenta according to the prescription given by the expression for the Landé g-factor (Sect. 7.4).
- The experimental demonstration of space quantisation by the Stern–Gerlach experiment (Sect. 7.5).
- The concept of electron shells in atoms and the understanding of the electronic structure of the periodic table (Chap. 8).
- The discovery of Pauli's exclusion principle and the requirement of four quantum numbers to describe the properties of atoms (Sect. 8.3).
- The discovery of electron spin as a distinctive quantum property of these particles – although inferred from angular momentum concepts, the intrinsic spin of the electron is not 'angular momentum' in the sense of rotational motion (Sect. 8.4).
- The confirmation of the existence of light quanta as a result of Compton's X-ray scattering experiments (Sect. 9.1).
- The discovery of Bose–Einstein statistics which differed very significantly from classical Boltzmann statistics (Sect. 9.2).
- The introduction of de Broglie waves associated with the momentum of the electron and the extension of the wave–particle duality to particles as well as to light quanta (Sect. 9.3).

The reader will immediately recognise that the above list is the product of hindsight, highlighting features which would have to be incorporated into a new theory of what was to become *quantum mechanics*. The old quantum theory could not accommodate these into a single coherent theory. In essence, the old quantum theory can be considered classical physics with the addition of the Bohr–Sommerfeld quantisation rules, namely,

$$\oint p_\phi \, \mathrm{d}\phi = nh \,, \tag{9.29}$$

as well as a set of selection rules to ensure that the observed optical and X-ray spectral features could be reproduced.

Besides the problems of coherence, there was the highly non-trivial problem that, although the properties of the hydrogen atom could be explained rather well by the Bohr–Sommerfeld model, the next element in the periodic table, helium with two electrons rather than one and twice the positive nuclear charge, defied explanation despite the strenuous efforts of many theorists. For heavier elements, the problems were no simpler.

But even more fundamentally, the stability of atoms, particularly the decay of the electron orbits because of the emission of electromagnetic radiation had no satisfactory solution. The old quantum theory got round this problem by simply requiring that accelerated electrons at the atomic level did not radiate electromagnetic radiation. The best that could be done was to assert that Bohr's stationary states genuinely were 'stationary' and at the atomic level the continuous loss of energy did not occur. Energy loss only occurred in transition between stationary states. Einstein had proposed how this could be formalised in his remarkable introduction of spontaneous and induced transition probabilities between stationary states and, as we will see, this was to be the touchstone for the development of approaches which would ultimately lead to the revolution of a genuine theory of quantum mechanics. Something had gone profoundly wrong with classical mechanics and dynamics at the atomic level – the solution of that problem was to prove to be truly revolutionary.

PART III

THE DISCOVERY OF QUANTUM MECHANICS

10 The collapse of the old quantum theory and the seeds of its regeneration

'In spite of its high-sounding name and its successful solutions of numerous problems in atomic physics, quantum theory, and especially the quantum theory of polyelectron systems, prior to 1925, was, from the methodological point of view, a lamentable hodge-podge of hypotheses, principles, theorems and computational recipes rather than a logical consistent theory. Every single quantum-theoretic problem had to be solved first in terms of classical physics; its classical solution had then to pass through the mysterious sieve of the quantum conditions or, as it happened in the majority of cases, the classical solution had to be translated into the language of quanta in conformance with the correspondence principle. Usually, the process of finding the 'correct solution' was a matter of skillful guessing and intuition, rather than of deductive or systematic reasoning.' (Jammer, 1989)

Although written with the benefit of hindsight, there is no doubt that, by the end of 1924, there was a major crisis in the attempts to create a system of 'quantum mechanics'[1] which could encompass all the features of atoms and their spectra. At the heart of the problem was the wave–particle duality first enunciated by Einstein in 1905 and reinforced by de Broglie's remarkable association of 'matter-waves' with electrons in 1924. As recorded by Jammer (1989),

'This state of affairs was well characterised by Sir William Bragg when he said that physicists are using on Mondays, Wednesdays, and Fridays the classical theory and on Tuesdays, Thursdays and Saturdays the quantum theory of radiation.'

The transition from the old quantum theory to the completed theory of quantum mechanics took place remarkably rapidly, over a period of only a few years, but it was not a simple story. Born remarked that it was not even 'a straight staircase upward', but rather a 'tangle of interconnected alleys.'

The story begins with one of the thorniest problems, the understanding of the physics of the dispersion of light by material media. Despite the apparently intractable nature of the problem, the attempts to unravel it were to lead to new approaches which would soon lead to a radically new description of physics at the atomic level.

10.1 Ladenburg, Kramers and the theory of dispersion

The dispersion problem is at the heart of the interaction between matter and radiation. Classically, the dispersion of electromagnetic waves results from the dependence upon frequency of the refractive index of the medium through which the waves propagate. The simplest example occurs in continuous media in which there is a linear dependence of the polarisation P of the medium upon the applied electric field strength E, $P = \chi \epsilon_0 E$, where χ is the electric susceptibility of the medium. It is straightforward to solve the first two of Maxwell's equations (1.5) and (1.6) for an incident electromagnetic wave. In a linear medium,

$$D = \epsilon_0 E + P = (1 + \chi)\epsilon_0 E = \epsilon \epsilon_0 E , \tag{10.1}$$

where ϵ is the relative permittivity of the material. Going through the standard procedure[2] of reducing (1.5) and (1.6) to a wave equation in E or H, the speed of propagation of the waves is $(\epsilon \epsilon_0 \mu_0)^{-1/2} = c/\sqrt{\epsilon}$. The refractive index of the material $n = \sqrt{\epsilon}$ is a readily measurable quantity. The term dispersion is used to describe the phenomena which occur when the refractive index is a function of frequency, resulting in the 'dispersion' or 'smearing out' of the frequency components of a wave-packet.

In the presence of an absorption line, there is a strong dependence of χ upon frequency and it had been established empirically that the variation of χ with frequency could be written in the form

$$\chi_i = \frac{e^2}{m_e} \frac{f_i}{\omega_i^2 - \omega^2} , \tag{10.2}$$

where ω_i is the central frequency of the absorption line and f_i is a constant, the significance of which will become apparent shortly.

A relation of this form can be derived from classical electromagnetic theory and, in fact, we have already almost derived it in Sect. 2.3.3 – the classical theory of dispersion was expounded by Paul Drude (1900). Formally, the polarisation P is the dipole moment per unit volume of the material,[3] in this case, induced by the electric field of the incident electromagnetic wave. For convenience, we repeat the formula (2.9) which describes the motion of an oscillator of natural angular frequency ω_0 under the influence of an incident electromagnetic wave:

$$\ddot{x} + \gamma \dot{x} + \omega_0^2 x = \frac{F}{m_e} . \tag{10.3}$$

For simplicity, we take the reduced mass of the oscillator to be the electron mass m_e. If the oscillator is accelerated by the E field of an incident wave, $F = eE_0 \exp(i\omega t)$. To find the response of the oscillator, a trial solution for x of the form $x = x_0 \exp(i\omega t)$ is adopted. Then

$$x_0 = \frac{eE_0}{m_e \left(\omega_0^2 - \omega^2 + i\gamma\omega \right)} . \tag{10.4}$$

The complex factor in the denominator means that the oscillator does not vibrate in phase with the incident wave. The dipole moment of the oscillator is $p = ex$ with respect to its rest position and so, for a single oscillator, multiplying through by e, we find

$$p = \frac{e^2 E_0}{m_e \left(\omega_0^2 - \omega^2 + i\gamma\omega \right)} . \tag{10.5}$$

Multiplying by the complex conjugate of (10.5) and taking the square root, we find the amplitude of the induced dipole moment,

$$|p| = \frac{e^2 E_0}{m_e \left[\left(\omega_0^2 - \omega^2 \right)^2 + \gamma^2 \omega^2 \right]^{1/2}}.$$

(10.6)

Thus, provided the frequency is not too close to the resonance frequency ω_0, the second term in square brackets in the denominator can be neglected and the expression for the electric susceptibility χ_0 of *a single oscillator* is

$$\chi_0 = \frac{e^2}{m_e \left(\omega_0^2 - \omega^2 \right)},$$

(10.7)

exactly the same as the relation (10.2). Summing over all the oscillators with angular frequencies ω_i in unit volume of the material, Drude's formula for the electric susceptibility is obtained,

$$\chi = \sum_i \frac{e^2}{m_e} \frac{f_i}{\omega_i^2 - \omega^2},$$

(10.8)

where the factor f_i was interpreted by Drude as the 'number of dispersion electrons' per atom. The same formula was derived by Pauli in his important review of 1926 where f_i was referred to as the 'strength' of the oscillator (Pauli, 1926). This variation of the dispersion in the vicinity of an absorption line was known as *anomalous dispersion*. If measurements of the anomalous dispersion were measured away from the line centre, the value of f_i could be measured for the oscillator.

The seeds of a new approach to the wave–particle duality were contained in a paper by Ladenburg (1921) who combined Drude's theory with Einstein's quantum theory of the emission and absorption of radiation (Einstein, 1916) discussed in Sect. 6.2. Consider a system of N oscillators each of mass m_e, charge e and oscillation frequency $\nu_0 = \omega_0/2\pi$ in thermal equilibrium in an enclosure. The classical average energy loss rate of a single oscillator is given by (2.4),

$$-\frac{dE}{dt} = \gamma E,$$

(10.9)

where $\gamma = \omega_0^2 e^2 / 6\pi \epsilon_0 c^3 m$. The radiation loss rate of the system of oscillators J_{cl} was written in the following form by Ladenburg,

$$J_{cl} = \frac{\overline{U} N}{\tau},$$

(10.10)

where $\tau = 1/\gamma$ and $\overline{U}$ is the mean energy of each oscillator. As shown by Planck, the mean energy of the oscillator is directly related to the mean energy density of radiation within the enclosure. In Sect. 2.3.3, this relation was derived for the case in which the motion of the electron could be modelled by three orthogonal oscillators and so for a single oscillator, (2.20) can be written,

$$u(\nu_0) = \frac{8\pi \nu_0}{3c^3} \overline{U} = \frac{2\omega_0^2}{3\pi c^3} \overline{U}.$$

(10.11)

Notice that $u(\nu_0)$ is the energy density of radiation per unit frequency interval. Therefore, substituting for $\overline{U}$ in (10.10), the classical rate of loss of energy of the oscillator in the enclosure can be written

$$J_{\text{cl}} = \frac{e^2}{4\epsilon_0 m_e} N u(\nu_0) . \tag{10.12}$$

This expression relates the rate of loss of energy directly to the spectral energy density $u(\nu_0)$ of the radiation field at frequency ν_0.

Ladenburg now tackled the same problem from the quantum theoretical point of view using the Einstein A and B coefficients discussed in Sect. 6.2:

$$A_m^n = \frac{8\pi h \nu^3}{c^3} B_m^n \quad \text{and} \quad B_m^n = B_n^m . \tag{10.13}$$

m labels the lower energy state ε_m and n the upper state ε_n and so $h\nu_0 = \varepsilon_n - \varepsilon_m$. The rate of absorption of energy by the set of oscillators is therefore

$$J_{\text{qu}} = h\nu_0 N_m B_m^n u(\nu_0) . \tag{10.14}$$

Ladenburg next used Bohr's correspondence principle to equate the classical energy loss rate to the quantum theoretical rate of absorption of energy, $J_{\text{cl}} = J_{\text{qu}}$, and so from (10.13),

$$N = N_m \frac{\epsilon_0 m_e c^3}{2\pi \nu_0^2 e^2} A_m^n . \tag{10.15}$$

According to Bohr, the correspondence principle should only be applied to transitions between states with large principal quantum numbers, but Ladenburg assumed that (10.15) should apply for all quantum transitions. Next, he identified N/N_m with f_i, the number of oscillators per atom derived in (10.8). We recall that classically Drude identified f_i as the number of dispersion electrons per atom. In Ladenburg's exposition, N_m is taken to be the number of atoms in the ground state. Therefore,

$$f_i = \frac{\epsilon_0 m_e c^3}{2\pi \nu_0^2 e^2} A_m^n . \tag{10.16}$$

Thus, Ladenburg had derived an expression for Einstein's spontaneous emission coefficient A_m^n in terms of measurable quantities since f_i can be found from the frequency dependence of the polarisation properties of the medium. Inserting (10.16) into (10.8), we find the relation between the polarisation and the spontaneous transition probabilities,

$$P = \chi E = \frac{\epsilon_0 E c^3}{8\pi^3} \sum_i \frac{A_m^n}{\nu_0^2(\nu_0^2 - \nu^2)} . \tag{10.17}$$

This expression can be thought of as describing the amplitude of the varying electric dipole moment of the medium which is to account for dispersion phenomena. But, implicitly Ladenburg had made a crucial conceptual advance, which was not explicitly stated in his paper. As noted by van der Waerden (1967),

'Ladenburg replaced the atom, so far as its interaction with the radiation field is concerned, by a set of harmonic oscillators with frequencies equal to the absorption frequencies ν_0 of the atom.'

This was explicitly recognised by Bohr, Kramers and Slater (1924) who referred to the model as consisting of a set of *virtual oscillators*.

The result (10.17) was generalised by Kramers (1924) for the case in which the lower state of the transition was not the ground state. In this case, he argued that (10.17) should include the contribution of virtual oscillators with energies $\varepsilon < \varepsilon_n$. To convert (10.17) into a more convenient form, let us label the energy level of the state we are interested in as i, higher energy states by k and lower energy states by k'. Then, $\nu_0 = \nu_{ik} = (\varepsilon_k - \varepsilon_i)/h$ and $\nu_{k'i} = (\varepsilon_i - \varepsilon_{k'i})/h$. Thus, (10.19) becomes

$$P = \chi E = \frac{\epsilon_0 E c^3}{8\pi^3} \sum_k \frac{A_i^k}{\nu_{ik}^2(\nu_{ik}^2 - \nu^2)} . \tag{10.18}$$

Kramers showed that a similar additional term to that in (10.17) should be included when the transitions are not to the ground state so that the full expression of the polarisation would read

$$P = \chi E = \frac{\epsilon_0 E c^3}{8\pi^3} \left[\sum_{\substack{k \\ E_k > E_i}} \frac{A_i^k}{\nu_{ik}^2(\nu_{ik}^2 - \nu^2)} - \sum_{\substack{k' \\ E_{k'} < E_i}} \frac{A_{k'}^i}{\nu_{k'i}^2(\nu_{k'i}^2 - \nu^2)} \right] . \tag{10.19}$$

Kramers realised that this second term had a somewhat curious significance, but that it had to be there. As he wrote in his paper to *Nature*,

> 'The reaction of the atom against the incident radiation can thus be formally compared with the action of a set of virtual harmonic oscillators inside the atom, conjugated with the different possible transitions to other stationary states.... one might introduce the following terminology: in the final state of the transition the atom acts as a "positive virtual oscillator" of relative strength $+f$; in the initial state, it acts as a negative virtual oscillator of strength $-f$. However unfamiliar this "negative dispersion" might appear from the point of view of the classical theory, it may be noted that it exhibits a close analogy with the "negative absorption" which was introduced by Einstein, in order to account for the law of temperature radiation on the basis of the quantum theory.'

The connection with the correspondence principle was rubbed home in the last sentence of his paper

> 'It may be remembered, however, that the presence of the second term in [10.19] is necessary if the classical theory can be applied in the limiting region where the motions in successive stationary states differ by only small amounts from each other.'

Kramers did not give the proof of (10.19) in his paper of 1924, but it was derived in detail in his paper with Heisenberg in the following year (Kramers and Heisenberg, 1925).

Without going through the details of that proof, we can understand why the second term on the right-hand side of (10.19) is necessary. It will be recalled that, in Einstein's derivation of Planck's formula using Einstein's A and B coefficients, it is essential to incorporate the induced emission term in order to obtain the correct form of the black-body spectrum in the long wavelength, or Rayleigh–Jeans, limit. If the term is omitted, the Wien, rather than the Planck, distribution is obtained. Now, according to the correspondence principle, the classical and quantum formalisms should coincide in the limit of large quantum numbers and this corresponds exactly to the Rayleigh–Jeans region of the Planck spectrum. This explains why the 'negative virtual oscillators' need to be included in (10.19). This important feature of the formalism was clarified in a paper by John van Vleck (1924). Again, the correspondence principle provides guidance about what the correct form of the quantum relations has to be.

10.2 Slater and the Bohr–Kramers–Slater theory

The next contribution to the 'tangle of interconnected alleys' was made by John C. Slater who completed his doctorate at Harvard University in 1923 and arrived in Copenhagen as a Sheldon Fellow of Harvard University in 1924. The source of Slater's insight came from the difficulty of reconciling the classical picture of electromagnetic waves with Einstein's concept of light quanta. In particular, how could exact conservation of energy and momentum be maintained when a continuous electromagnetic wave interacted with the discrete energy levels within atoms? As expressed by Jammer (1989),

> '. . . it was hard to understand how, for example, in a system composed of an electromagnetic radiation field, susceptible of only continuous changes of energy, and an aggregate of atoms, emitting and absorbing only discrete quanta of energy, the sum total of a continuous and of a discrete amount of energy could be a constant. . . . But there was an alternative: one could reject the energy principle as an exact law and regard it merely as a statistical law.'

This concept had been advocated by Darwin (1922) who remarked that,

> '. . . as pure dynamics has failed to explain many atomic phenomena, there seems no reason to maintain the exact conservation of energy, which is only one of the consequences of the dynamical equations.'

Darwin's paper of 1922 on the quantum theory of dispersion contained within it ideas which were to be taken over by Slater. In particular, according to Darwin's theory, when an electromagnetic wave interacts with an atom, the atom acquires a probability of emitting a spherical wave-train which interferes with the incident waves, the probability being a function of the intensity of the incident radiation.

Slater (1924) went much further, suggesting that, so long as an atom is in a stationary state, each atom can communicate with all other atoms through the action of a *virtual radiation field* originating from the *virtual oscillators* which have frequencies associated with the quantum transitions between stationary states – these are the same virtual oscillators which were introduced by Ladenburg and Kramers in their theory of dispersion. Here are Slater's own words.

> 'Any atom may, in fact, be supposed to communicate with other atoms all the time it is in a stationary state, by means of a virtual field of radiation originating from oscillators having the frequencies of possible quantum transitions and the function of which is to provide for the statistical conservation of energy and momentum by determining the probabilities for

quantum transitions. The part of the field originating from the given atom itself is supposed to induce a probability that that atom loses energy spontaneously, while radiation from external sources is regarded as inducing additional probabilities that it gain or lose energy, much as Einstein has suggested. The discontinuous transition finally resulting from these probabilities has no other significance than simply to mark the transfer to a new stationary state, and the change from the continuous radiation appropriate to the old state to that of the new.'

Slater's vision was that the resultant virtual field would determine the paths of the light quanta and also determine the probability that they propagate in a particular direction.

Bohr and Kramers discussed these ideas intensively with Slater on his arrival in Copenhagen. Bohr was still not convinced of the reality of light quanta, despite the evidence of Compton's experiments on the inelastic scattering of X-rays (Sect. 9.1). The results of these discussions was the paper by Bohr, Kramers and Slater (1924) which adopted many of Slater's ideas, but not those involving light quanta. They explicitly stated that the laws of energy and momentum conservation were only to refer to statistical averages, rather than to individual interactions. Two quotations will serve to illustrate the radical nature of the proposal.

'Further, we will assume that the occurrence of transition processes for the given atom itself, as well as for the other atoms with which it is in mutual communication, is connected with this mechanism by probability laws which are analogous to those which in Einstein's theory hold for the induced transitions between stationary states when illuminated by radiation. On the one hand, the transitions which in this theory are designated as spontaneous are, on our view, considered as induced by the virtual field of radiation which is connected with the virtual harmonic oscillators conjugated with the motion of the atom itself. On the other hand, the induced transitions of Einstein's theory occur in consequence of the virtual radiation in the surrounding space due to other atoms.'

'As regards the occurrence of transitions, . . . we abandon . . . any attempt at a causal connexion between the transitions in distant atoms, and especially a direct application of the principles of conservation of energy and momentum, so characteristic for the classical theories.'

Thus, not only has conservation of energy and momentum at the microscopic level been abandoned, causality has gone as well. There is no formal working out of these ideas in their paper, which Pais refers to as a proposal rather than a theory and also as being 'obscure in style' (Pais, 1991).

The Bohr, Kramers and Slater paper presented several experimental and theoretical challenges to the community. First of all, it was a challenge to the experimenters to demonstrate the validity of the conservation of energy and momentum at the microscopic atomic level. As Bohr, Kramers and Slater pointed out, the Compton scattering experiments were based upon statistical averages rather than on individual scattering events. Experimental evidence on the issue of the validity of the conservation laws was not long in coming. Bothe and Geiger carried out Compton scattering experiments using early coincidence techniques, in which they measured the arrival times of the scattered X-rays and the ejected K-shell electrons (Bothe and Geiger, 1924). The likelihood of the coincidences they observed occurring by chance was less than one part in 10^5. This was soon followed by cloud chamber experiments by Compton and Simon (1925) in which they determined the direction and time of ejection of the electron from the target. They verified that one recoil electron is produced on average for each scattered X-ray photon. Occasionally a secondary electron track was created by the X-ray and from these measurements, the geometry of the collision could be determined. They found the 'unequivocal answer' that the Compton collisions obeyed precisely the expectations of the laws of conservation of energy and momentum at the atomic level. As they stated

> '... the results do not appear to be reconcilable with the view of the statistical production of recoil and photo-electrons by Bohr, Kramers and Slater. They are, on the other hand, in direct support of the view that energy and momentum are conserved during the interaction of radiation and individual electrons.'

Thus, although it was evident that classical dynamics could not account for physics at the atomic level, the laws of conservation of energy and momentum still had to hold good.

Although the Bohr, Kramers and Slater proposal was shown to be incorrect in a remarkably short space of time, the paper itself was influential in that it rejected many of the fundamental tenets of classical physics. The paper was widely discussed and illustrated the necessity for quite different approaches to physics on the scale of atoms. Van der Waerden summarised the three radical elements of the paper as follows (van der Waerden, 1967):

1. Slater's concept of a virtual radiation field associated with the virtual oscillators of atoms;
2. The statistical conservations of energy and moment;
3. The statistical independence of the processes of emission and absorption of radiation.

The legacy of the paper was that the first postulate was to prove to be correct, whilst the second and third were in conflict with experiment. Following the intense discussions stimulated by the paper, Kramers went on to make his contributions to the understanding of the formula for the phenomena of dispersion, culminating in the major paper with Heisenberg which was discussed in the last section (Kramers and Heisenberg, 1925). In turn, these considerations were to prove to be crucial for Heisenberg when he made the first major advances in the development of a completely new approach to quantum physics.

Equally important was the fact that in these works the concept of atoms was gradually being displaced by a much more abstract concept of how physical processes should be envisioned at the atomic level. Jammer noted that the introduction of virtual oscillators and virtual radiation fields

> 'paved the way for the subsequent quantum-mechanical concept of probability as something endowed with physical reality and not merely a mathematical category of reasoning.'

Atoms were to be thought of as systems which consisted of virtual oscillators which were coupled to other atoms by virtual radiation fields associated with the oscillators. In the development which we are about to trace, the concept of atoms themselves became much more abstract. This was part of the penalty which had to be paid in coming up with a set of rules to replace the laws of classical dynamics.

10.3 Born and 'quantum mechanics'

The inadequacy of the old quantum theory to account for physics on the scale of atoms was now a central theme of the leading theoretical groups, nowhere more so than at Göttingen where Max Born and his colleagues were searching for new approaches to the problem of translating the structures of classical physics to the atomic domain. This change of attitude is reflected in Born's important paper of 1924 (Born, 1924). In the introduction to the paper he wrote,

> 'Since one knows that, in certain circumstances, atoms react to light waves completely "non-mechanically" (i.e. they are excited by quantum jumps), it is not to be expected that the interaction between electrons of one and the same atom should comply with the laws of classical mechanics; this disposes of any attempt to calculate the stationary orbits by using a classical perturbation theory complemented by quantum rules.'

He had been strongly impressed by the radical proposals by Bohr, Kramers and Slater and by Kramers' paper in which the dispersion formula had been derived. In particular, Kramers' derivation of the dispersion relation involved the interaction between light and the virtual oscillators and resulted in a quantum formulation for the process of dispersion while at the same time satisfying the correspondence principle in the long wavelength limit. The object of Born's paper was to extend this formalism to the case of the interactions of electrons within atoms. The ambition of the paper was to find out whether or not Kramers' approach to quantisation was founded on some more general properties of perturbed mechanical systems at the quantum level. In Born's words,

> 'What we shall do, is to bring the classical laws for the perturbation of a mechanical system, caused by the internal couplings or external fields, into one and the same form, which would very strongly suggest the formal passage from classical mechanics to a "quantum mechanics".'

For the first time, the term *quantum mechanics* appears in the literature.

Born begins by developing the classical theory of the perturbation of multiply periodic non-degenerate systems of the type discussed in Sect. 5.5. As discussed in Sect. 5.4, action–angle variables provide the natural system of coordinates for such motions. The perturbed Hamiltonian is written in the form

$$H = H_0 + \lambda H_1 \,, \tag{10.20}$$

where the perturbation H_1 is given by a Fourier series. These procedures were by now well-established in the literature, particularly because of the success of the formalism in the old quantum theory. It was also familiar from the perspective of the dynamics of astronomical systems where multiply periodic systems, such as the orbits of the planets, are subject to small perturbations. Having set up the formalism, Born shows that, if the perturbation is associated with incident electromagnetic waves, the classical expression for the dispersion is obtained.

Born now makes the key innovation of this paper by showing how to translate the formalism of classical mechanics to a 'quantum mechanics'. The argument depends upon two relations which we have already derived. The first is the use of action–angle variables as the natural system of coordinates for the description of multiply periodic motions, such as the orbits of electrons in atoms. As demonstrated in Sect. 5.5, the orbital frequency is given by (5.121),

$$\nu_i = \frac{\partial H}{\partial J_i} ,$$ (10.21)

where H is the Hamiltonian which is the total energy in this case. The second is the application of Bohr's correspondence principle according to which the classical and quantum relations should coincide for large quantum numbers and for small changes in the principal quantum number n. This was demonstrated for the hydrogen atom in Sect. 6.3. There, the key relation (6.18) can be written

$$\nu = \alpha \nu_e ,$$ (10.22)

where $\alpha = \Delta n$ and ν_e is the orbital frequency of the electron in the orbit with principal quantum number n, $\nu_e = m_e e^4 / 4\epsilon_0^2 h^3 n^3$. Classically, (10.22) has a natural interpretation in that $\alpha = 1$ corresponds to the orbital frequency of the electron and $\alpha = 2, 3, 4, \ldots$ correspond to the Fourier components of the orbital motion which can always be decomposed into a series of harmonics. Combining (10.21) and (10.22), the formula for the frequencies of the lines becomes

$$\nu_\alpha = \alpha \nu_e = \alpha \frac{dH}{dJ} .$$ (10.23)

Let us now write Bohr's quantum frequency condition for the action variable J as follows: $J = nh$ and so

$$h\nu(n, n - \tau) = E[J(n)] - E[J(n - \tau)], \quad \text{that is} \quad \nu(n, n - \tau) = \frac{H[J(n)] - H[J(n - \tau)]}{h} .$$
(10.24)

In the limit in which n is very large and $\tau = 1, 2, 3, \ldots$ small, a Taylor expansion of the last term on the right-hand side of (10.24) gives

$$\nu(n, n - \tau) = \frac{1}{h} \frac{dH}{dJ} \Delta J \quad \text{and, since} \quad \Delta J = h\tau, \quad \nu(n, n - \tau) = \tau \frac{dH}{dJ} .$$ (10.25)

Thus, the classical and quantum formulae (10.23) and (10.24) become the same in the limit of large quantum numbers, satisfying Bohr's correspondence principle. The last term in (10.25) can also be written

$$\nu(n, n - \tau) = \tau \frac{dH}{dJ} = \frac{\tau}{h} \frac{dH}{dn} .$$ (10.26)

Born's key insight was that the translation between the classical and quantum prescriptions involved replacing the *differential* $\tau \, dH/dn$, by the *difference* $H(n) - H(n - \tau)$. Born then postulates that this translation should apply more generally to all transitions between the classical and quantum formulations. Thus, for any arbitrary function $\phi(n)$ which defines the stationary state n, the differential $\tau \, d\phi/dn$ should be replaced by the difference $\phi(n) - \phi(n - \tau)$. Symbolically,

$$\tau \frac{d\phi(n)}{dn} \longleftrightarrow \phi(n) - \phi(n - \tau) .$$ (10.27)

This will play an important role in the future development of the structure of the theory. For convenience, Jammer (1989) refers to this equivalence as *Born's correspondence rule*.

Born then went on to show how, using this formulation, Kramers' quantum dispersion formula (10.19) could be derived. While Born was working on this paper, Heisenberg was his assistant at Göttingen and he contributed to the calculations carried out in the paper. As Born acknowledges in a footnote

> 'By happy coincidence, I was able to discuss the contents of this paper with Mr. Niels Bohr which contributed greatly to a clarification of the concepts. I am also greatly indebted to Mr. W. Heisenberg for much advice and help with the calculations.' (Born, 1924)

Heisenberg was only 22 years old at the time, but with his quick grasp of physical concepts and his technical expertise, he greatly benefitted from these studies with Born. He next spent the winter of 1924–1925 in Copenhagen working with Bohr and Kramers. It was there that he and Kramers worked out the full quantum theory of scattering and dispersion discussed in Sect. 10.1 (Kramers and Heisenberg, 1925). Their rigorous development of the theory started with the classical picture and then the transition to the quantum formulation was achieved by replacing the differential quotients $\partial/\partial J$ by the appropriate difference quotients, according to Born's correspondence rule.

Their result for the polarisation in the state i was given by an expression similar to (10.19) which we quoted above from Kramers' paper (Kramers, 1924), but with a significant difference in notation. Using (10.16) to write the polarisation in terms of f_{ki} rather than A_i^k, their expression can be written

$$P_i = \chi_i E = \frac{e^2 E}{4\pi^2 m_e} \left[\sum_{\substack{k \\ E_k > E_i}} \frac{f_{ki}}{(\nu_{ki}^2 - \nu^2)} - \sum_{\substack{k' \\ E_{k'} < E_i}} \frac{f_{ik'}}{(\nu_{ik'}^2 - \nu^2)} \right]. \qquad (10.28)$$

In the classical development f_i was the number of dispersing electrons in the state i, whereas in the new formulation the f_{ik} are not necessarily integers. They are better described as the *oscillator strengths*, the terminology introduced by Pauli.

Although the significance of the f_{ik}s has changed from the classical meaning, there is an important relation discovered by Kuhn (1925) and independently by Thomas (1925) concerning the sum of all the oscillator strengths of a given state i. Use is again made of the correspondence principle. Let us first simplify the notation by writing f_a for the f_{ki}, meaning absorption transitions from the state i and f_e for the $f_{ik'}$, meaning induced emission terms which must appear according to the analyses of Kramers and van Vleck. Consider now the limit in which the frequency ν is very much greater than ν_{ki} and $\nu_{ik'}$. Then, the quantum expression for the polarisation of an atom in a state i becomes

$$P_i = \chi_i E = -\frac{e^2 E}{4\pi^2 m_e \nu^2} \left[\sum_k f_a - \sum_{k'} f_e \right]. \qquad (10.29)$$

The total energy loss rate of an oscillator is given by Thomson's classical formula

$$-\left(\frac{dE}{dt} \right)_{rad} = \frac{\omega^4 e^2 |x_0|^2}{12\pi \epsilon_0 c^3} = \frac{\omega^4 |p_0|^2}{12\pi \epsilon_0 c^3}. \qquad (10.30)$$

From the classical analysis of the response of an oscillator to an incident electromagnetic wave, in the limit $\omega \gg \omega_0$ (Sect. 2.3.3),

$$|x_0|^2 = \frac{e^2 E_0^2}{m_e^2 \omega^4} ,$$

(10.31)

and so

$$-\left(\frac{dE}{dt}\right)_{rad} = \frac{e^4 E_0^2}{12\pi\epsilon_0 c^3 m_e^2} .$$

(10.32)

Now, let us work out the corresponding dispersion loss using the quantum formula (10.29) for a single state i within an atom for which the polarisation is $P_i \equiv p_0$,

$$-\left(\frac{dE}{dt}\right)_{rad} = \frac{\omega^4 |p_0|^2}{12\pi\epsilon_0 c^3} = \frac{e^4 E_0^2}{12\pi\epsilon_0 c^3 m_e^2}\left[\sum_k f_a - \sum_{k'} f_e\right]^2 .$$

(10.33)

Kuhn argued that, according to the correspondence principle, (10.32) and (10.33) should be the same in the classical limit. Therefore, the oscillator strengths for a given state i of an atom should obey the rule

$$\left[\sum_k f_a - \sum_{k'} f_e\right] = 1 .$$

(10.34)

This is the *sum rule of Kuhn and Thomas* and was to play an important role in the formulation of quantum mechanics.

10.4 Mathematics and physics in Göttingen

The developments described in this chapter were the last contributions prior to Heisenberg's epochal paper of 1925 which was to lead eventually to quantum mechanics as we know it today. The theoretical work was carried out principally in Göttingen and Copenhagen, under the guidance of a number of distinguished experimental and theoretical physicists. The Copenhagen Institute for Theoretical Physics was a very recent foundation, thanks to the efforts of Bohr, and very quickly became an international centre of excellence in all aspects of quantum physics. The most distinguished theorists visited Bohr regularly and were put through the most rigorous examinations of their researches, principally by Bohr himself.

The development of mathematical, theoretical and experimental research in Göttingen was very different.[4] There had been chairs of physics and mathematics in the University of Göttingen since the mid-eighteenth century. The great tradition of mathematics research was established by Carl Friedrich Gauss who became Professor of Mathematics in 1807. Gauss's vast contributions to mathematics need not be elaborated here, except to note that he strongly encouraged the application of mathematics to physical problems and made many important contributions. Wilhelm Weber was appointed to the Professorship of Physics in 1831 and was at the forefront of the theoretical attack upon the problems of electromagnetism.

Gauss was succeeded by a galaxy of distinguished mathematicians including Lejeune Dirichlet, Bernhard Riemann, Alfred Clebsch and Felix Klein – Arnold Sommerfeld was a student of Klein. In a separate appointment, David Hilbert became Director of the Mathematics Institute in 1895. Four new institutes were founded in 1905 to complement the existing Physics and Mathematics Departments and the Astronomical Observatory, the Director of which was Karl Schwarzschild. The new institutes were the Geophysical Institute, the Institute of Applied Electricity, the Institute of Applied Mechanics and the Institute of Applied Mathematics, the respective directors being Emil Weichert, Hermann Theodor Simon, Ludwig Prandtl and Carl Runge.

David Hilbert's appointment as Professor of Mathematics led to a remarkable flowering of research in mathematics, what Mehra and Rechenberg describe as

> 'a school of phenomenal brilliance, which had never been equalled before in the history of mathematics.'[5]

In 1900, Hilbert enunciated his 23 great problems of mathematics, which were to challenge the greatest mathematicians for the next century (Hilbert, 1900). Hilbert was not only interested in pure mathematics, however, but also in its application to physics and in 1902 he laid out his plans for a closer involvement of mathematics in physics. In 1903, he attracted his old friend from Königsberg, Hermann Minkowski, to the Chair of Mathematics but sadly Minkowski died in 1909 of appendicitis. This event interrupted Hilbert's plans for physics, but his endeavours were greatly helped by the endowment of the Wolfskehl Prize for the proof of Fermat's last theorem, which has already appeared in our story in connection with Bohr's Wolfskehl lectures of 1922 (Sect. 8.2). While Fermat's last theorem remained unsolved, the Mathematical Institute was allowed to use the interest on the capital, which amounted to about 5,000 marks per year, to endow the series of Wolfskehl lectures. Hilbert was encouraged by his colleagues to submit a solution to Fermat's last theorem, but Hilbert refused, remarking

> 'Why should I kill the goose that lays the golden egg?'

Hilbert and Klein wanted to keep abreast of developments in physics, remarking that 'Physics is much too hard for physicists'. As discussed in Sect. 8.2, Hilbert used the interest on the Wolfskehl endowment to invite distinguished lecturers in physics to Göttingen on a regular basis. Poincaré delivered the lectures in 1909, Lorentz in 1910 and Sommerfeld in 1912. The sixth of the 23 problems which Hilbert had formulated in his 1900 lecture was 'to establish the axioms of theoretical physics'.

In pursuit of this objective, he invited Einstein in the summer of 1915 to lecture on the development of the general theory of relativity. In late 1915, just weeks after the appearance of Einstein's great paper of 1915, Hilbert independently derived the field equations of general relativity (Hilbert, 1915). In fact, he went significantly further. In his axiomatic approach, he assumed that, as in general relativity, at each point in space-time 10 gravitational potentials $g_{\mu\nu}$ ($\mu, \nu = 1, 2, 3, 4$) are required to describe the gravitational field and four potentials A_μ ($\mu = 1, 2, 3, 4$) for the electromagnetic field. From these potentials, he formulated a Hamiltonian, a 'world-function' H, which depended on the $g_{\mu\nu}$ and their first and second spatial derivatives and the A_μ and their spatial derivatives. The first axiom asserted that the laws of physics should be determined by taking the variations of the integral $\int H \sqrt{g} \, dx^4$, where g is the determinant of the metric tensor $g_{\mu\nu}$. The second axiom was the 'axiom of general invariance' according to which H must be invariant with respect to arbitrary coordinate transformations. From these axioms, Hilbert showed that general relativity and a generalised form of Maxwell's equations could be derived. Klein went on to unify the different analyses of energy–momentum conservation in general relativity by Einstein, Hilbert and Lorentz. In a further brilliant analysis, Emmy Noether, a student of Hilbert and Klein, established the relationship between conservation laws and symmetry (Noether, 1918).

Hilbert and Klein took the view that this *field-theoretic approach* had solved Hilbert's problem 6 and provided the basis for the future development of theoretical physics. Although Einstein admired greatly Hilbert's deep understanding of mathematical physics, he was not an enthusiast for Hilbert's approach to field theory. Nonetheless, he continued to develop the field-theoretic approach and this resulted in his long, and in the end fruitless, endeavours to find a unified field theory. It would turn out that the breakthrough did not come from the Einstein–Hilbert approach, but from a quite different route.

In 1919, Klein retired from his position as Professor of Mathematics because of ill health and was succeeded by Richard Courant, who was not only an outstanding mathematician, but also an excellent organiser. He began a series of monographs on *Fundamental Topics of Mathematical Sciences in Monographs*. Significantly, No. 12 of the series was the first volume of the famous monograph *The Methods of Mathematical Physics* by Courant and Hilbert (1924). The book was based upon those aspects of Hilbert's lectures which found application in theoretical physics. This monograph set out systematically the mathematical tools needed to tackle the mathematical physics of quantum theory. Courant also developed ambitious plans for a mathematical institute similar to the institutes for the experimental sciences and his success in this endeavour resulted in the opening of the Mathematical Institute in 1929.

In physics, the older generation was represented by Rieche and Voigt but they supported the researches of the new generation of physicists which included Max Abraham, Johannes Stark and Walther Ritz. In 1920, Rieche retired and was replaced by Pohl. In the same year, Debye left for Zurich and by good fortune it proved possible to appoint simultaneously James Franck to the chair of experimental physics and Max Born to the chair of theoretical physics.

This remarkable battalion of mathematicians and physicists was about to tackle the quite different horizons opened up by Heisenberg's new insights. Once the way was shown, the strength of mathematical physics meant that the development of the theory proceeded rapidly. The glory years were to continue until 1933 when the Nazi prohibition on those of Jewish descent from holding university positions wreaked havoc with the brilliance of Göttingen mathematics and physics. From the Göttingen faculties alone, Hermann Weyl, Richard Courant, Edmund Landau, Emmy Noether, Max Born and James Franck all emigrated in the face of Nazi oppression.

11 The Heisenberg breakthrough

11.1 Heisenberg in Göttingen, Copenhagen and Helgoland

Werner Heisenberg studied under Sommerfeld in Munich and was present at the *Bohr Festspiele* held in Göttingen in 1922 (Sect. 8.2). Although aged only 20, he challenged Bohr's support of Kramers' analysis of the quadratic Stark effect, having studied the paper in detail for Sommerfeld's seminar in Munich. The result was a long walk with Bohr during which they discussed this topic and the more general problems of quantum physics. This encounter made a strong impression on Heisenberg. Much later Heisenberg stated:

> 'That discussion, which took us back and forth over Hainberg's wooded heights, was the first thorough discussion I can remember on the fundamental physical and philosophical problems of modern atomic theory, and it has certainly had a decisive influence on my later career. For the first time, I understood that Bohr's view of his theory was much more sceptical than that of many other physicists – for example, Sommerfeld – at that time, and that his insight into the structure of the theory was not a result of mathematical analysis of the basic assumptions, but rather of an intense occupation with the actual phenomena, such that it was possible for him to sense the relationships intuitively rather than derive them formally.'

Heisenberg spent the winter of 1922–1923 working in Göttingen as Born's assistant. The astronomers had made great progress in the use of perturbations techniques within the action–angle formulation of classical dynamics to study the gravitational perturbations of planetary orbits. Born and Heisenberg adapted these procedures to the case of the orbits of electrons in atoms, in particular, the problem of the two-electron helium atom. Although some qualitative features of the helium atom could be accounted for, the quantitative results did not agree with experiment (Born and Heisenberg, 1923a,b).

Born was so impressed by Heisenberg's abilities that, following Pauli's departure from Göttingen, he asked Sommerfeld if he would release Heisenberg to become his assistant once he had completed his doctoral dissertation in Munich. Sommerfeld agreed to this. Heisenberg completed his PhD dissertation on turbulence in Munich and returned to Göttingen in October 1923 as Born's assistant. Born has provided a portrait of Heisenberg at that time.

> 'He looked like a simple peasant boy, with short, fair hair, clear bright eyes and a charming expression. He took his duties as an assistant more seriously than Pauli and was a great help to me. His incredible quickness and acuteness of apprehension has always enabled him to do a colossal amount of work without much effort: he finished his hydrodynamical

thesis, worked on atomic problems partly alone, partly in collaboration with me, and helped me to direct my research students.' (van der Waerden, 1967)

In the course of his studies with Sommerfeld in Munich and with Born in Göttingen, Heisenberg perfected his use of their approaches and techniques, in which well-defined mechanical problems were tackled systematically according to the precepts of the old quantum theory. He had already encountered Bohr's quite different approach during the 1922 Wolfskehl lectures. Bohr had also been deeply impressed by Heisenberg's abilities and invited him to Copenhagen for a visit in March 1924. Bohr took a personal interest in Heisenberg's intellectual development which was deepened during a three-day walking tour through the island of Zealand. Whereas Sommerfeld and Born were interested in the solution of well-posed problems which were solved by rigorous mathematics, Bohr was much more interested in the underlying physical concepts which he related to the results of experiment. In this, he was much more akin to Einstein for whom physical concepts were the basis of theory which were then elaborated using the appropriate mathematical tools. Heisenberg appreciated that Bohr was thinking deeply and philosophically about the underlying physical concepts which had to be incorporated into a new system of quantum mechanics – Heisenberg's specific interests in the Zeeman effect and the problems of multi-electron atoms paled into insignificance in the light of Bohr's deeper concerns.

During early 1924, Bohr was preoccupied with the development of the Bohr–Kramers–Slater theory which sought to incorporate the concept of light quanta into the Bohr–Sommerfeld picture of the atom, at the expense of the exact conservation of energy and momentum in interactions between matter and radiation (Sect. 10.2) (Bohr *et al.*, 1924). At the same time, Kramers was developing his improved version of Ladenburg's theory of dispersion by replacing the atom by a collection of virtual oscillators (Sect. 10.1) (Kramers, 1924). The collaboration between Copenhagen and Göttingen intensified during this period and, during a visit by Bohr to Göttingen, he invited Heisenberg to visit for a more extended period during the winter semesters of 1924–1925. Heisenberg duly arrived in Copenhagen in September 1924.

Kramers, who was five years older than Heisenberg, had been a collaborator of Bohr's since 1916 and had become Bohr's principal assistant when the Institute for Theoretical Physics in Copenhagen was opened in 1920. Initially, Heisenberg was somewhat in awe of Kramers' sophistication and technical ability but they soon became good friends. During this longer visit, Heisenberg fully absorbed the 'Bohrian' approach and, in particular, became an adherent of the correspondence principle and its refinement in tackling the problems of quantum physics. It was during this period that Kramers and Heisenberg developed the much more complete quantum picture of the dispersion of light incorporating Born's correspondence rule (Sect. 10.3) (Kramers and Heisenberg, 1925). Also, during his visit, Heisenberg revisited the thorny problems of the anomalous Zeeman effect and the interpretation of complex spectra.

In April 1925, Heisenberg returned to Germany, taking first a well-deserved break in Southern Germany before taking up his position as a Privatdozent at Göttingen at the beginning of the next semester. With Heisenberg, Pascual Jordan and Friedrich Hund, Born had assembled a very powerful team for theoretical physics and spectroscopy in Göttingen.

In the meantime, Bohr had become more pessimistic about his recent researches. The Bohr–Kramers–Slater picture was no longer tenable in the light of the experiments of Bothe and Geiger (1924) and he lost faith in the models of multi-electron atoms and complex spectra, which were the basis of Heisenberg's researches at Copenhagen. As a result, Heisenberg changed the direction of his research to the study of the intensities of the spectral lines of the hydrogen atom. His reasoning was that hydrogen is a single electron system and so is not subject to the difficulties of helium or multi-electron atoms. His intention was to apply Born's correspondence rule for quantisation to the radiation of the electron in circular and elliptical orbits about the nucleus. To achieve this, he had to develop Fourier series in two dimensions for elliptical orbits and this programme rapidly became prohibitively complex. Instead, he turned to a simpler problem, the emission of a nonlinear oscillator. As we will discuss in Sect. 11.5, the nonlinear oscillator has the feature that higher harmonics of the fundamental frequency can be found by recursive formulae. Some of these ideas were foreshadowed in correspondence with Kronig in 1922.

By 1925, as the old quantum theory seemed to be irretrievably incapable of accounting for the experimental data, Heisenberg decided that a more radical approach was needed, inspired by the success of Ladenburg's, Kramers' and Born's quite different approach to the interaction of radiation with atoms. As discussed in Chap. 10, the orbits of electrons in atoms were dispensed with and replaced by virtual oscillators with frequencies corresponding to the observed lines in the spectra of atoms. In an inspired application of the correspondence principle, Heisenberg argued that the Born approach to quantisation should be applied, not only to quantities such as the frequencies of the virtual oscillator, but also to the *kinematics* of the electron itself. The great, and profound, insight was that the spatial position of the electron $x(t)$ should be subject to the Born correspondence rule. In other words, the classical kinematics of Galileo and Newton had to be replaced by their quantum theoretical counterparts. The breakthrough occurred in June 1925 when Heisenberg suffered a severe bout of hay fever. To recover, he took a break on the barren German island of Helgoland in the North Sea – with its offshore climate, the island is almost free of pollen. During the nine or ten days he was there, he worked out the mathematics of these ideas. On his way back to Göttingen, he discussed his new ideas with Pauli in Hamburg, and then in Göttingen wrote up his paper on the subject, sending it to Pauli for his urgent consideration. Although notoriously critical, Pauli was impressed. Born recognised the significance of what Heisenberg had achieved and submitted it for publication to the *Zeitschrift für Physik*. This paper marks the beginning of the major change of direction which was needed to develop a consistent theory of quantum mechanics. Let us investigate exactly what Heisenberg did.

11.2 *Quantum-theoretical re-interpretation of kinematic and mechanical relations* (Heisenberg, 1925)

Heisenberg's great paper of 1925 is very far from a final polished theory of quantum mechanics, but rather an exploration of a route to the resolution of the intractable problems faced by the old quantum theory (Heisenberg, 1925). Mehra and Rechenberg (1982b) point out the remarkable similarities to Einstein's great paper of 1905, *On the electrodynamics of moving bodies* (Einstein, 1905c), which established the new framework of special relativity and which abolished the concepts of absolute space and time. In contrast to Einstein's paper, which is a complete exposition of the underlying physics of the special theory of relativity, Heisenberg's paper is a work in progress.

Heisenberg's summary at the beginning of the paper is as follows:

> 'The present paper seeks to establish a basis for theoretical quantum mechanics founded exclusively upon relations between quantities which in principle are observable.'

He begins by reviewing the well-known problems of the old quantum theory and states unambiguously that

> 'the *Einstein–Bohr* frequency condition (which is valid in all cases) already represents such a complete departure from classical mechanics, or rather . . . from the kinematics underlying this mechanics, that even for the simplest quantum theoretical problems the validity of classical mechanics cannot be maintained.'

But he has a further concern. The old quantum theory is based upon calculations of the orbits and orbital frequencies of electrons in atoms and he contends that these are unobservable quantities. The quantities which are observable are the frequencies and intensities of emission and absorption lines – the positions of electrons in atoms are in principle unobservable. Much later, this concept would be incorporated into Heisenberg's uncertainty principle, but that was still a long way in the future. He goes on:

> 'In this situation, it seems sensible to discard all hope of observing hitherto unobservable quantities, such as the position and period of the electron, and to concede that the partial agreement of the quantum rules with experience is more or less fortuitous. Instead, it seems more reasonable to try to establish a theoretical quantum mechanics, analogous to classical mechanics, but in which only relations between observable quantities occur.'

Heisenberg later stated that he had been strongly influenced by Einstein's paper on special relativity of 1905 in his insistence upon considering only observable quantities. The crucial first part of Einstein's paper is entitled *Kinematical part* and demolishes Newton's concepts of absolute space and time. Space and time need to be defined in terms of measurable quantities with the result that simultaneity becomes relative, the key concept of the *relativity of simultaneity* (Rindler, 2001). By using the two principles of relativity, namely that the laws of physics should be form-invariant between inertial frames of reference and that the speed of light is a constant in all such frames, Einstein discovered operationally self-consistent definitions of space and time. To express this in another way, what had gone wrong with Newtonian physics was the description of the *kinematics* of particles, that is, the way in which points in four-dimensional space-time are described. The new kinematics of particles and light waves are fully derived in this first kinematical part of Einstein's great paper and are embodied in the Lorentz transformations.

Heisenberg was about to carry out a similar revolution in the concept of kinematics on the scale of atoms. He was to be guided by the insights gained from Kramers', Born's and his own studies of dispersion phenomena. In particular, he was to use Born's correspondence rule to create a quantum mechanics in which only relations between observables are permitted.

Heisenberg's paper is not the easiest to comprehend. As we will see, even physicists such as Enrico Fermi and Steven Weinberg[1] have found the content difficult and somewhat obscure. The problem is compounded by changes of notation which Heisenberg adopted in the course of the paper.[2] Another problem is that Heisenberg does not describe how the results he presents were obtained. These problems are addressed and resolved in an impressive paper by Ian Aitchison and his colleagues (2004) which will be used in the exposition which follows. In the following sections, I have translated Heisenberg's arguments into a single self-consistent notation which I hope does justice to the revolutionary content of the paper.

11.3 The radiation problem and the translation from classical to quantum physics

In the next part of the paper (Sect. 1), the basic postulates of the new approach are laid out, in particular, the way in which the correspondence rule is to be used to translate between classical and quantum physics. Heisenberg begins with the formula for the intensity of emission in classical electrodynamics. First of all, he notes that the standard results for dipole radiation are only the first step in the evaluation of higher order terms associated with quadrupole and higher multipoles of the acceleration history of the electron. These can all be determined by first taking the Fourier transform of the motion of the electron and then working out the intensity of radiation associated with each Fourier component. But, he has a deep insight – in his words,

'The point has nothing to do with electromagnetism but rather – and this seems particularly important – is of a purely kinematic nature. We may pose the question in its simplest form thus: If instead of a classical quantity $x(t)$ we have a quantum-theoretic quantity, what quantum-theoretic quantity will take the place of $x(t)^2$.'

We can appreciate his line of thinking by recalling the expression for the dipole radiation of an oscillator. In the simplest case of dipole radiation, the radiation rate is given by (2.1), namely

$$-\left(\frac{dE}{dt}\right)_{rad} = \frac{|\ddot{p}|^2}{6\pi\epsilon_0 c^3} = \frac{e^2|\ddot{r}|^2}{6\pi\epsilon_0 c^3}. \tag{11.1}$$

For the case of an oscillator performing simple harmonic oscillations with amplitude x_0 at angular frequency ω_0, $x = |x_0|\exp(i\omega_0 t)$ and then the *average* radiation loss rate is given by (2.3),

$$-\left(\frac{dE}{dt}\right)_{average} = \frac{\omega_0^4 e^2|x_0|^2}{12\pi\epsilon_0 c^3}. \tag{11.2}$$

Let us note some features of this expression which were important for the pioneers of quantum mechanics. The electric field strength at distance r from the accelerated charge in the far field limit is given by the expression

$$E_\theta = \frac{|\ddot{p}| \sin\theta}{4\pi\epsilon_0 c^2 r} = \frac{e|\ddot{x}| \sin\theta}{4\pi\epsilon_0 c^2 r} = \frac{e\omega_0^2|x_0| \sin\theta}{4\pi\epsilon_0 c^2 r} \ , \tag{11.3}$$

where $p = er$ is the dipole moment of the electron. The radiation loss rate (11.1) is obtained from (11.3) by integrating the Poynting vector flux $S = |E \times H| = E_\theta^2/Z_0$ of the emitted radiation over 4π steradians, that is, over $\frac{1}{2}\sin\theta \, d\theta$, where $Z_0 = (\mu_0/\epsilon_0)^{1/2}$ is the impedance of free space. The total radiation rate can be related to Einstein's coefficient A_m^n for spontaneous emission by the following argument. We recall that A_m^n is defined for the isotropic emission of the oscillator, whereas (11.2) corresponds to the dipole emission of a single oscillator. Just as we argued in Sect. 2.3.3, we obtain the isotropic result if we consider the total loss rate to be associated with three mutually perpendicular oscillators. Then, we find

$$h\nu_0 \, A_m^n = 3 \times \frac{\omega_0^4 e^2 |x_0|^2}{12\pi\epsilon_0 c^3} \quad ; \quad A_m^n = \frac{\omega_0^3 e^2 |x_0|^2}{2\epsilon_0 c^3 h} \ . \tag{11.4}$$

Note the key result that A_m^n is the *probability* of spontaneous emission and that it depends upon the *square of the amplitude* of the oscillator $|x_0|^2$. This is why Heisenberg asks the question, '... what quantum-theoretic quantity will take the place of $x(t)^2$?'

There is one further link between $|x_0|$ and electrodynamics. In the classical theory of the emission of electromagnetic waves, it is simplest to derive the expression for the electric field strength E of the wave from the vector potential A using the relations

$$A = \frac{\mu_0}{4\pi}\frac{e\dot{r}}{r} = \frac{\mu_0}{4\pi}\frac{ev}{r} \ , \quad E = -\frac{\partial A}{\partial t} = -\frac{e\ddot{r}}{4\pi\epsilon_0 c^2 r} = -\frac{e\dot{v}}{4\pi\epsilon_0 c^2 r} \ . \tag{11.5}$$

Thus, for an oscillator of angular frequency ω_0,

$$|E| = -\left|\frac{\partial A}{\partial t}\right| = \omega_0 |A| \ . \tag{11.6}$$

Thus, comparing (11.3) and (11.6), the magnitude of the vector potential is also proportional to $|x_0|$, the maximum displacement of the oscillator.

After this mild detour into classical electrodynamics, let us return to the development of Heisenberg's paper. He takes the radical point of view that what had gone wrong with the old quantum theory is that the spatial coordinates x_0 should be replaced by their quantum analogues. This is a dramatic step. As remarked by Jammer (1989),

'Heisenberg, influenced by both Sommerfeld and Bohr, considered now the possibility of 'guessing' – in accordance with the correspondence principle – not the solution of a particular quantum-theoretic problem but the very mathematical scheme for a new theory of mechanics. By integrating in this way the correspondence principle once and for all in the very foundations of the theory, he expected to eliminate the necessity of its recurrent application to every problem individually without jeopardising its general validity.'

A key difference which has to be built into this translation is that, according to classical electrodynamics, the radiation rate depends only upon one quantum number, the principal quantum number n which determines the orbital frequency of the electron – there may also be higher harmonics of the orbital frequency, but essentially the radiation depends only on orbital frequency. This is formalised by (10.23) in which the classical frequencies of the emission are given by

$$v(\alpha) = \alpha v_0 = \alpha \frac{dW}{dJ} , \tag{11.7}$$

where α is an integer, $\alpha = 1, 2, 3, \ldots$. In the old quantum theory, the orbit with principal quantum number n has action variable $J = nh$ and then (11.7) becomes

$$v(n, \alpha) = \alpha v(n) = \alpha \frac{dW}{dJ} = \alpha \frac{1}{h} \frac{dW}{dn} . \tag{11.8}$$

In contrast, Bohr's frequency relation necessarily depends upon the properties of two states through the relation

$$v(n, n - \tau) = \frac{1}{h}[W(n) - W(n - \tau)] , \tag{11.9}$$

where the Ws represent the binding energies of the stationary states and $\tau = 1, 2, 3, \ldots$.

There is also a contrast in the ways in which combinations of frequencies are added in classical and quantum theory. Classically, for the case of radiation of harmonics α and β associated with principal quantum number n, the combination rule is

$$v(n, \alpha) + v(n, \beta) = v(n, \alpha + \beta) . \tag{11.10}$$

This type of combination of frequencies occurs in the nonlinear coupling between the harmonics α and β. The corresponding quantum combination rule for the virtual oscillators associated with transitions from the stationary state n to $n - \tau'$ and then to $n - \tau' - \tau$ is *Ritz's combination principle* (Sect. 1.6) which Heisenberg writes:

$$v(n, n - \tau') + v(n - \tau', n - \tau' - \tau) = v(n, n - \tau' - \tau) , \tag{11.11}$$

$$v(n - \tau, n - \tau' - \tau) + v(n, n - \tau) = v(n, n - \tau' - \tau) . \tag{11.12}$$

Notice the important point that at the atomic level, the classical frequency rule corresponds to the frequency of the electron's orbit and its harmonics, whereas the quantum theoretical rule does not depend upon the orbital frequency, but upon the energy differences between the three states defined by n, $n - \tau'$ and $n - \tau' - \tau$; the associated frequencies in (11.11) and (11.12) are all observables. This completes the first part of Heisenberg's agenda of working only in terms of observable quantities.

The second part is to work out the intensities and polarisations of the radiation emitted in the quantum transitions. Classically, the motion of the electron in an orbit with principal quantum number n can be represented by a Fourier series expansion in harmonics of the orbital frequency,

$$x(n, t) = \sum_{\alpha} x(n, \alpha) \, e^{i\omega(n)\alpha t} , \tag{11.13}$$

where the sum extends over all positive and negative integral values of α, that is, $\sum_{\alpha}$ is taken to mean $\sum_{\alpha=-\infty}^{\alpha=+\infty}$. From the requirement that $x(n, t)$ must be real, it follows that $x(n, -\alpha) = x^*(n, \alpha)$, where $x^*(n, \alpha)$ is the complex conjugate of $x(n, \alpha)$.

To find the value of $|x^2|$ in the classical case, we multiply together the Fourier series for $x(n, t)$ and its complex conjugate, obtaining a double sum over the products of the components of $x(n, \alpha)$. Since the Fourier series extends from $-\infty$ to $+\infty$ and the αs are integers, we obtain the same result if we write

$$x(n, t) = \sum_{\alpha'} x(n, \alpha') \, e^{i\omega(n)\alpha' t} = \sum_{\alpha'} x(n, \alpha' - \alpha) \, e^{i\omega(n, \alpha' - \alpha)t} \,. \tag{11.14}$$

The complex conjugate of the last expression is $\sum_{\alpha'} x(n, \alpha - \alpha') \, e^{i\omega(n, \alpha - \alpha')t}$ and so we can write for $|x^2|$ the double sum over α' and α,

$$x^2(n, t) = \left[\sum_{\alpha'} x(n, \alpha') \, e^{i\omega(n)\alpha' t} \right] \times \left[\sum_{\alpha} x(n, \alpha - \alpha') \, e^{i\omega(n)(\alpha - \alpha')t} \right] \,. \tag{11.15}$$

Multiplying out the components, the double sum can be rewritten

$$x^2(n, t) = \sum_{\alpha} \left[\sum_{\alpha'} x(n, \alpha') x(n, \alpha - \alpha') \, e^{i\omega(n)\alpha' t} \, e^{i\omega(n)(\alpha - \alpha')t} \right] \,. \tag{11.16}$$

Because of the rule (11.10), it immediately follows that the product of exponents in (11.16) is $e^{i\omega(n)\alpha t}$ and so the double sum can be reduced to the form

$$x^2(n, t) = \sum_{\alpha} x^{(2)}(n, \alpha) \, e^{i\omega(n)\alpha t} \,, \tag{11.17}$$

where

$$x^{(2)}(n, \alpha) = \sum_{\alpha'} x(n, \alpha') x(n, \alpha - \alpha') \,. \tag{11.18}$$

From Heisenberg's perspective this is the key result since $x^2(n, t)$ can now be represented by the sum over Fourier components, or 'virtual oscillators', with frequencies $\alpha\omega(n)$ which are directly related to the intensities of radiation through expressions of the form (11.2), in other words, to the intensities and polarisations of the 'virtual oscillators'.

Heisenberg now seeks the quantum theoretical equivalent for $x(n, t)$. According to Bohr's correspondence principle, the quantum theoretical frequency $\nu(n, n - \tau)$ corresponds to $\nu(n, \alpha)$ and so by analogy the quantity $x(n, n - \tau)$ corresponds to the Fourier component $x(n, \alpha)$. This translation can be written symbolically

$$\nu(n, \alpha) \longleftrightarrow \nu(n, n - \tau) \quad \text{and so} \quad x(n, \alpha) \longleftrightarrow x(n, n - \tau) \,. \tag{11.19}$$

Notice that I am maintaining a clear distinction between the αs, which represent harmonics of the fundamental frequency ω_0, and τ which labels different stationary states of the electron in the atom. But this procedure immediately encounters the problem that there is no unique way in which $x(t)$ can be represented by a sum over the quantities $x(n, n - \tau)$ which depend upon two variables. Nonetheless, Heisenberg presses on with the following statement:

> 'However, one may regard the ensemble of quantities $x(n, n - \tau) \exp[i\omega(n, n - \tau)t]$ as a *representation* of the quantity $x(t)$ and then attempt to answer the above question: how is the quantity $x^2(t)$ to be represented?'

Consequently, Heisenberg seeks the quantum equivalent of (11.17) which can be written

$$x^2(n, t) = \sum_\tau x^{(2)}(n, n - \tau) \exp[i\omega(n, n - \tau)]) . \tag{11.20}$$

The next step is to find the quantum theoretical equivalent of (11.18) for $x^{(2)}(n, n - \tau)$. These will be the quantities which can be inserted into the classical formula for dipole radiation and so determine the intensities of the transitions with angular frequency $\omega(n, n - \tau)$.

Let us make the replacements (11.19) in (11.16). Then, the τ component of the outer sum, the quantity in square brackets in (11.16), becomes

$$\sum_{\tau'} x(n, n - \tau') x[n, n - (\tau - \tau')] \exp[i\omega(n, n - \tau')t] \exp[i\omega(n, n - (\tau - \tau'))t] .$$

$$\tag{11.21}$$

The problem with this replacement is that it does not result in terms with frequency $\omega(n, n - \tau)$. The quantum frequency addition rule (11.11) derived from Ritz's combination principle is

$$\omega(n, n - \tau') + \omega(n - \tau', n - \tau' - \tau) = \omega(n, n - \tau' - \tau) , \tag{11.22}$$

and so the sum in the exponential term in (11.21)

$$\omega(n, n - \tau') + \omega[n, n - (\tau - \tau')] \tag{11.23}$$

is clearly unacceptable. Heisenberg was therefore driven to the 'almost cogent' conclusion that the sum of the frequency factors had instead to satisfy the relation

$$\omega(n, n - \tau) = \omega(n, n - \tau') + \omega(n - \tau', n - \tau) . \tag{11.24}$$

Consequently, (11.21) had to be rewritten as follows

$$\sum_{\tau'} x(n, n - \tau') x(n - \tau', n - \tau) \exp[i\omega(n, n - \tau')t] \exp[i\omega(n - \tau', n - \tau)t]$$

$$= \left(\sum_{\tau'} x(n, n - \tau') x(n - \tau', n - \tau) \right) \exp[i\omega(n, n - \tau)t] . \tag{11.25}$$

The quantum theoretical expression for $x^{(2)}(n, n - \tau)$ therefore has to be

$$x^{(2)}(n, n - \tau) = \sum_{\tau'} x(n, n - \tau') x(n - \tau', n - \tau) , \tag{11.26}$$

rather than (11.18). This is the *new quantum theoretical multiplication rule which replaces the classical expression* (11.18). The quantities $x(n, n - \tau)$ can be thought of as *transition amplitudes* since the square of their moduli determine the transition probabilities between the states n and $n - \tau$.

Heisenberg immediately carried out the same procedure to determine $x^3(n)$ and also wrote down the expression for the product of two quantities x_n and y_n. If we write

$$x(n, \alpha) \leftrightarrow x(n, n - \tau) \exp[i\omega(n, n - \tau)t], \quad y(n, \alpha) \leftrightarrow y(n, n - \tau) \exp[i\omega(n, n - \tau)t],$$

$$\tag{11.27}$$

then, in general, $x_n y_n \neq y_n x_n$. This can be understood by writing out the product as a sum over the components τ' according to the prescription (11.25). Then, in general,

$$\sum_{\tau'} x(n, n - \tau') y(n - \tau', n - \tau) \neq \sum_{\tau'} y(n, n - \tau') x(n - \tau', n - \tau) . \qquad (11.28)$$

The new multiplication rule of quantum mechanics is *non-commutative* and this result seriously worried Heisenberg. It turned out, however, to be a key innovation of Heisenberg's paper.

11.4 The new dynamics

Undaunted by the puzzle of non-commuting variables, Heisenberg proceeded to the second section of his paper which deals with the mechanics of particles incorporating the new kinematics. According to the procedures of the old quantum theory, the analysis begins with the classical equation of motion, for example,

$$\ddot{x} + f(x) = 0 , \qquad (11.29)$$

and this equation is solved classically. The phase integral $J = \oint p \, dq = nh$ is then introduced to satisfy the quantum conditions. In the one-dimensional case, $p = m\dot{x}$, $dq = dx = \dot{x} \, dt$ and so

$$J = \oint m \dot{x}^2 \, dt = nh . \qquad (11.30)$$

Heisenberg now evaluates J according to classical physics and then translates the result to quantum physics using the new kinematics. First, x is written classically as a Fourier series, exactly as in (11.13):

$$x = x(n, t) = \sum_{\alpha} x(n, \alpha) \, e^{i\omega(n)\alpha t} . \qquad (11.31)$$

It follows that

$$m\dot{x} = m \sum_{-\infty}^{+\infty} x(n, \alpha) \, i\alpha\omega(n) \, e^{i\omega(n)\alpha t} . \qquad (11.32)$$

We now form $\dot{x}^2$, by multiplying $\dot{x}$ by its complex and integrating over a period of the motion $t = 2\pi/\omega(n)$. All the products of the series except those involving products such as $x(n, \alpha)x(n, -\alpha)$ are zero on integrating over one period. We also recall that $x(n, -\alpha)$ conjugate is the complex conjugate of $x(n, \alpha)$ and so the integral over one period of the motion is

$$J = \oint m \dot{x}^2 \, dt = 2\pi m \sum_{-\infty}^{+\infty} x(n, \alpha)x(n, -\alpha) \alpha^2 \omega(n) = 2\pi m \sum_{-\infty}^{+\infty} |x(n, \alpha)|^2 \alpha^2 \omega(n) . \qquad (11.33)$$

Now the quantum condition $J = nh$ is applied, but Heisenberg immediately takes the derivative of nh with respect to n, preparing the way for the application of Born's correspondence rule,

$$\frac{d}{dn}(nh) = \frac{d}{dn} \oint m \dot{x}^2 \, dt , \tag{11.34}$$

$$h = 2\pi m \sum_{-\infty}^{+\infty} \alpha \frac{d}{dn} \left[\alpha \omega(n) |x(n, \alpha)|^2 \right] . \tag{11.35}$$

Born's correspondence rule (10.27) tells us that differentials are to be replaced by differences according to the recipe

$$\alpha \frac{d\phi(n)}{dn} \longleftrightarrow \phi(n) - \phi(n - \tau) . \tag{11.36}$$

This had been generalised by Kramers and Heisenberg (1925) in their theory of dispersion as follows:

$$\alpha \frac{d\phi(n, \tau)}{dn} \longleftrightarrow \phi(n + \tau, n) - \phi(n, n - \tau) . \tag{11.37}$$

In making this translation, the angular frequency of the α harmonic and the amplitude $x(n, \alpha)$ are translated as

$$\phi(n + \tau, n): \quad \alpha\omega(n) \longleftrightarrow \omega(n, n + \tau) \quad ; \quad x(n, \alpha) \longleftrightarrow x(n, n + \tau), \tag{11.38}$$

$$\phi(n - \tau, n): \quad \alpha\omega(n) \longleftrightarrow \omega(n, n - \tau) \quad ; \quad x(n, \alpha) \longleftrightarrow x(n, n - \tau). \tag{11.39}$$

Consequently, (11.35) becomes

$$h = 4\pi m \sum_{0}^{+\infty} \left[|x(n, n + \tau)|^2 \omega(n, n + \tau) - |x(n, n - \tau)|^2 \omega(n, n - \tau) \right] , \tag{11.40}$$

where we have used the fact that x is real to convert the sum from $-\infty$ to $+\infty$ to twice that from 0 to $+\infty$ since $|x(n, \tau)|^2 = |x(n, -\tau)|^2$. Furthermore, the values of the xs can be determined absolutely since there is a requirement that, in the ground state n_0, no emission processes are possible to a lower energy state, in other words, $x(n_0, n_0 - \tau) = 0$ for $\tau > 0$.

These are remarkable calculations, as Heisenberg realised:

> 'Equations [(11.29)] and [(11.40)], if soluble, contain a complete determination not only of frequencies and energy values, but also of quantum theoretical transition probabilities. However, at present the actual mathematical solution can be obtained only in the simplest cases.'

The concept is that the equation of motion is solved in terms of Fourier components and then these values entered into the quantum theoretical relation (11.40) which incorporates the quantisation conditions. But much more is obtained since the quantities $x(n_0, n_0 - \tau)$ are directly related to transition probabilities. Heisenberg noted in a footnote that (11.40) can be simply reduced to the Kuhn–Thomas relation by relating the values of $x(n_0, n_0 - \tau)$ to Einstein's A coefficient for spontaneous emission and to the values of the oscillator strengths f_i. This can be demonstrated as follows.

We have already derived the relation between the Einstein A coefficient and f_i in (10.16), namely,

$$f_i = \frac{\epsilon_0 m_e c^3}{2\pi v_0^2 e^2} A_m^n \; . \tag{11.41}$$

We also showed in (11.4) how A_m^n can be related to $|x_0|^2$,

$$A_m^n = \frac{\omega_0^3 e^2 |x_0|^2}{2\epsilon_0 c^3 h} \; . \tag{11.42}$$

Inserting this value for A_m^n into (11.41), we find the relation

$$|x_0|^2 \omega_0 = \frac{h}{\pi m_e} f_i \; . \tag{11.43}$$

Now, we make use of the correspondence rule to write

$$[|x_0|^2 \omega_0]_a \longleftrightarrow |x(n, n+\tau)|^2 \omega(n, n+\tau) = \frac{h}{\pi m_e} f_a \; , \tag{11.44}$$

$$[|x_0|^2 \omega_0]_e \longleftrightarrow |x(n, n-\tau)|^2 \omega(n, n-\tau) = \frac{h}{\pi m_e} f_e \; . \tag{11.45}$$

Inserting these relations for absorption from and induced emission into the quantum state n into (11.40), Heisenberg demonstrated that the new formalism resulted in the Kuhn–Thomas formula (10.34).[3]

This simple reduction illustrates how the new formalism operates entirely in terms of observables, namely, the energies of the stationary states, the frequencies of emission and the transition probabilities between states.

11.5 The nonlinear oscillator

Heisenberg realised that his new formalism could only be applied to the simplest cases. He had already attempted to quantise the orbits of an electron in the hydrogen atom, but even this case proved to be prohibitively complicated. Instead, in the third section of his paper, he treats first the case of the nonlinear oscillator and then a rotator in which an electron orbits the nucleus at constant distance a. This is the most difficult section of the paper since Heisenberg gives no indication of how he obtained the results quoted. The paper by Aitchison and his colleagues (2004) makes a convincing case for the route Heisenberg probably followed to obtain these results and we will follow that exposition. The working out requires a great deal of algebraic manipulation, but, as we have already noted, Born commented that Heisenberg's

> '... incredible quickness and acuteness of apprehension has always enabled him to do a colossal amount of work without much effort.' (van der Waerden, 1967)

On first reading, the reader may prefer to jump to the results (11.93)–(11.97) – although not easy reading, the following calculations have the virtue of demonstrating Heisenberg's remarkable technical fluency.

The classical equation of motion of a nonlinear oscillator can be written

$$\ddot{x} + \omega_0^2 x + \lambda x^2 = 0 , \tag{11.46}$$

where λx^2 is the nonlinear term and λ is a small quantity. The advantage of treating this oscillator over a purely harmonic oscillator is that the Fourier series to represent its motion automatically contains a complete set of harmonics of the fundamental frequency ω_0. The form of Fourier series solution chosen by Heisenberg is as follows:

$$x = \lambda a_0 + a_1 \cos \omega t + \lambda a_2 \cos 2\omega t + \lambda^2 a_3 \cos 3\omega + \cdots + \lambda^{\tau-1} a_\tau \cos \tau \omega t + \cdots ,$$
$$\tag{11.47}$$

where the coefficients $a_0, a_1, a_2, \ldots, a_\tau, \ldots$ and ω are also expanded as power series in λ:

$$a_0 = a_0^{(0)} + \lambda a_0^{(1)} + \lambda^2 a_0^{(2)} + \cdots , \tag{11.48}$$

$$a_1 = a_1^{(0)} + \lambda a_1^{(1)} + \lambda^2 a_1^{(2)} + \cdots , \tag{11.49}$$

$$\cdots = \cdots ,$$

$$\omega = \omega^{(0)} + \lambda \omega^{(1)} + \lambda^2 \omega^{(2)} + \cdots \tag{11.50}$$

The nonlinear equation (11.46) cannot be solved in general and so a perturbation solution is sought in terms of power series in the small quantity λ. Inserting (11.47), (11.48) and (11.49) into (11.46), the recursion formulae for values of $a_0, a_1, a_2, \ldots$ are found by equating the constant terms and those separately in $\cos \omega t$, $\cos 2\omega t$, $\cos 3\omega t$, ... to zero. Ignoring terms in λ, it is straightforward to repeat Heisenberg's calculation and find the following recursion formulae for the first few terms of the series for the classical nonlinear oscillator:

$$[\text{constant}] \quad \lambda \left\{ \omega_0^2 a_0(n) + \tfrac{1}{2} a_1^2(n) \right\} = 0 ;$$

$$[\cos \omega t] \quad \left\{ -\omega^2 + \omega_0^2 \right\} = 0 ;$$

$$[\cos 2\omega t] \quad \lambda \left\{ [-(2\omega)^2 + \omega_0^2] a_2(n) + \tfrac{1}{2} a_1^2(n) \right\} = 0 ;$$

$$[\cos 3\omega t] \quad \lambda^2 \left\{ [-(3\omega)^2 + \omega_0^2] a_3(n) + a_1(n) a_2(n) \right\} = 0 . \tag{11.51}$$

Heisenberg now makes the transition from classical to quantum mechanics. When x is replaced by the $x(n, n - \tau) \exp[i\omega(n, n - \tau)]$, the first two terms of (11.46) become

$$[-\omega^2(n, n - \tau) + \omega_0^2] x(n, n - \tau) \exp[i\omega(n, n - \tau)] . \tag{11.52}$$

The third nonlinear term in x^2 has to be replaced by the quantum-theoretical multiplication rule (11.26) and so the term becomes

$$\lambda \sum_{\tau'} x(n, n - \tau') x(n - \tau', \tau' - \tau) \exp[i\omega(n, n - \tau)] . \tag{11.53}$$

Substituting (11.52) and (11.53) into (11.46), we obtain a recursion relation for the transition amplitude $x(n, n - \tau)$,

$$[-\omega^2(n, n - \tau) + \omega_0^2] x(n, n - \tau) + \lambda \sum_{\tau'} x(n, n - \tau') x(n - \tau', \tau' - \tau) = 0 . \tag{11.54}$$

Once again no general solution can be found and Heisenberg carries out the same perturbation analysis as for the classical case. He proposes that the translation should take the form:

$$\lambda a_0(n) \longleftrightarrow \lambda a_0(n, n) \ ;$$
$$a_1(n) \longleftrightarrow a(n, n - 1) \cos \omega(n, n - 1)t \ ;$$
$$\lambda a_2(n) \longleftrightarrow \lambda a(n, n - 2) \cos \omega(n, n - 2)t \ ;$$
$$\lambda^2 a_3(n) \longleftrightarrow \lambda^2 a(n, n - 3) \cos \omega(n, n - 3)t \ ;$$
$$\cdots \qquad\qquad \cdots$$
$$\lambda^{\alpha-1} a_\alpha(n) \longleftrightarrow \lambda^{\alpha-1} a(n, n - \alpha) \cos \omega(n, n - \alpha)t \ ;$$
$$\cdots \qquad\qquad \cdots \qquad\qquad\qquad (11.55)$$

In his paper, Heisenberg simply writes out the result of going through exactly the same analysis as in the classical case. Let us write down the appropriate translation to quantum language of (11.47)–(11.50). (11.47) becomes

$$x = \lambda a(n, n) + a(n, n - 1) \cos \omega(n, n - 1)t + \lambda a(n, n - 2) \cos \omega(n, n - 2)t$$
$$+ \lambda^2 a(n, n - 3) \cos \omega(n, n - 3)t + \cdots + \lambda^{\alpha-1} a(n, n - \alpha) \cos \omega(n, n - \alpha)t + \cdots$$
$$(11.56)$$

Once again, as in (11.48)–(11.50), the a coefficients and ω are expanded in a power series in λ so that the equivalent functions are:

$$a(n, n) = a^{(0)}(n, n) + \lambda a^{(1)}(n, n) + \lambda^2 a^{(2)}(n, n) + \cdots , \qquad (11.57)$$
$$a(n, n - 1) = a^{(0)}(n, n - 1) + \lambda a^{(1)}(n, n - 1) + \lambda^2 a^{(2)}(n, n - 1) + \cdots , \qquad (11.58)$$
$$\cdots = \cdots ,$$
$$\omega(n, n - \alpha) = \omega^0(n, n - \alpha) + \lambda \omega^{(1)}(n, n - \alpha) + \lambda^2 \omega^{(2)}(n, n - \alpha) + \cdots \qquad (11.59)$$

Now, the same type of manipulation involved in deriving (11.51) is applied to (11.46) and (11.56) to (11.59). In other words, recursion formulae for values of $a(n, n)$, $a(n, n - 1)$, $a(n, n - 2)$, ... are found by equating the constant terms and those separately in $\cos \omega(n, n - 1)t$, $\cos \omega(n, n - 2)t$, $\cos \omega(n, n - 3)t$, ... to zero. More details of the analysis are included in the paper by Aitchison *et al.* (2004) in which the recurrence relations for the transition amplitudes up to second order in λ are given. Exactly as before, the recurrence relations to zero order in λ, which Heisenberg quotes without proof, are as follows:

[constant] $\quad \lambda \left\{ \omega_0^2 \, a_0(n, n) + \frac{1}{4}[a^2(n + 1, n) + a^2(n, n - 1)] = 0 \right\} \ ; \quad (11.60)$

$[\cos \omega(n, n - 1)t] \quad -\omega^2(n, n - 1) + \omega_0^2 = 0 \ ; \qquad (11.61)$

$[\cos \omega(n, n - 2)t] \quad \lambda \left\{ [-\omega^2(n, n - 2) + \omega_0^2] a(n, n - 2) \right.$
$$\left. + \tfrac{1}{2} a(n, n - 1) a(n - 1, n - 2) = 0 \right\} \ ; \qquad (11.62)$$

$[\cos \omega(n, n - 3)t] \quad \lambda^2 \left\{ [-\omega^2(n, n - 3) + \omega_0^2] a(n, n - 3) \right.$
$$+ \tfrac{1}{2}[a(n, n - 1) a(n - 1, n - 3)]$$
$$\left. + \tfrac{1}{2}[a(n, n - 2) a(n - 2, n - 3)] = 0 \right\} \ . \qquad (11.63)$$

Next, we recall that the object of the exercise is to determine the quantity $x^{(2)}(n, n - \tau)$ which is given by (11.26). We therefore need to relate the quantities $a(n, n - \tau)$ derived above to the $x(n, n - \tau)$. In the classical case, $x_\alpha(n) = x^*_{-\alpha}(n)$ since $x(t)$ is real. The quantum theoretical analogue of this relation is

$$x(n, n - \tau) = x^*(n - \tau, n) \,. \tag{11.64}$$

This can be appreciated from the fact that the product of the associated exponentials $\exp[i\omega(n, n - \tau)] \times \exp[i\omega(n - \tau, n)] = 1$ according to the quantum rule for the addition of frequencies. In principle, $x(n, n - \tau)$ could be a complex number, but Heisenberg implicitly assumed that, by using the simple cosine expansion that they would the real. Consequently,

$$x(n, n - \tau) = x(n - \tau, n) \,. \tag{11.65}$$

We can now compare this with the series expansion (11.56) for $x(t)$, a typical term of which can be written as follows:

$$\lambda^{\tau-1} a(n, n - \tau) \cos \omega(n, n - \tau)t$$
$$= \frac{\lambda^{\tau-1}}{2} a(n, n - \tau) \{\exp[i\omega(n, n - \tau)] + \exp[-i\omega(n, n - \tau)]\} \,. \tag{11.66}$$

Since Bohr's frequency condition (11.9) requires $\omega(n, n - \tau) = -\omega(n - \tau, n)$ and $x(n, n - \tau) = x(n - \tau, n)$, this expression can be rewritten

$$\frac{\lambda^{\tau-1}}{2} a(n, n - \tau)\exp[i\omega(n, n - \tau)] + \frac{\lambda^{\tau-1}}{2} a(n - \tau, n)\exp[i\omega(n - \tau, n)] \,. \tag{11.67}$$

It is not surprising that the cosine expression corresponds to the sum of two exponentials in the series (11.20). Equating the terms in $\exp[i\omega(n, n - \tau)]$, it follows that

$$x(n, n - \tau) = \frac{\lambda^{\tau-1}}{2} a(n, n - \tau) \,. \tag{11.68}$$

Strictly speaking this equation only applies for positive values of τ. In general, the result is

$$x(n, n - \tau) = \frac{\lambda^{|\tau|-1}}{2} a(n, n - \tau) \qquad \tau \neq 0 \,. \tag{11.69}$$

The solutions of the equations of motion have been found in terms of $a(n, n - \tau)$ rather than $x(n, n - \tau)$ and so it is convenient to use (11.69) to translate (11.40) into this notation. Then,[4]

$$h = \pi m \sum_0^{+\infty} \left[|a(n, n + \tau)|^2 \omega(n, n + \tau) - |a(n, n - \tau)|^2 \omega(n, n - \tau) \right] \,. \tag{11.70}$$

We can now find the solutions for the transition amplitudes, the energies of the stationary states and the transition frequencies. We obtain many of the key results from the lowest order solutions in which all the λ terms are omitted. In this approximation, all the as and ωs acquire the superscript (0), $a^{(0)}$ and $\omega^{(0)}$. In this limit, (11.61) becomes

$$\omega^{(0)}(n, n - 1) = \omega_0 \,, \tag{11.71}$$

for all n. This makes sense since, if the perturbation term in λ is zero, the nonlinear oscillator becomes a linear oscillator with angular frequency ω_0. Now we use (11.71) to replace $\omega^{(0)}(n, n+1)$ and $\omega^{(0)}(n, n-1)$ in (11.70). Then,

$$\frac{h}{\pi m \omega_0} = [a^{(0)}(n, n+1)]^2 - [a^{(0)}(n, n-1)]^2 . \tag{11.72}$$

By inspection, we see that the solution of this difference equation is

$$[a^{(0)}(n, n-1)]^2 = \frac{h}{\pi m \omega_0}(n + \text{constant}) . \tag{11.73}$$

The value of the constant is found from Heisenberg's argument that there should be no transitions from the ground state to lower states, that is,

$$[a^{(0)}(0, -1)]^2 = 0 . \tag{11.74}$$

Consequently, the constant is zero. The solution for $a^{(0)}(n, n-1)$ can therefore be written

$$a^{(0)}(n, n-1) = \beta \sqrt{n} \quad \text{where} \quad \beta = (h/\pi m \omega_0)^{1/2} . \tag{11.75}$$

We now repeat the procedure for (11.60):

$$a^{(0)}(n, n) = -\frac{1}{4\omega_0^2} \left\{ [a^{(0)}(n+1, n)]^2 + [a^{(0)}(n, n-1)]^2 \right\} , \tag{11.76}$$

and so using (11.75),

$$a^{(0)}(n, n) = -\frac{\beta}{4\omega_0^2}(2n+1) . \tag{11.77}$$

Next, we repeat the procedure for (11.62). Then,

$$\left\{ -[\omega^{(0)}(n, n-2)]^2 + \omega_0^2 \right\} a^{(0)}(n, n-2) + \tfrac{1}{2}a^{(0)}(n, n-1) a^{(0)}(n-1, n-2) = 0 . \tag{11.78}$$

The frequency $\omega^{(0)}(n, n-2)$ must obey the frequency composition rule

$$\omega^{(0)}(n, n-2) = \omega^{(0)}(n, n-1) + \omega^{(0)}(n-1, n-2) , \tag{11.79}$$

and since, $\omega^{(0)}(n, n-1) = \omega_0$ for all n, $\omega^{(0)}(n, n-2) = 2\omega_0$. In fact, it is obvious from the frequency combination rule that repeated application gives the result

$$\omega^{(0)}(n, n-\tau) = \tau \omega_0 . \tag{11.80}$$

This result again makes a lot of sense. The harmonics of the nonlinear oscillator are simply integral multiples of the fundamental frequency ω_0 in the lowest approximation.

It is straightforward to show, by the same reasoning which led to (11.75) and (11.77), that

$$a^{(0)}(n, n-3) = \frac{\beta^3}{48\omega_0^4} \sqrt{n(n-1)(n-2)} , \tag{11.81}$$

and that in general, in this lowest order of approximation,

$$a^{(0)}(n, n-\tau) = A_\tau \frac{\beta^\tau}{\omega_0^{2(\tau-1)}} \sqrt{\frac{n!}{(n-\tau)!}} , \tag{11.82}$$

where A_τ is a numerical factor which depends upon τ.

The final task is to work out the energies of the stationary states of the oscillator. Heisenberg begins with the classical expression for the energy of a nonlinear oscillator. Following the usual procedure, we multiply (11.46) by $m\dot{x}$ and integrate with respect to time with the result

$$W = \tfrac{1}{2}m\dot{x}^2 + \tfrac{1}{2}m\omega_0^2 x^2 + \tfrac{1}{3}\lambda m x^3 . \tag{11.83}$$

He then quotes the answers according to classical and quantum physics without explaining how they are obtained. Aitchison and his colleagues conjecture entirely plausibly that he reasoned as follows. Heisenberg now had a formalism for taking products of transition amplitudes, (11.25) and (11.26), which are repeated here for convenience:

$$\sum_{\tau'} x(n, n - \tau')x(n - \tau', n - \tau) \exp[i\omega(n, n - \tau')t] \exp[i\omega(n - \tau', n - \tau)t]$$

$$= \left(\sum_{\tau'} x(n, n - \tau')x(n - \tau', n - \tau)\right) \exp[i\omega(n, n - \tau)t] . \tag{11.84}$$

Therefore, x^2 is represented by

$$\left(\sum_{\tau'} x(n, n - \tau')x(n - \tau', n - \tau)\right) \exp[i\omega(n, n - \tau)t] . \tag{11.85}$$

Correspondingly, $\dot{x}^2$ can be replaced by the product

$$\sum_{\tau'} i\omega(n, n - \tau')x(n, n - \tau')i\omega(n - \tau', n - \tau)x(n - \tau', n - \tau)$$

$$\times \exp[i\omega(n, n - \tau')t] \exp[i\omega(n - \tau', n - \tau)t]$$

$$= \sum_{\tau'} \omega(n, n - \tau')\omega(n - \tau, n - \tau')x(n, n - \tau')x(n - \tau', n - \tau) \exp[i\omega(n, n - \tau)t] , \tag{11.86}$$

where we have used the relation $\omega(n, m) = -\omega(m, n)$, according to Bohr's frequency relation. Thus, the expression for the energy of the nonlinear oscillator is of the form

$$W = W(n, n - \tau) \exp[i\omega(n, n - \tau)t]) . \tag{11.87}$$

Heisenberg fully appreciated the fact that energy will only be conserved if W is time-independent, meaning that the terms in which $\tau \neq 0$ must be zero,

$$W(n, n - \tau) = 0, \quad \text{if} \quad \tau \neq 0 . \tag{11.88}$$

This provided a key test of the validity of the new quantum formalism.

Let us first evaluate W given the requirement (11.88). We are interested in the lowest order solutions and so we can neglect terms in λ and its higher powers. Therefore, we can neglect the term in $\tfrac{1}{3}\lambda m x^3$ in the expression for the total energy. Likewise, in the expansion (11.56), the only terms which do not depend upon λ are those in $a(n, n - 1)\cos\omega(n, n - 1)t$, in other words, transitions in which n changes only by one. Therefore, the only terms which survive in (11.87) are those in $W(n, n)$, $W(n, n - 2)$ and $W(n, n + 2)$. In fact, Aitchison and his colleagues show that $W(n, n - 2)$ and $W(n, n + 2)$ both sum to zero and so the only surviving terms are those in $W(n, n)$. Therefore, we need only evaluate $W(n, n)$ for which $\tau = 0$ and so

$$W = \tfrac{1}{2}m\dot{x}^2 + \tfrac{1}{2}m\omega_0^2 x^2$$

$$= \tfrac{1}{2}m \sum_{\tau'} \omega(n, n - \tau')\omega(n, \, n - \tau')x(n, n - \tau')\, x(n - \tau', n)$$

$$+ \tfrac{1}{2}m\omega_0^2 \sum_{\tau'} x(n, n - \tau')\, x(n - \tau', n)\,. \tag{11.89}$$

To find the total energy, we sum over the only surviving terms in τ', $\tau' = \pm 1$. The ωs can now be awarded superscripts (0) since all terms in λ and its higher orders are omitted in the lowest order. Therefore,

$$W = \tfrac{1}{2}m \left\{ [\omega^{(0)}(n, n - 1)\omega^{(0)}(n, \, n - 1)x(n, n - 1)\, x(n - 1, \, n)] \right.$$

$$\left. + [\omega^{(0)}(n, n + 1)\omega^{(0)}(n, \, n + 1)x(n, n + 1)\, x(n + 1, \, n)] \right\}$$

$$+ \tfrac{1}{2}m\omega_0^2 \left\{ x(n, n - 1)\, x(n - 1, \, n) + x(n, n + 1)\, x(n + 1, \, n) \right\}\,. \tag{11.90}$$

From (11.65), $x(n, n + 1) = x(n + 1, \, n)$ and we can now replace the xs by the $a^{(0)}$s according to (11.69). Also, because the transitions are all equally spaced in frequency, $\omega^{(0)}(n, n - 1) = \omega^{(0)}(n, n + 1) = \omega_0$. Therefore,

$$W = \tfrac{1}{2}m\omega_0^2 \left\{ \tfrac{1}{4}[a^{(0)}(n, n - 1)]^2 + \tfrac{1}{4}[a^{(0)}(n, n + 1)]^2 \right\}$$

$$+ \tfrac{1}{2}m\omega_0^2 \left\{ \tfrac{1}{4}[a^{(0)}(n, n - 1)]^2 + \tfrac{1}{4}[a^{(0)}(n, n + 1)]^2 \right\}\,. \tag{11.91}$$

Finally, we insert the solutions for the $a^{(0)}$s from (11.75) which become

$$a^{(0)}(n, n - 1) = \beta\sqrt{n} \quad \text{and} \quad a^{(0)}(n + 1, n) = \beta\sqrt{n + 1} \quad \text{where} \quad \beta = (h/\pi m\omega_0)^{1/2}\,. \tag{11.92}$$

The final result is

$$W = \frac{h\omega_0}{2\pi}\left(n + \tfrac{1}{2}\right)\,. \tag{11.93}$$

Heisenberg immediately notes that this expression for the energy of the harmonic oscillator differs from the 'classical' result $W = nh\omega_0/2\pi$ because of the inclusion of what is now referred to as the 'zero point energy' $\tfrac{1}{2}h\omega_0/2\pi$.

The above analysis is only to zero-order in the small parameter λ. Heisenberg does not proceed to evaluate the higher order corrections of the nonlinear oscillator (11.46), but rather treats the simpler example

$$\ddot{x} + \omega_0^2 x + \lambda x^3 = 0\,, \tag{11.94}$$

the reason being that solutions only contain 'odd' terms:

$$x = a_1 \cos \omega t + \lambda a_3 \cos 3\omega t + \lambda^2 a_5 \cos 5\omega t + \cdots \tag{11.95}$$

Preserving terms up to λ^2, he finds the result

$$W = \frac{h\omega_0}{2\pi}\left(n + \tfrac{1}{2}\right) + \lambda \frac{3\left(n^2 + n + \tfrac{1}{2}\right)h^2}{32\pi^2\omega_0^2 m} - \lambda^2 \frac{h^3}{512\pi^3\omega_0^5 m^2}\left(17n^3 + \tfrac{51}{2}n^2 + \tfrac{59}{2}n + \tfrac{21}{2}\right)\,. \tag{11.96}$$

He also makes the important parenthetical remark,

'(I could not prove in general that all periodic terms actually vanish, but this was the case for all terms evaluated)'

but gives no details about which terms he tested. As discussed in the context of (11.87) and (11.88), a key requirement of the theory is to ensure that energy is conserved. Aitchison and his colleagues (2004) demonstrate in Appendix B of their paper how these calculations can be performed using Heisenberg's approach and demonstrate that up to order λ the terms $W(n, n - \alpha)$ are indeed zero, if $\alpha \neq 0$. They also carry out the equivalent analysis which led to (11.96) for the case analysed by Heisenberg with the nonlinear term λx^2 and find the result to order λ^2,

$$W = \frac{h\omega_0}{2\pi} \left(n + \tfrac{1}{2}\right) - \frac{5\lambda^2 h^2}{48\pi^2 m\omega_0^4} \left(n^2 + n + \tfrac{11}{30}\right) . \tag{11.97}$$

11.6 The simple rotator

The final part of Sect. 3 of Heisenberg's paper concerns the case of an electron in a circular orbit of radius a about the nucleus. This turns out to be a much simpler calculation. The classical quantum condition is $\oint p \, dq = nh$ and in this case $p = mva$, the angular momentum, and $dq = d\theta$:

$$nh = \oint mva \, d\theta = \oint m\omega a^2 \, d\theta . \tag{11.98}$$

Differentiating with respect to n

$$h = \frac{d}{dn}(2\pi ma^2\omega) . \tag{11.99}$$

Using Kramer's and Heisenberg's version of Born's correspondence rule, this expression translates into

$$h = 2\pi m[a^2\omega(n + 1, n) - a^2\omega(n, n - 1)] . \tag{11.100}$$

Just as in the case of the nonlinear oscillator, the solution by inspection is

$$\omega(n, n - 1) = \frac{h(n + \text{constant})}{2\pi ma^2} . \tag{11.101}$$

Again, since there should be no amplitude for transitions from the ground state to lower states, the constant is zero and so

$$\omega(n, n - 1) = \frac{hn}{2\pi ma^2} . \tag{11.102}$$

The energy of the electron is $W = \tfrac{1}{2}mv^2$ and so, using the same procedures as in Sect. 11.5, the energy is

$$W = \frac{m}{2}a^2 \frac{\omega^2(n, n - 1) + \omega^2(n + 1, n)}{2} = \frac{h^2}{8\pi^2 ma^2} \left(n^2 + n + \tfrac{1}{2}\right) . \tag{11.103}$$

As Heisenberg points out, this expression again satisfies the condition $\omega(n, n-1) = (2\pi/h)[W(n) - W(n-1)]$. This result can also be written in the suggestive form

$$W = \frac{h^2}{8\pi^2 I}\left[n(n+1) + \tfrac{1}{2}\right], \qquad (11.104)$$

where $I = ma^2$ is the moment of inertia of the electron in its orbit about the nucleus.[5]

The importance of this calculation for Heisenberg was that the expression (11.104) was in excellent agreement with the band spectra measured by Adolf Kratzer (1922). Kratzer had made a detailed analysis of the cyanide spectroscopic bands and found that he had to introduce half-integral quantum numbers to account for details of these rotational spectra. Heisenberg fully appreciated that his new quantum formalism automatically resulted in half-integral quantisation.

The last paragraph of Heisenberg's paper reads as follows:

> 'Whether a method to determine quantum-theoretical data using relations between observable quantities, such as that proposed here, can be regarded as satisfactory in principle, or whether this method after all represents far too rough an approach to the physical problem of constructing a theoretical quantum mechanics, an obviously very involved problem at the moment, can be decided only by a more intensive mathematical investigation of the method which has been very superficially employed here.'

11.7 Reflections

Heisenberg's caution in his concluding paragraph contrasts dramatically with its revolutionary content. As we will see, the theory was rapidly taken up by the leading theorists and converted into much more accessible form. There can be no doubt that Born fully appreciated the importance of what Heisenberg had achieved and made the correct decision to forward it to the *Zeitschrift für Physik* for immediate publication as soon as he had studied it in detail.

Heisenberg was not happy with the paper, which contrasted with his previous papers in which well-posed problems were solved using established mathematical procedures. In his discussions with Pauli, he considered the 'positive' part of the paper to be 'poor'. As remarked by Mehra and Rechenberg (1982b)

> 'Now, instead of a complete formulation of a consistent atomic theory, the future quantum mechanics, Heisenberg felt that he was able to present only a step in formulating its foundations, but by no means the final solution. ... He had just been able to write down a few formal equations, which were difficult to handle mathematically and even more difficult to interpret physically.'

What is staggering about the paper is how many ideas were to prove to be absolutely correct. Let us review these achievements.

1. By far the most significant and original idea was the concept that what had gone wrong with the old quantum theory was its kinematical content.
2. This concept was coupled with the innovation that the quantum theory had to be written in terms of the properties of atomic systems which were measurable.
3. This led to the concept, modelled on classical electromagnetism, that the spatial coordinates were to be replaced by 'transition amplitudes', the squares of which would determine the probability of such transitions taking place.
4. Guided by the correspondence principle, the product law for the transition amplitudes was found to be non-commutative.
5. In application to the specific problems which could be treated with his new formalism, the quantum expression for the energy levels of an oscillator and a simple rotator had a 'zero-point energy' of $h\nu/2$.
6. In contrast to the kinematics, which was changed radically, the dynamics of particles and oscillators remained 'Newtonian'.
7. As pointed out by Aitchison and his colleagues (2004), many of the results derived by Heisenberg have their exact analogues in the completed version of quantum mechanics. For example, (11.54) illustrates a basic rule of quantum mechanics that a transition amplitude for a transition from one state to another is found by summing over all possible intermediate states.

Despite its problems, Heisenberg's paper is undoubtedly one of the great papers in theoretical physics. Almost immediately, it opened up routes to more powerful mathematical approaches. Heisenberg could only have achieved so much under the joint influences of the intuitive approach of Bohr and the more mathematical approaches of Born and Sommerfeld.

12 Matrix mechanics

12.1 Born's reaction

In his reminiscences, Born recounted his memories of these exciting days (Born, 1978):

> 'Meanwhile Heisenberg pursued some work of his own, keeping its idea and purpose somewhat dark and mysterious. Towards the end of the summer semester, in the first days of July 1925, he came to me with a manuscript and asked me to read it and decide whether it was worth publishing... He added that though he had tried hard, he could not make any progress beyond the simple considerations contained in his paper, and he asked me to try myself, which I promised...
>
> His most audacious step consists in the suggestion of introducing the transition amplitudes of the coordinates q and momenta p in the formulae of mechanics...
>
> I was most impressed by Heisenberg's considerations, which were a great step forward in the programme which we had pursued...
>
> After having sent Heisenberg's paper to *Zeitschrift für Physik* for publication, I began to ponder about his symbolic multiplication, and was soon so involved in it that I thought the whole day and could hardly sleep at night. For there was something fundamental behind it... And one morning... I suddenly saw the light: Heisenberg's symbolic multiplication was nothing but matrix calculus, well known to me since my student days from the lectures of Rosanes at Breslau.'

At that time, matrices and matrix algebra were regarded as the province of the mathematicians and there were few examples of their application in physics. Heisenberg was certainly unaware of matrices, as he himself confessed. In fact, Born was one of the few physicists who had used matrices in his earlier researches with von Kármán on the theory of crystal lattices (Born and von Kármán, 1912). He had therefore been familiar with matrix algebra, with infinite matrices and the techniques of transforming infinite quadratic forms to principal axes. Born had moved away from these pursuits, but now he recalled his earlier activities.

What Born recognised was that Heisenberg's new multiplication rule (11.26) for transition amplitudes corresponded to the multiplication rule for matrices, namely,

$$x^{(2)}(n, n - \tau) = \sum_{\tau'} x(n, n - \tau') x(n - \tau', n - \tau) . \tag{12.1}$$

In addition, it was well known that the multiplication rule for matrices is non-commutative, exactly the feature of his scheme which had caused Heisenberg so much concern. What Born did was to change slightly Heisenberg's notation, by setting the transition amplitude

$q(n, n + \tau) \equiv q(n, m)$ and then regarding $q(n, m)$ as the elements of a matrix $\mathbf{q}$. Then, when the matrix product $\mathbf{pq}$ of two matrices $\mathbf{p}$ and $\mathbf{q}$ is taken, it is not necessarily equal to $\mathbf{qp}$.

The calculation which Born carried out can be appreciated from Heisenberg's product rule (11.28) for the product of two transition amplitudes $x(n, n - \tau)$ and $y(n, n - \tau)$:

$$x(n, n - \tau)\, y(n, n - \tau)\, \mathrm{e}^{\mathrm{i}\omega(n,n-\tau)t} \equiv \sum_{\tau'} x(n, n - \tau')\, y(n - \tau', \, n - \tau)\, \mathrm{e}^{\mathrm{i}\omega(n,n-\tau)t} \,. \quad (12.2)$$

Born immediately made the significant step of defining a quantised 'momentum amplitude' through the definition

$$p \equiv p(n, n - \tau) = m\dot{x}(n, n - \tau)\,, \quad (12.3)$$

by analogy with Heisenberg's transition amplitudes. Thus, if $x = x(n, n - \tau) = \sum_{\tau} x(n, n - \tau)\, \mathrm{e}^{\mathrm{i}\omega(n,n-\tau)t}$, then

$$p = m\dot{x}(n, n - \tau) = m \sum_{\tau} x(n, n - \tau)\, \mathrm{i}\omega(n, n - \tau)\, \mathrm{e}^{\mathrm{i}\omega(n,n-\tau)t} \,. \quad (12.4)$$

In so doing, Born had immediately introduced the pair of canonically conjugate space and momentum variables in their quantum guise, p and q. Therefore, using the rule (12.2), the product px is given by

$$px = m \sum_{\tau'} x(n, n - \tau')\, \mathrm{i}\omega(n, n - \tau')\, \mathrm{e}^{\mathrm{i}\omega(n,n-\tau')t} x(n - \tau', n - \tau)\, \mathrm{e}^{\mathrm{i}\omega(n-\tau',n-\tau)t} \quad (12.5)$$

$$= m \sum_{\tau'} x(n, n - \tau')\, \mathrm{i}\omega(n, n - \tau')\, x(n - \tau', n - \tau)\, \mathrm{e}^{\mathrm{i}\omega(n,n-\tau)t} \,, \quad (12.6)$$

where $\sum_{\tau'}$ means 'take the sum over all integral values of τ' from $-\infty$ to $+\infty$'. Now, let us evaluate this sum for the case $\tau = 0$, corresponding to the time-invariant value of the product of p and q and to the diagonal elements of the matrix product. Then,

$$px = m \sum_{\tau'} x(n, n - \tau')\, \mathrm{i}\omega(n, n - \tau')\, x(n - \tau', n)\,, \quad (12.7)$$

and so,

$$px - xp = m \sum_{\tau'} x(n, n - \tau')\, \mathrm{i}\omega(n, n - \tau')\, x(n - \tau', n)$$
$$- m \sum_{\tau'} x(n, n - \tau')\, \mathrm{i}\omega(n - \tau', n)\, x(n - \tau', n)\,. \quad (12.8)$$

We now use the rules $\omega(n, n - \tau') = -\omega(n - \tau', n)$ from Bohr's frequency condition and $x(n, n - \tau') = x^*(n, n - \tau')$ because the x should be real. Let us first sum over all values of $\tau' > 0$. Then, (12.8) becomes

$$px - xp = 2m\mathrm{i} \sum_{\tau'>0} \omega(n, n - \tau')\, |x(n - \tau', n)|^2 \,. \quad (12.9)$$

We obtain a similar result if we sum over all values of $\tau' < 0$, but let us now write $\tau' = -\tau'(+)$, where the $\tau'(+)$ are positive integers. Then,

$$px - xp = 2mi \sum_{\tau' < 0} \omega(n, n - \tau') |x(n - \tau', n)|^2$$

$$= \sum_{\tau'(+) > 0} \omega(n, n + \tau'(+)) |x(n + \tau'(+), n)|^2 . \qquad (12.10)$$

Combining (12.9) and (12.10), we evidently obtain the result

$$px - xp = 2mi \sum_{\tau' > 0} \left[\omega(n, n - \tau') |x(n - \tau', n)|^2 - \omega(n + \tau', n) |x(n + \tau', n)|^2 \right].$$

$$(12.11)$$

This expression can now be compared with Heisenberg's quantum relation (11.40),

$$h = 4\pi m \sum_0^{+\infty} \left[|x(n, n + \tau)|^2 \omega(n, n + \tau) - |x(n, n - \tau)|^2 \omega(n, n - \tau) \right] . \quad (12.12)$$

It immediately follows that

$$px - xp = \frac{h}{2\pi i} . \qquad (12.13)$$

Born appreciated the deep significance of this result – the non-commutativity of the product $px - xp$ was directly related to quantum processes at the atomic level through the appearance of Planck's constant in (12.13).

He was well aware of the fact that the rule (12.13) only applied to the diagonal elements of the matrix product, but he guessed that all the off-diagonal terms had to be zero. Then, the non-commutativity rule could be written

$$\mathbf{px} - \mathbf{xp} = \frac{h}{2\pi i} \mathbf{I} , \qquad (12.14)$$

where $\mathbf{I}$ is the unit matrix, but he was unable prove this conjecture. He needed help.

12.2 Born and Jordan's matrix mechanics

Born's first reaction was to ask Pauli to collaborate with him in developing the mathematical physics of matrix algebra as applied to quantum mechanics, but Pauli was dismissive of the suggestion – he did not believe in non-commutativity and preferred to let Heisenberg continue the line of research he had pioneered. He was also opposed to Born's formal mathematical approach to quantum physics. He told Born

> 'Yes, I know, you are fond of tedious and complicated formalisms. You will spoil Heisenberg's physical ideas by your futile mathematics.'

Instead, Born turned to Pascual Jordan, another remarkably talented product of the Göttingen school of mathematical physics. Jordan had attended Courant's courses on mathematics and had assisted him in the preparation of his classic text with Hilbert *The Methods of Mathematical Physics* (Courant and Hilbert, 1924). As a result of this work, he had some familiarity with the theory of matrices, although he did not claim any particular speciality in the field. He had already helped Born in the preparation of his review of the dynamics of crystal lattices (Born, 1923) and written a paper with Born in which van Vleck's theory of dispersion was extended to aperiodic motions (Born and Jordan, 1925b). He immediately took up Born's challenge and within a few days proved Born's conjecture concerning the non-commutativity of (12.14) using matrix methods. Born and Jordan now took up the challenge of attempting to work out a fully self-consistent system of quantum mechanics using matrix methods. Born was exhausted and suffered a minor breakdown during the succeeding month and so much of the analysis was carried out by Jordan alone. In a remarkably short space of time, they made huge progress towards the first complete exposition of quantum mechanics using matrix methods. Their important paper was submitted only 60 days after Heisenberg's (Born and Jordan, 1925b).

Their agenda is clearly set out in the introduction to their paper

> '... it is in fact possible, starting with the basic premises given by Heisenberg, to build up a closed theory of quantum mechanics which displays strikingly close analogies with classical mechanics, but at the same time preserves characteristic features of quantum phenomena.'

Born's objective was to rewrite all the equations of classical physics in matrix notation so that the key concept of non-commutativity would automatically be incorporated in the new quantum mechanics. Let us review the remarkable contents of Born and Jordan's paper.

12.2.1 Matrix algebra

Matrices and matrix algebra had been highly developed by the mathematicians but had, as yet, found few applications in mathematical physics. They appear in Chapter 1 of Courant and Hilbert's *The Methods of Mathematical Physics* and in Bôcher's *Algebra* (Bôcher, 1911), which were consulted by Jordan.[1] In Chap. 1 of Born and Jordan's paper, the elementary operations of matrix calculations were reviewed. We recall only the essential features of these calculations, emphasising how Born and Jordan had to develop the standard procedures.

First, they used a slightly different notation from the usual use of suffices to identify the matrix elements, the objective being to make the notation similar to that used by Heisenberg, so that, as noted above $x(n, n - \tau) \equiv x(n, m)$. We will adopt the following notation:

$$\mathbf{a} = [a(nm)] = \begin{bmatrix} a(00) & a(01) & a(02) & \dots \\ a(10) & a(11) & a(12) & \dots \\ a(20) & a(21) & a(22) & \dots \\ \dots & \dots & \dots & \dots \end{bmatrix}. \tag{12.15}$$

Then the standard texts show how many of the operations are similar to those of ordinary algebra, an important exception being the rule for matrix multiplication. Thus,

- The equality of two matrices:

$$\mathbf{a} = \mathbf{b} \quad \text{means} \quad a(nm) = b(nm) \, . \tag{12.16}$$

- Matrix addition:

$$\mathbf{a} = \mathbf{b} + \mathbf{c} \quad \text{means} \quad a(nm) = b(nm) + c(nm) \, . \tag{12.17}$$

- Matrix multiplication:

$$\mathbf{a} = \mathbf{bc} \quad \text{means} \quad a(nm) = \sum_{k=0}^{k=\infty} b(nk)c(km) \, . \tag{12.18}$$

This is the rule Born recognised in Heisenberg's new multiplication rule for transition amplitudes. Powers are defined by repeated matrix multiplication.

- The associative rule for matrix multiplication and the distributive rule for combined matrix multiplication and addition are

$$(\mathbf{ab})\mathbf{c} = \mathbf{a}(\mathbf{bc}) \, ; \tag{12.19}$$

$$\mathbf{a}(\mathbf{b} + \mathbf{c}) = \mathbf{ab} + \mathbf{ac} \, . \tag{12.20}$$

- Non-commutativity of matrix multiplication means that, in general,

$$\mathbf{ab} \neq \mathbf{ba} \, . \tag{12.21}$$

If it turns out that $\mathbf{ab} = \mathbf{ba}$, the matrices $\mathbf{a}$ and $\mathbf{b}$ are said to commute.

- The unit matrix $\mathbf{I}$ is defined by

$$\mathbf{I} \equiv [\delta(n, m)] \qquad \begin{cases} \delta(n, m) = 0 & \text{if } n \neq m \\ \delta(n, n) = 1 & \text{if } n = m \, . \end{cases} \tag{12.22}$$

It follows that

$$\mathbf{aI} = \mathbf{Ia} = \mathbf{a} \, . \tag{12.23}$$

- The reciprocal matrix is defined by

$$\mathbf{a}^{-1}\mathbf{a} = \mathbf{aa}^{-1} = \mathbf{I} \, , \tag{12.24}$$

provided the determinant of $\mathbf{a}$ is not zero. As shown in the standard textbooks, the elements of $\mathbf{a}^{-1}$ are found by forming the adjoint matrix (adj $\mathbf{a}$), the elements of which are the transposed cofactors of $a(nm)$. Then, $\mathbf{a}(\text{adj } \mathbf{a}) = (\text{adj } \mathbf{a})\mathbf{a} = |\mathbf{a}|\mathbf{I}$.

- Differentiation with respect to a parameter t. If the elements of the matrices $\mathbf{a}$ and $\mathbf{b}$ are functions of a parameter t, then differentiation of an element of the product $\mathbf{ab}$ is given by

$$\frac{\mathrm{d}}{\mathrm{d}t} \sum_k a(nk)\, b(km) = \sum_k \left\{ \dot{a}(nk)\, b(km) + a(nk)\, \dot{b}(km) \right\} \, . \tag{12.25}$$

This can be rewritten

$$\frac{\mathrm{d}}{\mathrm{d}t}(\mathbf{ab}) = \dot{\mathbf{a}}\mathbf{b} + \mathbf{a}\dot{\mathbf{b}} \, . \tag{12.26}$$

Note the importance of the order of the differentiations because of the non-commutativity of the matrix product.

- Repeated application of (12.26) results in the following rule

$$\frac{d}{dt}(\mathbf{x}_1\mathbf{x}_2\ldots\mathbf{x}_n) = \dot{\mathbf{x}}_1\mathbf{x}_2\ldots\mathbf{x}_n + \mathbf{x}_1\dot{\mathbf{x}}_2\ldots\mathbf{x}_n + \cdots + \mathbf{x}_1\mathbf{x}_2\ldots\dot{\mathbf{x}}_n . \tag{12.27}$$

- Finally, we can define functions of matrices, for example,

$$\mathbf{f}_1(\mathbf{y}_1,\ldots\mathbf{y}_m;\mathbf{x}_1,\ldots\mathbf{x}_n) = 0 ,$$

$$\ldots\ldots\ldots\ldots\ldots\ldots\ldots\ldots \tag{12.28}$$

$$\mathbf{f}_n(\mathbf{y}_1,\ldots\mathbf{y}_m;\mathbf{x}_1,\ldots\mathbf{x}_n) = 0 .$$

These results were all in the literature and they were quickly used by Jordan to demonstrate the correctness of Born's conjecture (12.14). For the case of the one-dimensional nonlinear harmonic oscillator, Jordan rewrote the equation of motion as follows:

$$\ddot{\mathbf{q}} + \omega_0^2\mathbf{q} + \lambda\mathbf{q}^n = 0 , \qquad n = 2, 3\ldots \tag{12.29}$$

Multiplying (12.29) by $m\mathbf{q}$, first from the right, then from the left and subtracting these matrix relations, Jordan found the result

$$m(\ddot{\mathbf{q}}\mathbf{q} - \mathbf{q}\ddot{\mathbf{q}}) = 0 , \tag{12.30}$$

where 0 is the infinite zero matrix. We now use (12.26) to write

$$m\frac{d(\dot{\mathbf{q}}\mathbf{q})}{dt} = m\frac{d\dot{\mathbf{q}}}{dt}\mathbf{q} + m\dot{\mathbf{q}}\dot{\mathbf{q}} = m\ddot{\mathbf{q}}\mathbf{q} + m\dot{\mathbf{q}}^2 . \tag{12.31}$$

Similarly,

$$m\frac{d(\mathbf{q}\dot{\mathbf{q}})}{dt} = m\dot{\mathbf{q}}^2 + m\mathbf{q}\ddot{\mathbf{q}} . \tag{12.32}$$

Substituting (12.31) and (12.32) into (12.30), we find

$$m\frac{d}{dt}(\dot{\mathbf{q}}\mathbf{q} - \mathbf{q}\dot{\mathbf{q}}) = 0 . \tag{12.33}$$

Finally, in the spirit of Born's extension of the concept of transition amplitudes to momenta, we write $\mathbf{p} = m\dot{\mathbf{q}}$, and so

$$\frac{d}{dt}(\mathbf{p}\mathbf{q} - \mathbf{q}\mathbf{p}) = 0 . \tag{12.34}$$

Now, the off-diagonal terms of the product matrices $\mathbf{pq}$ and $\mathbf{qp}$ all contain exponential time-varying terms of the form $\exp[(2\pi i/h) \times (E_n - E_m)t]$ where E_n and E_m are the energies of the stationary states n and m and so the only way in which (12.34) can be satisfied is if the off-diagonal terms are all zero. This result, and the speed with which Jordan obtained it, greatly pleased Born. They agreed to develop the full theory of matrix mechanics collaboratively.

As remarked by Jammer (1989),

'The equations of motion have shown that $(\mathbf{pq} - \mathbf{qp})$ is a diagonal matrix. That all diagonal elements are equal to $h/2\pi i$ has been a consequence of the correspondence principle. Moreover, since [(12.14)], as Born soon realised, is the only fundamental equation in which h appears, the introduction of Planck's constant into quantum mechanics has likewise been a consequence of the correspondence principle.'

Reflecting on these calculations, Born (1978) later remarked,

'I shall never forget the thrill I experienced when I succeeded in condensing Heisenberg's ideas on quantum conditions into the mysterious equation $(\mathbf{pq} - \mathbf{qp}) = (h/2\pi\,\mathrm{i})\mathbf{I}$.'

12.2.2 Matrix dynamics

With these new insights, Born and Jordan set about rewriting the laws of dynamics in matrix form. While the results obtained above could be obtained from the standard textbooks, the next steps involved the development of a formalism corresponding to the differentiation of one matrix with respect to another. Again, care has to be taken since the variables do not commute under matrix multiplication. Therefore, each matrix product has to be written out and the correct order of differentiation preserved. For example, if $\mathbf{y}$, $\mathbf{x}_1$, $\mathbf{x}_2$ and $\mathbf{x}_3$ are matrices and if, say,

$$\mathbf{y} = \mathbf{x}_1^2\mathbf{x}_2\mathbf{x}_1\mathbf{x}_3, \quad \text{then} \quad \frac{\partial \mathbf{y}}{\partial \mathbf{x}_1} = \mathbf{x}_1\mathbf{x}_2\mathbf{x}_1\mathbf{x}_3 + \mathbf{x}_2\mathbf{x}_1\mathbf{x}_3\mathbf{x}_1 + \mathbf{x}_3\mathbf{x}_1^2\mathbf{x}_2\,, \quad (12.35)$$

where we have used the permutation rule that $\mathbf{x}_1\mathbf{x}_1\mathbf{x}_2\mathbf{x}_1\mathbf{x}_3 = \mathbf{x}_1\mathbf{x}_2\mathbf{x}_1\mathbf{x}_3\mathbf{x}_1 = \mathbf{x}_1\mathbf{x}_3\mathbf{x}_1^2\mathbf{x}_2$ to bring each $\mathbf{x}_1$ to the front of the queue for differentiation. These and further rules for matrix manipulation completed Sect. 2 of Chap. 1 of Born and Jordan's paper.

Chapter 2 introduces the laws of dynamics in matrix form. Immediately, they state that the dynamical system is to be defined by a spatial coordinate $\mathbf{q}$ and momentum $\mathbf{p}$ according to the definitions

$$\mathbf{q} = \left[q(nm)\,\mathrm{e}^{2\pi\,\mathrm{i}\,\nu(nm)t}\right]\,, \qquad \mathbf{p} = \left[p(nm)\,\mathrm{e}^{2\pi\,\mathrm{i}\,\nu(nm)t}\right]\,. \qquad (12.36)$$

Whereas Heisenberg had developed his arguments starting from Newton's laws of motion, Born and Jordan replaced these by the more powerful methods of Hamiltonian dynamics which had considerable success in the old quantum theory. Like Heisenberg, they only considered systems of one degree of freedom, one of their intentions being to demonstrate that this more rigorous formal approach could reproduce Heisenberg's results and so create a formally self-consistent theory of quantum mechanics in which the dynamical entities were matrices.

Immediately, they noted that, in general, the $q(nm)$ and $p(nm)$ are complex numbers. As a result, these matrices must be *Hermitian matrices* in which, on transposition, each element goes over into its complex conjugate. As a result, it follows that

$$q(nm)\,q(mn) = |q(nm)|^2 \quad \text{and} \quad \nu(nm) = -\nu(mn)\,. \qquad (12.37)$$

Then, in the case of Cartesian coordinates, they identified $|q(nm)|^2$ as a measure of the probability of the transitions $n \leftrightarrow m$.

Next, they translated the Hamiltonian equations of classical mechanics, which we discussed in Sect. 5.4.3, into matrix form. The Hamiltonian (5.64) can be written in matrix form as

$$H = \frac{1}{2m}\mathbf{p}^2 + \mathbf{U}(\mathbf{q}) . \tag{12.38}$$

Then, the equations of motion can be written in canonical form

$$\begin{cases} \dot{\mathbf{q}} = \dfrac{\partial \mathbf{H}}{\partial \mathbf{p}} , \\[2mm] \dot{\mathbf{p}} = -\dfrac{\partial \mathbf{H}}{\partial \mathbf{q}} . \end{cases} \tag{12.39}$$

They then showed that this formulation leads back to Heisenberg's quantisation rule (11.40) and to the Thomas–Kuhn expression (10.34).

Having set out the basic elements of matrix mechanics, Born and Jordan next state,

> 'The content of the previous paragraphs furnishes the basic rules of the new quantum mechanics in their entirely. All other laws of quantum mechanics, whose general validity is to be verified, must be *derivable* from these basic tenets. As instances of such laws to be proved, the law of conservation of energy and the Bohr frequency condition primarily enter into consideration.'

The law of conservation of energy requires that $\dot{\mathbf{H}} = 0$ and consequently, as we have shown, the Hamiltonian matrix $\mathbf{H}$ must be diagonal. Then, Heisenberg identified the diagonal elements $H(nn)$ as the energies of the stationary states with quantum number n and so Bohr's frequency condition followed immediately:

$$h\nu(nm) = H(nn) - H(mm) , \tag{12.40}$$

and the energy of the nth state is

$$W_n = H(nn) + \text{constant} . \tag{12.41}$$

Born and Jordan show formally how this comes about.

Chapter 3 of their paper concerns the application of their new approach to harmonic and anharmonic oscillators, the intention being to derive Heisenberg's result with the new matrix formalism. Suffice to say that they succeeded, but without the need to introduce some of the assumptions made by Heisenberg. For example, Heisenberg had implicitly assumed that transitions only occurred between neighbouring states, whereas the new formalism required the quantum transitions to correspond to changes of quantum number $\Delta n = \pm 1$. The final Chap. 4 of their paper concerned an early application of these concepts to electromagnetic radiation. A significant feature of this last section was the demonstration that the squares of the amplitudes of the matrix elements representing the dipole moment of the transition determine the transition probabilities.

As already noted, Born was recovering from exhaustion during the elaboration of these concepts and, whilst he was the instigator of the programme and introduced the concepts of the matrix multiplication rule and the basic quantum relation (12.14), the detailed elaboration of the rules of matrix mechanics and dynamics was largely Jordan's work. Once the paper was submitted, the collaboration with Heisenberg resumed and resulted in the definitive paper describing the full theory of matrix mechanics by Born, Heisenberg and Jordan (1926).

12.3 Born, Heisenberg and Jordan (1926) – the Three-Man Paper

It is remarkable that the 'Three-Man Paper' ended up being as coherent as it did since there was actually remarkably little face-to-face contact between the three authors. After the completion of his revolutionary paper of 1925, Heisenberg spent nearly a month and a half in Munich and a walking tour in the Alps. Then, rather than return to Göttingen, he went back to Copenhagen and only returned to Göttingen in late October. Born returned to Göttingen following his period of recuperation and then departed for his trip as a 'foreign lecturer' to the Massachusetts Institute of Technology on 31 August 1926. Thus, much of the work on the paper, which was received for publication by the *Zeitschrift für Physik* on 15 November 1926, was carried out by correspondence. Equally remarkable is the fact that this long paper *On quantum mechanics II* was a remarkably complete exposition of the theory of matrix mechanics as applied to quantum physics and dealt with many of the incompletenesses of Heisenberg's and Born and Jordan's earlier papers (Born *et al.*, 1926).

The fact that the theory was developed so rapidly can be attributed to a number of factors. Although Heisenberg confessed that he had not appreciated that non-commutativity is a key feature of the mathematics of matrices, as soon as he heard of Born and Jordan's successes, he was an instant convert and quickly learned and applied these tools to the various problems he had worked on. Born and Jordan were delighted that he joined their collaboration since he brought to the research quite different skills and approaches to those of Born and Jordan. The second feature of the collaboration was that all three were experts in the classical theory of multiply periodic systems, the perturbation schemes used and the theory of canonical transformations. These classical techniques now had to be translated into the language of matrix mechanics. The third piece of good fortune was that the theory of infinite quadratic forms had been developed by Hilbert, Born's teacher, 20 years earlier. Born not only had a deep understanding of these methods, but also realised that the matrix theory of canonical transformations was identical with the theory of principal-axes transformations which would turn out to be the route to the calculation of the energy levels in atomic systems. A further bonus was that Hilbert's formalism, as extended by Ernst Hellinger, enabled discrete and continuous states of atomic systems to be treated in a unified manner.

Mehra and Rechenberg provide a succinct summary of the achievements of the Three-Man Paper:

> 'The endeavours of Born, Heisenberg and Jordan led to the development of the theory of matrix mechanics, which was applicable to all types of multiply periodic systems, to non-degenerate and degenerate ones, and in principle even to aperiodic systems. In addition, the authors realised that the matrix equations had a simpler structure than the corresponding classical equations. Thus, for example, the set of Hamilton–Jacobi partial differential equations of classical mechanics was replaced by a set of algebraic equations; their solution established a unitary transformation of the quantum-mechanical Hamiltonian matrix to a diagonal matrix. Now the discussion of conservation laws also appeared to be considerably more elementary.' (Mehra and Rechenberg, 1982c)

As discussed by van der Waerden (1967) and Mehra and Rechenberg (1982c), the ideas were developed in a flurry of correspondence between the three authors, as well as with Pauli and Bohr. Let us give some impression of the contents of the 'final' version of matrix mechanics.

12.3.1 Basic theory – systems with one degree of freedom

The introduction was written by Heisenberg and he immediately summarised the advances since his own paper (Heisenberg, 1925) and Born and Jordan's Paper I (Born and Jordan, 1925b):

> 'The present paper sets out to develop further a general quantum-theoretical mechanics whose physical and mathematical basis has been treated in two previous papers by the present authors. It was found possible to extend the above theory to systems having several degrees of freedom (Chapter 2), and by the introduction of "canonical transformations" to reduce the problem of integrating the equations of motion to a known mathematical formulation. From this theory of canonical transformations we are able to derive a perturbation theory (Chapter 1, §4) which displays close similarity to classical perturbation theory. On the other hand we were able to trace a connection between quantum mechanics and the highly-developed mathematical theory of quadratic forms of infinitely many variables (Chapter 3).'

But, they recognised that the new quantum mechanics is not visualisable as was the old quantum theory:

> 'Admittedly, such a system of quantum-theoretical relations between observable quantities, when compared with the quantum theory employed hitherto, would labour under the disadvantage of not being directly amenable to a geometrically visualizable interpretation, since the motions of the electrons cannot be described in terms of the familiar concepts of space and time.'

Despite this,

> 'If one reviews the fundamental differences between classical and quantum theory, differences which stem from the basic quantum theoretical postulates, then the formalism proposed . . . , if proved to be correct, would appear to represent a system of quantum mechanics as close to that of classical theory as could reasonably be hoped.'

Chapter 1 begins by setting out the basic formalism of matrix calculus for systems having one degree of freedom and re-introducing Born's equation (12.14),

$$\mathbf{px} - \mathbf{xp} = \frac{h}{2\pi i}\mathbf{I}. \tag{12.42}$$

They remark on the fact that this

> '. . . is the only one of the basic formulae in the quantum mechanics proposed here which contains Planck's constant h. It is satisfying that the constant h already enters into the basic tenets of the theory at this stage in so simple a form.'

This basic relation leads to a more general expression. They show that, if $\mathbf{f(pq)}$ is any function of $\mathbf{p}$ and $\mathbf{q}$, then

$$\mathbf{fq} - \mathbf{qf} = \frac{\partial \mathbf{f}}{\partial \mathbf{p}} \frac{h}{2\pi i} \, , \tag{12.43}$$

$$\mathbf{pf} - \mathbf{fp} = \frac{\partial \mathbf{f}}{\partial \mathbf{q}} \frac{h}{2\pi i} \, . \tag{12.44}$$

They demonstrate this by showing that if the formulae (12.43) and (12.44) are valid for each of the functions φ and ψ, then they must also be valid for the combinations $\varphi + \psi$ and $\varphi \cdot \psi$. The first of these is trivial, while the second follows from a simple calculation. Setting $\mathbf{f} = \varphi \cdot \psi$, (12.43) becomes

$$(\varphi \cdot \psi)\mathbf{q} - \mathbf{q}(\varphi \cdot \psi) = \varphi(\psi \mathbf{q} - \mathbf{q}\psi) + (\varphi \mathbf{q} - \mathbf{q}\varphi)\psi$$
$$= \left(\varphi \frac{\partial \psi}{\partial \mathbf{p}} + \frac{\partial \varphi}{\partial \mathbf{p}} \psi \right) \frac{h}{2\pi i} = \frac{\partial(\varphi \cdot \psi)}{\partial \mathbf{p}} \frac{h}{2\pi i} \, . \tag{12.45}$$

A similar treatment follows for $\mathbf{p}(\varphi \cdot \psi) - (\varphi \cdot \psi)\mathbf{p}$ in (12.44). Therefore, since (12.43) and (12.44) hold for $\mathbf{p}$ and $\mathbf{q}$ they must also be valid for every function $\mathbf{f}$ which can be expressed as a power series in $\mathbf{p}$ and $\mathbf{q}$.

Next, the canonical equations (12.39) are introduced and energy conservation and Bohr's frequency condition derived from the basic postulates. Central to the development is the frequency combination principle according to which

$$\nu(nm) + \nu(mk) = \nu(nk) \, , \tag{12.46}$$

which leads to the expression

$$\nu(nm) = \frac{(W_n - W_m)}{h} \, , \tag{12.47}$$

where the W_ns are the energy levels or 'energy terms'. These can be converted into a diagonal matrix $\mathbf{W}$ according to the definition

$$\mathbf{W} = [W(nm)] = [\delta_{nm} W_n] = \begin{cases} W_n \text{ for } n = m \, , \\ 0 \text{ for } n \neq m \, . \end{cases} \tag{12.48}$$

Now, for any quantum theoretical matrix $\mathbf{a}$, the elements of the matrix take the form

$$\mathbf{a} = [a(nm) e^{2\pi i \nu(nm)t}] \, , \tag{12.49}$$

and so the time derivative of $\mathbf{a}$ is

$$\dot{\mathbf{a}} = [2\pi i \nu(nm) a(nm)] \, , \tag{12.50}$$

where the exponential time dependence has been dropped since it cancels through all the formulae. We can now use (12.48) to write the energy terms as follows:

$$\mathbf{Wa} = \left[\sum_k W(nk) a(km) \right] = \left[\sum_k \delta_{nk} W_k a(km) \right] = [W_n a(nm)] \, , \tag{12.51}$$

$$\mathbf{aW} = \left[\sum_k a(nk) W(km) \right] = \left[\sum_k a(nk) \delta_{km} W_m \right] = [W_m a(nm)] \, . \tag{12.52}$$

Inserting these relations into (12.47) and (12.50), we find the general result for any quantum theoretical quantity **a**,

$$\dot{\mathbf{a}} = \frac{2\pi i}{h}(\mathbf{Wa} - \mathbf{aW}) .$$ (12.53)

Next, we need to show that this formalism is consistent with Hamilton's equations in matrix form. In (12.53), we set successively $\mathbf{a} = \mathbf{q}$ and $\mathbf{a} = \mathbf{p}$ and substitute for $\dot{\mathbf{q}}$ and $\dot{\mathbf{P}}$ in (12.39). Then, we use (12.43) and (12.44) with $\mathbf{f} = \mathbf{H}$ to determine $\partial\mathbf{H}/\partial\mathbf{q}$ and $\partial\mathbf{H}/\partial\mathbf{p}$. The result is the pair of equations

$$\mathbf{Wq} - \mathbf{qW} = \mathbf{Hq} - \mathbf{qH} \quad ; \quad \mathbf{Wp} - \mathbf{PW} = \mathbf{Hp} - \mathbf{PH} ,$$ (12.54)

which can be rewritten

$$(\mathbf{W} - \mathbf{H})\mathbf{q} - \mathbf{q}(\mathbf{W} - \mathbf{H}) = 0 \quad ; \quad (\mathbf{W} - \mathbf{H})\mathbf{p} - \mathbf{p}(\mathbf{W} - \mathbf{H}) = 0 .$$ (12.55)

Thus, $(\mathbf{W} - \mathbf{H})$ commutes with **p** and **q** and hence with every other function of $(\mathbf{p}, \mathbf{q})$. In particular, it commutes with the energy function, or Hamiltonian, **H**,

$$(\mathbf{W} - \mathbf{H})\mathbf{H} - \mathbf{H}(\mathbf{W} - \mathbf{H}) = 0 .$$ (12.56)

Hence, from (12.53), we find

$$\dot{\mathbf{H}} = 0 .$$ (12.57)

This result proves that energy is conserved and that **H** is a diagonal matrix, $H(nm) = \delta_{nm}H_n$. Consequently, from the first expression in (12.54),

$$q(nm)(H_n - H_m) = q(nm)(W_n - W_m) \quad ; \quad \frac{(H_n - H_m)}{h} = \nu(nm) .$$ (12.58)

Now, Born and his colleagues invert the argument with important consequences for the further development of the theory. If now they assume energy conservation and the frequency condition (12.58) and that the energy function **H** is an analytic function of variables **P** and **Q**, then provided

$$\mathbf{PQ} - \mathbf{QP} = \frac{h}{2\pi i}\mathbf{I} ,$$ (12.59)

the canonical equations

$$\dot{\mathbf{Q}} = \frac{\partial\mathbf{H}}{\partial\mathbf{P}} , \quad \dot{\mathbf{P}} = -\frac{\partial\mathbf{H}}{\partial\mathbf{Q}}$$ (12.60)

always apply. This leads naturally to one of the most important features of their scheme of quantum mechanics, the concept of *canonical transformations*. The above arguments suggest that a canonical transformation from variables $(\mathbf{P}, \mathbf{Q})$ to $(\mathbf{p}, \mathbf{q})$ can be written

$$\mathbf{pq} - \mathbf{qp} = \mathbf{PQ} - \mathbf{QP} = \frac{h}{2\pi i}\mathbf{I} .$$ (12.61)

Heisenberg and Born both worked on the nature of the transformations from variables $(\mathbf{P}, \mathbf{Q})$ to $(\mathbf{p}, \mathbf{q})$. Born's knowledge of matrix algebra enabled him to come up with the most general form of transformation which was of the form

$$\mathbf{P} = \mathbf{S}\mathbf{p}\mathbf{S}^{-1}, \quad \mathbf{Q} = \mathbf{S}\mathbf{q}\mathbf{S}^{-1}. \tag{12.62}$$

By the same reasoning as above, these transformations must also hold good for sums and products of $\mathbf{p}$ and $\mathbf{q}$ and so more generally,

$$\mathbf{f}(\mathbf{P}, \mathbf{Q}) = \mathbf{S}\mathbf{f}(\mathbf{p}, \mathbf{q})\mathbf{S}^{-1}. \tag{12.63}$$

The importance of this result is most simply explained in the authors' own words:

> 'The importance of the canonical transformation is due to the following theorem: If any pair of values $\mathbf{p}_0$, $\mathbf{q}_0$ be given which satisfy [(12.60)], then the problem of integrating the canonical equations for the energy function $\mathbf{H}(\mathbf{p}, \mathbf{q})$ can be reduced to the following: A function $\mathbf{S}$ is to be determined, such that when
>
> $$\mathbf{p} = \mathbf{S}\mathbf{p}_0\mathbf{S}^{-1}, \quad \mathbf{q} = \mathbf{S}\mathbf{q}_0\mathbf{S}^{-1}, \tag{12.64}$$
>
> the function
>
> $$\mathbf{H}(\mathbf{p}, \mathbf{q}) = \mathbf{S}\mathbf{H}(\mathbf{p}_0, \mathbf{q}_0)\mathbf{S}^{-1} = \mathbf{W}, \tag{12.65}$$
>
> becomes a diagonal matrix. Equation [(12.65)] is the analogue to the Hamilton partial differential equation, and in a sense stands for the action function.'

This is the key result – the diagonal elements of the matrix $\mathbf{W}$ are the stationary energy terms of the system. The problem has been reduced to the determination of the transformation of $\mathbf{H}(\mathbf{p}, \mathbf{q})$ to diagonal form and the procedures for doing this were already in the mathematical literature. The remainder of Chap. 1 of their paper was devoted to showing how the new formalism could deal with perturbation theory and time-variability of the energy function $\mathbf{H}$.

12.3.2 Systems with an arbitrary number of degree of freedom and Hermitian forms

The second chapter deals with the extension of the formalism to several degrees of freedom, $f > 1$ and involves replacing the two-dimensional matrices by $2f$-dimensional matrices corresponding to the $2f$-dimensional manifold of stationary states:

$$\mathbf{q}_k = [q_k(n_1 \ldots n_k, m_1 \ldots m_f)] \quad , \quad \mathbf{p}_k = [p_k(n_1 \ldots n_k, m_1 \ldots m_f)]. \tag{12.66}$$

The equations of motion corresponding to (12.39) can be taken over in the form

$$\dot{\mathbf{q}}_k = \frac{\partial \mathbf{H}}{\partial \mathbf{p}_k} ; \dot{\mathbf{p}}_k = -\frac{\partial \mathbf{H}}{\partial \mathbf{q}_k}. \tag{12.67}$$

In addition, the commutation relations have to be extended as follows:

$$\begin{cases} \mathbf{p}_k\mathbf{q}_l - \mathbf{q}_l\mathbf{p}_k = \dfrac{h}{2\pi \mathrm{i}}\delta_{kl}, \\ \mathbf{p}_k\mathbf{p}_l - \mathbf{p}_l\mathbf{p}_k = 0, \\ \mathbf{q}_k\mathbf{q}_l - \mathbf{q}_l\mathbf{q}_k = 0. \end{cases} \tag{12.68}$$

They demonstrated how all the results described in Sect. 12.3.1 can be extended naturally to the $2f$-dimensional case. Also in Chap. 2, they showed how the scheme can deal with degenerate as well as non-degenerate quantum systems.

Chapter 3 was written by Born. He had the advantage over Heisenberg and Jordan in having in-depth knowledge of the mathematical advances carried out by Hilbert and Hellinger which turned out to be exactly the tools needed to complete the formulation of matrix mechanics. As Born expressed it,

> 'But behind the formalism of this perturbation theory there lurks a very simple, purely algebraic connection.... Apart from the deeper insight into the mathematical structure of the theory, we thereby gain the advantage of being able to use the methods and results developed earlier in mathematics.'

Born's remark refers to his realisation that the transformation of matrices can be regarded as equivalent to a system of linear transformations for bilinear forms. The reason is that, to every matrix $\mathbf{a} = [a(nm)]$, there corresponds a *bilinear form* defined by

$$A(xy) = \sum_{nm} a(nm)x_n y_m \,, \tag{12.69}$$

of the two series of variables $x_1, x_2, \ldots, x_n$ and $y_1, y_2, \ldots, y_n$. One of the features of the matrix elements of $\mathbf{a}$ is that they should be Hermitian so that the squares of the amplitudes are real numbers. In the standard theory, Hermitian matrices are defined such that the transposed matrix $\mathbf{a}^*$ is equal to its complex conjugate $\tilde{\mathbf{a}}$, that is, $\tilde{\mathbf{a}} = \mathbf{a}^*$. In terms of bilinear forms, their Hermitian properties are defined by $a(mn) = a^*(nm)$. Then, if we write $y_n = x_n^*$, it follows that

$$A(xx^*) = \sum_{nm} a(nm)x_n x_m^* \tag{12.70}$$

is real.

The importance of these considerations is that the theory of bilinear forms had been pioneered by Hilbert and his studies were brought together in his book *Grundzüge einer allgemeinen Theorie der linearen Integralgleichungen* (Hilbert, 1912). In particular, he showed that, for a finite number of variables, it is always possible to carry out an orthogonal transformation of a bilinear form to a sum of squares, the procedure known as the *transformation to principal axes*, specifically,

$$A(xx^*) = \sum_{n} W_n y_n y_n^* \,. \tag{12.71}$$

Rewriting this expression in terms of matrices, there exists an orthogonal matrix $\mathbf{v}$ such that

$$\mathbf{v}\tilde{\mathbf{v}}^* = \mathbf{I} \quad \text{and} \quad \mathbf{va}\tilde{\mathbf{v}}^* = \mathbf{vav}^{-1} = \mathbf{W} \,, \tag{12.72}$$

where $\mathbf{W} = (W_n \delta_{nm})$ is a diagonal matrix. Born knew that the problem of the diagonalisation of Hermitian forms was equivalent to an 'eigenvalue problem' and he called the solutions for W_n the *eigenvalues* of the set of linear equations

$$W x_k - \sum H(kl)x_l = 0 \,, \tag{12.73}$$

a typical eigenvalue equation. As van der Waerden remarks, for Born the eigenvalues were mathematical tools for determining the energy levels of the atomic system. Only with the somewhat different approach of Schrödinger was it appreciated that the eigenvectors $(x_1, x_2, \ldots)$ determine the properties of the stationary states of the atom.

Born and his colleagues also appreciated that, in addition to enabling the energy levels of the atomic systems to be determined, the formalism of Hilbert and Hellinger also enabled continuous spectra to be treated. Continuous spectra are solutions of the same equations of motion, but the orthogonality relations now need to be written in terms of differential spectra rather than discrete spectral lines. As they wrote in their paper,

> 'The simultaneous appearance of both continuous and line spectra as solutions of the same equations of motion and the same commutation relations seemed to us to represent a particularly significant feature of the new theory.... there nevertheless are characteristic distinctions, both mathematically and physically, between continuous and discrete spectra, corresponding to the differences between Fourier series and Fourier integrals in classical theory.'

They illustrate the differences by considering the classical analogues of periodic and aperiodic motions. In the case of multiply periodic systems such as those considered in the old quantum theory, a Fourier series $a(\nu)$ can be associated with oscillations of the form $\exp(2\pi i \nu t)$. In contrast, for aperiodic motions, it is necessary to work in terms of Fourier integrals where $\varphi(\nu)\, d\nu$ takes the place of $a(\nu)$. In the quantum mechanical case, the quantities $q(kl)$ are replaced by the differential quantities $q(k, W)\, dW$ or $q(W, W')\, dW\, dW'$, depending on whether one or both indices lie in the continuous region, where we denote states in the continuum by the Ws. Thus, the quantities to be subject to the orthogonality conditions are differential energies rather than total energies.

It turned out that Ernst Hellinger, an old friend of Born's, had completed his doctoral dissertation at Göttingen in 1907 under Hilbert's supervision on precisely the topic of bilinear forms which could not be represented by discrete quantities $q(kl)$ (Hellinger, 1909). In cases in which $\sum_{mn} H(mn) x_m x_n^*$ cannot be converted into the expression $\sum_n W_n y_n y_n^*$ by an orthogonal transformation, it was assumed that a representation including a continuum spectrum exists:

$$\sum_{mn} H(mn) x_m x_n^* = \sum_n W_n y_n y_n^* + \int W(\varphi) y(\varphi) y^*(\varphi)\, d\varphi\,, \qquad (12.74)$$

in which the variables x_n are connected to the variables y_n and $y(\varphi)$ by an orthogonal transformation. Hellinger had shown that the corresponding orthogonality conditions for any two interval Δ_1 and Δ_2 of the continuous spectrum can be written

$$\sum_k \int_{\Delta_1} x_k(W')\, dW' \int_{\Delta_2} x_k(W'')\, dW'' = \int_{\Delta_{12}} d\varphi(W) = \phi\left(W^{(2)}\right) - \phi\left(W^{(1)}\right)\,, \quad (12.75)$$

where Δ_{12} is the interval common to both Δ_1 and Δ_2 and $W^{(2)}$ and $W^{(1)}$ are the endpoints of Δ_{12}. If there is no overlap between Δ_1 and Δ_2, the right-hand side of (12.75), $\phi(W^{(2)}) - \phi(W^{(1)})$, is zero. The orthogonal matrix for transformation to principal axes now takes the

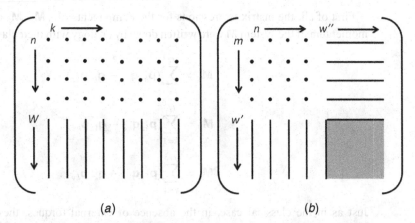

Fig. 12.1 (a) A schematic representation of the 'orthogonal matrix' **S** showing the discrete values x_{kn} and the continuous distribution $x_k(W)\,\mathrm{d}W$. (b) Matrix representing the discrete and continuum values of the momentum and coordinate matrices **p** and **q** according to matrix mechanics (Born *et al.*, 1926).

form

$$\mathbf{S} = [x_{kn}, x_k(W)\,\mathrm{d}W] \tag{12.76}$$

and is represented schematically in Fig. 12.1a.

These matrices can then be used to reduce the momentum and coordinate matrices to diagonal form by the orthogonal transformations

$$\mathbf{p} = \mathbf{S}\mathbf{p}_0\mathbf{S}^{-1}, \quad \mathbf{q} = \mathbf{S}\mathbf{q}_0\mathbf{S}^{-1}, \tag{12.77}$$

resulting in four types of elements for **p**, which are illustrated schematically in the matrix shown in Fig. 12.1b:

$$\begin{cases} p(mn) & = \sum_{kl} x_{km}^* p^0(kl) x_{ln}, \\ p(m, W)\,\mathrm{d}W & = \sum_{kl} x_{km}^* p^0(kl) x_l(W)\,\mathrm{d}W, \\ p(W, n)\,\mathrm{d}W & = \sum_{kl} x_k^*(W)\,\mathrm{d}W \cdot p^0(kl) x_{ln}, \\ p(W', W'')\,\mathrm{d}W'\,\mathrm{d}W'' & = \sum_{kl} x_k^*(W')\,\mathrm{d}W\, p^0(kl) x_l(W'')\,\mathrm{d}W''. \end{cases} \tag{12.78}$$

These correspond to four types of transition: top left, from ellipse to ellipse; top right, from ellipse to hyperbola; bottom left, from hyperbola to ellipse; bottom right, from hyperbola to hyperbola. In the language of atomic physics, these correspond to bound–bound, bound–free, free–bound and free–free transitions.

12.3.3 The quantisation of angular momentum

The fourth and final chapter of the Three-Man Paper concerned physical applications of the theory. The first part of the chapter is devoted to the laws of conservation of momentum and angular momentum and the determination of selections rules for transitions between stationary states.

First of all, the matrix expressions for the components $\mathbf{M}_x$, $\mathbf{M}_y$, $\mathbf{M}_z$ of the total angular momentum of the system $\mathbf{M}$ were written down by analogy with their classical counterparts.

$$\mathbf{M}_x = \sum_{k=1}^{f/3}(\mathbf{p}_{ky}\mathbf{q}_{kz} - \mathbf{q}_{ky}\mathbf{p}_{kz}) \,, \tag{12.79}$$

$$\mathbf{M}_y = \sum_{k=1}^{f/3}(\mathbf{p}_{kz}\mathbf{q}_{kx} - \mathbf{q}_{kz}\mathbf{p}_{kx}) \,, \tag{12.80}$$

$$\mathbf{M}_z = \sum_{k=1}^{f/3}(\mathbf{p}_{kx}\mathbf{q}_{ky} - \mathbf{q}_{kx}\mathbf{p}_{ky}) \,. \tag{12.81}$$

Just as in the classical case, in the absence of external torques, these components are conserved, but this now follows from the commutation properties of the $\mathbf{p}$s and $\mathbf{q}$s. Similarly, total linear momentum and the components of the linear momentum are conserved in the absence of forces

$$\mathbf{p} = \sum_{k=1}^{f/3}\mathbf{p}_k = \text{constant} \,, \quad \mathbf{p}_x = \sum_{k=1}^{f/3}\mathbf{p}_{kx} = \text{constant} \,, \quad \dots \tag{12.82}$$

From the commutation relations (12.68), they immediately derived the fundamental quantum mechanical relation

$$\mathbf{M}_x\mathbf{M}_y - \mathbf{M}_y\mathbf{M}_x = \frac{h}{2\pi\mathrm{i}}\mathbf{M}_z \,. \tag{12.83}$$

The total angular momentum $\mathbf{M}$ is a diagonal matrix and its square commutes with the z-component of the angular momentum:

$$\mathbf{M}^2\mathbf{M}_z - \mathbf{M}_z\mathbf{M}^2 = 0 \,. \tag{12.84}$$

The elements of the diagonal total angular momentum matrix $\mathbf{M}^2$ were shown to be $j(j + 1)(h/2\pi)^2$ where j is an integer or half-integer. Finally, the selection rules for transitions were shown to be

$$m \to m+1 \quad \text{or} \quad m \quad \text{or} \quad m-1 \,, \tag{12.85}$$

$$j \to j+1 \quad \text{or} \quad j \quad \text{or} \quad j-1 \,, \tag{12.86}$$

where the js and ms are integers or half-integers and $m \le j$. These are precisely the rules which had been derived empirically by Landé and his colleagues. The formalism also enabled the intensities and polarisations of the lines to be determined.

After a brief discussion of the Zeeman effect, the paper concludes with an analysis by Jordan of the statistics of the wave fields emitted by an ensemble of harmonic oscillators. Using the new formalism, he was able to recover Einstein's expression (3.41) for the fluctuations in black-body radiation

$$\overline{\Delta^2} = h\nu\overline{E} + \frac{\overline{E^2}}{z_\nu V} \,, \tag{12.87}$$

where $z_\nu \, d\nu$ is the number of eigenvibrations, or normal modes, in the frequency interval ν to $\nu + d\nu$.

12.3.4 Reflections

The Three-Man Paper was a quite remarkable achievement. The three authors were entering entirely new areas of theoretical physics using new mathematical tools. The extraordinary feature of their work is that they successfully worked out all the essential features of the full theory of non-relativistic quantum mechanics. But this was achieved at a price – the mathematics was known to relatively few physicists and the physical content of the theory was not yet fully understood. The authors were well aware of this. In his introduction, Heisenberg made every effort to endow the theory with physical meaning, but there was still some way to go. Fortunately, they were in continuous correspondence with their theoretical colleagues, in particular, with Pauli and Weyl. Once the clues were provided, they rapidly produced independent analyses which resulted in the same set of commutation relations. Still, the community of physicists at large was wary of the complexity of the new theory until Pauli used the new matrix methods to derive the energy levels and selection rules for the hydrogen atom.

12.4 Pauli's theory of the hydrogen atom

Pauli was sceptical about matrix mechanics, still railing against what he perceived to be Born's excessively formal mathematical approach to the problems of quantum physics. He felt that the 43-year old Born should leave the problems of quantum physics to the new generation of young physicists – in his opinion, the new physics was *Knabenphysik*, young man's physics. Even by October 1925, he was still opposed to the Göttingen approach, as he wrote to Ralph Kronig:

> 'One must first seek to liberate Heisenberg's mechanics from Göttingen's deluge of formal learning and better expose its physical essence.'

Heisenberg was not to take this lying down. Having seen Pauli's letter to Kronig, he responded:

> 'With respect to both of your last letters I must preach you a sermon, and beg your pardon for proceeding in Bavarian: It really is a pigsty that you cannot stop indulging in a slanging match. Your eternal reviling of Copenhagen and Göttingen is a shrieking scandal.... When you reproach us that we are such big donkeys that we have never produced anything new in physics, it may well be true. But then, you are also an equally big jackass because you have not accomplished it either...'

This provoked Pauli into action, showing the Göttingen physicists that he could beat them at their own game. Born, Heisenberg and Jordan were well aware that the 'crown jewels' would be the demonstration that the new theory of matrix mechanics could account for the Balmer spectrum of the hydrogen atom. Bohr's theory of the hydrogen atom was now regarded, even by Bohr himself, as an 'accident' – according to the new perspective, it was meaningless to talk about electron orbits since they did not correspond to observables. The Three-Man Paper, however much it struck right at the heart of quantum mechanics, did not produce the answer, much to their frustration.

The problem they encountered was that the matrix formalism could not be applied directly to the conjugate of the angular momentum matrix **M**. Classically, the conjugate of the action variable J is the angle variable ϕ, but there was no matrix corresponding to ϕ. In the second section of his paper, Pauli explained the problem:

> '... we must first ... develop the requisite rules for simultaneously operating with matrices **x**, **y** and **z** of the Cartesian coordinates of the electron ..., the matrix **r** of the magnitude of the radius vector, and their time derivatives. The present version of the laws of the new quantum mechanics requires that we avoid the introduction of a polar angle ϕ. Since this is not confined within finite limits, it cannot, namely, be formally represented as a matrix in the same way as the above-mentioned coordinates, which execute librations[2] in classical mechanics.'

Pauli quickly mastered the techniques of matrix mechanics and then discovered an ingenious route for circumventing the problems associated with the polar angle ϕ. He had just completed a major survey of quantum physics for the *Handbuch der Physik* and in that review he described the approach to the analysis of orbits in an inverse-square law field by his colleague at Hamburg, Wilhelm Lenz (Pauli, 1926). Lenz avoided using the angle ϕ by introducing another constant of the motion, the vector A, which was defined by

$$A = \frac{4\pi\epsilon_0}{Ze^2 m_0}[M \times p] + \frac{r}{r}, \tag{12.88}$$

where m_0 is the mass of the electron, p its linear momentum, M its angular momentum vector and r the vector from the focus to a point on the ellipse. The significance of A can be appreciated from the following calculation. Taking the scalar product of A with r,

$$A \cdot r = \frac{4\pi\epsilon_0}{Ze^2 m_0}[M \times p] \cdot r + \frac{r \cdot r}{r},$$

$$|A||r|\cos\phi = \frac{4\pi\epsilon_0}{Ze^2 m_0}[p \times r] \cdot M + |r|,$$

$$\frac{4\pi\epsilon_0}{Ze^2 m_0}\frac{|M|^2}{|r|} = 1 - |A|\cos\phi, \tag{12.89}$$

since $M = r \times p$. This is exactly the expression for the elliptical orbit of an electron in an inverse-square electrostatic field written in pedal, or (r, ϕ), form as in (5.1),[3]

$$\frac{\lambda}{r} = 1 + \varepsilon\cos\phi \quad \text{where} \quad \lambda = \frac{4\pi\epsilon_0|M|^2}{Ze^2 m_0}, \tag{12.90}$$

and ε is the eccentricity of the ellipse. This result provides a physical interpretation of the meaning of the vector A. Comparison of (12.89) and (12.90) shows that the magnitude of A is the eccentricity ε of the ellipse and that the vector points from one focus of the ellipse along the major axis towards the other focus, as can be appreciated from the geometry of Fig. 5.1.

Next, (12.90) can be written in terms of the energy W of the orbit. According to classical dynamics, the total energy of the elliptical orbit under the influence of the inverse-square law of electrostatics is

$$W = T + U = -\frac{U}{2} = -\frac{Ze^2}{8\pi\epsilon_0 a} \quad \text{where} \quad a = \frac{\lambda}{(1-\varepsilon^2)} . \tag{12.91}$$

T and U are the kinetic and electrostatic potential energies respectively and a is the length of the semi-major axis of the ellipse. This result was derived in Sect. 5.2 as expression (5.22). Substituting the value of λ from (12.90),

$$\left(1 - \varepsilon^2\right) = -W|M|^2 \frac{32\pi^2\epsilon_0^2}{Z^2 e^4 m_0} , \tag{12.92}$$

that is,

$$1 - |A|^2 = -W|M|^2 \frac{32\pi^2\epsilon_0^2}{Z^2 e^4 m_0} . \tag{12.93}$$

The importance of this result for Pauli was the fact that the orbits of the ellipses in the old quantum theory were confined to a plane and were characterised by two variables, the action–angle variables J and ϕ. Now, he had two independent vectors which are constants of the motion, M and A, which do not suffer from the problems of the angle variable. In addition, (12.93) depended only upon the values of the two constants of the motion and the energy of the orbit. Pauli realised that the energies of the stationary states could be found by converting this classical formalism into the new matrix mechanics and then the energy levels of the hydrogen atom could be determined according to the prescription of Born, Heisenberg and Jordan.

In an extraordinary *tour de force*, he now set about converting the vectors M and A into matrix form, using his own very considerable ability at 'formal mathematics' as well as ingenuity and insight in making the problem of the hydrogen atom soluble. An outline of his calculation is as follows.[4]

First of all, the angular momentum matrix was defined as $\mathbf{M} = m_e(\mathbf{r} \times \mathbf{v})$ and was shown to commute with the Hamiltonian matrix of the one-electron atom. Hence, the angular momentum was a constant of the motion. Next, from (12.88) he defined the matrix corresponding to A as

$$\mathbf{A} = \frac{4\pi\epsilon_0}{Ze^2 m_0} \frac{1}{2} [\mathbf{M} \times \mathbf{p} - \mathbf{p} \times \mathbf{M}] + \frac{\mathbf{r}}{r} . \tag{12.94}$$

Then, he established the commutation relations for $\mathbf{M}$ and $\mathbf{A}$

$$(\mathbf{M} \times \mathbf{M}) = -\frac{h}{2\pi i}\mathbf{M} , \tag{12.95}$$

$$(\mathbf{A} \times \mathbf{M}) = -\frac{h}{2\pi i}\mathbf{A} , \tag{12.96}$$

$$(\mathbf{A} \times \mathbf{A}) = \frac{h}{2\pi i}\left(Z^2 e^4 m_0\right)^{-1} 2\mathbf{H} \cdot \mathbf{M} , \tag{12.97}$$

where **H** is the Hamiltonian of the system. The matrix expression corresponding to (12.93) became

$$1 - \mathbf{A}^2 = -\frac{16\pi^2 \epsilon_0^2}{Z^2 e^4 m_0} 2\mathbf{H} \left(\mathbf{M}^2 + \frac{h^2}{4\pi^2} \mathbf{I} \right) . \tag{12.98}$$

The objective was now to find the eigenvalues associated with the solutions of (12.98) but Pauli was well aware of the fact that they would be degenerate. The same problem occurred in the old quantum theory in which the energies of ellipses with the same semi-major axes were all the same. There was, however, a standard procedure for relieving the degeneracy, which had already been used by Born and Pauli in a paper of 1922, and that was to perturb the Hamiltonian by the addition of a term of the from $\lambda \mathbf{H}_1$ which would split the degenerate energy levels. To achieve the complete separation of the levels, Pauli included two types of perturbation in $\mathbf{H}_1$, a non-Coulomb radial field and a magnetic field in the z-direction. As the small parameter λ tends to zero, the solution for the degenerate case is recovered.

Pauli knew about the rules of quantisation of angular momentum described in Sect. 12.3.3 and used these to obtain the same result as Born, Heisenberg and Jordan (1926) that the eigenvalues of the matrix $\mathbf{M}^2$ were $j(j+1)(h/2\pi)$. The determination of the elements of the matrix **A** remained a challenge which Pauli solved using the rules for the relative intensities of the Zeeman components which had previously been derived by Hönl (1925) and Goudsmit and Kronig (1925). Finally, Pauli obtained the result that the absolute values of the energy levels of the hydrogen atom were

$$W_n = H(j, m; j, m) = \frac{Z^2 e^4 m_0}{8\epsilon_0^2 h^2 (j_{\max} + 1)^2} . \tag{12.99}$$

Setting $j_{\max} + 1 = n$, Bohr's formula for the energy levels of the hydrogen atom (4.25) was recovered:

$$W_n = \frac{Z^2 e^4 m_0}{8\epsilon_0^2 h^2 n^2} . \tag{12.100}$$

In coming to this spectacular result, Pauli obtained other key features of the hydrogen atom as well. He showed that the values of j and m had to be integers and that the maximum value of the angular momentum quantum number $j_{\max} = n - 1$, where $n = 1$ is the ground state. Thus, unlike the old quantum theory, the lowest energy state of the hydrogen atom has zero angular momentum. This is a pure quantum mechanical result, quite unlike the lowest energy state of the Bohr atom. This had the effect of getting rid of the empirical restrictions which had to be placed upon the allowed orbits in the old quantum theory, for example, the exclusion of linear orbits in which the electron passes through the nucleus (Sect. 5.2). It will be recognised that Pauli had derived many of the essential features of the quantum mechanical model of the hydrogen atom.

In addition, the same new quantum conditions resolved the problem of the motion of the electron in the hydrogen atom under the influence of crossed electric and magnetic fields. According to the old quantum theory, the perturbation of the energy of the state W_n in the presence of crossed electric and magnetic fields was

$$\Delta W_{n_1, n_2} = \left(\frac{n}{2} - n_1 \right) \omega_1 h + \left(\frac{n}{2} - n_2 \right) \omega_2 h , \tag{12.101}$$

where $\omega_1 = \nu_L + \nu_s$, $\omega_2 = |\nu_L - \nu_s|$, ν_L is the Larmor frequency and ν_s is the precession frequency associated with the Stark effect. Pauli showed that according to the new quantum theory, the quantum number n should be replaced by $j_{max} = n - 1$, again eliminating linear orbits in which the electron passes through the nucleus. In the old quantum theory, these orbits had to be excluded arbitrarily since they violated the adiabatic principle.

12.5 The triumph of matrix mechanics and its incompleteness

Pauli's achievement in solving the problem of the hydrogen atom according to matrix mechanics was universally applauded as a triumph for the new quantum mechanics. Pauli had carried out the calculations described in Sect. 12.4 in only three weeks and communicated the results to Heisenberg, who wrote on 3 November 1925,

> 'I need not tell you how delighted I am about the new theory of hydrogen and how pleasantly surprised I am about the speed with which you produced this theory.'

Bohr learned about Pauli's new results from Kramers soon after this date and immediately wrote to Pauli:

> 'To my great joy I heard from Kramers that you have succeeded in deriving the Balmer formula.'

Pauli responded by sending Bohr and his colleagues details of his calculations which impressed them greatly. Bohr wrote to Pauli on 5 December 1925:

> 'Kramers, Kronig and I, who have just gone once again with the greatest pleasure through your beautiful calculation of the hydrogen spectrum, send you many friendly greetings from Tisvelde.'

These remarkable calculations convinced most physicists that the new matrix mechanics had the power to account for quantum phenomena in a self-consistent fashion.

The new quantum mechanics was however unfamiliar to the physicist-in-the-street and there were relatively few problems which could be successfully addressed by the formalism as it stood. The helium atom was still too hard and the spin of the electron still had to be incorporated into the scheme of quantum mechanics.

One other notable success was the matrix mechanics of diatomic molecules which was carried out by Lucie Mensing, a graduate student of Wilhelm Lenz in Hamburg. The two papers of Born, Heisenberg and Jordan (1925b; 1926) had already considered the cases of the nonlinear oscillator and the rotator in quantum mechanics and so she was able to use much of the material already in these papers to address the quantisation of the vibrational and rotational stationary states of diatomic molecules (Mensing, 1926). She found that the energy states of the diatomic molecule were given by the expression

$$W_{n,j} = U_0 + \frac{h^2}{8\pi^2 m_0 a^2} j(j+1) + h\left(n + \tfrac{1}{2}\right)\left[\nu_0 + \beta j(j+1)\right]$$
$$+ \alpha h^2 \left[n(n+1) + \tfrac{11}{30}\right] + \cdots \tag{12.102}$$

This expression has a similar form to the classical expression for the vibrational and rotational frequencies of a diatomic molecule, but with some important differences:

- The second term on the right-hand side represents the rotational states of the molecule with the classical j^2 being replaced by the quantum mechanical $j(j+1)$.
- The third term corresponds to the classical term $hn(\nu_0 + \beta j^2)$ and describes the harmonic oscillations of the nuclei relative to their centre of mass and the mixed rotational-vibrational modes.
- The fourth term describes the anharmonicity of the oscillations and replaces the classical term $h\alpha n^2$.
- Particularly important was the inclusion of the factor $(n + \frac{1}{2})$ in the third term. The term one-half corresponds to the zero point energy of the oscillator and its presence resolved the discrepancy between the experimental values of the vibrational modes of the molecule and the classical theory. To account for this discrepancy half-integral quantum numbers had been introduced, but Mensing's calculation now showed that this was unnecessary. The only allowed transitions were associated with $\Delta j = \pm 1$ for the rotational terms and $\Delta n = 0, \pm 1, \pm 2, \ldots$ for the vibrational terms.

The successes of matrix mechanics was undeniable, but there was still a long way to go before the full richness of quantum phenomena could be encompassed by a fully self-consistent theory. Matrix mechanics was only the first step along the route to the modern theory of quantum mechanics.

Dirac's quantum mechanics

Göttingen and Copenhagen were the undoubted capitals of the new discipline of quantum mechanics. The expertise in experimental and mathematical physics and in pure mathematics made Göttingen the epicentre of the revolution which was taking place in the mathematical physics of quanta. Whilst this was to remain the case for the next few years, other actors soon appeared on the scene who were to contribute to Born's 'tangle of interconnected alleys'. What was truly remarkable was how quickly the different approaches to the problems of quantum theory were developed and the rapid assimilation of all of them into a coherent and self-consistent theory of quantum mechanics. Whilst the theory itself was completed relatively quickly, the understanding of its physical content was to take many more years.

The new players on the scene included Paul Dirac at Cambridge, Erwin Schrödinger in Vienna and Norbert Wiener at the Massachusetts Institute of Technology. Each of them brought quite new approaches to the development of quantum theory – their innovations were to supersede the matrix mechanics of Born, Heisenberg and Jordan, but there can be no doubt that the success of that theory indicated clearly the route ahead. They were however to involve the introduction of new mathematical techniques into the description of quantum phenomena.

13.1 Dirac's approach to quantum mechanics

Paul Dirac was trained as an electrical engineer at Bristol University, but he had a very strong mathematical bent. He was a solitary character who was notoriously quiet and self-effacing. He simply worked things out on his own. He was interested in problems which could be treated on a strict mathematical basis and looked for beauty in the mathematics needed to describe nature. On the one hand, his training as an engineer had taught him to be somewhat pragmatic about the mathematics necessary to address any particular problem. As he wrote in his memoirs of 1967,

> 'I think if I had not had this engineering training, I should not have had any success with the kind of work I did later on, because it was really necessary to get away from the point of view that one should only deal with exact equations, and that one should deal only with results which could be deduced logically from known exact laws which one accepted, in which one had implicit faith. Engineers were concerned only with getting equations which were useful for describing nature. . . .

And that led me of course to the view that this outlook was really the best outlook to have. We wanted a description of Nature. We wanted to find the equations which would describe Nature, and the best we could hope for was usually approximate equations, and we would have to reconcile ourselves to an absence of strict logic . . . ' (Dirac, 1977)

On the other hand, the theory had to be 'beautiful'. In the same memoir, he writes about his meeting with Schrödinger:

'. . . of all the physicists that I met, I think Schrödinger was the one that I felt to be most closely similar to myself. I found myself getting into agreement with Schrödinger more readily than with anyone else. I believe the reason for this was that Schrödinger and I both had a very strong appreciation of mathematical beauty, and this appreciation of mathematical beauty dominated all our work. It was a sort of faith with us that any equations which described the fundamental laws of Nature must have great mathematical beauty in them. It was like a religion with us. It was a very profitable religion to hold, and can be considered as the basis for much of our success.' (Dirac, 1977)

13.2 Dirac and *The fundamental equations of quantum mechanics* (1925)

Picking up the story from the end of Chap. 11, Heisenberg left the decision about whether or not to publish his revolutionary paper of 1925 to Born, since he had to leave for Cambridge where he was to deliver a lecture to the Kapitsa Club on 28 July 1925. These gatherings had been instituted by Piotr Kapitsa and at them the latest developments in physics were discussed by the staff and research students of the Cavendish Laboratory and other cognate departments. Dirac was a member of the club. The title of Heisenberg's talk was *Termzoologie und Zeemanbotanik* and was principally concerned with his recent researches into the anomalous Zeeman effect. Dirac had only vague memories of the meeting, but some discussion of Heisenberg's most recent work on the fundamentals of quantum mechanics took place either during the lecture or in the informal discussions afterwards. Ralph Fowler, who was Dirac's supervisor, was impressed by what Heisenberg described about this new work and asked Heisenberg to send copies of the proofs of his paper when they became available. This duly happened a few weeks later and Fowler sent the proofs to Dirac, who had returned home to Bristol for part of the vacation, seeking his opinion on Heisenberg's work.

In Dirac's words,

'I received it (the proofs) either at the end of August or the beginning of September . . . At first I was not very impressed by it. It seemed to me to be too complicated. I just did not see the main point of it, and in particular his determination of quantum conditions seemed to me too far fetched, so I just put it aside as being of no interest. However, a week or ten days later I returned to this paper of Heisenberg's and studied it more closely. And then I suddenly realised that it did provide the key to the whole solution of the difficulties which we were concerned with.

My previous work had been all concerned with studying individual states . . . Heisenberg brought out the quite new idea that one had to consider quantities associated with two stationary states rather than one.'

But this scarcely does justice to the profound impact Heisenberg's paper had upon Dirac. He quickly homed in on Heisenberg's discovery that the variables involved at the quantum level do not commute. Dirac's great achievement in his paper of 1925 was to recast Heisenberg's insights into the language of Hamiltonian dynamics. As expressed by Jammer,

'In a few weeks' time he achieved his objective and thus established one of the most profound and useful relations between quantum mechanics and the classical Hamilton–Jacobi formulation of mechanics.' (Jammer, 1989)

13.2.1 Reformulating non-commutability

Like Born, Dirac appreciated that the deep insights in Heisenberg's paper were the introduction of quantities which depended upon two variables rather than one and the fact that these quantities were non-commuting, the feature which greatly disturbed Heisenberg. Born's analysis led to the development of matrix mechanics which was the subject of Chap. 12. Quite independently and entirely on his own, Dirac discovered a rather different and, in the end, more powerful approach to the reformulation of quantum mechanics. An important aspect of Dirac's approach was that he firmly believed that, whatever the correct formulation of the theory, it should be derivable from Hamiltonian mechanics, a subject in which he was already an expert. Like most theorists at the time, he had become fluent in the use of the techniques of Hamilton–Jacobi theory and of action-angle variables in the old quantum theory. As he remarked much later,

'At that time, I was expecting some kind of connection between the new mechanics and Hamiltonian dynamics (as Hamiltonian dynamics was used so much with Sommerfeld's development of the Bohr theory) and it seemed to me that this connection should show up with large quantum numbers.'

Notice again Bohr's correspondence principle at work.

To appreciate what Dirac did, let us recall the Heisenberg prescription for taking the product of the quantities x and y, what Dirac refers to as the 'Heisenberg product'. Following Jammer (1989), consider the one-dimensional classical case, that is, a system of one degree of freedom, in which x and y are functions of the action–angle variables J and w. As in (11.13), for a multiply periodic system, $x(n, t)$ and $y(n, t)$ can be written as Fourier series:

$$x = x(n, t) = x(J, w) = \sum_\tau x_\tau(J) \exp(2\pi i \tau w), \tag{13.1}$$

$$y = y(n, t) = y(J, w) = \sum_\tau y_\tau(J) \exp(2\pi i \tau w), \tag{13.2}$$

where, according to the old quantum theory, $J = nh$. In Heisenberg's scheme, the quantities $x_\tau(J)$ and $y_\tau(J)$ correspond to the quantum quantities $x(n, n - \tau)$ and $y(n, n - \tau)$. Dirac was interested in the limit of large quantum numbers $n \gg \tau$, expecting that in this limit he would find the classical analogue of $xy - yx$. The $(n, n - \tau - \sigma)$ component of $xy - yx$ is given by the difference of Heisenberg products

$$(xy - yx)_{n,n-\tau-\sigma} = [x(n, n - \tau)y(n - \tau, n - \tau - \sigma) - y(n, n - \sigma)x(n - \sigma, n - \tau - \sigma)]$$
$$\times \exp[i\omega(n, n - \tau - \sigma)t]. \tag{13.3}$$

Concentrating on the terms in the first pair of square brackets on the right-hand side of (13.3), this expression can be rewritten by subtracting and adding $x(n - \sigma, n - \tau - \sigma)y(n - \tau, n - \tau - \sigma)$ to each of the terms so that

$$x(n, n - \tau)y(n - \tau, n - \tau - \sigma) - y(n, n - \sigma)x(n - \sigma, n - \tau - \sigma)$$
$$= [x(n, n - \tau) - x(n - \sigma, n - \sigma - \tau)]y(n - \tau, n - \tau - \sigma)$$
$$- [y(n, n - \sigma) - y(n - \tau, n - \tau - \sigma)]x(n - \sigma, n - \tau - \sigma). \tag{13.4}$$

Now, working backwards, these Heisenberg products can be converted into their classical equivalents for large quantum numbers $n \gg \tau + \sigma$ using the equivalences $x(n, n - \tau) \to x_\tau(J)$ and $y(n, n - \sigma) \to y_\sigma(J)$. Then,

$$\begin{cases} x(n, n - \tau) - x(n - \sigma, n - \sigma - \tau) & \to x_\tau(J) - x_\tau(J - h\sigma) = \dfrac{\partial x_\tau(J)}{\partial J}h\sigma, \\[2mm] y(n, n - \sigma) - y(n - \tau, n - \tau - \sigma) & \to y_\sigma(J) - y_\sigma(J - h\tau) = \dfrac{\partial y_\tau(J)}{\partial J}h\tau. \end{cases} \tag{13.5}$$

Therefore, (13.4) becomes

$$\frac{\partial x_\tau(J)}{\partial J}h\sigma y_\sigma(J) - \frac{\partial y_\tau(J)}{\partial J}h\tau x_\tau(J). \tag{13.6}$$

But, from the definitions of the σ and τ components of x and y,

$$\begin{cases} \dfrac{\partial}{\partial w}[y_\sigma(J)\exp(2\pi i\sigma w)] & = 2\pi i[\sigma y_\sigma(J)\exp(2\pi i\sigma w)], \\[2mm] \dfrac{\partial}{\partial w}[x_\tau(J)\exp(2\pi i\tau w)] & = 2\pi i[\tau y_\tau(J)\exp(2\pi i\tau w)]. \end{cases} \tag{13.7}$$

Combining (13.6) and (13.7), it follows that the nm component of the quantum mechanical expression for $xy - yx$ corresponds to

$$\frac{h}{2\pi i}\sum_{\tau+\sigma=n-m}\left\{\frac{\partial}{\partial J}[x_\tau\exp(2\pi i\tau w)]\frac{\partial}{\partial w}[y_\sigma\exp(2\pi i\sigma w)]\right.$$
$$\left. -\frac{\partial}{\partial J}[y_\sigma\exp(2\pi i\sigma w)]\frac{\partial}{\partial w}[x_\tau\exp(2\pi i\tau w)]\right\}, \tag{13.8}$$

where we have dropped the arguments (J) in x_τ and y_σ. The result is that the quantity $xy - yx$ is identical to the quantity

$$\frac{h}{2\pi i}\left(\frac{\partial y}{\partial J}\frac{\partial x}{\partial w} - \frac{\partial x}{\partial J}\frac{\partial y}{\partial w}\right), \tag{13.9}$$

which is the classical expression for the *Poisson bracket* of the quantities x and y, which are functions of the canonical variables J and w.

There is a delightful story about how Dirac realised the significance of Poisson brackets during the exciting days of October 1925. Dirac had a strict rule about relaxing on Sunday afternoons by taking country walks around Cambridge. In his memoirs, he writes:

'It was during one of the Sunday Walks in October 1925 when I was thinking very much about this $uv - vu$, in spite of my intention to relax, that I thought about Poisson brackets... I did not remember very well what a Poisson bracket was. I did not remember the precise formula for a Poisson bracket and only had some vague recollections. But there were exciting possibilities there and I thought I might be getting on to some big new idea.

Of course, I could not [find out what a Poisson bracket was] right out in the country. I just had to hurry home and see what I could then find out about Poisson brackets. I looked through my notes and there was no reference there anywhere to Poisson brackets. The text books which I had at home were all too elementary to mention them. There was nothing I could do, because it was Sunday evening then and the libraries were all closed. I just had to wait impatiently through that night without knowing whether this idea was any good or not but still I think my confidence grew during the course of the night. The next morning, I hurried along to one of the libraries as soon as it was open and then I looked up Poisson brackets in Whittaker's *Analytic Dynamics* [(1917)] and I found that they were just what I needed. They provided the perfect analogy with the commutator.'

13.2.2 Poisson brackets

Like so many topics in analytic dynamics which have appeared in this story, Poisson introduced what are now known as *Poisson brackets* in the course of a study of an n-body problem in astronomy, namely, the motion of a planet about the Sun in the presence of the $(n - 1)$ other planets. His problem was to solve the $3n$ Euler–Lagrange equations of motion (5.59) in the presence of the time-varying perturbation Ω of the gravitational potential due to the other planets. The full significance of Poisson's pioneering studies was only fully appreciated later in the century when Jacobi independently discovered the power of Poisson brackets. Jacobi referred to Poisson's combination rule for Poisson brackets[1] as 'the most profound discovery of Monsieur Poisson' and 'the most important theorem in dynamics' (Jacobi, 1841).

The Poisson bracket[2] of the functions g and h is defined to be

$$[g, h] = \sum_{i=1}^{n} \left(\frac{\partial g}{\partial p_i} \frac{\partial h}{\partial q_i} - \frac{\partial g}{\partial q_i} \frac{\partial h}{\partial p_i} \right), \tag{13.10}$$

where g and h are functions of p_i and q_i, the generalised, or canonical, momentum and position coordinates of the n degrees of freedom of the system. In general, we can write the time variation of, say g, as

$$\dot{g} = \sum_{i=1}^{n} \left(\frac{\partial g}{\partial q_i} \dot{q}_i + \frac{\partial g}{\partial p_i} \dot{p}_i \right). \tag{13.11}$$

Hamilton's equations (5.69) enable us to replace $\dot{q}$ and $\dot{p}$ by the derivatives of the Hamiltonian with respect to p_i and q_i and so

$$\dot{g} = \sum_{i=1}^{n} \left(\frac{\partial g}{\partial q_i} \frac{\partial H}{\partial p_i} - \frac{\partial g}{\partial p_i} \frac{\partial H}{\partial q_i} \right) = [H, g]. \tag{13.12}$$

Therefore, Hamilton's equations can be rewritten

$$\dot{q}_i = [H, q_i], \qquad \dot{p}_i = [H, p_i].$$ (13.13)

What Dirac appreciated was that the Poisson brackets have exactly the commutation properties needed to describe the non-commutation features of the difference of Heisenberg products $xy - yx$. Thus, in (13.10), if we set $g = q_i$ and $h = q_j$, or $g = p_i$ and $h = p_j$, we obtain the results

$$[q_i, q_j] = 0, \qquad [p_i, p_j] = 0,$$ (13.14)

because of the independence of the q_is and p_is. Furthermore, if $j \neq k$,

$$[p_j, q_k] = 0.$$ (13.15)

If, however, $g = p_k$ and $h = q_k$, then

$$[p_k, q_k] = 1, \qquad [q_j, p_k] = -1.$$ (13.16)

Pairs of quantities with zero Poisson brackets are said to *commute*, whereas those with Poisson brackets equal to unity are *non-commuting*. Pairs of quantities with Poisson brackets equal to unity are said to be *canonically conjugate*. From (13.13), it follows that any quantity which commutes with the Hamiltonian H does not change with time. In particular, H is a constant of the motion since it commutes with itself.

But the Poisson brackets had other key properties. Most important was the fact that *all* sets of dynamical variables Q_k and P_k which are obtained from q_k and p_k by a canonical transformation which leaves the Hamilton equations of motion unaltered, also leave their Poisson brackets unchanged. We can summarise (13.14)–(13.16) by the expressions

$$[q_k, p_i] = \delta_{ki}, \qquad [q_k, q_i] = [p_k, p_i] = 0,$$ (13.17)

where δ_{ki} is the Kronecker symbol meaning $\delta_{ki} = 1$ if $k = i$ and zero otherwise. Then, if Q_k and P_k are obtained from q_k and p_k by a canonical transformation, it follows that

$$[Q_k, P_i] = \delta_{ki}, \qquad [Q_k, Q_i] = [P_k, P_i] = 0.$$ (13.18)

Furthermore, the Poisson brackets of any two functions F_1 and F_2 remain invariant under all such transformations,

$$\sum_{i=1}^{n} \left(\frac{\partial F_1}{\partial p_i} \frac{\partial F_2}{\partial q_i} - \frac{\partial F_1}{\partial q_i} \frac{\partial F_2}{\partial p_i} \right) = \sum_{i=1}^{n} \left(\frac{\partial F_1}{\partial P_i} \frac{\partial F_2}{\partial Q_i} - \frac{\partial F_1}{\partial Q_i} \frac{\partial F_2}{\partial P_i} \right).$$ (13.19)

These and many other properties of Poisson brackets were included in Whittaker's *Analytic Dynamics* and Dirac immediately set about rewriting Heisenberg's insight in terms of Hamiltonian dynamics. The great advance was that Dirac could now write down the quantisation conditions in terms of classical Hamiltonian dynamics, using the equivalence of the difference of Heisenberg products to Poisson brackets. From (13.9), this equivalence can be written

$$xy - yx = \frac{h}{2\pi i} \left(\frac{\partial x}{\partial w} \frac{\partial y}{\partial J} - \frac{\partial y}{\partial w} \frac{\partial x}{\partial J} \right) \equiv \frac{h}{2\pi i} [x, y],$$ (13.20)

where the differentiations are taken with respect to the canonical coordinates J and w. Thus, for the special cases of canonical coordinates p and q, it follows that

$$q_r q_s - q_s q_r = 0, \quad p_r p_s - p_s p_r = 0, \quad q_r p_s - q_s p_r = \delta_{rs} \frac{ih}{2\pi}. \qquad (13.21)$$

There is another key feature of Dirac's approach to quantum mechanics. Notice that the procedure embeds quantum mechanics into the very heart of Hamiltonian dynamics. Just as Heisenberg appreciated that he had applied quantum concepts to space itself, so Dirac had done the same thing by treating the momentum and space coordinates on the same footing and introducing Bohr's quantisation condition into the foundations of Hamiltonian mechanics. As expressed by Jammer (1989)

> '... Dirac absorbed, so to speak, the correspondence principle as an integral part of the very foundation of his theory and, like Heisenberg, disposed thereby of the necessity of resorting to Bohr's principle each time a problem had to be solved.'

13.2.3 *The fundamental equations of quantum mechanics* (1925)

Dirac quickly wrote his paper with the above title and showed it to Fowler who fully appreciated its importance. Fowler communicated the paper to the Royal Society for rapid publication in the *Proceedings of the Royal Society* and it was received on 7 November 1925 – it was published on 1 December 1925. Dirac's perspective was clearly laid out in the introduction:

> 'In a recent paper, Heisenberg [(1925)] puts forward a new theory which suggests that it is not the equations of classical mechanics that are in any way at fault, but that the mathematical operations by which physical results are deduced from them require modification. *All* the information supplied by the classical theory can thus be made use of in the new theory.'

After summarising Heisenberg's innovations, Dirac rewrites the Heisenberg product as follows,

$$xy(nm) = \sum_k x(nk)\, y(km). \qquad (13.22)$$

Born had made the same simplification and realised that (13.22) represented matrix multiplication. Dirac took the more general view that (13.22) should be the multiplication rule for the quantum variables x and y without specifying at this stage exactly what they were. They had however to obey the addition rule

$$\{x + y\}(nm) = x(nm) + y(nm), \qquad (13.23)$$

as in the case of matrix addition. But now, Dirac takes a different route. He considers the rules for differentiation of the quantum variables x and y which are functions of the parameter v. As discussed by Mehra and Rechenberg, he now applied his knowledge of projective geometry to the problem of defining the differentiation dx/dv of the quantum variable x. His key insight was that there should be a linear relation between dx/dv and

x. This is consistent with the fact that both dx/dv and x can be represented by Fourier series of the form (13.1). Therefore, he assumed that the (nm) component of dx/dv could be written as

$$\frac{dx}{dv} = \sum_{n'm'} a(nm; n'm')x(n'm') .$$ (13.24)

He then argued that quantum differentiation should satisfy the following conditions:

$$\frac{d}{dv}(x + y) = \frac{d}{dv}x + \frac{d}{dv}y ,$$ (13.25)

$$\frac{d(xy)}{dv} = \frac{d}{dv}x \cdot y + x \cdot \frac{d}{dv}y ,$$ (13.26)

where the order of multiplication of the variables in (13.26) has to be preserved. Dirac now carried out an analysis of the constraints on the process of quantum differentiation with respect to v assuming that the rules (13.24) and (13.25) had to be maintained. From this analysis, he concluded that the most general form of quantum differentiation was

$$\frac{dx}{dv} = xa - ax ,$$ (13.27)

where the new quantum variable a has components $a(nm)$. Thus, the differential of the quantum variable x with respect to any parameter v is expressed as the *difference* of the Heisenberg products of the quantum variables x and a. This key result showed that the differential equations of classical mechanics were to be replaced by algebraic equations involving the addition and multiplication of quantum variables. Dirac was now in a position to apply the full apparatus of classical Hamiltonian dynamics to quantum mechanics.

Particularly notable was his derivation of Bohr's quantisation condition according to the new formalism. Just as in the case of matrix mechanics, 'diagonal' terms of the form $C(nn)$ correspond to constants of the motion. Therefore, the Hamiltonian terms $H(nn)$ correspond to the energies of the stationary states of the system. From (13.13), it follows that

$$\dot{x} = [H, x] .$$ (13.28)

Combining this with the fundamental relation (13.20),

$$xH - Hy = \frac{ih}{2\pi}[x, H] ,$$ (13.29)

it follows that

$$x(nm)H(mm) - H(nn)x(nm) = \frac{ih}{2\pi}\dot{x}(nm) = -\frac{h}{2\pi} \cdot \omega(nm)x(nm) ,$$ (13.30)

since the time dependence of $x(nm)$ is entirely determined by the term $\exp[i\omega(nm)t]$. Hence,

$$h\nu(nm) = \frac{h}{2\pi}\omega(nm) = H(nn) - H(mm) = E_n - E_m ,$$ (13.31)

where E_n and E_m are the energies of the stationary states n and m.

Fowler fully appreciated the deep significance of Dirac's paper and this was corroborated by the response of Heisenberg, to whom Dirac had sent a manuscript copy of his paper. On 20 November 1925, Heisenberg wrote to Dirac as follows:

> 'I have read your extraordinarily beautiful paper on quantum mechanics with the greatest interest, and there can be no doubt that all your results are correct as far as one believes at all in the newly proposed theory. ... [Dirac's paper was] also really better written and more concentrated than our attempts here [in Göttingen].'

Heisenberg particularly liked the derivation of Bohr's frequency condition, which used rather general principles rather than a specific model of atomic systems. Dirac was well aware that Born, Heisenberg and Jordan had developed their matrix mechanical approach to quantum mechanics and so he immediately turned all his energies to perfecting his own rather different approach.

13.3 Quantum algebra, q- and c-numbers and the hydrogen atom

Dirac had stressed the formal algebraic nature of the quantum variables in his paper of 1925, but he now formalised the concept by introducing a distinction between the quantum variables, which he called q-numbers, and the ordinary numbers of arithmetic, which he called c-numbers. As he wrote, 'q stands for *quantum*, or maybe *queer*, and c for *classical*, or maybe *commuting*' (Dirac, 1977). He makes this distinction clear in the introduction to his paper (Dirac, 1926d).

> 'The fact that the variables used for describing a dynamical system do not satisfy the commutative law means, of course, that they are not numbers in the sense of the word previously used in mathematics. To distinguish the two kinds of numbers, we shall call the quantum variables q-numbers and the numbers of classical mathematics which satisfy the commutative law c-numbers ...
>
> At present one can form no picture of what a q-number is like. One cannot say that one q-number is greater or lesser that another ... One knows nothing of the processes by which the numbers are formed except that they satisfy all the ordinary laws of algebra, excluding the commutative law of multiplication ... '

In his reminiscences, he wrote

> 'Now, I did not know anything about the real nature of q-numbers. Heisenberg's matrices I thought were just an example of q-numbers; maybe q-numbers were something more general. ... I proceeded to develop a theory in which I felt free to make any assumption I wanted to, unless they led immediately to an inconsistency. I did not bother at all about finding a precise mathematical nature for q-numbers, or any kind of precision in dealing with them.' (Dirac, 1977)

In this paper and three succeeding papers of 1926 (Dirac, 1926b,c,e), Dirac put the emphasis upon the new algebra necessary to accommodate Heisenberg's rules for quantum multiplication, the last paper published in the *Proceedings of the Cambridge Philosophical*

Society having the title *On quantum algebra* (Dirac, 1926b). We recall the remark at the beginning of his paper of 1925 that

> '[it is] not the equations of classical mechanics that are in any way at fault, but that the mathematical operations by which physical results are deduced from them require modification.'

In other words, it is not the forms of the laws of motion which are at fault – the wrong algebra is being used to carry out the mathematical operations.

Dirac had exactly the right training for making this imaginative leap in the dark. From the beginning he had a strong mathematical bent with a particular interest in geometry. After his engineering degree at Bristol University, he studied for a second degree in mathematics, which he completed in two rather than three years A particularly important influence was Peter Fraser, an outstanding mathematics lecturer, from whom Dirac learned the importance of rigorous mathematics, as opposed to the non-rigorous tools commonly used by engineers and physicists. He also learned projective geometry which was to have an important influence on his thinking.

At Cambridge, his supervisor was Ralph Fowler, the leading quantum theorist in Cambridge who stimulated his interest in atomic problems. At the same time, he participated regularly in the Saturday afternoon geometrical tea parties held by Henry Baker, the Lowndean Professor of Astronomy and Geometry. All Baker's students were required to attend these parties and there were lectures and discussions of geometrical topics after tea – Dirac presented his first lecture at one of these.

Baker's masterpiece was his six-volume *Principles of Geometry*, published from 1922 to 1925, volume one of which was to be of special significance for Dirac (Baker, 1922). In his memoirs, Dirac wrote:

> '. . . I was always very much interested in geometry . . . You can divide all mathematicians into two classes, those whose main interest is geometry, those whose main interest is algebra. . . . Now, a good mathematician has to be a master of both geometry and of algebra, and he has to be able to pass from one to the other quite freely according to the nature of the problem that he is working on. . . . my preference was strongly on the side of geometry, and has always remained so.'

Later, he remarks

> '[Projective geometry] was a most useful tool for research, but I did not mention it in my published work. . . . I felt that most physicists were not familiar with it. When I obtained a particular result, I translated it into an analytic form and put down the argument in terms of equations. That was an argument which any physicist would be able to understand without having had this special training.'

Although projective geometry had applications in relativity theory, no-one had considered that it would have importance for quantum theory. Dirac, however, appreciated a key result contained in Baker's book, namely, that there existed a complete set of laws for 'non-commuting' numbers, similar to those of ordinary commuting numbers.

Section III of Chap. 1 of Baker's first volume describes the elements of *symbolic algebra* needed to represent geometrical operations (Baker, 1922). He uses the term *symbol* rather than variable since the algebraic rules should apply for a wide range of mathematical entities, for example, vectors, matrices and so on – he gives examples of various types of symbols on pp. 68–69. These examples include 2×2 matrices with complex components which were eventually to become Dirac spin matrices. Specifically, he demonstrates with these examples that 'the equation $ab - ba$ is not generally true'. Baker stated the basic axiom as follows:

> 'The symbols which we first introduce are, speaking in general terms, subject to all the laws of ordinary algebra, except the commutative law of multiplication. They are not necessarily, and will not be finally, arranged in order of magnitude, so that, in their entirety, they are wider than the real numbers of arithmetic.'

It is remarkable that Dirac takes over, almost word-for-word, the foundations of Baker's symbolic algebra as the basis for his rules of quantum algebra. Pages 62–69 of Baker's book indicate clearly the source of Dirac's insights and of his translation of the symbols of projective geometry into the variables of quantum algebra.

Thus, if z_1, z_2 and z_3 are q-numbers, they obey the rules of ordinary algebra as follows:

$$z_1 + z_2 = z_2 + z_1 \,,$$
$$(z_1 + z_2) + z_3 = z_1 + (z_2 + z_3) \,,$$
$$(z_1 z_2) z_3 = z_1 (z_2 z_3) \,,$$
$$z_1(z_2 + z_3) = z_1 z_2 + z_1 z_3 \,, \qquad (z_1 + z_2) z_3 = z_1 z_3 + z_2 z_3 \,.$$

If
$$z_1 z_2 = 0 \,, \quad \text{either} \quad z_1 = 0 \quad \text{or} \quad z_2 = 0 \,;$$

but
$$z_1 z_2 \neq z_2 z_1 \,,$$

in general, except when z_1 or z_2 is a c-number.

In Sect. 2, the rules of Poisson bracket manipulations are adapted for q-numbers. In particular, the full apparatus of Hamiltonian mechanics can be taken over into his quantum algebra of q-numbers. Thus, if Q_r and P_r are canonical variables, which are functions of q-numbers, and these obey the rules for Poisson brackets

$$[Q_r, P_s] = \delta_{rs} \,, \qquad [Q_r, Q_s] = [P_r, P_s] = 0 \,, \tag{13.32}$$

then, just as in Hamiltonian mechanics, Q_r and P_r can be transformed to another set of canonical variables q_r and q_s by relations of the form

$$Q_r = b q_r b^{-1} \,, \qquad P_r = b p_r b^{-1} \,, \tag{13.33}$$

where b is a q-number. Notice the strong similarity to the canonical transformations (12.64) introduced by Born and his colleagues in their matrix treatment of quantum algebra. Interestingly, Dirac remarks that 'these formulae do not appear to be of great practical value', whereas they were central to Born's procedures for the transformation to principal axes, that is, the diagonalisation of the matrices from which the energies of the stationary states could be determined. By the same arguments, the equations of motion could be written

$$\dot{x} = [x, H] \,, \tag{13.34}$$

where the Hamiltonian H is now a q-number. Dirac does not define what q-numbers are until he is forced to confront the theory with experiment. He does however concede that, in the case of multiply periodic motion, they can be represented by a set of harmonic components of the form $x(nm) \exp i\omega(nm)t$, where $x(nm)$ and $\omega(nm)$ are c-numbers. Then $\dot{x}$ has components $i\omega(nm) x(nm) \exp i\omega(nm)t$.

Before proceeding to the application of this algebra to quantum mechanics, Dirac develops a number of algebraic theorems which illustrate the care which has to be taken in manipulating q-numbers, in particular, the order in which the operations are carried out. Thus, for example, he establishes the results that

$$\frac{1}{xy} = \frac{1}{y} \cdot \frac{1}{x} \quad ; \quad \frac{d}{dt}\left(\frac{1}{x}\right) = -\frac{1}{x}\dot{x}\frac{1}{x} \,. \tag{13.35}$$

The binomial expansion of $(1 + x)^n$ is the same as in ordinary algebra where n is a c-number and the exponential series can be defined as the same power series as in ordinary algebra. Generally, however, $e^{x+y} \neq e^x e^y$, unless x and y commute.

Dirac now needed to show that the new formalism could be used to carry out useful computations which could be compared with known results in quantum physics. This was not a trivial task since there were relatively few problems which could be tackled with the state of development of quantum algebra at the time. In his paper entitled *Quantum mechanics and a preliminary investigation of the hydrogen atom* (Dirac, 1926d), Dirac tackled the determination of the energy levels of the hydrogen atom. To do this, he had to convert the classical theory of multiply periodic systems into the language of q-numbers and quantum algebra. He started with the advantage that he was already an expert on the mathematical procedures of action–angle variables and their application to multiply periodic systems. Section 4 of the paper carries out this translation. On the basis of his new quantum algebra, he set out the postulates of the quantum theory of multiply periodic dynamical systems where the action–angle variables J_r and w_r are now q-numbers and so have to obey the commutation relations.

Dirac showed that, using his new formalism he could work out the frequency of the orbits of electrons in the hydrogen atom and Bohr's quantum relation, noting that the quantities involved were q-numbers rather than the c-numbers of the old quantum theory. In addition, he derived Heisenberg's multiplication rule for the quantities x and y,

$$xy(n, n - \gamma) = \sum_\alpha x(n, n - \alpha)y(n - \alpha, n - \gamma) \,. \tag{13.36}$$

Sections 5, 6 and 7 are concerned with solving the orbital equations for the electron in the hydrogen atom, but now using all the rules of quantum algebra. This analysis required considerable care but resulted in a Hamiltonian of the form

$$H = \frac{1}{2m}\left(p_r^2 + \frac{k_1 k_2}{r^2}\right) - \frac{e^2}{4\pi\epsilon_0 r} \,, \tag{13.37}$$

where k_1 and k_2 are q-numbers given by $k_1 = k + \frac{1}{2}h$ and $k_2 = k - \frac{1}{2}h$, k being the q-number which describes the angular momentum, which is the conjugate momentum associated with the angle variable θ.

After a detailed analysis, Dirac obtained the expression for the energies of the stationary states of the hydrogen atom and the transition frequencies of the hydrogen atom in the form,[3]

$$v = \frac{H(P + nh) - H(P)}{h} = \frac{m_e e^4}{8\epsilon_0^2 h^3}\left[\frac{1}{P^2} - \frac{1}{(P + nh)^2}\right], \qquad (13.38)$$

where $P = mh$ and m takes integral values. This is precisely the formula for the lines in the frequency spectrum of the hydrogen atom.

Dirac's was no less of a *tour de force* than Pauli's. They had both completed calculations of very considerable complexity and, by quite different approaches – Pauli by applying the concepts of matrix mechanics and Dirac by invoking his rules of quantum algebra – arrived at the same result which agreed with experiment. In his correspondence with Heisenberg, Dirac had been informed that Pauli had already solved the problem of the hydrogen atom using matrix mechanics,[4] but this did not worry Dirac particularly. From his point of view, this was only a 'reality check', that quantum algebra was not just a theoretical construct, but a version of quantum mechanics which could account for the results of experiment.

13.4 Multi-electron atoms, *On quantum algebra* and a PhD dissertation

Dirac's objective all along had been to use the formalism of quantum algebra to tackle the problems of multi-electron atoms and this he proceeded to carry out, the end result being the paper with the unpromising title *The elimination of the nodes in quantum mechanics*. Once again, the inspiration had come from celestial mechanics and the problem of the orbits of point masses acting under gravity. As expressed by Dirac,

> 'In the classical treatment of the dynamical problem of a number of particles or electrons moving in a central field of force and disturbing one another, one always begins by making the initial simplification, known as the elimination of the nodes, which consists in obtaining a contact transformation from the Cartesian co-ordinates and momenta of the electrons to a set of canonical variables, of which all except three are independent of the orientation of the system as a whole, while these three determine the orientation. In the absence of an external field of force, the Hamiltonian, when expressed in terms of the new variables, must be independent of these three, which simplifies the equations of motion.' (Dirac, 1926e)

The objective of the paper was to determine the necessary contact transformation to enable the programme to be carried out. First of all, Dirac derived the properties of the q-numbers representing linear and angular momentum in his quantum algebra and obtained independently the rules for the quantisation of angular momentum which had already been established by Born, Heisenberg and Jordan (1926) (see Sect. 12.3.3). Next, the appropriate forms of the action–angle variables are determined. Once this was achieved, the transformation equations were determined, first for a single electron, then for a system of two electrons and finally for systems with more than two electrons. In the last case, the multi-electron atom was modelled by the 'core plus valance electron picture' so that the problem was reduced to a 'two-electron' problem (see Sect. 7.4).

The paper is once again a technical *tour de force*. Of particular interest is Dirac's treatment of the anomalous Zeeman effect in Sect. 9. He uses the standard atomic model of an electron plus core and, following the experimental evidence for the anomalous value of the gyromagnetic ratio (Sects. 7.5.2), gave the core twice the ratio of magnetic moment to angular momentum as compared with the standard Lorenz value (7.39). With this assumption, he derived an expression for the Landé g factor

$$g = \left(1 + \frac{1}{2} \frac{j_1 j_2 - k_1 k_2 + k_1' k_2'}{j_1 j_2}\right), \tag{13.39}$$

which agreed precisely with Landé's empirical expression discussed in Sect. 7.4 and given by (7.31). Finally, he showed that the theory also gave Kronig's results for the relative intensities of the multiplets and their components in a weak magnetic field.

The fourth paper was his short contribution *On quantum algebra* to the *Mathematical Proceedings of the Cambridge Philosophical Society* (Dirac, 1926b). This was a formal description of the algebra of q-numbers stated axiomatically and concisely. It summarises in one paper the necessary formalities for the application of quantum algebra to physics at the quantum level.

Dirac submitted his PhD dissertation with the title *Quantum mechanics* in May 1926, including in it the four papers discussed in this chapter (Dirac, 1926a). This was a remarkable achievement in that it provided a complete and self-consistent theory of quantum mechanics, all discovered in the previous nine months. The examiners included Arthur Eddington and they were very impressed by the dissertation. In June 1926, Eddington took the unusual step of writing personally to Dirac to congratulate him warmly on his remarkable achievement (Farmelo, 2009).

This brought to a close Dirac's first burst of creativity in developing the fundamentals of quantum theory. Soon, he was to add the quite different approach pioneered by Erwin Schrödinger to his armoury of weapons for attacking the problems of quantum theory.

Schrödinger and wave mechanics

On 13 March 1926, the first of six papers on wave mechanics by Erwin Schrödinger was published in the *Annalen der Physik* with the title *Quantisation as an eigenvalue problem (Part 1)*[1] (Schrödinger, 1926b). The startling first paragraph reads:

> 'In this paper, I wish to consider, first, the simple case of the hydrogen atom (non-relativistic and unperturbed), and show that the customary quantum conditions can be replaced by another postulate, in which the concept of "whole numbers", merely as such, is not introduced. Rather, when integralness does appear, it arises in the same natural way as it does in the case of *node-numbers* of a vibrating string. The new conception is capable of generalisation, and strikes, I believe, very deeply at the true nature of quantum rules.'

These papers were the fruits of an extraordinary burst of creativity on Schrödinger's part which resulted from his interactions with Einstein in the latter part of 1925. Central to these exchanges were de Broglie's remarkable researches which culminated in his famous PhD dissertation and published papers of 1924. These events have already be recounted in Chap. 9. The subsequent developments which led to Schrödinger's discovery of the equation which bears his name will be taken up in Sect. 14.2, but let us first understand more about Schrödinger's background.

14.1 Schrödinger's background in physics and mathematics

14.1.1 Education and career up to 1925

Unlike Heisenberg, Jordan, Pauli and Dirac, Erwin Schrödinger was not one of the young Turks who developed *Knabenphysik*, young man's physics. In 1926, they were all aged about 25, while Schrödinger was 38, more or less of the same generation as Born who was five years his senior. Schrödinger was born in 1887 in Vienna, which had become a leading centre for experimental and theoretical physics in the latter years of the nineteenth century. In 1866, Josef Stefan had become director of the Physical Institute of the University of Vienna where his doctoral students included Ludwig Boltzmann, Marian Smoluchowski and Johann Josef Loschmidt. In turn, Stefan and Boltzmann were the teachers of Friedrich Hasenöhrl who was to succeed Boltzmann as Professor of Theoretical Physics at the University of Vienna. Hasenöhrl was the only Austrian to participate in the 1911 Solvay Conference on quanta. Following his tenure of the Chair of Physics at Prague, Ernst Mach returned to the University of Vienna in 1895 as Professor of Philosophy. The stage was set for the vehement debates between Boltzmann and Mach on the reality of the existence of atoms.

Schrödinger began his university studies in 1906 as the community of Viennese scientists was coming to terms with the tragedy of Boltzmann's suicide in September of that year. Schrödinger studied physics at the University of Vienna between 1906 and 1910 under Franz Exner in experimental physics and Hasenöhrl in theoretical physics. He was recognised as an outstanding student and, following a year's military service, became an assistant at Exner's Physical Institute in 1911. Having obtained his Habilitation in January 1914, he was appointed a Privatdozent at the University of Vienna where he carried out a wide range of studies in both experimental and theoretical physics.

During the First World War, Schrödinger participated in the war effort as a commissioned officer in the Austrian fortress artillery. Tragically, Hasenöhrl was killed during the hostilities in October 1915. Later, on receiving the Nobel Prize for Physics in 1933, Schrödinger wrote in his autobiographical notes,

> 'At that time, Hasenöhrl was killed in battle, and I feel that otherwise today his name would now stand in place of mine.' (Schrödinger, 1935)

In 1917, still on military duty, he returned to Vienna as an instructor in meteorology for anti-aircraft officer-candidates. He was able to continue his researches during the War years, producing 10 scientific papers on a variety of themes. In 1919, he completed what was to prove to be his last paper on experimental physics – from this time onwards, his scientific papers were entirely of a theoretical nature.

With the collapse and breakup of the Austro-Hungarian empire at the end of the First World War and the poor prospects of obtaining a position in Austria, Schrödinger decided that his future lay in Germany and in the spring of 1920 he became the assistant to Max Wien, the brother of Wilhelm (Willy) Wien, at the University of Jena. This was the beginning of a period of continuous transfer between one university and another as he moved up the academic ladder. In September 1920, he was appointed an extraordinary Professor of Theoretical Physics at the *Technische Hochschule* in Stuttgart and then in the spring of 1921 to an ordinary professorship at the University of Breslau. Finally, in October 1921 he was called to the Professorship of Theoretical Physics at the University of Zürich where he was to remain for the next six years – during this period he carried out his pioneering researches in wave mechanics.

14.1.2 Scientific accomplishments up to 1925

Schrödinger had received a very thorough training in experimental and theoretical physics at the University of Vienna. The most influential of his teachers was Hasenöhrl whose excellent lectures included the foundations of analytic mechanics, the dynamics of deformable bodies, special attention being paid to the solution of partial differential equations and eigenvalue problems, as well as Maxwell's equations, electromagnetic theory, optics, thermodynamics and statistical mechanics. It is striking that Schrödinger's interests in physics spanned a very wide range of topics, whilst the problems of quantum physics occupied a relatively minor role in his early work. He published papers on the kinetic theory of magnetism, on dielectrics and anomalous dispersion, atmospheric electricity and the origin of penetrating radiation. The last topic preceded the discovery in 1912 of the cosmic radiation by Victor Hess, who was first assistant to Stefan Meyer, director of the *Institut für Radiumforschung* in Vienna (Hess, 1913). Schrödinger had worked on the height dependence of the penetrating radiation assuming it originated in the Earth – Hess showed conclusively that the radiation was extraterrestrial and made reference to Schrödinger's paper in his article.

More germane to the present story was Schrödinger's research on the atomic structure of solids. Following the pioneering researches on the specific heat capacities at low temperatures by Einstein and Debye, the lattice theory of the structure of solids had been the subject of important papers by Born and von Kármán (Born and von Kármán, 1912). This work was elaborated theoretically by Schrödinger in his first important paper under the title *On the dynamics of elastically coupled point systems* (Schrödinger, 1914). The theory required the use of advanced methods in the eigenvalue theory for infinite systems, but this posed little difficulty for him because of his training in theoretical physics under Hasenöhrl. In his courses, Hasenöhrl had treated

> 'the higher theoretical schemes of mechanics as well as the problem of eigenvalues in continuum physics.' (Schrödinger, 1935)

During the War years, Schrödinger studied Smoluchowski's theory of fluctuations and general relativity. Towards the end of the War, he wrote a substantial review of atomic and molecular specific heats which necessarily involved the introduction of quantisation to understand their low temperature behaviour (Schrödinger, 1919). He continued his broad interests in physics and other disciplines such as the theory of colours and colour perception. He maintained an interest in quantum problems, for example, in his paper on penetrating orbits which was discussed in Sect. 8.2 and which dates from his short stay at Stuttgart (Schrödinger, 1921), but he was not one of the main protagonists of quantum theory. It therefore came as a complete surprise when Schrödinger discovered wave mechanics which at first sight was a completely different approach from the technically complex methods developed in Göttingen, Copenhagen and Cambridge. Let us trace the steps which led to Schrödinger's dramatic discovery of his wave equation.

14.2 Einstein, De Broglie and Schrödinger

The deep significance of Bose's famous paper of 1924 was fully appreciated by Einstein. It described a new type of statistics for counting photons which resulted in an elegant derivation of the spectrum of black-body radiation. Einstein realised that the counting procedures for indistinguishable particles could be applied equally well to atoms as to photons. The procedures described in Sec. 9.2 led to the expression (9.7) for the number of particles in the state k

$$n_k = \frac{g_k}{e^{\alpha + \beta \varepsilon_k} - 1} , \tag{14.1}$$

where the constants α and β are to be determined and g_k is the degeneracy of the energy level with energy ε_k for photons or atoms. On thermodynamic grounds, $\beta = 1/kT$ and α is determined by fixing the number of photons or atoms. For black-body radiation, α is set equal to zero, meaning that the photon number is not conserved, but that the number density of photons is matched to the total energy under the black-body spectrum – the properties of the black-body spectrum are uniquely defined by the single parameter T, the thermodynamic temperature of the radiation.

Einstein carried out an exactly parallel analysis for the atoms of a monatomic gas, but now the numbers of particles had to be conserved and so α had to be a non-zero constant. The same statistical procedures as in Sect. 9.2 (Einstein, 1924) were used. The total number of available states in momentum space is given by (9.9), but now the momenta of the atoms of the gas are given by $p^2 = 2mE$, rather than by $p = h\nu/c$. Therefore, the volume of phase space for a monatomic gas is

$$V \, dp_x \, dp_y \, dp_z = V 4\pi p^2 \, dp \,, \tag{14.2}$$

where $p = (2mE)^{1/2}$, $dp = \frac{1}{2}(2mE)^{-1/2} \times 2m \, dE$ and V is the physical volume within which the gas is contained. Since the elementary volume of momentum space is h^3, the number of available states with energy in the interval E to $E + dE$ in the volume V is

$$g_k = \frac{V \, 4\pi p^2 \, dp}{h^3} \,, \tag{14.3}$$

and so the number of particles in the energy interval E to $E + dE$ is

$$N(E) \, dE = \frac{g_k}{e^{\alpha + \beta E} - 1} = \frac{4\pi V (2m^3 E)^{1/2}}{h^3 \left(e^{\alpha + \beta E} - 1\right)} \, dE \,. \tag{14.4}$$

In the limit of very low densities and high temperatures, (14.4) becomes the standard Boltzmann distribution and so $\beta = 1/kT$. We can therefore write

$$N(E) \, dE = \frac{g_k}{\left(Be^{E/kT} - 1\right)} = \frac{4\pi V (2m^3 E)^{1/2}}{h^3 \left(Be^{E/kT} - 1\right)} \, dE \,, \tag{14.5}$$

where $B = e^\alpha$.

Einstein fully appreciated the remarkable properties of this new distribution (Einstein, 1924, 1925). B cannot be less than one or else the number of particles could become negative and so $B \geq 1$. When B is very close to 1, the number of particles in low energy states can become very large. In fact, the effect is much more dramatic than this, as Einstein demonstrated in his second paper. A version of Einstein's analysis is given in the endnotes.[2] What had gone wrong was that the continuum approximation (14.3), in which g_i was replaced by $N(E) \, dE$, does not take account of the fact that the zero energy quantised state can contain a finite number of particles. The quantisation of the states, which results in a ground state with zero energy, has to be included in the calculation. At low enough temperatures, $T < T_B$, where

$$T_B = \frac{2\pi\hbar^2}{mk} \left(\frac{N}{2.612V}\right)^{2/3} \,, \tag{14.6}$$

the particles accumulated in the zero energy state and, as $T \to 0$, all the particles would be condensed into the same zero energy state. This is the origin of the phenomenon of the *Bose–Einstein condensation*. The gas is split into two parts, the 'normal phase' in which the particles are distributed among the excited energy states and the 'condensed phase' in which all the remaining atoms of the gas are in the quantised ground state. At low enough energies, essentially all the particles would be in this condensed phase. Although it was not appreciated at the time, this was the first realisation of a phase transition using the techniques of statistical mechanics of identical particles.[3]

Einstein appreciated fully that Bose's statistics were very different from classical Boltzmann statistics. The key difference is that the particles can no longer be considered to be statistically independent. As explained in Sect. 9.2, since the particles are indistinguishable, each possible configuration is counted only once. In Boltzmann statistics the configurations [A|B] and [B|A] are considered as separate realisations of the distribution of distinguishable particles, whereas according to Bose–Einstein statistics, if A and B are indistinguishable, they are counted only once. This induces a statistical correlation between the particles. In his second paper on the Bose–Einstein gas, Einstein was quite clear about the strange nature of the new statistics (Einstein, 1925):

> '[It] therefore expresses indirectly a certain hypothesis about a mutual dependence of the molecules, which for the present is of a totally enigmatic nature and which just creates the same statistical probability for the cases which are defined here as *complexions*.'

There were two other suggestive features of the distribution to which Einstein drew attention. First, the new distribution was consistent with Nernst's heat theorem, or the third law of thermodynamics, according to which the entropy of all gases tends to zero at zero temperature. This is indeed the case for a Bose–Einstein distribution, but not for a classical Boltzmann distribution.

Secondly, just as in the case of black-body radiation, the fluctuations in the number density of particles is composed of two parts. The fractional fluctuation for photons is given by (3.41) and a similar relation is found for the monatomic gas,

$$\overline{\left(\frac{\Delta_\nu}{n_\nu}\right)^2} = \frac{1}{n_\nu} + \frac{1}{z_\nu}, \tag{14.7}$$

where n_ν is the mean number of atoms and z_ν is the number of cells in phase space with energies in the interval E to $E + dE$. The first term is the usual statistical fluctuation in the number of particles according to a Poisson distribution, $\Delta N/N \approx N^{-1/2}$. The second term arises from fluctuations associated with the superposition of random waves, as was demonstrated in Sect. 3.6.2. Again, Einstein fully appreciated the significance of this result:

> 'It arises in the case of radiation from interference fluctuations. We may also interpret it for gases in a corresponding way by associating with the gas, in a suitable manner, a ray phenomenon and then computing the interference fluctuations of the latter.' (Einstein, 1925)

He went on to remark that

'I pursue this interpretation further, since I believe that here we have to do with more than a mere analogy.'

In December 1924, Einstein had been sent a copy of de Broglie's doctoral thesis by Langevin, who had been one of its examiners. Einstein was profoundly impressed, referring to it as 'a very notable publication'. In his paper of 1925, he suggested that the type of matter waves proposed by de Broglie enabled him to interpret the fluctuation term in (14.7) for the Bose–Einstein gas and repeated de Broglie's suggestion that experiments might be carried out to search for the interference of matter waves in beams of particles, although he realised that the effect would be extremely small. The subsequent story of the discovery of electron diffraction by Davisson and George Thomson was recounted in Sect. 9.4, but these results were not established at the time when Schrödinger began to take matter waves seriously.

Schrödinger's correspondence with Einstein began in February 1925 when he was studying the various expressions for the entropy of an ideal gas at low temperatures. Having studied Einstein's first paper on the equation of state of an ideal gas, he wrote to Einstein about his concern that the particle distribution according to Bose–Einstein statistics was inconsistent with the Boltzmann distribution. Einstein responded by stating that his concern was indeed valid, but there was nothing wrong with his calculation. As Einstein wrote,

'Your reproach is not unjustified, although I have not made a mistake in my paper. In Bose statistics, which I use, the quanta or molecules are not considered as being *mutually independent objects*.'

Schrödinger delayed replying until November 1925 when he confessed that he had not appreciated the originality of Einstein's paper. In particular, he was intrigued by the new light it shed upon quantum degeneracy.

In the same letter, Schrödinger makes his first reference to de Broglie's thesis (de Broglie, 1924a). He had been intrigued by Einstein's reference to it in his second paper on Bose–Einstein statistics (Einstein, 1925) and only obtained a copy of it in late summer 1925. He discussed the thesis with Pieter Debye, his colleague at Zurich, who suggested that Schrödinger present a colloquium on de Broglie's ideas. This duly took place in late November or early December 1925. According to Felix Bloch,

'When he had finished, Debye casually remarked that he thought this way of talking was rather childish. As a student of Sommerfeld, he had learned that, to deal properly with waves, one had to have a wave equation. It sounded quite trivial and did not seem to make a great impression, but Schrödinger evidently thought a bit more about the idea afterwards.' (Bloch, 1976)

Just a few weeks later, Schrödinger gave another colloquium in which he is reported to have begun with the following words:

'My colleague Debye suggested that one should have a wave equation; well, I have found one.'

This was the discovery of *Schrödinger's wave equation* for the hydrogen atom.

14.3 The relativistic Schrödinger wave equation

In seeking a wave equation to describe de Broglie's matter waves, Schrödinger began by attempting to find an appropriate relativistic wave equation. The reason is clear from the fact that, as demonstrated in Sect. 9.3, de Broglie used extensively relativistic arguments to find a self-consistent description of the phase relations of the matter waves and the kinematics of the electron. These first attempts at the derivation of the relativistic wave equation were never published, but the argument can be traced in Schrödinger's notebooks and a three-page memorandum he wrote on the eigenvibrations of the hydrogen atom. This first attempt was to be superseded by his first non-relativistic paper on wave mechanics (Mehra and Rechenberg, 1987).

Schrödinger was seeking a time-independent wave equation which would have the standard form

$$\Delta \psi + k^2 \psi = 0 \,, \tag{14.8}$$

where $k = 2\pi/\lambda$ is the wavenumber of the wave. $u = \omega/k$ is the phase velocity of the wave and, in his dissertation, de Broglie had shown that u is the superluminal velocity $u = c^2/v$, where v is the speed of the electron (Sect. 9.3). Schrödinger therefore made the following identification:

$$u = \frac{c^2}{v} = \frac{\gamma m_e c^2}{\gamma m_e v} = \frac{E}{p} = \frac{h\nu}{\gamma m_0 v} \,. \tag{14.9}$$

In place of de Broglie's relation $E = h\nu = \gamma m_e c^2$, Schrödinger includes the electrostatic potential energy so that

$$h\nu = \gamma m_0 c^2 - \frac{e^2}{4\pi \epsilon_0 r} \,. \tag{14.10}$$

These expressions were rearranged to eliminate the velocity v of the electron so that

$$\frac{v}{c} = \left[1 - \frac{m_e c^2}{\left(h\nu + \frac{e^2}{4\pi \epsilon_0 r} \right)^2} \right]^{1/2} \,, \qquad \gamma = \left(\frac{h\nu + \frac{e^2}{4\pi \epsilon_0 r}}{m_e c^2} \right) \,. \tag{14.11}$$

Then, the phase velocity of the electron is

$$u = c \frac{h\nu/m_e c^2}{\left[\left(\frac{h\nu}{m_e c^2} + \frac{e^2}{4\pi \epsilon_0 m_e c^2 r} \right)^2 - 1 \right]^{1/2}} \,. \tag{14.12}$$

Using the relation $k = 2\pi\nu/u$ and substituting for u into the wave equation (14.8) using (14.11), Schrödinger found the following wave equation,

$$\Delta \psi + \frac{4\pi^2 m_e^2 c^2}{h^2} \left[\left(\frac{h\nu}{m_e c^2} + \frac{e^2}{4\pi \epsilon_0 m_e c^2 r} \right)^2 - 1 \right] \psi = 0 \,. \tag{14.13}$$

As we will see, this equation is of the same form as the non-relativistic wave equation.

Schrödinger was now on home ground and knew exactly how to treat (14.13) as an eigenfunction problem. All he had to do was to follow the standard procedure of separating variables in spherical harmonic coordinates, namely $\psi(r, \theta, \phi) = R(r)\,\Theta(\theta)\,\Phi(\phi)$, and apply the appropriate boundary conditions. The functions $\Theta(\theta)$ and $\Phi(\phi)$ could be expressed by the standard angular functions for $m \geq 0$

$$Y_l^m(\theta, \phi) = \Theta(\theta)\,\Phi(\phi) = (-1)^m \left[\frac{(2l+1)}{4\pi} \frac{(l-m)!}{(l+m)!} \right]^{1/2} P_l^m(\cos\theta)\exp(im\phi), \quad (14.14)$$

where $P_l^m(\cos\theta)$ are the standard spherical harmonics and l is a positive integer.[4] For values $m < 0$, the spherical harmonics are

$$Y_l^{-|m|}(\theta, \phi) = (-1)^{|m|}\left[Y_l^{|m|}(\theta, \phi) \right]^*, \quad (14.15)$$

where the asterisk denotes the complex conjugate of $Y_l^m(\theta, \phi)$. The radial equation could be written as an ordinary differential equation in the following form,

$$\frac{\mathrm{d}^2 R}{\mathrm{d}r^2} + \frac{2}{r}\frac{\mathrm{d}R}{\mathrm{d}r} + \left(-A + \frac{2B}{r} - \frac{C}{r^2} \right) R = 0. \quad (14.16)$$

Schrödinger, after a bit of effort, solved this equation and found quantised energy levels, but the solution was not quite right. He was well aware of the fact that one of the triumphs of the old quantum theory had been Sommerfeld's relativistic model for the hydrogen atom which was discussed in Sect. 5.3. The issue is illustrated by the analysis of Mehra and Rechenberg (1987) who show that the solutions for the constants A and B can be written in equivalent form in terms of the quantities A' and B' for both the Sommerfeld and Heisenberg solutions as follows:

$$\text{Sommerfeld}: \quad \frac{2\pi}{h}\frac{B'}{\sqrt{-A'}} = n_r + \sqrt{k^2 - \alpha^2}\,,$$

$$\text{Schrödinger}: \quad \frac{2\pi}{h}\frac{B'}{\sqrt{-A'}} = n_r + \sqrt{\left(k+\tfrac{1}{2}\right)^2 - \alpha^2} - \tfrac{1}{2}\,.$$

The constants A' and B' determine the energies of the stationary states of the electron in the hydrogen atom and Sommerfeld's expression (5.36) was in excellent agreement with experiment. The extra factors of $\frac{1}{2}$ in Schrödinger's expression spoiled that agreement and would require the introduction of half-integral quantum numbers. This was a discouragement for Schrödinger and he set the problem aside for a brief period while he completed other pieces of writing. Although it was not understood at the time, what had gone wrong was that, as soon as a relativistic theory of the hydrogen atom is adopted, the spin of the electron has to be included. This was, however, a significant piece of analysis – it was the first time a wave equation for the electron in the atom had been introduced into quantum physics. Notice that de Broglie's waves were propagating waves whereas Schrödinger had converted the problem into one of standing waves, like the vibrations of a violin string under tension.

14.4 *Quantisation as an Eigenvalue Problem (Part 1)*

14.4.1 Preliminaries

Schrödinger spent the Christmas break of 1925–1926 at the Villa Herwig at Arosa where he intended to relax and enjoy the skiing. His mind was, however, consumed by his recent researches at the expense of what should have been a period of relaxation. As he noted in a letter to Willy Wien of 27 December 1925,

> 'At the moment I am plagued by a new atomic theory... I believe I can write down a vibrating system – constructed in a comparatively natural manner and not by *ad hoc* assumptions – which has as its eigenfrequencies the term frequencies of the hydrogen atom.'

His notebooks show that the first derivation of the non-relativistic wave equation for the hydrogen atom was formulated during that holiday. In fact, the derivation was no more than a straightforward simplification of the relativistic wave equation discussed in Sect. 14.3. In the non-relativistic case, (14.10) becomes

$$h\nu = m_e c^2 + \frac{1}{2} m_e v^2 - \frac{e^2}{4\pi \epsilon_0 r} , \tag{14.17}$$

and the phase velocity of the wave (14.9) becomes

$$u = \frac{E}{p} = \frac{h\nu}{m_e v} . \tag{14.18}$$

As before, eliminating v between (14.17) and (14.18), a simpler expression for the phase velocity is obtained,

$$u = \frac{h\nu}{m_e v} = \frac{h\nu}{\sqrt{2m_e \left(h\nu - m_e c^2 + \frac{e^2}{4\pi \epsilon_0 r} \right)}} , \tag{14.19}$$

and so substituting into the wave equation (14.8),

$$\Delta \psi + \frac{8\pi^2 m_e}{h^2} \left(h\nu - m_e c^2 + \frac{e^2}{4\pi \epsilon_0 r} \right) \psi = 0 . \tag{14.20}$$

This wave equation could again be solved by separation of variables in spherical polar coordinates, $\psi(r, \theta, \phi) = R(r)\,\Theta(\theta)\,\Phi(\phi)$, as in Sect. 14.3, resulting in the same form of ordinary differential equation for $R(r)$,

$$\frac{d^2 R}{dr^2} + \frac{2}{r}\frac{dR}{dr} + \left(-A + \frac{2B}{r} - \frac{C}{r^2} \right) R = 0 , \tag{14.21}$$

but now with much simpler expressions for the coefficients A, B and C:

$$A = \frac{8\pi^2 m_e}{h^2}(m_e c^2 - h\nu) , \qquad B = \frac{\pi m_e e^2}{\epsilon_0 h^2} , \qquad C = l(l+1) , \tag{14.22}$$

where l is an integer corresponding to the azimuthal quantum number in the old quantum theory. The spherical harmonic solution (14.14) forces the condition that the following expression for A, B and C should only take integral values, n_r,

$$\frac{B}{\sqrt{A}} - \sqrt{C + \tfrac{1}{4}} + \tfrac{1}{2} = n_r \ . \tag{14.23}$$

Inserting the above values for A, B and C, he found

$$\frac{B}{\sqrt{A}} = \sqrt{\frac{m_e e^4}{8\epsilon_0^2 h^2 (m_e c^2 - h\nu)}} = (n_r + l) = n \ . \tag{14.24}$$

Squaring both sides and inserting the values of A and B, we find

$$\frac{R_\infty h}{m_e c^2 - h\nu} = n^2 \ , \tag{14.25}$$

which can be reorganised as follows:

$$E = h\nu = m_e c^2 - \frac{R_\infty h c}{n^2} \ , \tag{14.26}$$

where R_∞ is the Rydberg constant, $R_\infty = m_e e^4 / 8\epsilon_0^2 h^3 c$. This was a startling result – the energies of the standing waves were exactly those of the stationary states of the hydrogen atom and the differences in energies between them would result in Bohr's formula for the lines observed in the hydrogen spectrum.

Schrödinger still had a lot to do. In particular, he still had to solve (14.21) for the radial dependence of ψ upon radius. He had with him Schlesinger's textbook *Introduction to the Theory of Differential Equations* (1900) and struggled to find the solution. In fact, Courant and Hilbert's *Methods of Mathematical Physics, Volume 1* had appeared in 1924 and gave a solution to a similar equation in terms of Laguerre polynomials – the actual solution is obtained using associated Laguerre polynomials, which appear in Part 3 of Schrödinger's papers of 1926 with the same title *Quantisation as an eigenvalue problem* (Schrödinger, 1926e).

14.4.2 The non-relativistic theory of the hydrogen atom

Within a few weeks, Schrödinger completed the first of his great series of six papers and it was received by the *Annalen der Physik* on 27 January 1926. The ideas and approach were refined with the striking feature that he devotes most of the arguments to the mathematical requirements of the wave equation he had discovered, rather than relating the formalism to the physics of the hydrogen atom, or quantum phenomena in general. He explains the reason for this approach in the third section of his paper.

> 'It is, of course, strongly suggested that we should try to connect the function ψ with some *vibration process* in the atom, which would more nearly approach reality than the electronic orbits, the real existence of which is being very much questioned today. I originally intended to found the new quantum conditions in this more intuitive manner, but finally gave them the above neutral mathematical form, because it brings more clearly to light what is really essential. The essential thing seems to me to be, that the postulation

of "whole numbers" no longer enters into the quantum rules mysteriously, but that we have traced the matter a step further back, and found the "integralness" to have its origin in the finiteness and single-valuedness of a certain space function.'

Schrödinger was to defer the more physical arguments to the second paper of the series.

His first step was to rederive the wave equation (14.20) in a somewhat more formal manner, starting from the Hamilton–Jacobi differential equation (5.86),

$$H\left(q, \frac{\partial S}{\partial q}\right) = E , \qquad (14.27)$$

which was derived in Sect. 5.4.4. He points out that the solutions for S are normally found in terms of the *sum* of functions, each being a function of only one independent variable q_i. This was illustrated in the case of Sommerfeld's analysis of the elliptical orbits in the old quantum theory in Sect. 5.5. Rewriting (5.98) in our present notation in Cartesian coordinates,

$$H = \frac{1}{2m}\left[\left(\frac{\partial S}{\partial x}\right)^2 + \left(\frac{\partial S}{\partial y}\right)^2 + \left(\frac{\partial S}{\partial z}\right)^2\right] - \frac{Ze^2}{4\pi\epsilon_0 r} = E . \qquad (14.28)$$

Now, in his analysis of the wave equation (14.20), Schrödinger had shown that the solution should be sought in terms of the products of independent functions as illustrated by the separation of variables which led to (14.21). In contrast, the analysis of Sect. 5.5 showed that the solution is written as the sum of independent functions of the orthogonal coordinates r, θ, ϕ. In Cartesian notation, (5.99) would be written

$$S = S_x(x) + S_y(y) + S_z(z) . \qquad (14.29)$$

The solution is to write S as the logarithm of some function ψ such that $S = K \ln \psi$. Then, (14.27) can be written

$$H\left(q, \frac{K}{\psi}\frac{\partial \psi}{\partial q}\right) = E . \qquad (14.30)$$

He states that he is to apply this formalism to the non-relativistic case of the hydrogen atom. Then comes the key to the new approach:

'We now seek the function ψ, such that for any arbitrary variation of it the integral of the said quadratic form, taken over the whole of coordinate space, is stationary, ψ being everywhere real, single-valued, finite, and continuously differentiable up to second order. *The quantum conditions are replaced by this variation problem.*'

Now, (14.28) can be rewritten

$$H = \frac{K^2}{2m}\left[\left(\frac{1}{\psi}\frac{\partial \psi}{\partial x}\right)^2 + \left(\frac{1}{\psi}\frac{\partial \psi}{\partial y}\right)^2 + \left(\frac{1}{\psi}\frac{\partial \psi}{\partial z}\right)^2\right] - \frac{Ze^2}{4\pi\epsilon_0 r} = E , \qquad (14.31)$$

or, converting this into an appropriate form for finding stationary solutions for ψ,

$$J = \left(\frac{\partial \psi}{\partial x}\right)^2 + \left(\frac{\partial \psi}{\partial y}\right)^2 + \left(\frac{\partial \psi}{\partial z}\right)^2 - \frac{2m}{K^2}\left(E + \frac{Ze^2}{4\pi\epsilon_0 r}\right)\psi^2 = 0 . \qquad (14.32)$$

This is the quadratic form which is to be subject to arbitrary variations over the whole of coordinate space. The stationary solution is found by requiring $\delta J = 0$, that is,

$$\delta J = \delta \iiint \mathrm{d}x\, \mathrm{d}y\, \mathrm{d}z \left[\left(\frac{\partial \psi}{\partial x}\right)^2 + \left(\frac{\partial \psi}{\partial y}\right)^2 + \left(\frac{\partial \psi}{\partial z}\right)^2 - \frac{2m}{K^2}\left(E + \frac{Ze^2}{4\pi \epsilon_0 r}\right)\psi^2 \right] = 0,$$
(14.33)

where the integral is taken over all space. Schrödinger then states that

'From this we find in the usual way,'

$$\tfrac{1}{2}\delta J = \oint_S \left(\frac{\partial \psi}{\partial n}\right)\delta\psi\, \mathrm{d}A - \iiint \mathrm{d}x\, \mathrm{d}y\, \mathrm{d}z\, \delta\psi \left[\nabla^2 \psi + \frac{2m}{K^2}\left(E + \frac{Ze^2}{4\pi \epsilon_0 r}\right)\psi^2 \right] = 0,$$
(14.34)

where $\mathrm{d}n$ is the differential element of distance perpendicular to the element of area $\mathrm{d}A$. We outline in the endnotes how this expression is derived using the tools developed in Sect. 5.4.2.[5] For arbitrary variations $\delta\psi$, both terms must be zero and so it follows that

$$\nabla^2 \psi + \frac{2m}{K^2}\left(E + \frac{e^2}{4\pi \epsilon_0 r}\right)\psi = 0,$$
(14.35)

with the requirement that the first integral of (14.34) should be zero when taken over 4π steradians at infinity,

$$\oint_S \left(\frac{\partial \psi}{\partial n}\right)\delta\psi\, \mathrm{d}A = 0.$$
(14.36)

Schrödinger had already derived an equation of similar form starting from his interpretation of de Broglie's concept of matter waves which resulted in (14.20). Comparing (14.20) and (14.35), it immediately follows that $K = h/2\pi$. Therefore,

$$\nabla^2 \psi + \frac{8\pi^2 m}{h^2}\left(E + \frac{e^2}{4\pi \epsilon_0 r}\right)\psi = 0.$$
(14.37)

This is the definitive version of *Schrödinger's time-independent wave equation* for the hydrogen atom.[6] Most of the rest of the paper is concerned with a careful analysis of the mathematical properties of the wave equation (14.37).

First of all, Schrödinger recognised that, because of the spherical symmetry of the problem, spherical polar coordinates were the natural system to adopt. He sought solutions in these coordinates by separation of variables, $\psi = R(r)\,\Theta(\theta)\,\Phi(\phi)$, the functions $Y(\theta, \phi) = \Theta(\theta)\,\Phi(\phi)$ being the usual spherical harmonics given by (14.14) and (14.15). The key feature of these functions is that, in order that the spherical harmonics are single-valued, l can only take positive integral values $l \geq 0$ while m can only take integral values in the range $-l \leq m \leq l$. The spherical harmonics form a complete, orthogonal set of functions so that any distribution on a sphere can be decomposed into the sum of spherical harmonics.

The separation of variables results in the following equation for $R(r)$,

$$\frac{\mathrm{d}^2 R}{\mathrm{d}r^2} + \frac{2}{r}\frac{\mathrm{d}R}{\mathrm{d}r} + \left[\frac{8\pi^2 m_e E}{h^2} + \frac{8\pi^2 m_e e^2}{h^2 r} - \frac{l(l+1)}{r^2} \right] R = 0,$$
(14.38)

where $l = 0, 1, 2, 3 \ldots$ Schrödinger now carries out a detailed analysis of the properties of acceptable solutions of this equation, paying particular attention to the issues of singularities at $r = 0$ and $r = \infty$. The solutions are found in terms of Laplace transformations and are characterised by the quantity

$$\frac{m_e e^2}{\sqrt{-8\epsilon_0^2 h^2 m_e E}}. \tag{14.39}$$

He then establishes the following results – the italics are Schrödinger's:

1. *[(14.38)] has, for every positive E, solutions which are everywhere single-valued, finite, and continuous; and which tend to zero as $1/r$ at infinity under continual oscillations.*
2. *For negative values of E which do not satisfy the condition* [that (14.39) takes integral values] *our variation problem has no solution.*
3. *For negative values of E, our variation problem has solutions, if and only if, E satisfies the condition*

$$\frac{m_e e^2}{\sqrt{-8\epsilon_0^2 h^2 m_e E}} = n, \tag{14.40}$$

where $n = 1, 2, 3, 4, \ldots$
4. *There are no solutions of the problem if $l \geq n$. Only values smaller than n (and there is always one such at our disposal) can be given to the integer l, which denotes the order of the surface harmonic appearing in the equation.*

The energies of the stationary states can now be found from (14.40) and are

$$E_n = -\frac{m_e e^4}{8\epsilon_0^2 n^2 h^2}. \tag{14.41}$$

These are precisely the energies of the stationary states of the electron in the hydrogen atom.

Schrödinger fully appreciated the significance of these calculations. Immediately after his derivation of the energies of the stationary states of the hydrogen atom, he explains the significance of the quantum numbers – I have translated the quantum numbers into modern usage.

> 'Our n is the principal quantum number. $l + 1$ is analogous to the azimuthal quantum number. The splitting up of this number through a closer definition of the surface harmonic can be compared with the resolution of the azimuthal quantum number into an "equatorial" and "polar" quantum. These numbers *here* define the system of nodes-lines on the sphere. Also the "radial quantum number" $n - l - 1$ gives exactly the number of the "node-spheres", for it is easily established that the function $[R(r)]$ has exactly $n - l - 1$ positive real roots. The positive E-values correspond to the continuum of hyperbolic orbits, to which one may ascribe, in a certain sense, the radial quantum number ∞.'

He also notes that the amplitudes of the standing wave solutions for $R(r)$ tend to zero at radii greater than a_n/n, where a_n is the corresponding Bohr–Sommerfeld value for the semi-major axis of the classical elliptical orbits.

14.4.3 Reflections

Schrödinger realised the importance of these calculations in providing insight into the nature of quantisation. Close to the end of his paper, he writes:

> 'It is hardly necessary to emphasise how much more congenial it would be to imagine that at a quantum transition the energy changes from one form of vibration to another, than to think of a jumping electron. The changing of the vibration form can take place continuously in space and time, and it can readily last as long as the emission process lasts empirically.'

He had made the remarkable discovery that quantisation can be introduced into the description of quantum phenomena through the essential requirement that the wavefunction ψ should be everywhere *real, single-valued, finite, and continuously differentiable up to second order*. As he emphasised, the somewhat arbitrary quantum conditions of the old quantum theory are replaced by these constraints on the properties of the wavefunction.

The paper gives the impression of being written in the white heat of inspiration – Schrödinger realises that there are many loose ends to be tied up, but he was driven to publish his discovery in an almost unvarnished form. He fully acknowledges the importance of de Broglie's papers:

> 'Above all, I wish to mention that I was led to these deliberations in the first place by the suggestive papers of M. Louis de Broglie (de Broglie, 1924a), ... '

It is perfectly understandable that he should give the highest priority to publishing his wave equation, without going into a more thorough examination of his intellectual breakthrough. Still working at breakneck speed, his next paper *Quantisation as an eigenvalue problem (Part 2)* was received by the *Annalen der Physik* on 23 February 1926. Part 2 might more logically have preceded Part 1 since it provides a deeper insight into how the wave equation can be formulated as an extension of the principles of Fermat and Hamilton. The analysis of that paper is our next task.

14.5 *Quantisation as an eigenvalue problem (Part 2)*

Schrödinger's second paper was concerned with providing a more formal basis for his wave equation and providing further examples of its application (Schrödinger, 1926c). The attempt to place Lagrangian mechanics and physical optics on the same formal basis had been pioneered by Hamilton, inspired by Lagrange's approach to classical mechanics and Fresnel's success in developing the wave theory of optics to describe diffraction and interference phenomena. Hamilton's objective was no less than to provide a single unified theory which would describe both the motion of particles and the propagation of light rays (Hamilton, 1833). He achieved this by showing that his characteristic function S, introduced in Sect. 5.4, could be applied to both the dynamics of particles and the paths of light rays.

14.5.1 The fundamentals of undulatory mechanics

For light rays, *Fermat's principle of least time* states that the path of a light ray is that which minimises the time between the source of light and its detection through a medium in which the refractive index n varies with position

$$\delta \int \frac{n}{c}\, dl = 0\,, \tag{14.42}$$

where c is the speed of light and dl is the element of distance. Snell's law of refraction, $n_1 \sin \theta_1 = n_2 \sin \theta_2$, follows immediately from the geometrical application of the principle, where n_1 and n_2 are the refractive indices of the medium and the angles θ_1 and θ_2 are the angles of incidence and refraction of the ray relative to the normal to the surface.

The analogue of Fermat's principle for mechanical systems can be readily derived from *Maupertuis' principle*, historically the first of the variational principles, and a variant of the *principle of least action*. In the present instance, we are only interested in cases in which the time does not appear explicitly so that neither the Lagrangian nor the Hamiltonian depend upon time, in other words, the systems involve conservative fields of force. In terms of general coordinates, the Hamiltonian $H(p,q) = E = $ constant.

Landau and Lifshitz (1976) demonstrate that, under these conditions, Maupertuis' principle can be written

$$\delta S_0 = 0 \quad \text{where} \quad S_0 = \int \sum_i p_i\, dq_i \,, \tag{14.43}$$

where p_i and q_i are the generalised coordinates introduced in Sect. 5.4.3, with $p_i = \partial \mathcal{L}/\partial \dot{q}_i$. Schrödinger uses the principle in its simplest form in which p_i is the Cartesian momentum $p = mv$ and q the position coordinate. Then the trajectory of a particle is given by that path for which the minimisation condition is

$$\delta \int \boldsymbol{p} \cdot \frac{dl}{dt}\, dt = \delta \int \boldsymbol{p} \cdot \boldsymbol{v}\, dt = \delta \int 2T\, dt = 0\,, \tag{14.44}$$

where T is the kinetic energy.[7] Expression (14.44) was to be the starting point for Schrödinger's next attack on the fundamentals of what he called in this paper *undulatory mechanics*.

Starting from Maupertuis' principle (14.44), dt can be replaced by dl/v where v is the velocity of the particle. Since the kinetic energy of the particle is $\frac{1}{2}mv^2$, Maupertuis' principle becomes

$$\delta \int mv\, dl = 0\,. \tag{14.45}$$

But, we can write v in terms of the energy E and the potential energy U as $\frac{1}{2}mv^2 + U = E$ and so the minimisation procedure becomes

$$\delta \int [2m(E - U)]^{1/2}\, dl = 0\,. \tag{14.46}$$

According to Hamilton, the minimisation procedure (14.46) should be the same as Fermat's principle of least time (14.42). Therefore, the equivalence between the phase velocity v of the wave and the motion of a particle can be written symbolically

$$v_{\text{wave}} = \frac{c}{n} \quad \Rightarrow \quad \frac{C}{[2m(E-U)]^{1/2}} , \tag{14.47}$$

where C is a constant. Jammer (1989) refers to this equivalence as *Hamilton's optical-mechanical analogy*.

Hamilton took the analogy further. Hamilton's action function is defined to be $S = \int_{t_0}^{t} \mathcal{L} \, dt$, where $\mathcal{L} = (T - U)$ is the Lagrangian, and defines an *action surface* $S(x, y, z, t) =$ constant. The properties of the action surface are directly related to the dynamics of the particles of the system. From the action function, the following relations can be derived, as demonstrated in the endnotes.[8] The momentum and total energy of the particle are given by

$$\boldsymbol{p} = \nabla S \quad \text{and} \quad \frac{\partial S}{\partial t} = \mathcal{L} - pv = -E , \tag{14.48}$$

where E is the constant total energy for motion under conservative forces. The corresponding equations for the wavefront of a wave of the form $\exp(i\phi) = \exp[i(\boldsymbol{k} \cdot \boldsymbol{r} - \omega t)]$ are

$$\boldsymbol{k} = \nabla \phi \quad \text{and} \quad \frac{\partial \phi}{\partial t} = -\omega , \tag{14.49}$$

where ϕ is the phase factor of the wave and $\omega = 2\pi v$. Comparing (14.48) and (14.49), the surfaces of constant action of a system of particles are the exact analogues of the surfaces of constant phase for optical waves. In addition, the wavevector $\boldsymbol{k}$ is the analogue of the momentum $\boldsymbol{p}$ and the angular frequency ω the analogue of the energy E of the particle. These insights, published by Hamilton between 1828 and 1837 (Hamilton, 1931), were far ahead of their time. The formal analogy was apparent, but there was no obvious interpretation of the velocity v as a wave velocity. With the exception of few theorists, such as Felix Klein, Hamilton's optical-mechanical analogy was neglected until Schrödinger brought the concept to the forefront of quantum theory in 1926.

In his second paper, Schrödinger emphasises that Fermat's principle in optics and Hamilton's principle in mechanics provide the classical equivalence of ray optics and particle mechanics. Fermat's principle makes no reference to the wave nature of light. In contrast, the variational procedures for wave motion were deeply embedded in Fresnel's approach to wave optics and so there should be a parallel wave equivalence for particle motion. Classical ray optics works extremely well, provided the scale of the system is large compared with the wavelength of light. It breaks down, however, when the wavelength is of the order of scale of the system, giving rise to the characteristic diffraction and interference phenomena. Schrödinger postulates that the same should be true for particle mechanics. Classical mechanics works very well, provided the scale of the system is very much greater than the typical scale over which quantum phenomena are observed. On the atomic scale, however, the wave properties of matter cannot be neglected if the scale of the system is of the same order as the de Broglie wavelength, $\lambda = h/p$, where p is the momentum of the particle.

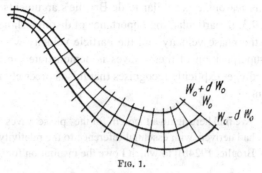

FIG. 1.

Fig. 14.1 Schrödinger's visualisation of the motion of the action surface defined by $S = \int_{t_0}^{t} \mathcal{L} \, dt$, where $\mathcal{L} = T - U$ is the Lagrangian (Schrödinger, 1926c).

Schrödinger first discusses the dynamics of the particle in terms of the motion of the action surface and shows that its normal velocity is

$$v = \frac{E}{\sqrt{2m(E - U)}}. \tag{14.50}$$

His pictorial representation of the motion of the action surface is shown in Fig. 14.1. We can derive this relation directly from the equivalences discovered by Hamilton between the velocity of the action surface and that of a wavefront, (14.49) and (14.50) respectively. Thus, the phase velocity of the wave or action surface is

$$v_{\text{ph}} = \left| \frac{\omega}{k} \right| = \frac{\partial \phi / \partial t}{\nabla \phi} \equiv \frac{\partial S / \partial t}{\nabla S} = \frac{E}{p} = \frac{E}{\sqrt{2m(E - U)}}. \tag{14.51}$$

The problem which faced Hamilton and the nineteenth century theorists was that (14.50) is not the velocity of the particle which is

$$v_{\text{part}} = \frac{\sqrt{2m(E - U)}}{m}. \tag{14.52}$$

This was the principal reason why Hamilton's insights were neglected.

Schrödinger realised, however, that, according to undulatory mechanics, the expression (14.51) corresponded to a dispersion relation for standing waves associated with the stationary states of the electron in the atom. The energy of the stationary state is $E = \hbar\omega$ and so (14.52) becomes

$$\frac{\omega}{k} = \frac{\hbar\omega}{\sqrt{2m(\hbar\omega - U)}} \; ; \quad k = \frac{1}{\hbar}\sqrt{2m(\hbar\omega - U)}. \tag{14.53}$$

Hence the group velocity $v_{\text{gr}} = d\omega/dk$, which is the velocity of a wave-packet, is given by

$$\frac{dk}{d\omega} = \frac{1}{v_{\text{gr}}} = \frac{m}{\sqrt{2m(\hbar\omega - U)}} \; ; \quad v_{\text{gr}} = \frac{\sqrt{2m(E - U)}}{m}. \tag{14.54}$$

This is precisely the expression for the velocity of the particle. Note also that $v_{\text{ph}} v_{\text{gr}} = E/m$.

This reasoning is similar to de Broglie's arguments which were developed in detail in Sect. 9.3. In particular, the importance of distinguishing the de Broglie waves which move with the phase velocity and the particle velocity which travels at the group velocity of the superposition of these waves is demonstrated in the expressions (9.15) and (9.16). Schrödinger explicitly recognises that this is precisely the insight provided by de Broglie, writing

> 'We find here again a theorem for the "phase waves" of the electron, which M. de Broglie had derived, with essential reference to the relativity theory, in those fine researches [(de Broglie, 1924b)] to which I owe the inspiration for this work.'

Schödinger devotes considerable attention and caution to the formal underpinning of undulatory mechanics according to the precepts of Hamilton–Jacobi theory and then adopts the simplest means of formulating the wave equation. The standard wave equation for ψ can be written

$$\nabla^2 \psi - \frac{1}{v_{\text{ph}}^2} \ddot{\psi} = 0 \,. \tag{14.55}$$

Time-independent solutions are sought in which the wavefunction ψ depends upon time as $\exp(i\omega t) = \exp(i 2\pi \nu t)$ and so

$$\nabla^2 \psi + \frac{\omega^2}{v_{\text{ph}}^2} \psi = 0 \,. \tag{14.56}$$

Now substituting (14.52) for v_{ph} and recalling that $E = h\nu$, we find

$$\nabla^2 \psi + \frac{8\pi^2 m}{h^2}(E - U)\psi = 0 \,, \quad \text{or} \quad \nabla^2 \psi + \frac{8\pi^2 m}{h^2}(h\nu - U)\psi = 0 \,. \tag{14.57}$$

These equations are exactly the same as the wave equation (14.37) derived in his first paper. Schrödinger is duly cautious about the uniqueness of this equation, but argues on grounds of simplicity that it will be adopted as the wave equation for describing quantum phenomena. He also notes that, as a partial differential equation, very large numbers of solutions are possible. Although the quantum relation $E = h\nu$ has been introduced, the quantisation of the energy levels of atomic systems is not bolted on arbitrarily, but is determined by the boundary conditions needed to satisfy the requirements that the function ψ must be single-valued, finite and continuous throughout configuration space. He had applied these concepts in Part 1 and now he gives further examples of the power of these methods.

14.5.2 Applications

The solution of Schrödinger's wave equation for the hydrogen atom was a truly impressive feat and now he extended the concepts to the problems which had already been treated with success by Born, Heisenberg and Jordan (1926). By the time he was working on Part 2 of his series of papers, he had discovered Courant and Hilbert's *Methods of Mathematical Physics* (1924) and was bowled over by its contents. In an enthusiastic letter to Wien of 22 February 1926, he writes:

'Time is flying. Each second or third day brings with it a small novelty – *it* works, not I, and that "It" is the magnificent classical mathematics and Hilbert's mathematics, the wonderful edifice of eigenvalues. These unfold everything so clearly before us, that all we have to do is take it, without any labour and bothering; because the correct method is provided in time, as soon as one needs it, completely automatically. I am so happy to have escaped from terrible mechanics, including its action and angle variables and the perturbation theory, which I have never really understood. Now everything becomes *linear*, everything can be superposed; one computes as easily and comfortably as in good old acoustics. Even perturbation theory [in the new mechanics] is not more complicated than [considering] the forced vibrations of a string.'

This remarkable coincidence of the appearance of Courant and Hilbert's book and the immediate application of eigenfunctions and eigenvalues to wave mechanics was one of the reasons for Schrödinger's amazingly rapid progress in 1926.

The Planck oscillator

The first new application was to the harmonic oscillator, referred to as the Planck oscillator because of its fundamental role in Planck's pioneering paper of 1900. Writing the kinetic and potential energy terms for the one-dimensional harmonic oscillator as $T = \frac{1}{2}m\dot{x}^2$ and $U = \frac{1}{2}m\omega_0^2 x^2$, the wave equation becomes

$$\frac{\mathrm{d}^2\psi}{\mathrm{d}x^2} + \frac{8\pi^2 m}{h^2}\left(E - \frac{1}{2}m\omega_0^2 x^2\right)\psi = 0 \,. \tag{14.58}$$

Changing the notation to $a = 8\pi^2 mE/h^2$ and $b = 4\pi^2 m^2 \omega_0^2/h^2$, this equation becomes

$$\frac{\mathrm{d}^2\psi}{\mathrm{d}x^2} + (a - bx^2)\psi = 0 \,. \tag{14.59}$$

Changing variables to $y = xb^{1/4}$, the equation is reduced to a standard form

$$\frac{\mathrm{d}^2\psi}{\mathrm{d}y^2} + \left(\frac{a}{\sqrt{b}} - y^2\right)\psi = 0 \,. \tag{14.60}$$

Eigenfunction solutions for this equation had been presented in Courant and Hilbert's *Methods of Mathematical Physics* (1924). The eigenvalues can only take the values $a/\sqrt{b} = 1, 3, 5, \ldots, (2n + 1), \ldots$ and the eigenfunctions are

$$\psi(y) = \mathrm{e}^{-y^2/2}\, H_n(y) \,, \tag{14.61}$$

where $H_n(y)$ are the orthogonal Hermite polynomials, the first few of which are

$$H_0(y) = 1 \qquad\qquad H_1(y) = 2y$$
$$H_2(y) = 4y^2 - 2 \qquad\qquad H_3(y) = 8y^3 - 12y$$
$$H_4(y) = 16y^4 - 48y^2 + 12 \qquad\qquad \cdots$$

From the definitions of a and b, the eigenvalues are therefore

$$\frac{a}{\sqrt{b}} = \frac{2E}{h\nu_0} = 1, 3, 5, \ldots (2n - 1), \ldots \,, \tag{14.62}$$

that is,

$$E = \left(n + \tfrac{1}{2}\right) h\nu_0 = \left(n + \tfrac{1}{2}\right) \hbar\omega_0 \,. \tag{14.63}$$

This striking relation is identical to Heisenberg's results obtained from the matrix mechanics approach to quantum phenomena. Schrödinger was also able to derive the forms of the eigenfunctions, to which we will return in the next section. He was well aware of the fact that the solution included the *zero point energy*, $\tfrac{1}{2}h\nu_0$, and that it is observed in measurements of the frequencies of band edges (see Sects. 11.5 and 12.5).

Rotator with a fixed axis

In this case, the energy is entirely in the kinetic energy of rotation of the rotator and the only variable is the phase ϕ of the angle of rotation. For this rotational motion, the equivalence with the linear motion of the electron in a harmonic oscillator is

$$\tfrac{1}{2}m\dot{x}^2 \equiv \tfrac{1}{2}I\dot{\phi}^2 \,, \quad m \equiv I \,, \quad x \equiv \phi \,. \tag{14.64}$$

Therefore, the wave equation becomes

$$\frac{d^2\psi}{d\phi^2} + \frac{8\pi^2 I}{h^2} E\psi = 0 \,. \tag{14.65}$$

This is a simple harmonic equation with solution

$$\psi(\phi) = \frac{\sin}{\cos} \left[\left(\frac{8\pi^2 I}{h^2}\right)^{1/2} \phi \right] \,. \tag{14.66}$$

The quantisation arises from the requirement that the wavefunction be single-valued and continuous and so $(8\pi^2 I/h^2)^{1/2} = 1, 2, 3, \ldots$, and the quantised energy levels of the rotator are

$$E_n = \frac{n^2 h^2}{8\pi^2 I} \,, \tag{14.67}$$

in agreement with previous quantum arguments.

Rigid rotator with free axes

In this case, the motion of the rotator is constrained such that the ends of any diameter lie on the surface of a sphere. Therefore, the wave equation has to be written in spherical polar coordinates. The kinetic energy in the θ and ϕ directions is written in terms of the angular momentum about the i_θ and i_ϕ directions. Since $T = \tfrac{1}{2}m(v_\theta^2 + v_\phi^2)$ and the angular momenta about these axes are $L_\theta = mr\,v_\theta$ and $L_\phi = mr\sin\theta\,v_\phi$,

$$T = \frac{1}{2I}\left(L_\theta^2 + \frac{L_\phi^2}{\sin^2\theta}\right) \,. \tag{14.68}$$

Therefore, the wave equation becomes

$$\nabla^2\psi = \frac{1}{\sin\theta}\frac{\partial}{\partial\theta}\left(\sin\theta\frac{\partial\psi}{\partial\theta}\right) + \frac{1}{\sin^2\theta}\frac{\partial^2\psi}{\partial\phi^2} + \frac{8\pi^2 I E}{h^2}\psi = 0, \tag{14.69}$$

where there is no radial dependence since the circumference of the rotator is constrained to lie on the surface of a fixed sphere – the Laplacian operator can only depend upon the polar angles θ and ϕ. The solution is sought by the usual procedure of separating variables, $\psi = \Theta(\theta)\,\Phi(\phi)$, and then the quantisation condition enters from the requirement that the functions $\Theta(\theta)$ and $\Phi(\phi)$ are continuous and single valued on the surface of the sphere. These conditions lead to the requirement that only discrete values of the angular quantum numbers are possible so that the energy eigenvalues are

$$E_l = \frac{l(l+1)h^2}{8\pi^2 I}\,, \tag{14.70}$$

where $l = 0,\ 1,\ 2,\ 3,\ \dots$ Notice that we are using modern standard notation for the angular momentum quantum number l, rather than Schrödinger's n. This expression will be recognised as the standard expression for the quantised energy levels of, for example, a diatomic molecule.[9]

The non-rigid rotator – the diatomic molecule

In the final part of his paper, Schrödinger tackles the problem of the rotating and vibrating molecule. The problem now involves six degrees of freedom as well as harmonic coupling between the two atoms of the diatomic molecule. We simply quote Schrödinger's final result for the quantised energy levels:

$$E = E_t + \frac{l(l+1)h^2}{8\pi^2 I}\left(1 - \frac{\epsilon}{1+3\epsilon}\right) + (n + \tfrac{1}{2})h\nu_0\sqrt{1+3\epsilon}\,, \tag{14.71}$$

where $n = 0,\ 1,\ 2,\dots$ and $l = 0,\ 1,\ 2,\dots$ The small quantity

$$\epsilon = \frac{l(l+1)h^2}{16\pi^4\nu_0^2 I^2} \tag{14.72}$$

is the ratio of the rotational to vibrational energy of the molecule. E_t is the molecule's translational energy. The second and third terms correspond to the quantised rotational and vibrational energies and are the familiar terms with small corrections represented by the small quantity ϵ for the coupling between the modes. Schrödinger recognised that this calculation does not take account of the important deviations from a harmonic potential for a more realistic model of interatomic forces. For this, perturbation solutions of the wave equation are needed and the appropriate procedures were to be developed in Part 3 of Schrödinger's series of papers.

14.6 Wave-packets

Schrödinger's third paper, published on 9 July 1926, was a short note to *Die Naturwissenschaften* concerning the representation of an oscillator in undulatory mechanics by a wave-packet, defined by the superposition of the eigenfunctions of the harmonic oscillator

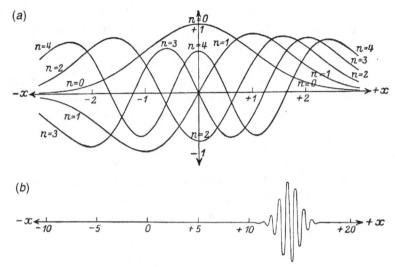

Fig. 14.2 (*a*) The first five eigenfunctions of a harmonic oscillator $\psi_n = e^{x^2/2}\, H_n(x)$. The functions decay exponentially outside the range $-3 \leq x \leq 3$ shown in the diagram. (*b*) The wave-packet associated with the Planck, or harmonic, oscillator as represented by the superposition of the wavefunctions $\psi_n(x)$ (Schrödinger, 1926a).

(Schrödinger, 1926a). He summarises the results of the calculations for the wavefunctions of the harmonic oscillator as follows:

$$\psi_n = e^{-y^2/2}\, H_n(y)\, e^{i\omega_n t} , \tag{14.73}$$

where $\omega_n = (n + \frac{1}{2})\omega_0$ and we have used the notation of Sect. 14.5.2. $H_n(y)$ are the Hermite polynomials and $y = x\sqrt{2\pi m\omega_0/h}$. The functions (14.73) are normalised by multiplying by $(2^n n!)^{-1/2}$ and these are referred to as Hermite's orthogonal functions – the first five normalised wavefunctions are displayed in the range $-3 \leq y \leq +3$ in Fig. 14.2a, which is taken from Schrödinger's paper. Outside this range the wavefunctions decrease exponentially to zero.

To create a wave-packet, Schrödinger adopts the complete set of eigenfunctions and assumes that they are of large amplitude $A \gg 1$. Then, he chooses the wave-packet to be described by the following function

$$\psi = \sum_{n=0}^{\infty} \left(\frac{A}{2}\right)^n \frac{\psi_n}{n!} = e^{i\omega_0 t/2} \sum_{n=0}^{\infty} \left(\frac{A}{2} e^{i\omega_0 t}\right)^n \frac{1}{n!} e^{-y^2/2} H_n(y) . \tag{14.74}$$

The significance of (14.74) is that Hermite's orthogonal functions $(2^n n!)^{-1/2}\psi_n$ are weighted by the factor $A^n/\sqrt{2^n n!}$. The latter function can be compared with the function $z^n/n!$ which, for large values of n has a sharp maximum at $n = z$. Thus, by weighting the eigenfunctions with this factor, a narrow range of values of n is selected about the value $n = A^2/2$. This choice of weighting also has the advantage that the series (14.74) can be summed exactly since

$$\sum_{n=0}^{\infty} \frac{s^n}{n!} e^{-y^2/2} H_n(y) = \exp\left(-s^2 + 2sy - \frac{y^2}{2}\right) . \tag{14.75}$$

Hence,

$$\psi(y, t) = \exp\left(i\omega_0 t/2 - \frac{A^2}{4}e^{i2\omega_0 t} - \frac{y^2}{2}\right). \tag{14.76}$$

Taking the real part of (14.77), the evolution of the wave-packet is

$$\psi = \exp\left[\frac{A^2}{4} - \tfrac{1}{2}(y - A\cos\omega_0 t)^2\right]\cos\left[\omega_0 t/2 + (A\sin\omega_0 t)\cdot\left(y - \frac{A}{2}\cos\omega_0 t\right)\right]. \tag{14.77}$$

This is the remarkable final result of Schrödinger's calculation, one which was to usher in a new approach to understanding the nature of physics at the quantum level. The solution is illustrated in Fig. 14.2b. Schrödinger explains carefully the significance of each term in (14.77). The first term in large square brackets is a Gaussian error-curve which is centred at the position $y = A\cos\omega_0 t$. The width of the distribution is of order unity and so is small compared with the amplitude of the oscillation of the error-curve along the y-axis. If we now revert to the x-coordinate, the amplitude a of the oscillation is

$$a = A\sqrt{h/2\pi m\omega_0}, \tag{14.78}$$

and so the classical energy of the oscillator, if the particle is assumed to have mass m, is

$$E_{\text{vib}} = \tfrac{1}{2}\omega_0^2 a^2 m = \frac{A^2}{2}\hbar\omega_0 = n\hbar\omega_0, \tag{14.79}$$

exactly the mean energy of the oscillator which has the average quantum number n of the group. Thus, the wave-packet represents the oscillation of a particle of mass m in the quantum state n.

The second term in large square brackets in (14.78) represents the modulation of the Gaussian error-curve by the 'carrier' signal, shown in Fig. 14.2b. The wave-packet oscillates back and forward about $y = 0$, exactly as in the case of a harmonic oscillator. Schrödinger notes the important point that, unlike normal wave-packets in which the waveform is eventually broadened because of dispersion, there is no dispersion of the wave-packet according to (14.77). We recall that the sum is an exact solution of the problem of the oscillator and so there are no higher order corrections in the case of the Planck oscillator. Another elementary example of the non-dispersive propagation of a wave-packet in wave mechanics occurs in the case of the representation of a particle moving at constant speed v, as illustrated in the endnote.[10]

The solution (14.78) for half an oscillation cycle is shown in Fig. 14.3 for the case $A = 20$. This diagram illustrates how, for large quantum numbers n, the behaviour of the wave-packet exactly mimics that of an oscillator of mass m. In this representation, time $t = 0$ corresponds to the maximum value of x which occurs at $\cos\omega_0 t = 1$, so that $\sin\omega_0 t = 0$. In this case, the second term in the second square bracket is zero and there are no 'beats' in the profile of the wave-packet. On the other hand, when $\cos\omega_0 t = 0$ and $\sin\omega_0 t = 1$, the profile is modulated by the function $\cos Ax = \cos 20x$. The overall mean locus of the wave-packet in Fig. 14.3 is half a cosine wave centred on $t = 0$. In complete analogy with classical mechanics, Schrödinger remarks that the

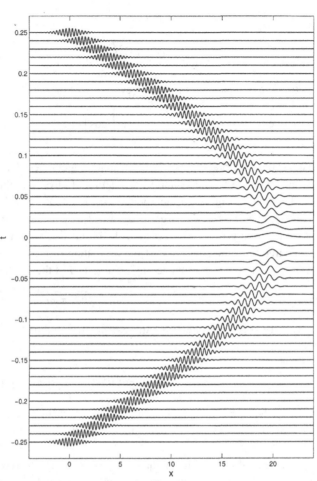

Fig. 14.3 The evolution of the form of a wave-packet with $A = 20$ over half a period of oscillation of a quantised harmonic oscillator according to (14.77) (Schrödinger, 1926a). The diagram was kindly created by Dr. David Green.

'variability of the 'corrugations' is to be conceived as depending on velocity, and, as such, is completely intelligible from all general aspects of undulatory mechanics – but I do not wish to discuss this further at present.'

14.7 *Quantisation as an eigenvalue problem (Part 3)*

Having discovered Courant and Hilbert's *Methods of Mathematical Physics*, Schrödinger redoubled his efforts. The next task was to develop perturbation theory, a major goal being the application of these techniques to account for the Stark broadening of the Balmer lines of hydrogen. This had been one of the great triumphs of the old quantum theory, thanks to the work of Epstein (1916a,b) and Schwarzschild (1916).

14.7.1 Courant and Hilbert (1924)

Courant and Hilbert's remarkable textbook on the *Methods of Mathematical Physics* provided a complete summation of all the tools Schrödinger needed to find solutions of his wave equation. The titles of the chapters themselves provide an indication of how appropriate the mathematical methods were for the solution of wave equations – 'III Linear Integral Equations, IV The Calculus of Variations, V Vibration and Eigenvalue Problems, VI Application of the Calculus of Variations to Eigenvalue Problems, VII Special Functions Defined by Eigenvalue Problems'. For many purposes, Schrödinger simply had to translate the language of Courant and Hilbert into what he now called *wave mechanics*. As he remarked in the introduction of this paper, Part 3 of the series,

> 'The method is essentially the same as that used by Lord Rayleigh in investigating the vibrations of a string with *small inhomogeneities* in his *Theory of Sound* [(Strutt (Lord Rayleigh), 1894)]. This was a particularly simple case, as the differential equation of the unperturbed problem had *constant* coefficients, and only the perturbing terms were arbitrary functions along the string. A complete generalisation is possible not merely with regard to these points, but also for the specially important case of *several* independent variables, *i.e.* for *partial* differential equations, in which *multiple [eigenvalues]* appear in the unperturbed problem, and where the addition of a perturbing term causes the *splitting up* of such values and is of the greatest interest in well-known spectroscopic questions (Zeeman effect, Stark effect, Multiplicities).'

The key mathematical tools were the use of Hermitian operators and the techniques developed by Sturm and Liouville to treat perturbation solutions of particular types of second-order differential equations. The concepts of eigenfunctions and eigenvalues were already well known, as indicated by the reference to Rayleigh's analyses in the context of sound waves, but now they had to be applied to Schrödinger's wave equation. The key mathematical features of the eigenfunctions involved in quantum mechanics are described in textbooks on quantum mechanics and are as follows:

1. If L is a differential operator, the eigenvalue equation is

$$Lu(x) = \lambda u(x) . \tag{14.80}$$

The region Ω over which the problem is to be solved has to be prescribed and appropriate boundary conditions adopted. The key feature of the solutions needed for quantum mechanics is that L should be *Hermitian*, meaning that

$$\int_\Omega u^*(\boldsymbol{x})\, Lv(\boldsymbol{x})\, \mathrm{d}^3x = \left[\int_\Omega v^*(\boldsymbol{x})\, Lu(\boldsymbol{x})\, \mathrm{d}^3x \right]^* , \tag{14.81}$$

where the asterisk means the complex conjugate and u and v are arbitrary functions which satisfy the boundary conditions. The Hermitian behaviour of the wavefunctions is exactly equivalent to the properties of Hermitian matrices, discussed in Sect. 12.3. In the case of matrices, the process of determining the eigenvalues is part of the operation of converting the matrices into diagonal form.

2. If L is a Hermitian operator, the eigenvalues are real, just as in the case of Hermitian matrices.

3. The eigenfunctions of a Hermitian differential operator, associated with different eigenvalues, are orthogonal, meaning that

$$uv = \int_\Omega u^*(x)\, v(x)\, \mathrm{d}^3 x = 0 , \quad \text{if } u \neq v . \tag{14.82}$$

4. Under very general conditions, the set of eigenfunction solutions forms a complete orthogonal, normalised set, or *orthonormal* set, so that, just as in the case of Fourier series, *any* well behaved function of x can be synthesised by an infinite series of these eigenfunctions. Thus,

$$f(x) = \sum_n c_n u_n(x) . \tag{14.83}$$

Since the u_n are orthonormal, the coefficients c_n can be readily found,

$$u_m \cdot f = \sum_n c_n\, u_m \cdot u_n = \sum_n c_n\, \delta_{mn} = c_m . \tag{14.84}$$

A differential operator of special importance is that associated with the *Sturm–Liouville equation*. The operator is

$$L(y) = p\frac{\mathrm{d}^2 y}{\mathrm{d}x^2} + \frac{\mathrm{d}p}{\mathrm{d}x}\frac{\mathrm{d}y}{\mathrm{d}x} - qy = \frac{\mathrm{d}}{\mathrm{d}x}\left[p\frac{\mathrm{d}y}{\mathrm{d}x} \right] - qy , \tag{14.85}$$

where $y \equiv y(x)$ is the dependent function, $p \equiv p(x)$, $\mathrm{d}p/\mathrm{d}x$ and $q \equiv q(x)$ are continuous functions of x and $p \geq 0$. If $L(y) = 0$, this becomes a linear, homogeneous, second-order differential equation and it can be readily shown to be Hermitian. Generalising for some arbitrary *weighting function* $\rho(x)$ which is a continuous function of x and which never becomes negative or zero, the set of eigenfunctions found by setting $L(y) = E\rho(x)y(x)$ and satisfying the boundary conditions is complete and orthonormal. The solutions of the homogeneous equation result in the eigenfunctions u_k and eigenvalues E_k for the stationary states in, for example, the hydrogen atom. Note that, with the inclusion of the weighting function, the normalisation condition for the eigenfunctions is

$$\int \rho(x)\, u_i(x)\, u_k(x)\, \mathrm{d}x = \delta_{ij} = \begin{cases} 1 & \text{if } i = j \\ 0 & \text{if } i \neq j. \end{cases} \tag{14.86}$$

The next step is to find the solutions when the system is subject to a small perturbation. We begin by assuming that the unperturbed solution for the eigenfunctions u_k and eigenvalues E_k are known and then ask how these change when a small perturbing term $-\lambda r(x)y$ is added to the wave equation so that it become

$$L[y] - \lambda r(x)y + E\rho(x)y(x) = 0 . \tag{14.87}$$

λ is assumed to be a small quantity and $r(x)$ is an arbitrary continuous function of x. It is therefore expected that the solutions will only result in small changes of the coefficients q in (14.85). The continuity aspect of the solutions was a key feature for Schrödinger. In the introduction to the paper, he writes

'[The perturbation method] is based upon the important *property of continuity* possessed by [eigenvalues] and [eigenfunctions], principally, for our purpose, upon their *continuous* dependence on the coefficients of the differential equation, and less upon the extent of the domain... and... the boundary conditions... are generally the same for the unperturbed and perturbed problems.'

Again, Courant and Hilbert provide the complete solution. For small values of λ, the energies and eigenfunctions of the perturbed eigenstates are slightly different from their unperturbed values and so Schrödinger writes,

$$E_k^* = E_k + \lambda \epsilon_k \, ; \quad u_k^* = u_k(x) + \lambda v_k(x) \, . \tag{14.88}$$

Substituting these relations into (14.87) and recalling that the unperturbed eigenfunction for u_k satisfies the unperturbed wave equation, we find

$$L[v_k] = E_k \rho v_k = (r - \epsilon_k \rho) u_k \, . \tag{14.89}$$

This is now an inhomogeneous equation for v_k. Courant and Hilbert showed that the eigenfunction equation for v_k only has solutions if the right-hand side of (14.89) is orthogonal to the associated solution of the homogeneous equation. Therefore,

$$\int (r - \epsilon_k \rho) u_k^2 \, dx = 0 \, , \quad \epsilon_k = \frac{\int r u_k^2 \, dx}{\int \rho u_k^2 \, dx} \, . \tag{14.90}$$

If the functions u_i have been normalised, then

$$\epsilon_k = \int r u_k^2 \, dx \, . \tag{14.91}$$

Thus, the perturbed energy of the eigenstate k has been found, without determining the perturbed eigenfunction v_k. This result is the exact equivalent of that found in classical mechanics, namely, that the energy perturbation is in the first approximation equal to the perturbing function averaged over the unperturbed motion.

Finally, to find the function v_k, the inhomogeneous equation is solved in terms of the complete set of eigenfunctions $u_i(x)$,

$$v_k(x) = \sum_{i=1}^{\infty} \gamma_{ki} u_i(x) \, , \tag{14.92}$$

with the result that

$$\gamma_{ki} = \frac{c_{ki}}{E_k - E_i} = \frac{\int r u_k u_i \, dx}{E_k - E_i} \quad \text{if } i \neq j \, . \tag{14.93}$$

In this expression,

$$c_{ik} = \int (r - \epsilon_k \rho) u_k u_i \, dx = \begin{cases} \int r u_k u_i \, dx & \text{for } i \neq j \\ 0 & \text{for } i = k \, . \end{cases} \tag{14.94}$$

Hence, the perturbed eigenfunction and energy eigenvalue of the perturbed state are

$$u_k^*(x) = u_k(x) + \lambda \sum_{i=1}^{\infty} {}' \frac{u_i(x) \int r u_k u_i \, dx}{E_k - E_i} \, ; \quad E_k^* = E_k + \lambda \int r u_k^2 \, dx \, , \tag{14.95}$$

where the dash on the summation means that the term $i = j$ should be omitted from the sum.

Following his careful exposition of the fundamentals of perturbation theory, Schrödinger extends the procedures to several independent variables, resulting in a set of partial, rather than ordinary, differential equations.

14.7.2 The Stark effect

Schrödinger immediately applies the new formalism to the Stark effect, in which the perturbation is associated with the influence of a uniform electric field F upon the electron in the atom. Hence, the perturbed Schrödinger wave equation becomes

$$\nabla^2 \psi + \frac{8\pi^2 m}{h^2}\left(E + \frac{e^2}{4\pi\epsilon_0 r} - eFz\right)\psi = 0 . \tag{14.96}$$

Here, we have preserved Schrödinger's notation in which the electric field strength is written as F to avoid confusion with the energy E of the stationary states. He now solves the perturbation problem in two different ways.

He first uses the method of Epstein and Schwarzschild in which they transformed to the parabolic coordinates used in celestial mechanics and solved the wave equation in that coordinate system (see Sect. 7.2 and the relations (7.2)). The coordinate transformations adopted by Schrödinger are:

$$x = \sqrt{\lambda_1 \lambda_2} \cos\phi ; \quad y = \sqrt{\lambda_1 \lambda_2} \sin\phi ; \quad z = \tfrac{1}{2}(\lambda_1 + \lambda_2) . \tag{14.97}$$

With this transformation to $\lambda_1, \lambda_2, \phi$ coordinates, Schrödinger's wave equation can be written in self-adjoint, or Hermitian, form and the solution found by separation of variables, $\psi = \Lambda_1(\lambda_1)\,\Lambda_2(\lambda_2)\,\Phi(\phi)$. First of all, the unperturbed solution is found using the associated Laguerre polynomials, which Schrödinger had found in Courant and Hilbert's book, to describe the wavefunctions of the stationary states. Then, the solution for the perturbed wavefunctions under the influence of a uniform electric field is obtained. The perturbed stationary states have energies

$$E = -\frac{m_e e^4}{8\epsilon_0^2 h^2 n^2} - \frac{3h^2 \epsilon F}{2\pi m_e e}\, n(k_2 - k_1) , \tag{14.98}$$

where k_1 and k_2 are parabolic quantum numbers, corresponding exactly to the quantum numbers n_1 and n_2 introduced in the classical Epstein–Schwarzschild result (7.7) in the old quantum theory. The principal quantum number is $n = n_1 + n_2 + n_3$ as in Sect. 7.2. In addition, Schrödinger noted that, just as in Heisenberg's theory, there are stationary states with zero orbital quantum number and so the *pendulum orbits*, which had to be excluded on heuristic grounds in the old quantum theory, do not exist.

But Schrödinger did not leave matters there – he wanted to evaluate the intensities of the lines observed in the Stark effect as well. Here, he made use of his discovery of the equivalence between wave mechanics and matrix mechanics so that the matrix elements, which he had shown could be derived from wave mechanics, were identified with the dipole moments associated with transitions between stationary states (see Chap. 15).

Finally, he returned to his calculation of the stationary states in the presence of an electric field and recast the whole calculation in terms of (r, θ, ϕ) coordinates, what he referred to as the *method of Bohr*. Following a lengthy calculation, he demonstrated that exactly the same result as (14.99) was obtained, remarking that

> 'On the whole, we must admit that in the present case the method of secular perturbations [second approach] is considerably more troublesome than the direct application of a system of separation [first approach].'

14.8 *Quantisation as an eigenvalue problem (Part 4)*

The fourth paper of the series, received by the *Annalen der Physik* on 21 June 1926, concerned the development of the time-dependent form of Schrödinger's wave equation. In the first three papers of the series, Schrödinger had successfully developed the wave mechanical formalism for time-independent wave phenomena, but now he needed to extend the procedures to time-dependent phenomena such as, for example, particle scattering and transitions between stationary states in which the eigenvalues change from their initial to their final states. He achieved this by returning to the original form of the wave equation and understanding how it could be modified when, for example, the potential term is time-varying.

In time-independent form, the wave equation

$$\nabla^2 \psi - 2m \frac{(E-V)}{E^2} \frac{\partial^2 \psi}{\partial t^2} = 0 \quad \text{becomes} \quad \nabla^2 \psi + \frac{8\pi^2 m}{h^2}(E-V)\psi = 0 \,, \quad (14.99)$$

where the time dependence of the wavefunction is assumed to be of the form

$$\psi \propto \text{real part of } \left(e^{\pm 2\pi i E t / h} \right) \,, \quad (14.100)$$

and $E = h\nu = \hbar\omega$. It follows from this relation that

$$\frac{d\psi}{dt} = \pm \frac{2\pi i E}{h} \left(e^{\pm 2\pi i E t / h} \right) = \pm \frac{2\pi i E}{h} \psi; \quad \frac{d^2 \psi}{dt^2} = -\frac{4\pi^2 E^2}{h^2} \left(e^{\pm 2\pi i E t / h} \right) = -\frac{4\pi^2 E^2}{h^2} \psi \,.$$

$$(14.101)$$

Schrödinger's aim was to eliminate the energy E of the eigenstate from the wave equation so that it could become time-variable. He noted that the first differential of ψ in (14.101) determines the quantity $E\psi$ in terms of $d\psi/dt$ and so he makes this substitution into (14.99):

$$\nabla^2 \psi - \frac{8\pi^2 m}{h^2} V \psi = \pm \frac{4\pi i m}{h} \frac{\partial \psi}{\partial t} \,. \quad (14.102)$$

Schrödinger recognised that, by making this substitution, he required the wavefunction to be complex, but he had a prescription for overcoming this problem.

> 'We will require the complex wavefunction ψ to satisfy one of these two equations [14.102]. Since the conjugate complex function $\overline{\psi}$ will then satisfy the other equation, we may take the real part of ψ as the real wave function (if we require it).'

The necessity of introducing complex numbers into quantum mechanics to describe the time dependence of the wavefunction can be understood from elementary arguments starting from de Broglie's relation, as described in the endnotes.[11]

Having established the time-dependent wave equation, Schrödinger was in a position to tackle a wide variety of problems in quantum physics. In the remainder of Part 4, he concentrates upon the problem of the theory of dispersion, a problem which had been successfully tackled by Kramers and Heisenberg (1925) (see Sect. 10.3). He takes the incident radiation to be described as a perturbation to the potential V associated with the electric field of the incident waves. Thus, the potential can be written

$$V = V_0 + A(x) \cos(2\pi \nu t) , \qquad (14.103)$$

where $A(x)$ is the perturbation to the potential due to incident electric field F of the light waves and can be written $-F \sum_i e_i z_i$. Then, the time-dependent Schrödinger equation becomes

$$\nabla^2 \psi - \frac{8\pi^2 m}{h^2}(V_0 + A \cos 2\pi \nu t)\psi = \pm \frac{4\pi i m}{h} \frac{\partial \psi}{\partial t} . \qquad (14.104)$$

Schrödinger proceeds to solve this equation, first by finding the unperturbed solution and then by treating the time-varying incident field as a perturbation of that solution. We will not go through that analysis but simply note the form of the perturbed wavefunction,

$$\psi = u_k(x) \exp\left(\frac{2\pi i E_k t}{h}\right)$$
$$+ \frac{1}{2} \sum_{n=1}^{\infty} a'_{kn} u_n(x) \left[\frac{\exp\left(2\pi i t/h\right)(E_k + h\nu)}{E_k - E_n + h\nu} + \frac{\exp\left(2\pi i t/h\right)(E_k - h\nu)}{E_k - E_n - h\nu}\right] . \qquad (14.105)$$

The first term on the right-hand side of the expression is the *free vibration* of the system. The second term shows the effect of the perturbation caused by the incident radiation. Note that this formula excludes the resonant case in which $h\nu = E_k - E_n$. The important point is that, just as in Kramers and Heisenberg's analysis, there are two terms corresponding to the cases of induced absorption and emission of radiation from states with energy $h\nu$ displaced from $E_k - E_n$.

The aim of the calculation was to determine the electric dipole moment, or polarisation, of the medium under the influence of the incident radiation field and Schrödinger adopted the

> '*heuristic hypothesis* [that] the field scalar ψ represents the electric density as a function of space coordinates and the time, *if x stands for only three space coordinates*, i.e. if we are dealing with the problem of one electron.'

Then, the integral of $\psi\overline{\psi}$ multiplied by the charge of the particle taken over all the coordinates of the system is taken to represent the distribution of electric charge. Integrating over all particles of the system, the induced electric dipole moment associated with the incident radiation could be found. Specifically, if the classical expression for the dipole moment is $M_y = \sum e_i y_i$, the resultant electric dipole moment is

$$\int M_y \, \psi\overline{\psi} \rho \, dx , \qquad (14.106)$$

where ρ is the weighting function which ensures that the wavefunctions are self-adjoint. Schrödinger found that his expression for the dispersion of the medium was of similar form to that deduced by Kramers and Heisenberg, but also was an improvement of their expression. In later sections of the paper, he goes on to consider the resonant and degenerate cases.

Schrödinger appreciated that these revolutionary new procedures opened up the route to treating a vast range of problems in atomic physics. In his words,

> 'By superposing the perturbations due to a constant electric or magnetic field and a light wave, we obtain magnetic and electric double refraction, and the magnetic rotation of the plane of polarisation. Resonance radiation in a magnetic field also comes under this heading, ... Further, we can treat the action of an α-particle or electron flying past the atom in this way, if the encounter is not too close for the perturbation of each of the two systems to be calculable from the undisturbed motion of the other. All these questions are mere matters of calculation as soon as the [eigenvalues] and [eigenfunctions] of the unperturbed systems are known.'

14.9 Reflections

Schrödinger's was a quite astonishing achievement. To quote Jammer (1989):

> 'Schrödinger's brilliant paper was undoubtedly one of the most influential contributions ever made in the history of science. It deepened our understanding of atomic phenomena, served as a convenient foundation for the mathematical solution of problems in atomic physics, solid state physics and, to some extent, also in nuclear physics, and finally opened new avenues for thought. In fact, the subsequent development of non-relativistic quantum theory was to no small extent merely an elaboration and application of Schrödinger's work.'

The impact upon the community of physicists was instant for here was a system of wave mechanics which was founded upon well tried and tested procedures in analytic dynamics. Eigenvalues and eigenfunctions were the staple diet of the classical theorist and here they found new applications in physics at the atomic level. The techniques were immediately adopted by physicists such as Enrico Fermi and many others who found the matrix mechanics of Born, Heisenberg and Jordan and Dirac's q-numbers obscure and difficult to apply to real physical problems. In contrast, they felt at home with Schrödinger's transparent scheme of wave mechanics.

But, there was still a long way to go. We have intentionally discussed only five of the six papers which Schrödinger published in 1926. We missed out his paper on the reconciliation of the Born–Heisenberg–Jordan approach to quantum problems by matrix mechanics with wave mechanics – this is the subject of the next chapter. Furthermore, spin has not yet been incorporated into the infrastructure of quantum mechanics. Schrödinger fully appreciated that spin had to be included in his scheme of wave mechanics and this would involve an extension of the methods he had so successfully deployed so far. A central role would be played by a deeper understanding of the role of linear operators.

The physics community was now faced with two theories of quantum phenomena which could scarcely have differed more radically from one another and yet both had achieved remarkable successes in explaining precisely the same physical phenomena – the spectral lines of the hydrogen atom, the zero point energy of quantum systems, the quantisation of the harmonic oscillator, the quantum rotator and the Stark effect. Furthermore, both theories could account for the experimental data, unlike the predictions of the old quantum theory. Perhaps these quite different approaches are not so surprising when it is appreciated that matrix and wave mechanics started from the diametrically opposite poles of the wave–particle duality.

At the heart of the *Heisenberg approach*[1] was the fundamental role played by the non-commutative behaviour of the quantum variables and the quantisation of both the momentum and spatial variables. To accommodate these features, a new mathematical calculus had been invented from the realisation that matrices followed precisely the correct algebraic rules. The elaboration of this scheme led to the concept of the energy levels of a quantum system being associated with the diagonalisation of matrices using the eigenvalue procedure. As Jammer remarks, the theory

> '. . . defied any pictorial representation; it was an *algebraic* approach which, proceeding from the observed discreteness of spectral lines, emphasised the element of *discontinuity*; in spite of its renunciation of classical description in space and time it was ultimately a theory whose basic conception was the *corpuscle*.'

In complete contrast, *Schrödinger's approach* was firmly based upon de Broglie's insight into the wave properties of particles and the need to describe these by a wave equation. Again, quoting Jammer,

> 'Schrödinger's [approach]. . . was based on the familiar apparatus of differential equations, akin to the classical mechanics of fluids and suggestive of an easily visualisable representation: it was an *analytical* approach which, proceeding from a generalisation of the laws of motion, stressed the element of *continuity*, and, as its name indicates, it was a theory whose basic concept was the *wave*.'

Neither Heisenberg nor Schrödinger were initially particularly impressed by the other's approach. Heisenberg wrote to Pauli that

> 'The more I ponder about the physical part of Schrödinger's theory, the more disgusting it seems to me,'

while Schrödinger stated that

'I was discouraged, if not repelled, by what appeared to me a rather difficult method of transcendental algebra, defying any visualisation.'

Despite these reservations, a number of authors, including Schrödinger, sought to understand how the different approaches could be reconciled. There were general similarities, for example, in the use of eigenvalues in matrix algebra and in the solutions of the Schrödinger wave equation. Schrödinger was in the vanguard of this reconciliation when he discovered what he referred to as 'a formal mathematical identity' of wave and matrix mechanics. But this was only one of a number of insights into the mathematics necessary to accommodate both theories in a coherent picture – Lanczos, Born, Wiener, Pauli and Eckart all made key contributions to the reconciliation of the two approaches. These endeavours led to a deepening of the understanding of the content of the new theories and set the scene for the development of the modern theory of quantum mechanics. Many of these developments took place almost simultaneously and independently – they are all part of Born's 'tangle of interconnected alleys'. Let us begin with Schrödinger and then introduce the insights of the other pioneers.

15.1 Schrödinger (1926d)

Schrödinger interrupted his series of papers *Quantisation as an eigenvalue problem* between Parts 2 and 3 to write his paper *On the relation between the quantum mechanics of Heisenberg, Born, and Jordan, and that of Schrödinger* which was received by the *Annalen der Physik* on 18 March 1926 (Schrödinger, 1926d). He had been seeking an accommodation between the two theories, but as late as 22 February 1926 he still had failed to find a solution. Then, suddenly, in a matter of a couple of weeks, he found the answer he was looking for. In addition to demonstrating the equivalence of the theories, Schrödinger's paper was a stout defence of the wave mechanical approach to quantum physics, claiming that it contained all the mathematical apparatus needed to account for the results obtained by matrix mechanics with the additional virtue of being visualisable.

Schrödinger went straight to the heart of matrix mechanics, starting with Born's fundamental relation (12.14), from which all else followed

$$\mathbf{pq} - \mathbf{qp} = \frac{h}{2\pi i}\mathbf{I}, \tag{15.1}$$

where **p** and **q** are the matrices associated with the momentum and position variables respectively and **I** is the unit matrix.

Immediately, Schrödinger translates (15.1) into the language of operator calculus. As we will discover, all the methods of reconciling matrix and wave mechanics involved a deepening appreciation of the role of operators in quantum mechanics. Their significance lay in the fact that it was already well-known in the mathematical literature that operator calculus is non-commutative. Thus, if $\widehat{A}$ and $\widehat{B}$ are operators, the operator $\widehat{AB}$ is not, in general, the same as the operator $\widehat{BA}$, in other words, such operators do not commute. These and many more of the key properties of operators had been surveyed by Salvatore Pincherle in an important article of 1906 entitled *Functional operators and equations* published in the *Encyclopädie der mathematischen Wissenschaften* (Pincherle, 1906). We will find that the importance of operator calculus began to be appreciated by all those who sought to reconcile the different approaches to quantum mechanics and the essential tools were already formulated in the section of Pincherle's review entitled *The elements of operator calculus*.

It is worth quoting Schrödinger's words to understand how he approached the reconciliation of the two theories.

'The starting point in the construction of matrices is given by the simple observation that Heisenberg's peculiar calculating laws for functions of the *double* set of n quantities $q_1, q_2, \ldots, q_n$; $p_1, p_2, \ldots, p_n$ (position- and canonically conjugate momentum coordinates) agree exactly with the rules, which *ordinary analysis* makes *linear differential operators* obey in the *single* set of n variables $q_1, q_2, \ldots, q_n$. So, the *co-ordination* has to occur in such a manner that each p_l in the *function* is to be replaced by the operator $\partial/\partial q_l$. Actually, the operator $\partial/\partial q_l$ is exchangeable with $\partial/\partial q_m$, where m is arbitrary, but with q_m only, if $m \neq n$. The operator, obtained by interchange and subtraction when $m = l$, namely,

$$\frac{\partial}{\partial q_l} q_l - q_l \frac{\partial}{\partial q_l} \, , \tag{15.2}$$

when applied to any arbitrary function of the qs, *reproduces* the function, that is, the operator gives identity. This simple fact will be reflected in the domain of matrices as Heisenberg's interchange rule.'

This is a key point in the argument. If we write an arbitrary function of q_i as $\psi(q_1, q_2 \ldots q_n) \equiv \psi(q)$, then

$$\left[\frac{\partial}{\partial q_l} q_l - q_l \frac{\partial}{\partial q_l} \right] \psi(q) = \frac{\partial}{\partial q_l} [q_l \psi(q)] - q_l \frac{\partial \psi(q)}{\partial q_l}$$
$$= \psi(q) + q_l \frac{\partial \psi(q)}{\partial q_l} - q_l \frac{\partial \psi(q)}{\partial q_l} = \psi(q) \, , \tag{15.3}$$

confirming Schrödinger's statement that the operator is an identity operator.

With these words, Schrödinger begins his own development of the rules of operator calculus. He illustrates the rules by an example in which the function is described by a power series in the ps and qs, a single term of which might be

$$F(q_k, p_k) = f(q_1, \ldots, q_n) p_r p_s p_t \, g(q_1, \ldots, q_n) p_{r'} \, h(q_1, \ldots, q_n) \, p_{r''} p_{s''} \ldots \tag{15.4}$$

Then, to convert to the operator formalism, he replaces the ps, for example p_r, by the operator $K \, \partial/\partial q_r$, emphasising that the variables need to be 'well-arranged'. By this, he means that, just as in the case of matrix multiplication, the order of the products of operators must be carefully adhered to – the corresponding differentiations must be carried out in a strict order. Thus, (15.4) is converted into an operator which Schrödinger writes as $[F, \cdot]$:

$$[F, \cdot] = f(q_1, \ldots, q_n) K^3 \frac{\partial^3}{\partial q_r \, \partial q_s \, \partial q_t} \, g(q_1, \ldots, q_n) \, K \, \frac{\partial}{\partial q_{r'}} \, h(q_1, \ldots, q_n) \, K^2 \frac{\partial^2}{\partial q_{r''} \, \partial q_{s''}} \ldots$$
$$\tag{15.5}$$

If the operator $[F, \cdot]$ acts upon the function $u(q_1, \ldots q_n)$, some new function $[F, u]$ is created. The operators must obey the multiplicative rule that, if G is another well-arranged operator, applying G to $[F, u]$, a new function $[GF, u]$ is created, meaning that first the operator $[F, \cdot]$ acts upon u and then $[G, \cdot]$ operates upon $[F, u]$. In general, $[GF, u]$ is not the same as $[FG, u]$.

Next, Schrödinger associates with the operator $[F, \cdot]$ a matrix F^{kl} through the introduction of a complete orthonormal set of functions for $l \leq k$, $l \leq \infty$. Writing x for the set of variables $q_1, q_2, \ldots, q_n$ and $\int \mathrm{d}x$ for the integration over all q-space, the normalised orthonormal set of functions is

$$u_1(x)\sqrt{\rho(x)}, \ u_2(x)\sqrt{\rho(x)}, \ u_3(x)\sqrt{\rho(x)}, \ldots \text{ad inf.}, \qquad (15.6)$$

with the normalisation conditions

$$\int \rho(x)\, u_i(x)\, u_k(x)\, \mathrm{d}x = \begin{cases} 0 & \text{for } i \neq k, \\ 1 & \text{for } i = k, \end{cases} \qquad (15.7)$$

where $\rho(x)$ is the weight function introduced to ensure that the normalised functions are self-adjoint. Schrödinger defines the elements of the matrix F^{kl} by the following integral[2]

$$F^{kl} = \int \rho(x)\, u_k(x)\, [F, u_l(x)]\, \mathrm{d}x . \qquad (15.8)$$

In his words,

'... a matrix element is computed by *multiplying* the function of the orthogonal system denoted by the *row*-index (whereby we understand always u_i, not $u_i\sqrt{\rho}$) by the "density function" ρ, and by the result arising from using our operator on the orthogonal function corresponding to the *column*-index, and then by *integrating* the whole over the domain.'

Schrödinger identifies q_l with the 'scalar operator' q_l and p_l with the differential operator $K\partial/\partial q_l$. Thus, matrix elements can be associated with q_l and p_l as follows:

$$q_l^{ik} = \int \rho(x)\, u_i(x)\, q_l\, u_k(x)\, \mathrm{d}x , \qquad (15.9)$$

$$p_l^{ik} = K \int \rho(x)\, u_i(x)\, \frac{\partial u_k(x)}{\partial q_l}\, \mathrm{d}x . \qquad (15.10)$$

Next, this formalism is used to derive Born's quantum condition (15.1). The rule (15.8) can be used to derive the ik component of the matrix associated with the operator $[F, \cdot] = p_l q_l - q_l p_l$

$$(p_l q_l - q_l p_l)^{ik} = \int \rho(x)\, u_i(x)\, [F, u_k(x)]\, \mathrm{d}x , \qquad (15.11)$$

$$= K \int \left\{ \rho(x)\, u_i(x)\frac{\partial}{\partial q_l}\, [q_l\, u_k(x)] - \rho(x)\, u_i(x)\, q_l \frac{\partial}{\partial q_l} u_k(x) \right\} \mathrm{d}x , \quad (15.12)$$

$$= K \int \rho(x)\, u_i(x)\, u_k(x)\, \mathrm{d}x = \begin{cases} 0 & \text{for } i \neq k, \\ K & \text{for } i = k. \end{cases} \qquad (15.13)$$

because of the fact that (15.2) is an identity operator and because of the orthogonality of the set of functions $u_i(x)$ according to (15.7). This is exactly Born's *quantum relation* (15.1) if we set $K = h/2\pi i$. This calculation represents the introduction of the momentum operator

$$p_i \equiv \frac{h}{2\pi i} \frac{\partial}{\partial q_i} ,$$
(15.14)

into quantum physics.

Schrödinger shows that every function $F(p, q)$ can be translated into an operator $[F, \cdot]$ which in turn can be translated into a matrix F^{kl} using (15.8). Furthermore, these matrices obey all the rules of matrix mechanics required by Born, Heisenberg and Jordan (1926). As expressed by Jammer (1989),

> 'Any wave-mechanical equation could therefore consistently be translated into a matrix equation, the operation of F on the wave function ψ, corresponding to the application of the matrix (F^{ij}) on the column vector (a_k) whose components are the Fourier coefficients of ψ.'

Schrödinger was again in full flight – next he takes the complete set of orthonormal functions to be the eigenfunctions of the wave equation, which could be written as an operator equation

$$[H, \psi] = E\psi ,$$
(15.15)

where $[H, \cdot]$ is the operator associated with the Hamiltonian of the system. In fact, (15.15) is just his wave equation as can be seen as follows:

$$H = \frac{1}{2m}(p_x^2 + p_y^2 + p_z^2) + U(x, y, z) \equiv \frac{1}{2m}(p_x p_x + p_y p_y + p_z p_z) + U(x, y, z) .$$
(15.16)

Converting this well-arranged function into an operator equation using $p_x \equiv (h/2\pi i)\,\partial/\partial x$, $p_y \equiv (h/2\pi i)\,\partial/\partial y$, $p_z \equiv (h/2\pi i)\,\partial/\partial z$, and then substituting into (15.16), we find

$$-\frac{h^2}{8\pi^2 m}\left(\frac{\partial^2 \psi}{\partial x^2} + \frac{\partial^2 \psi}{\partial y^2} + \frac{\partial^2 \psi}{\partial z^2}\right) + U(x, y, z)\psi = E\psi ,$$

$$\nabla^2 \psi + \frac{8\pi^2 m}{h^2}\left[E - U(x, y, z)\right]\psi = 0 ,$$
(15.17)

exactly Schrödinger's time-independent wave equation (14.37). This calculation is the 'standard' method of deriving the Schrödinger wave equation using operator techniques.

Next, Schrödinger had to establish the rules of differentiation of the operator F with respect to the position q_l and momentum p_l coordinates. Just as in the case of Born, Heisenberg and Jordan's matrix mechanics, the definitions need some care. In his operator notation, Schrödinger showed that the appropriate forms are

$$\left[\frac{\partial F}{\partial q_l}, \cdot\right] = \frac{1}{K}\left[p_l F - F p_l, \cdot\right] ,$$
(15.18)

$$\left[\frac{\partial F}{\partial p_l}, \cdot\right] = \frac{1}{K}\left[F q_l - q_l F, \cdot\right] ,$$
(15.19)

where the p_l and q_l are variables, *not* operators. Born, Heisenberg and Jordan had shown that Hamilton's equations of motion (12.67) could be written in matrix form, the ik elements of the matrices being

$$\left(\frac{\partial q_l}{\partial t}\right)^{ik} = \left(\frac{\partial H}{\partial p_l}\right)^{ik}, \quad \left(\frac{\partial p_l}{\partial t}\right)^{ik} = -\left(\frac{\partial H}{\partial q_l}\right)^{ik}, \tag{15.20}$$

where $l = 1, 2, 3, \ldots n$ and $i, k = 1, 2, 3, \ldots$ ad inf. Now, according to (12.50), the time derivative of the matrix element q_l^{ik} associated with the pair of stationary states i and k is given by

$$2\pi i \, \nu(ik) q_l^{ik} = 2\pi i \, (\nu_i - \nu_k) q_l^{ik}, \tag{15.21}$$

where the frequencies ν_i and ν_k have been associated with the states i and k respectively. Now identifying $[F, \cdot]$ with the Hamiltonian operator $[H, \cdot]$ and using (15.19) and the first equation of (15.20), we find

$$(\nu_i - \nu_k) q_l^{ik} = \frac{1}{h}(Hq_l - q_l H)^{ik}. \tag{15.22}$$

Schrödinger now chooses the complete set of eigenfunctions associated with his wave equation (15.15) as the basis for the determination of the matrix elements and so

$$H^{ik} = E_l \int \rho(x) u_k(x) u_l(x) \, dx = \begin{cases} E_i & \text{for } i = k, \\ 0 & \text{for } i \neq k. \end{cases} \tag{15.23}$$

This expression is then used to work out the terms $(Hq_l)^{ik}$ and $(q_l H)^{ik}$. We recall that we need to sum over all the possible intermediate states m to find the value of, say, $(Hq_l)^{ik}$. Note that this procedure can be traced back to Heisenberg's rule (11.18) that the 'classical' Fourier term $x^{(2)}(n, \alpha) = \sum_{\alpha'} x(n, \alpha') x(n, \alpha - \alpha')$ should be converted into the matrix format of Born, Heisenberg and Jordan (1926) as $x^{ik} = \sum_m x^{im} x^{mk}$. Thus,

$$(Hq_l)^{ik} = \sum_m H^{im} q_l^{mk}. \tag{15.24}$$

But, according to the rules (15.23), the only non-zero member of the infinite series is that for which $i = m$, and for this case $H^{ii} = E_i$. Therefore,

$$(Hq_l)^{ik} = E_i q_l^{ik}. \tag{15.25}$$

Performing the same analysis for $(q_l H)^{ik}$

$$(q_l H)^{ik} = \sum_m q_l^{im} H^{mk} = E_k q_l^{ik}. \tag{15.26}$$

It immediately follows from (15.25) and (15.26) that, substituting these values into (15.22),

$$(\nu_i - \nu_k) = \frac{1}{h}(E_i - E_k), \tag{15.27}$$

relating the frequencies of transition to the differences of energies of the stationary states. Schrödinger was triumphant:

> 'Hence the solution of the whole system of matrix equations of Heisenberg, Born and Jordan is reduced to the natural boundary value problem of a linear partial differential equation. If we have solved the boundary value problem, then by the use of [15.8] we can calculate by differentiations and quadratures every matrix element we are interested in.'

These were considerable achievements and demonstrated the equivalence of matrix and wave mechanics. Strictly speaking, Schrödinger had demonstrated that he could translate wave mechanics into matrix mechanics and obtain all the results derived by Born, Heisenberg and Jordan. It was not so obvious that the translation would work completely symmetrically in the opposite direction, namely, did matrix mechanics necessarily imply wave mechanics, or did it contain additional features beyond those which could be described by wave mechanics? A number of authors were hot on the trail.

15.2 Lanczos (1926)

In fact, Schrödinger was not the first to seek a reformulation of the matrix mechanics of Born and Jordan (1925b) in terms of continuous functions. As soon as their paper of 1925 was published, Kornel (Cornelius) Lanczos showed that their matrix formulation of quantum mechanics could be written in terms of the kernels found in *integral equations*. This did not necessarily advance the cause of the new quantum mechanics among the community of physicists since, as remarked by Jammer (1989):

> '... the use of integral equations with which physicists were – and still are – much less familiar than with differential equations, the absence of any specific example or new result, and certainly also the fact that the publication almost coincided with that of Schrödinger's first communication, explain the relatively cool reception Lanczos's paper was given.'

Lanczos had already used integral equations in his studies of general relativity in the weak field limit and had become familiar with the techniques through Hilbert's pioneering study of 1912 (Hilbert, 1912) and the relevant chapters of Courant and Hilbert's *Methods of Mathematical Physics* (1924).

A key feature of Heisenberg's paper of 1925 was that the functions depended upon pairs of variables, say m and n. Lanczos appreciated that the *kernel function* $K(s, \sigma)$ found in integral equations also depended upon two points s and σ in a coordinate space of arbitrary dimensions. If the kernel is symmetrical, a system of eigenfunctions $\phi^i(s)$, with $s = 1, 2, 3, \ldots$ can be associated with the solution of the integral equation

$$\phi(s) = \lambda \int K(s, \sigma) \phi(\sigma) \, d\sigma , \qquad (15.28)$$

where λ is the corresponding eigenvalue. The system of eigenfunctions is orthonormal, complete and satisfies the condition

$$\int \phi^i(s) \phi^k(s) \, ds = \begin{cases} 1 & \text{for } i = k , \\ 0 & \text{for } i \neq k . \end{cases} \qquad (15.29)$$

Hence the function $f(s, \sigma)$ can be expanded in terms of the $\phi^i(s)$ as

$$f(s, \sigma) = \sum_i a_i(\sigma) \phi^i(s) . \qquad (15.30)$$

Then, the $a_i(\sigma)$ can be expanded in terms of the same infinite sequence of eigenfunctions $\phi^k(\sigma)$ and so

$$f(s, \sigma) = \sum_{i,k} a_{ik}\, \phi^i(s)\, \phi^k(\sigma)\,. \tag{15.31}$$

Lanczos identified the matrix $\mathbf{a} \equiv [a_{ik}]$ with the matrices of Born and Jordan's matrix mechanics. He further demonstrated how the complete system of matrix mechanics in Born and Jordan's paper could be expressed in terms of the kernels of linear integral equations. He concluded,

> 'Because of the mutual unique relation existing between the matrices and the kernel functions used in our representation, it does not matter at all for the formal treatment of the [quantum mechanical] problem, whether one writes down the fundamental equations in the form of integral equations – as we have done here – or whether one starts immediately from the constituent coefficients and applies the matrix equation.'

The similarities with Schrödinger's paper are apparent and he acknowledged Lanczos's contributions in a footnote to his paper. There is no question, however, about the greater impact of Schrödinger's paper which was couched in much more accessible mathematical terms and was also part of a sequence of ground-breaking insights into the nature of the new quantum mechanics.

15.3 Born and Wiener's operator formalism

In Sect. 12.3, we left Max Born on his way to the Massachusetts Institute of Technology, having completed the 'Three-Man Paper' with Heisenberg and Jordan. Born had met Norbert Wiener in 1924 when the latter was a visitor to Göttingen. An exchange programme was set up by Courant and Wiener, one of the first fruits of this collaboration being Born's appointment as a 'foreign lecturer' at MIT for the fall term of 1925–1926. Born's lectures were centred upon the new matrix mechanics, but he was well aware of the shortcomings of what he, Heisenberg and Jordan had achieved. They had found it impossible to deal with aperiodic motions, including linear motion in a straight line, and, as described in Sect. 12.4, there was no simple equivalence for the action–angle variables which were the natural language of the old quantum theory. Furthermore, there were real mathematical problems with Born's assumptions about the properties of unbounded matrices. Despite these concerns, the theory agreed well with the experimental results and eliminated the intractable problems of the old quantum theory. Born fully realised that an extension of matrix mechanics was needed and Wiener turned out to be exactly the person to do this.

Wiener was a mathematical prodigy who was to make many fundamental contributions to stochastic processes, in particular to the mathematical elaboration of fundamental topics in electronic engineering, electronic communication and control systems. Just before Born arrived in Cambridge, Massachusetts, Wiener had written a comprehensive account of the operational calculus, in particular, a rigorous study of the concept of operators, including

Volterra's integral transformations and Pincherle's transformations of one power series to another (Wiener, 1926). In his autobiography, Wiener wrote:

> 'When Professor Born came to the United States he was enormously excited about the new basis Heisenberg had just given for the quantum theory of the atom. Born wanted a theory which would generalise these matrices .... The job was a highly technical one, and he counted on me for aid. ... I had the generalisation of matrices already at hand in the form of what is known as operators. Born had a good many qualms about the soundness of my method and kept wondering if Hilbert would approve of my mathematics. Hilbert did, in fact, approve of it, and operators have since remained an essential part of quantum theory.' (Wiener, 1956)

As Jammer (1989) notes, the mathematical problems associated with the formulation of the new quantum mechanics fostered a spirit of collaboration between the pure mathematicians and the physicists, among the first fruits of this collaboration being the joint paper by Born and Wiener which was received by the *Zeitschrift für Physik* on 5 January 1926 (Born and Wiener, 1926). Jammer provides a clear description of the route Born and Wiener followed in formally introducing operators into quantum mechanics – we will follow his presentation here. The main topic of Wiener's paper of 1926 had been a generalisation of the Fourier integral which made it possible to apply integral operators to both analytic and non-analytic functions and this found immediate application in Born and Wiener's extension of matrix mechanics.

Consider the Fourier series for the two functions $y(t)$ and $x(t)$ of time t:

$$y(t) = \sum_m y_m \exp\left(2\pi i W_m t / h\right) , \tag{15.32}$$

$$x(t) = \sum_n x_n \exp\left(2\pi i W_n t / h\right) . \tag{15.33}$$

The aim of the calculation is to develop an operator expression relating $y(t)$ and $x(t)$, written symbolically $y(t) = qx(t)$, where q will turn out to be an integral operator. The x_n are found by multiplying (15.33) by $\exp\left(-2\pi i W_k t / h\right)$ and integrating over all t from $-\infty$ to $+\infty$. Then, the only non-zero term is that for which $k = n$, so that

$$x_n = \lim_{T=\infty} \frac{1}{2T} \int_{-T}^{+T} x(s) \exp\left(-2\pi i W_n s / h\right) \, ds . \tag{15.34}$$

Now, let us adopt the matrix transformation

$$y_m = \sum_n q_{mn} x_n . \tag{15.35}$$

Then, the expression for $y(t)$ becomes

$$y(t) = \sum_m y_m \exp\left(2\pi i W_m t / h\right) = \sum_{m,n} q_{mn} x_n \exp\left(2\pi i W_m t / h\right) , \tag{15.36}$$

$$= \lim_{T=\infty} \frac{1}{2T} \int_{-T}^{+T} \sum_{m,n} q_{mn} x(s) \exp\left[2\pi i \left(W_m t - W_n s\right) / h\right] \, ds . \tag{15.37}$$

If we now write

$$q(t, s) = \sum_{m,n} q_{mn} \exp\left[2\pi i\left(W_m t - W_n s\right)/h\right] ,\qquad(15.38)$$

we can write

$$y(t) = \lim_{T=\infty} \frac{1}{2T} \int_{-T}^{+T} q(t, s) x(s)\, ds .\qquad(15.39)$$

Thus, $x(s)$ is converted into $y(t)$ by the integral operator

$$q = \lim_{T=\infty} \frac{1}{2T} \int_{-T}^{+T} ds\, q(t, s)\ldots ,\qquad(15.40)$$

where it is understood that the function $x(s)$ is to be taken inside the integral. They then introduced the definition of a linear operator:

'An operator is a rule in accordance with which we may obtain from a function $x(t)$ another function $y(t)$.... It is linear if

$$q[x(t) + y(t)] = qx(t) + qy(t) .'\qquad(15.41)$$

Having established the procedure for obtaining the relation $y(t) = qx(t)$, Born and Wiener next consider the operator $Dq \equiv \partial q/\partial t$. From (15.39), we find

$$y(t) = Dq\, x(t) = \lim_{T=\infty} \frac{1}{2T} \int_{-T}^{+T} \frac{\partial q(t, s)}{\partial t} x(s)\, ds .\qquad(15.42)$$

Now, using the relation (15.38) for $q(t, s)$, differentiation gives

$$\frac{\partial q(t, s)}{\partial t} = \frac{2\pi i}{h} \sum_{m,n} q_{mn} W_m \exp\left[2\pi i\left(W_m t - W_n s\right)/h\right] .\qquad(15.43)$$

Just as (15.38) defines the relation between the operator q and the matrix q_{mn}, so (15.43) provides the relation between the operator Dq and the matrix element $(Dq)_{mn}$, namely

$$(Dq)_{mn} = \left(\frac{2\pi i}{h} W_m q_{mn}\right) .\qquad(15.44)$$

Next, consider the operator $qD = q\partial/\partial t$. Then,

$$y = q \frac{\partial x(t)}{\partial t} = \lim_{T=\infty} \frac{1}{2T} \int_{-T}^{+T} q(t, s) \frac{\partial x(s)}{\partial t}\, ds .\qquad(15.45)$$

Carrying out a partial integration, this integral becomes

$$y = q \frac{\partial x(t)}{\partial t} = -\lim_{T=\infty} \frac{1}{2T} \int_{-T}^{+T} x(t) \frac{\partial q(t, s)}{\partial s}\, ds .\qquad(15.46)$$

Now, using the relation (15.38) for $q(t, s)$, differentiation gives

$$\frac{\partial q(t, s)}{\partial s} = -\frac{2\pi i}{h} \sum_{m,n} q_{mn} W_n \exp\left[2\pi i\left(W_m t - W_n s\right)/h\right] .\qquad(15.47)$$

We find the relation between the operator qD and the matrix element $(qD)_{mn}$, namely,

$$(qD)_{mn} = \left(\frac{2\pi i}{h} W_n q_{mn}\right) .\qquad(15.48)$$

Hence, the matrix element mn associated with the operator $(Dq - qD)$ has elements

$$(Dq - qD)_{mn} = \frac{2\pi i q_{mn}}{h} (W_m - W_n) .$$ (15.49)

Since $h\nu_{mn} = W_m - W_n$, it follows that

$$(Dq - qD)_{mn} = 2\pi i \nu_{mn} q_{mn} .$$ (15.50)

Born and Wiener recognised that the right-hand side of (15.50) corresponded to the derivative of the mn component of the matrix with respect to time, as demonstrated by (12.49) and (12.50), and so they identified the right-hand side with $\dot{q}_{mn}$. The corresponding operator equation becomes

$$Dq - qD = \dot{q} .$$ (15.51)

This operator equation can be compared with the corresponding matrix equation (12.53). Given the clear correspondence between the matrix and operator formalisms, Born and Wiener went on to identify the operator form of the commutation relation as

$$pq - qp = \frac{h}{2\pi i} 1 ,$$ (15.52)

where 1 is the unit operator, and the canonical equations

$$\dot{q} = \frac{\partial H(pq)}{\partial p} , \quad \dot{p} = -\frac{\partial H(pq)}{\partial q} ,$$ (15.53)

as operator equations in which the operators p and q are Hermitian. This provided a complete equivalence between the operator and matrix mechanics formalisms.

With this new formalism, they identified the energy operator with $(h/2\pi i)D$. In addition, they solved the problems of the quantised harmonic oscillator and linear motion, thus demonstrating the ability of the scheme to deal with both periodic and aperiodic motion.

What is intriguing is that they did *not* make the simple identification of the momentum operator p through the relation $p = (h/2\pi i)\, \partial/\partial q$, which would have led directly to wave mechanics and Schrödinger's wave equation. As Born lamented,[3]

> 'We expressed the energy as d/dt and wrote the commutation law for energy and time as an identity by applying $[t(d/dt) - (d/dt)t]$ to a function of t; it was absolutely the same for q and p. But we did not see that. And I never will forgive myself, for had we done this, we would have had the whole wave mechanics from quantum mechanics at once, a few months before Schrödinger.'

15.4 Pauli's letter to Jordan

Born sent a copy of his paper with Wiener to Heisenberg in January 1926. Heisenberg then forwarded a copy to Pauli as well as a copy of the paper by Lanczos. Pauli was characteristically scathing about the formal mathematics developed by these authors and developed his own version of operator methods, which we will not enter into here. More significantly, in early February, Sommerfeld had asked him to look out for the publication of Schrödinger's paper on wave mechanics which Sommerfeld described as involving a 'totally crazy method', but which reproduced many of the results of matrix mechanics.

Pauli worked on Schrödinger's paper over the Easter holidays while he was visiting Copenhagen. By early April, he had achieved what he later described as the 'complete clarification of the connection of Schrödinger's theory and quantum mechanics'. The argument was set out in a letter of 12 April 1926 to Pascual Jordan.[4] Pauli was strongly impressed by the fact that Schrödinger had been able to recover precisely his results for the hydrogen atom and also by the fact that the two theories seemed to be based upon quite different hypotheses and mathematical techniques. The first part of the letter to Jordan simply reformulates de Broglie's insights and the derivation of Schrödinger's wave equation. The second half of the letter concerns Pauli's reconciliation of the two theories.

For simplicity, Pauli considers only the one-dimensional case of the Schrödinger wave equation which he writes

$$\frac{d^2\psi}{dx^2} + \frac{8\pi^2 m_0}{h^2}\left[E - E_{\text{pot}}(x)\right]\psi = 0 . \tag{15.54}$$

As Schrödinger had shown, for $E < E_{\text{pot}}$, the equation only has solutions for specific energy eigenvalues E_1, E_2, E_3, ... with corresponding eigenfunctions ψ_1, ψ_2, ... and these form a complete orthonormal set with

$$\int_{-\infty}^{\infty} \psi_n\,\psi_m\,dx = \begin{cases} 0 & \text{for } n \neq m , \\ 1 & \text{for } n = m . \end{cases} \tag{15.55}$$

It is simplest to quote the central insights of the letter in Pauli's own words:

'Now one considers in particular the expansion of $x\psi_n$:

$$x\psi_n(x) = \sum_m x_{nm}\psi_m(x) \quad \text{with} \quad x_{nm} = \int_{-\infty}^{+\infty} x\,\psi_n\,\psi_m\,dx . \tag{15.56}$$

One also puts

$$(p_x)_{nm} = \frac{ih}{2\pi}\int_{-\infty}^{+\infty} \frac{\partial\psi_n}{\partial x}\psi_m\,dx_i \quad \text{with} \quad \frac{ih}{2\pi}\frac{\partial\psi_n}{\partial x} = \sum_m (p_x)_{nm}\,\psi_m(x) . \tag{15.57}$$

... Now, if $x_{nm} = x_{mn}$ is real, $(p_x)_{nm} = -(p_x)_{mn}$ is purely imaginary. It can be shown without difficulty, that the matrices for x and p_x thus defined satisfy the equations of the Göttingen mechanics. Namely,

$$\mathbf{p}_x\mathbf{x} - \mathbf{x}\mathbf{p}_x = \frac{h}{2\pi i}\mathbf{I} \quad \text{and} \quad \mathbf{H} = \frac{1}{2m_0}\mathbf{p}_x^2 + \mathbf{E}_{\text{pot}}(x) , \tag{15.58}$$

with $\mathbf{H}$ representing the diagonal Hamiltonian matrix whose elements E_n provide the energy eigenvalues. From the rule of multiplication it follows that the matrix belonging to any function $F(x)$ of x is just given by the coefficients

$$F_{nm} = \int_{-\infty}^{+\infty} F(x)\,\psi_n\,\psi_m\,dx . \tag{15.59}$$

I shall not write out the calculation in detail; you will be able to verify the assertion easily.'

Pauli fully appreciated the great power of Schrödinger's approach which could be extended straightforwardly to incorporate perturbation theory and so to the solution of numerous problems including the intensities of the spectral lines in the Zeeman effect. Two further quotations from this letter are particularly noteworthy:

> 'In principle, in the Göttingen theory as well as in de Broglie's statement of the quantum problem, no description of the electron in the atom in space and time is given.'

> 'It seems that one also sees now, how from the point of view of quantum mechanics the contradistinction between 'point' and 'set of waves' fades away in favour of something more general.'

By 18 April 1926, Schrödinger had learned about Pauli's achievements and wrote to Sommerfeld the same day:

> 'With Pauli I have exchanged a couple of letters. He really is a phenomenal person. How he has discovered everything again? In a tenth of the time which I needed for it!'

15.5 Eckart and the operator calculus

In 1925, the 23-year old Carl Eckart won a fellowship to work at the California Institute of Technology. He had the good fortune to attend the lectures which Born presented on his latest research with Wiener into operator methods in quantum mechanics. These proved to be the inspiration for his synthesis of the wave and matrix mechanical approaches to quantum theory. Following Born's lectures, he made a thorough study of the operator formalism and, in his words,

> 'The result was that I... was completely familiar with what is now known as the Schrödinger operator (the energy operator) before Schrödinger's papers appeared in Pasadena.'

Then the papers of Lanczos and Schrödinger were published in March 1926. Eckart realised that these were fully consistent with the Born–Wiener operator approach. What he appreciated was the following:

> 'The Lanczos quantum theory requires a set of orthogonal functions, by means of which the matrices of Born and Jordan are to be determined... Schrödinger had published a quantum principle which leads directly to a set of orthogonal functions... If these are interpreted as the functions entering into the Lanczos theory, the matrices of the Born–Jordan theory are readily obtained.' (Eckart, 1926b)

Eckart published his paper in the *Proceedings of the National Academy of Sciences* to establish the precedence of his ideas and these were elaborated in much more detail in his paper published in the *Physical Review* later the same year (Eckart, 1926a).

The motivation for these studies was the fact that, although the new Born–Wiener scheme of operator dynamics provided a more rigorous foundation for quantum mechanics, it was still difficult to find solutions for more than a few of the simpler problems of atomic physics. Eckart's objective was to sythesise the approaches of Born and Wiener, Dirac and Schrödinger into a single operator calculus in which *all* the quantities appearing in the equations are operators. Again, all the necessary tools were already available in Pincherle's review (Pincherle, 1906), but now they had to be reformulated for applications in atomic physics.

In the first part of his paper in the *Physical Review*, Eckart rewrote the equations of classical mechanics entirely in the language of operators. Thus, Hamilton's equations now became

$$\frac{dP_j}{dT} = -\frac{\partial H(P, Q)}{\partial Q_j} \; ; \quad \frac{dQ_j}{dT} = \frac{\partial H(P, Q)}{\partial P_j} \; , \tag{15.60}$$

where the quantities, P, Q, T and H are all operators. In his first paper, Eckart noted that,

> 'In translating ordinary equations into operator notation, nothing new is introduced. The transition involves merely a change of mental focus. Instead of concentrating the attention on the numerical quantities, it is directed to the operations of combining them.'

The words are similar to those already quoted by Dirac:

> '...the classical equations are to be retained formally without alteration...only the operations by which the quantities involved are combined are to be altered.'

Let us outline what Eckart did. His operator calculus made use of the results already obtained by Dirac and by Born and Wiener. To do this, he adopted general linear operators which obeyed Born's condition (15.41) and defined an operator D_x by the following equation,

$$D_x X - X D_x = [1\otimes] \; , \tag{15.61}$$

where the right-hand side denotes the unit operator. This expression ensures that the operators are non-commutative. Then, following the rules of matrix calculus of Born and Jordan and the q-number calculus of Dirac, the following algebraic rules for operators were established:

$$\frac{dQ}{dX} = D_x Q - Q D_x \; , \tag{15.62}$$

$$\frac{d}{dX}(Q + P) = \frac{dQ}{dX} + \frac{dP}{dX} \; , \tag{15.63}$$

$$\frac{d}{dX}(QP) = \frac{dQ}{dX}P + Q\frac{dP}{dX} \; . \tag{15.64}$$

The definition (15.61) and the rule of differentiation (15.62) could be generalised to a number of independent operator variables $Q_1, Q_2, \ldots, Q_n$ so that

$$D_j Q_i - Q_i D_j = [\delta_{ij}\otimes] \; , \tag{15.65}$$

$$\frac{dF}{dQ_i} = D_j F - F D_j \; . \tag{15.66}$$

Then, the commutator relations for the quantum operators can be written

$$
\begin{cases}
P_i Q_j - Q_j P_i = \left[\dfrac{h}{2\pi \mathrm{i}} \delta_{ij} \otimes \right], \\[2mm]
Q_i Q_j - Q_j Q_i = 0, \\[2mm]
P_i P_j - P_j P_i = 0,
\end{cases}
\tag{15.67}
$$

and the dynamical equations (15.60) become

$$
\begin{cases}
\dfrac{\mathrm{d}P_j}{\mathrm{d}T} = \dfrac{\mathrm{d}}{\mathrm{d}t} P_j - P_j \dfrac{\mathrm{d}}{\mathrm{d}t} = -\dfrac{\partial H(P, Q)}{\partial Q_j}, \\[3mm]
\dfrac{\mathrm{d}Q_j}{\mathrm{d}T} = \dfrac{\mathrm{d}}{\mathrm{d}t} Q_j - Q_j \dfrac{\mathrm{d}}{\mathrm{d}t} = \dfrac{\partial H(P, Q)}{\partial P_j}.
\end{cases}
\tag{15.68}
$$

Next, the scheme had to be converted into a form which would allow numerical quantities to be compared with experiment. Eckart used the result which could be derived from Born and Wiener's paper, namely,

$$
\frac{1}{\psi} P_j \psi = p_i, \qquad \frac{1}{\psi} Q_j \psi = q_i,
\tag{15.69}
$$

or, more generally,

$$
\frac{1}{\psi} F \psi = f,
\tag{15.70}
$$

where F is any function of P_j and Q_j. For stationary states, Born and Wiener had used the result that the time dependence of the eigenfunctions takes the form $W_n \propto \exp(2\pi \mathrm{i} W_n t / h)$ and so, using (15.70) for the Hamiltonian operator $H(P_j Q_j)$, it follows that

$$
\frac{1}{\psi_n} H(P_j Q_j) \psi_n = W_n.
\tag{15.71}
$$

Eckert noted immediately that

> '[15.71] will be seen [to be] the equation published by Schrödinger in another form... which, in addition to defining ψ_n, serves to distinguish a certain discrete sequence of values of W from all others.'

To complete the argument, Eckart had to define the operators associated with Q_i and P_i and he chose the position operator Q_j to mean multiplication by the scalar quantity q_j and P_j to be the canonically conjugate momentum operator

$$
P_j = \frac{h}{2\pi \mathrm{i}} \frac{\partial}{\partial q_i}.
\tag{15.72}
$$

Inserting these definitions into (15.71), Schrödinger's wave equation follows directly and hence the solution of the energies of the stationary states of the hydrogen atom. To complete the picture, Eckart showed how to find the matrix elements $Q_j(nk)$ associated with the general operator F,

$$
Q_j f = \sum F_n Q_j(nk) \psi_k.
\tag{15.73}
$$

He concluded his paper with the remark that:

> 'The method of obtaining the matrices which has just been outlined differs very little from Lanczos's interpretation of the matrix calculus which is thus included in the present calculus.'

This work, submitted to the *Physical Review* on 7 June 1926, but only published in October 1926, completed the formal identity of matrix and wave mechanics and demonstrated the power of operator techniques in quantum mechanics.

15.6 Reconciling quantum mechanics and Bohr's quantisation of angular momentum – the WKB approximation

An issue which perplexed Bohr and his colleagues was: why was the simple Bohr model of the atom so successful when it now appeared that the quantisation of orbits was associated with boundary conditions which had to be imposed upon permissible solutions of the Schrödinger wave equation at infinity? The answer was soon provided by the independent analyses of Gregor Wentzel (1926a), Kramers (1926) and Léon Brillouin (1926), who used what is now known as the *WKB approximation*.[5]

Except for a few special cases, analytic solutions of Schrödinger's equation are not available, but the WKB approximation enables approximate solutions to be found in cases in which the potential varies slowly with position x. By 'slow', we mean that the changes in potential are small on the scale of the de Broglie wavelength. These were the cases discussed independently by Wentzel, Kramers and Brillouin. In the case in which there is no spatial variation of the potential function, the result corresponding to the solution in classical mechanics is found. In the case in which the potential is slowly varying, the solution to first order in the small quantity $h/2\pi i$ results in Sommerfeld's quantum relation

$$\oint p \, dx = nh \, . \tag{15.74}$$

In the later analyses by Kramers and his colleagues (Niessen, 1928; Kramers and Ittmann, 1929), it was shown that the relation is in fact

$$\oint p \, dx = \left(n + \tfrac{1}{2}\right) h \, . \tag{15.75}$$

Enrico Persico (1938) gives a simple derivation of the result (15.75) starting from Schrödinger's equation, making exactly the same approximations as in the papers by Wentzel (1926a), Kramers (1926) and Brillioun (1926). First, Persico writes the one-dimensional Schrödinger wave equation for an oscillator in the usual form

$$\frac{d^2\psi}{dx^2} + \frac{8\pi^2 m}{h^2}(E - U)\psi = 0 \, . \tag{15.76}$$

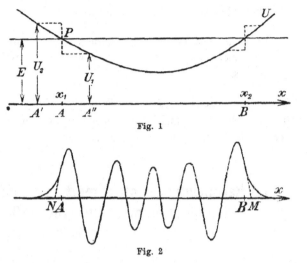

Fig. 15.1   Fig. 1 (top) illustrates the potential function of a harmonic oscillator with various values of U and E described in the text. Fig. 2 (bottom) shows a solution of the Schrödinger wave equation for a harmonic oscillator for principal quantum number $n = 8$. Both diagrams are from Persico (1938).

The potential U and the solution of Schrödinger's wave equation as a function of x are illustrated in Figs. 1 and 2 of Fig. 15.1 which are taken from Persico's paper. It can be seen that the oscillations of the wavefunction are much more rapidly varying than the change in the potential U. To a good approximation, in the region AB away from the immediate vicinity of the points A and B at which $E = U$, the angular frequency of oscillation is

$$\omega(x) = \sqrt{\frac{8\pi^2 m}{h^2}(E - U)}\,, \qquad (15.77)$$

being the solution of

$$\frac{\mathrm{d}^2 \psi}{\mathrm{d}x^2} + \omega^2(x)\psi = 0\,. \qquad (15.78)$$

If U were indeed independent of x, the solution for $\psi(x)$ would be

$$\psi(x) = C \sin(\omega x + \theta)\,. \qquad (15.79)$$

A better approximation would be to seek a solution of the form

$$\psi(x) = F(x)\sin[S(x)]\,, \qquad (15.80)$$

where $F(x)$ is a slowly varying function of x. Inserting this trial solution into (15.78), we find

$$(F'' - FS'^2 + \omega^2 F)\sin S + (2F'S' + FS'')\cos S = 0\,, \qquad (15.81)$$

where the dashes mean differentiation with respect to x. To satisfy this equation, the expressions in front of both the sine and cosine terms must be zero and so we obtain two conditions,

$$F'' - FS'^2 + \omega^2 F = 0 ; \quad 2F'S' + FS'' = 0 .$$
(15.82)

If F is a slowly varying function of x, $|F''| \ll \omega^2 |F|$, and so the first expression in (15.82) becomes

$$\omega^2 = S'^2 .$$
(15.83)

The solution for S is then

$$S = \int_{x_1}^x \omega \, dx + \theta ,$$
(15.84)

where θ is a constant.

The second expression in (15.82) can be integrated immediately to give

$$F = \frac{c}{\sqrt{S'}} = \frac{c}{\sqrt{\omega}} ,$$
(15.85)

using the result (15.83). Hence, the wave function in the region of the oscillations can be approximated by

$$\psi = \frac{c}{\sqrt{\omega}} \sin\left(\int_{x_1}^x \omega \, dx + \theta\right) .$$
(15.86)

Now, it is clear that the solution (15.86) is not valid at the points A and B, at which $E - U = 0$. Dealing with A first, Persico introduces the trick of replacing the parabolic function U with a 'step', indicated by the dashed lines in the vicinity of A in Fig. 1 of Fig. 15.1. In the interval A to A'', the potential takes the constant value U_1 and in the interval A to A', the constant value U_2. The potentials U_1 and U_2 are chosen such that

$$U_2 - E = E - U_1 .$$
(15.87)

Therefore, in the interval A to A'', the frequency of oscillation takes the constant value

$$\omega_1 = \sqrt{\frac{8\pi^2 m}{h^2}(E - U_1)} .$$
(15.88)

Inserting this value into (15.86), the wavefunction measured from A is

$$\psi = \frac{c}{\sqrt{\omega_1}} \sin[\omega_1(x - x_1) + \theta] .$$
(15.89)

Now consider the wavefunction in the interval A' to A. In this region, $U_2 > E$ and so the wave equation becomes

$$\psi'' - \frac{8\pi^2 m}{h^2}(U_2 - E)\psi = \psi'' - \frac{8\pi^2 m}{h^2}(E - U_1)\psi = \psi'' - \omega_1^2 \psi = 0 , \quad (15.90)$$

the first equality resulting from our choice of the values of U_1 and U_2 in (15.87). The solution of (15.90) is an exponentially decreasing function of decreasing x in the interval AA' with exponent ω_1,

$$\psi(x) = a e^{\omega_1 x} .$$
(15.91)

Now we need to join together the solutions (15.89) and (15.91) at A. As usual, for any wave, the function and its first derivative must be continuous at A and so

$$ae^{\omega_1 x_1} = \frac{c}{\sqrt{\omega_1}} \sin\theta \;, \quad a\omega_1 e^{\omega_1 x_1} = \frac{c}{\sqrt{\omega_1}}\, \omega_1 \cos\theta \;. \tag{15.92}$$

Dividing the first by the second equation in (15.92), we find $\tan\theta = 1$, $\theta = \pi/4$. Hence the approximate solution for ψ in the interval A to B is

$$\psi(x) = \frac{c}{\sqrt{\omega}} \sin\left(\int_{x_1}^{x} \omega \, dx + \frac{\pi}{4}\right) \;. \tag{15.93}$$

This is the WKB solution found by Kramers and his colleagues. Notice the important point that the first maximum of the wavefunction occurs at phase $\pi/4$ from x_1 in the positive x-direction.

Similar considerations apply to the wavefunction in the vicinity of B, the point x_2 lying $\pi/4$ beyond the last maximum. Therefore, the total phase difference between A and B, or x_1 and x_2, is

$$\int_{x_1}^{x_2} \omega \, dx = n\pi + \frac{\pi}{2} = \left(n + \frac{1}{2}\right)\pi \;, \tag{15.94}$$

where n is found by counting the number of nodes between A and B.

Now, classically, the momentum of the electron is $p = \pm\sqrt{2m(E - U)}$ and so is directly related to the angular frequency of the wavefunction through (15.77), $p = h\omega/2\pi$. If we now take the integral of $p \, dx$ through one complete oscillation of the harmonic oscillator, that is, from x_1 to x_2 and back to x_1, we find

$$\oint p \, dx = 2 \int_{x_1}^{x_2} p \, dx = \frac{h}{\pi} \int_{x_1}^{x_2} \omega \, dx \;, \tag{15.95}$$

and hence, from (15.94),

$$\oint p \, dx = \left(n + \frac{1}{2}\right) h \;, \tag{15.96}$$

precisely with Bohr–Sommerfeld quantisation condition, including the term $\frac{1}{2}h$.

We recall that there is no mention of the motion of the electron in the atom according to either wave or matrix mechanics, a point emphasised in the quotation by Pauli at the end of Sect. 15.4: 'no description of the electron in the atom in space and time is given'. It seems best to regard Bohr's quantisation of angular momentum as a lucky coincidence which opened up the route to the old and new quantum physics.

15.7 Reflections

By mid-1926, it was becoming clear that the operator calculus was the way ahead for the study of quantum problems, stimulated by the remarkable insights of Born, Dirac, Eckart, Heisenberg, Jordan, Schrödinger and Wiener, to mention only those whose work has been highlighted in this chapter. Born was unstinting in his praise for Schrödinger's achievement. In his obituary of Schrödinger, he referred to his papers of 1926 as 'of a grandeur unsurpassed in theoretical physics' (Born, 1961b). Planck was equally effusive:

> '[Schrödinger's wave equation] plays the same part in modern physics as do the equations established by Newton, Lagrange and Hamilton in classical mechanics.' (Planck, 1931b)

The next few years saw an extraordinary flowering of quantum mechanics as more and more flesh was put on the skeleton of the operator calculus.

16 Spin and quantum statistics

The discovery of the spin of the electron by Uhlenbeck and Goudsmit was a major advance in the understanding of physics at the atomic level. Its discovery coincided with the development of both matrix and wave mechanics and its incorporation into the scheme of quantum mechanics and statistics led to deeper understanding of the underlying structure of quantum mechanics. Almost immediately, Heisenberg and Jordan used the new scheme of matrix mechanics to derive the expression for the g-factor which Landé had derived empirically from a very close study of the anomalous Zeeman effect. An important consequence of these developments was that the different approaches of matrix and wave mechanics were brought together. In particular, the discovery of spin as a new quantum number suggested the possibility of understanding systems containing more than one electron. Heisenberg's analysis of the helium atom was to pave the way for the full incorporation of spin into quantum mechanics and quantum statistics.

16.1 Spin and the Landé g-factor

The story of the discovery of the spin of the electron by Uhlenbeck and Goudsmit (1925a) was told in Sect. 8.5. As discussed in that section, their discovery was based upon empirical studies of the regularities observed in the anomalous Zeeman effect, inspired by the intricate analyses of Landé. Although based originally upon the classical concept of a rotating electron, electron spin is a purely quantum mechanical property intrinsic to the electron. Opinions were strongly divided about the validity of the concept, Pauli taking a strongly negative position, while Bohr, Heisenberg and Jordan took a more positive view. The challenge taken up by Heisenberg was to find a quantum mechanical solution for the anomalous Zeeman effect using the concept of a spin-$\frac{1}{2}$ particle within the context of their recently completed matrix formalism.

Uhlenbeck and Goudsmit's innovations were to postulate that the electron has an intrinsic spin with spin quantum numbers $s = \pm \frac{1}{2}$ and that the gyromagnetic ratio associated with electron spin should be twice that associated with the electron's orbital motion. We recall that the gyromagnetic ratio for orbital motion of the electron $\mu_e(\text{orb})$ is the ratio of the magnetic moment associated with that motion to its angular momentum. For electron spin, the gyromagnetic ration $\mu_e(\text{spin})$ was postulated to be twice that value. Thus,

$$\mu_e(\text{orb}) = \frac{e}{2m_e} \boldsymbol{L} \; ; \quad \mu_e(\text{spin}) = \frac{e}{m_e} \boldsymbol{s} \; , \tag{16.1}$$

where L and s are the angular momentum vectors associated with the orbital motion and electron spin respectively. Then, the Hamiltonian H for an electron in a uniform magnetic field B can be written

$$H = H_0 + H_1 + H_2 + H_3 \,. \tag{16.2}$$

Heisenberg took the various contributions to H to be as follows:

- The term H_0 is the energy associated with the non-relativistic motion of a spinless electron about the nucleus in the absence of an external magnetic field.
- The second term H_1 represents the interaction energy between the orbital angular momentum L and spin s of the electron with the external magnetic field B,

$$H_1 = \frac{e}{2m_e} B \cdot (L + 2s) \,. \tag{16.3}$$

- The term H_2 represents the energy associated with the coupling between the induced magnetic field B_i observed in the frame of reference of the moving electron and the magnetic moment of the electron, what is termed *spin–orbit coupling*. For a single electron orbiting a nucleus of charge Ze, the electric field strength at radius r from the nucleus is

$$E = \frac{Ze}{4\pi\epsilon_0 r^3} r \tag{16.4}$$

and so the induced, or internal, magnetic flux density observed by the electron in its rest frame is

$$B_i = -\frac{v \times E}{c^2} = -\frac{Ze(v \times r)}{4\pi\epsilon_0 c^2 r^3} = \frac{ZeL}{4\pi\epsilon_0 m_e c^2}\left(\frac{1}{r^3}\right) \,. \tag{16.5}$$

Hence, the spin–orbit coupling is expected to have the form

$$H_2 = -\mu_e \cdot B_i = \frac{Ze^2}{4\pi\epsilon_0 m_e^2 c^2}\overline{\left(\frac{1}{r^3}\right)}(L \cdot s) \,, \tag{16.6}$$

where the overline means the value of r^{-3} averaged over the orbit of the electron. Note that more generally, when the effects of shielding by other electrons in the atom are taken into account, the electric field will not be of inverse-square form but can be written

$$E = -\frac{dV(r)}{dr} \,, \tag{16.7}$$

where $V(r)$ is the mean radial electrostatic potential. Associated with the interaction energy there is a precession of the axis of the magnetic moment of the electron about the internal magnetic field, just as in the case of the classical analysis of the Zeeman effect (Sec. 4.1),

$$\Omega_{LS} = \frac{eB}{m_e} = \frac{Ze^2}{4\pi\epsilon_0 m_e^2 c^2}\overline{\left(\frac{1}{r^3}\right)} L \,. \tag{16.8}$$

As compared with (4.10), however, (16.8) has included the factor of 2 associated with the gyromagnetic ratio of the spin of the electron. Notice also that the precession frequency is proportional to the interaction energy, $\Omega_{LS} \propto H_2$.

- The final term H_3 represents the relativistic corrections to the Hamiltonian which was taken from Sommerfeld's analysis of the classical relativistic hydrogen atom (Sommerfeld, 1916a).

Despite the less than encouraging views of Pauli, in November 1925 Heisenberg set about putting the above Hamiltonian through what he referred to as the 'matrix mill' in order to find the stationary states and line splittings associated with the anomalous Zeeman effect. Disappointingly, he almost reproduced Landé's formula for the anomalous Zeeman effect, but the crucial spin–orbit coupling term resulted in a factor of 2 discrepancy from Landé's expression, a result which cast doubt on the whole scheme.

In early January 1926, Heisenberg became aware of the new operator formulation of quantum mechanics of Born and Wiener (1926) which opened up the route for extending matrix mechanics to the more general operator formalism. He was then able to re-evaluate the problem using action–angle variables, but nonetheless, the stubborn factor of 2 remained, causing general disappointment among the proponents of electron spin.

The solution was, however, at hand thanks to the insight of Llewellyn Thomas who had arrived recently at Bohr's Institute in Copenhagen as a visiting graduate student. Thomas re-examined the relativistic transformations between the frame of the electron and the external frame and discovered that the expression (16.6) was not the complete story. Thomas was aware of the fact that there is an addition kinematic effect associated with the orbital motion of a vector, such as the spin vector of the electron, according to the special theory of relativity. Thomas was aware of such a precession which had been worked out by de Sitter in the context of the relativistic precession of the Moon and which is analysed in Eddington's book *The Mathematical Theory of Relativity* (Eddington, 1924). As Thomas expressed it in his short paper to *Nature* (Thomas, 1926),

> '[According to the above argument], the precession of the spin axis . . . is its precession in a system of coordinates (2) in which the centre of the electron is momentarily at rest. System (2) is obtained from system (1), in which the electron is moving and the nucleus at rest, by a Lorentz transformation with the velocity v. If the acceleration of the electron is [a], and the system (3) is obtained from system (1) by a Lorentz transformation with velocity $v + a \, dt$, then the precession which an observer at rest with respect to the nucleus would observe, and which should be summed to give the secular precession, is that precession which would turn the direction of the spin axis at time t in (2) into its direction at time $t + dt$ in (3) if both directions were regarded as directions in (1). To a first approximation system (3) is obtained from system (2) by a Lorentz transformation with the velocity $a \, dt$ together with a rotation $(1/2c^2)[v \times a] \, dt$.'

This purely kinematic effect results in an additional contribution to the precession, and hence interaction energy, of the electron which amounts to

$$\mathbf{\Omega}_{\mathrm{T}} = -\frac{1}{2c^2}[v \times a] \,, \tag{16.9}$$

and since, to a first approximation, $m_e a = eE$, it follows that

$$\mathbf{\Omega}_{\mathrm{T}} = -\frac{1}{2}\frac{Ze^2}{4\pi\epsilon_0 m_e^2 c^2}\overline{\left(\frac{1}{r^3}\right)} L \,. \tag{16.10}$$

Thus, the effective internal field is only half the value given by (16.6) and can account completely for the discrepant factor of 2. After considerable debate, even Pauli was converted and the paper on the quantum mechanical explanation for the anomalous Zeeman effect was published by Heisenberg and Jordan in June 1926 (Heisenberg and Jordan, 1926). Rechenberg has written in his summary of the history of quanta and quantum mechanics that the explanation of the anomalous Zeeman effect was one of the greatest triumphs of matrix mechanics (Rechenberg, 1995).

16.2 Heisenberg and the helium atom

Following the departure of Kramers to take up the chair of theoretical physics at Utrecht, Heisenberg succeeded to the post of university lecturer and assistant to Bohr in Copenhagen in May 1926. It was well-known that the old quantum theory had insuperable problems in dealing with atoms with more than one electron and Heisenberg was well aware of these, having already worked on the theory of the helium atom on the basis of the Bohr–Sommerfeld theory of atomic structure. By 1926, however, the problem was ripe for a renewed attack. The discoveries of electron spin by Uhlenbeck and Goudsmit (1925a) and Pauli's exclusion principle (1925) as well as the success with which Heisenberg and Jordan had accounted for the anomalous Zeeman effect using the operator enhancement of matrix mechanics suggested a new approach to the helium problem. At the same time, Heisenberg rapidly assimilated Schrödinger's wave mechanics which provided a simpler means of determining the relevant matrix elements.

The well-known problem associated with the spectra of helium and the alkaline-earth metals, such as magnesium and calcium, was that there are two apparently separate and independent sets of lines present in their optical spectra. In the case of helium, the corresponding separate term diagrams were referred to as belonging to *para* and *orthohelium* (Fig. 16.1). Both sets of energy levels were not so different from the terms appearing in the hydrogen term diagram. The parahelium lines were all singlets whereas the orthohelium series consisted of very narrow triplets – the energy levels of the ortho terms are slightly more tightly bound than those of the corresponding para terms. Originally, it was thought that helium was in fact a mixture of two gases, para and orthohelium.

According to Goudsmit's recollections,[1] he and Bohr discussed the problem of the helium spectrum in February 1926 during a visit to Copenhagen, not long after the discovery of spin. He recalled that

> 'By looking at the helium spectrum Bohr understood right away that if you turned over one of the two spins [of the electrons in the helium atom], the energy suddenly would be entirely different.'

Goudsmit attempted to explain the large energy difference between the ground states of ortho and parahelium in terms of a magnetic interaction between the magnetic moments of the two electrons, but could not obtain as large a difference as observed.

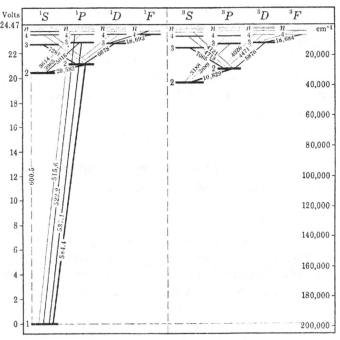

Fig. 16.1 The term diagram for helium (Herzberg, 1944). The term diagram on the left is for parahelium and is characterised by the two electrons having opposite spins and so $S = 0$, corresponding to singlet states. The term diagram on the right is for orthohelium in which the electrons have parallel spins so that $S = 1$ and the states are triplet states. Transitions between the singlet and triplet states are forbidden.

Heisenberg realised that the solution to the problem lay in applying Pauli's exclusion principle to the helium atom (see Sect. 8.4). On a postcard written to Pauli towards the end of April 1926, Heisenberg wrote

> 'we have found a rather decisive argument that your exclusion of equivalent orbits [of two electrons in an atom] is connected with the singlet–triplet separation. ... Consider the energy written as a function of the transition probabilities. Then, a large difference results if one ... has transition to $1S$, or if, according to your ban [exclusion principle], one puts them equal to zero. That is, para- and ortho-[helium] do have different energies, independently of the interaction between magnets.'

This provided the spur to Heisenberg to begin calculating. According to the Pauli exclusion principle, two electrons with opposite spins could occupy the ground state and would correspond to the parahelium series. On the other hand, if the spins were parallel, one of the electrons could remain in the ground state but the other would have to occupy a less tightly bound orbit, well above the ground state. Central to the analysis was the fact that the two electrons in the helium atom are identical particles and this could not be ignored. In the scheme of matrix mechanics, neglecting the weak interaction between the magnetic moments of the electrons, Heisenberg adopted a model in which the Hamiltonian matrix $\mathbf{H}^0$ was composed of two terms, $\mathbf{H}^a$ and $\mathbf{H}^b$, each term referring to the motion of the electron under the influence of the combined Coulomb field of the nucleus and the two electrons.

The charge experienced by an electron in the unshielded Coulomb field of the nucleus would be $2e$, or e if the nuclear charge were shielded by the other electron. The key point realised by Heisenberg was that the exchange of the a and b electrons would leave the energies of the stationary states unaltered.

Heisenberg proceeded with the calculation of the stationary states of the helium atom treating the Coulomb repulsion between the electrons as a first-order perturbation. The result was the discovery of two separate symmetric and antisymmetric solutions for the stationary states, transitions being allowed within each set of separate energy levels, but not between them. Initially, Heisenberg identified the symmetric solutions as the ones to be adopted and considered this choice to be consistent with Pauli's exclusion principle and with Bose–Einstein statistics. Much later, in an interview in 1963, he confessed to his confusion about the nature of the statistics to be associated with the electrons in atoms in 1926. He stated:

'For a long time, I continued to mix up Bose–Einstein and Fermi–Dirac statistics. I did not know Fermi–Dirac statistics at that time; I knew only the Pauli exclusion principle. I was always confused between Bose–Einstein statistics and the Pauli exclusion principle which produce different ways of counting states. When I wrote the equations for two identical electrons, there were two solutions, one symmetrical and the other anti-symmetrical. First I thought that I had to take the anti-symmetrical solution to obtain the Bose statistics and that must be the one that gave the Pauli principle. Later, I saw that it was the other way around. One must take the symmetrical solution to get Bose statistics, and the antisymmetrical solution to get Pauli's exclusion principle.' (Heisenberg, 1963)

Heisenberg proceeded to carry out the analysis to find the energy levels of the helium atom and, in doing so, made use of Schrödinger's wave mechanics to evaluate the perturbation terms. The Hamiltonian for the helium atom can be written

$$H = \frac{p_1^2}{2m_e} + \frac{p_2^2}{2m_e} - \frac{2e^2}{4\pi\epsilon_0 r_1} - \frac{2e^2}{4\pi\epsilon_0 r_2} + \frac{e^2}{4\pi\epsilon_0 r_{12}}, \qquad (16.11)$$

where r_1 and r_2 are the distances of the electrons from the nucleus and r_{12} is the distance between the electrons. For the unperturbed helium atom, in which the interaction between the electrons as represented by the last term in (16.11) is omitted, Schrödinger's equation is separable and wavefunction solutions of the form $\psi = \phi_n^1 \phi_m^2$ can be found, where ϕ_n^1 and ϕ_m^2 are the wavefunctions for electrons 1 and 2. Heisenberg found that, when he carried out the perturbation analysis of the two-electron system including the electrostatic repulsion between the electrons, the symmetric and antisymmetric solutions were

$$\psi^\bullet = \frac{1}{\sqrt{2}}(\phi_n^1 \phi_m^2 + \phi_m^1 \phi_n^2), \quad \psi^\times = \frac{1}{\sqrt{2}}(\phi_n^1 \phi_m^2 - \phi_m^1 \phi_n^2). \qquad (16.12)$$

Heisenberg 'used the steamroller' approach to find the energy levels of the helium atom, approximating the radial distribution of the electric potential by the function

$$V(r) = -\frac{Ze^2}{4\pi\epsilon_0 r} + f(r), \text{ where } f(r) = \begin{cases} \dfrac{e^2}{4\pi\epsilon_0 r_0} & \text{for } 0 \leq r \leq r_0 \\[2mm] \dfrac{e^2}{4\pi\epsilon_0 r} & \text{for } r_0 \leq r \leq \infty. \end{cases} \qquad (16.13)$$

With this approximation, Heisenberg was able to find a reasonable quantitative explanation of the energy levels in the para and ortho states of the helium atom. In the second part of his paper he included the small additional changes associated with electron spin, which included the correct splitting of the singlet and triplet states. Heisenberg's papers were published in August and October 1926 (Heisenberg, 1926a,b). In the second paper, Heisenberg acknowledged the value of Schrödinger's approach to the evaluation of the energy levels and transition probabilities. He wrote

> 'I used Schrödinger's formalism for help with the mathematics. It was clear to me that in order to calculate the shift of the levels in the helium atom, matrix elements were needed, and that they could be calculated quite well from Schrödinger's scheme. Such a calculation in matrix mechanics would have been difficult.'

This major advance in understanding the properties of the helium atom were refined by a number of authors, leading ultimately to an accurate estimate of the ionisation potential of the helium atom by Kellner (1927). These calculations were also to lead to the development of the concept of *exchange forces*, the quantum mechanical force associated with the fact that two identical electrons cannot occupy the same quantum state. This would ultimately lead to Heisenberg's application of the concept of exchange forces to the theory of ferromagnetism (Heisenberg, 1928), but first we need to get to grips with the statistical properties of spin-$\frac{1}{2}$ particles, Fermi–Dirac statistics.

16.3 Fermi–Dirac statistics – the Fermi approach

The young Enrico Fermi visited Göttingen and Leiden as a research fellow in 1923–1925 and became familiar with the issues of quantum physics. In 1925 he was appointed a lecturer in physics at the University of Florence and then to a new professorship of theoretical physics at the University of Rome in 1926. His early interest in quantum physics had been in the equation of state of an ideal gas, in particular, Nernst's heat theorem according to which the heat capacities of all substances tend to zero as the temperature tends to zero, contrary to the expectations of classical statistical mechanics. The papers by Bose and Einstein appeared in 1924 and 1925 (see Sect. 9.2) and Fermi certainly knew about them (Bose, 1924; Einstein, 1924, 1925). They had treated light quanta and the atoms of an ideal gas as indistinguishable particles and argued that the procedure of allocating these entities to the available states in phase space should avoid duplications, provided the particles were indistinguishable.

Fermi was even more strongly impressed by Pauli's recent enunciation of the exclusion principle for the electrons in the atom (Pauli, 1925), namely, that only one electron can occupy a particular quantum state with a given set of atomic quantum numbers (Sect. 8.4). In the standard formulation of statistical mechanics, the available states within an enclosure are enumerated by counting the available states in momentum space, which is equivalent to fitting a finite, but very large, number of wavelengths within its reflecting walls – the result is an essentially continuous distribution of energies and momenta of the particles. Fermi now proposed reducing the size of the enclosure so that it included only a single atom and then applying Pauli's exclusion principle to that volume. But unlike Pauli, who applied his rule to the electrons in the atom, Fermi proposed applying the same type of exclusion principle to the atoms of a perfect gas. Fermi wrote,

> '. . . one obtains a degeneracy [of the gas] of the expected order of magnitude, if one chooses the vessel to be so small as to contain, on average, only one molecule. . . . We shall demonstrate . . . that the application of the Pauli rule allows us to present a complete, consistent theory of the degeneracy of ideal gases.' (Fermi, 1926b)

Fermi's first publication on his new quantum statistics appeared in Italian in the *Rendiconti del Reale Accademia Lincei* on 7 February 1926 and then in German in the *Zeitschrift für Physik* on 11 May 1926 (Fermi, 1926a,b). Fermi's model consisted of N gas molecules of mass m located within a central 'elastic' potential U given by the expression

$$U = 2\pi^2 \nu^2 m r^2 , \tag{16.14}$$

where ν is the frequency of oscillation of the molecules about the centre $r = 0$. Considering only the translational motion of the atoms or molecules in the x-, y- and z-directions, the energies in the three independent coordinates are quantised so that the energy of each molecule is

$$w = h\nu (s_1 + s_2 + s_3) = sh\nu , \tag{16.15}$$

where s_1, s_2 and s_3 take integral values $0 \leq s_i \leq s$. It is straightforward to show that the number of different ways Q_s of obtaining s for this range of s_1, s_2 and s_3 is

$$Q_s = \frac{(s+1)(s+2)}{2} . \tag{16.16}$$

At absolute zero temperature, all the states would be occupied, with one molecule with zero energy, three with energy $h\nu$, six with energy $2h\nu$ and so on.

At temperatures greater than absolute zero, the atoms or molecules are distributed among the available states and Einstein had shown how this calculation is carried out in his paper on the Bose–Einstein distribution (Einstein, 1924, 1925). If there are N molecules and the total energy E is to be distributed among them, then

$$\sum_s N_s = N, \quad \text{and} \quad h\nu \sum_s s N_s = E , \tag{16.17}$$

with the requirement that $N_s \leq Q_s$. This becomes an exercise in the calculus of variations, exactly paralleling Einstein's analysis. The number of possible ways of distributing N_s atoms among Q_s possible locations is given by the usual permutation formula $\binom{Q_s}{N_s}$ and so the total number of possible ways of distributing the particles is the product of these factorials for all the energy states, subject to the constraints (16.17),

$$P = \binom{Q_0}{N_0} \binom{Q_1}{N_1} \binom{Q_2}{N_2} \dots \tag{16.18}$$

Using Stirling's formula for large factorials, $N! \approx N^N$, and taking logarithms, the problem reduces to finding the most likely value of

$$\ln P = \sum_s \ln \binom{Q_s}{N_s} = - \sum_s \left(N_s \ln \frac{N_s}{Q_s - N_s} + Q_s \ln \frac{Q_s - N_s}{Q_s} \right) , \tag{16.19}$$

subject to the constraints (16.17). Using the standard procedure of the method of undetermined multipliers, the result is

$$\frac{N_s}{Q_s - N_s} = \alpha \exp(-\beta s) \quad \text{or} \quad N_s = \frac{Q_s}{\exp(A + \beta s) + 1} , \tag{16.20}$$

where α, β and A are constants. This result can be compared with Einstein's expression (9.7),

$$n_k = \frac{g_k}{e^{\alpha + \beta \varepsilon_k} - 1} , \tag{16.21}$$

showing the characteristic difference between Fermi–Dirac and Bose–Einstein statistics, the plus sign in the denominator of the former and the minus sign of the latter.

Although the constants A and β could be determined from the statistical formula $S = k \ln P$ and the thermodynamic relation $T = dS/dS$, Fermi was not certain about the applicability of the thermodynamic laws at extremely low temperatures and so instead derived the constants from the asymptotic value of the particle distribution at very large distances from the origin. At these distances, the density of particles becomes very low and the energy distribution tends to a Maxwellian velocity distribution. In this way, he found that the constant $\beta = hv/kT$ and determined the value of A. Of particular significance was his determination of the pressure of the gas in the extreme low-temperature limit, at which the pressure is independent of temperature,

$$p = \frac{1}{20} \left(\frac{6}{\pi} \right)^{2/3} \frac{h^2 n^{5/3}}{m} , \tag{16.22}$$

the form of which can be derived from elementary physical arguments.[2] In addition, there is a zero point energy per particle which amounts to

$$E = \frac{3}{40} \left(\frac{6}{\pi} \right)^{2/3} \frac{h^2 n^{2/3}}{m} . \tag{16.23}$$

Finally, the specific heat capacity at low temperatures is

$$C_V = \left(\frac{2\pi^2}{\sqrt{3}} \right)^{4/3} \frac{mk^2 T}{h^2 n^{2/3}} . \tag{16.24}$$

Thus, C_V becomes zero as the temperature tends to zero, as required by Nernst's heat theorem.

16.4 Fermi–Dirac statistics – the Dirac approach

Meanwhile, back in Cambridge, Dirac continued his personal, rather lonely, agenda for the development of quantum mechanics, but profiting from correspondence with his colleagues in Göttingen and Copenhagen, particularly with Heisenberg. The latter recognised that Dirac's quantum algebra was a more powerful approach than matrix mechanics. He drew Dirac's attention to Schrödinger's papers and sought his advice and assistance in reconciling the very different approaches of matrix and wave mechanics. Heisenberg was well aware of Dirac's remarkable mathematical skills and he was not disappointed by the outcome. Dirac's q-number formulation of the rules of quantum mechanics had the advantage that he did not need to state explicitly what the q-number was, but could use the formalism for whatever mathematical object best suited his purposes for the problem at hand. As he stated in the introduction to his paper *On the theory of quantum mechanics* (Dirac, 1926f),

> 'One can build up a theory [of atomic systems] without knowing anything about the dynamical variables except the algebraic laws that they are subject to, and can show that they may be represented by matrices whenever a set of uniformising variables for the dynamical system exists. . . . It can be shown however . . . that there is no set of uniformising variables for a system containing more than one electron, so that the theory cannot progress very far on these lines.'

In his letters to Dirac, Heisenberg described the reconciliation of matrix and wave mechanics and Dirac fully appreciated the advantages of Schrödinger's theory. Characteristically, he set about a systematic exploration of wave mechanics, recasting it in his own language.[3] The q-numbers of his theory could be taken to be differential operators, the pairs of canonically conjugate position and momentum variables q_r and p_r and time and energy variables t and W being defined by

$$q_r , \quad p_r = -\mathrm{i}\frac{h}{2\pi}\frac{\partial}{\partial q_r} , \tag{16.25}$$

$$t , \quad W_r = -\mathrm{i}\frac{h}{2\pi}\frac{\partial}{\partial t} . \tag{16.26}$$

Furthermore, he showed that with these definitions, the classical Hamilton–Jacobi equation reduces to Schrödinger's wave equation and so the whole apparatus of wave mechanics followed naturally from his scheme of quantum mechanics. The relation between Schrödinger's eigenfunctions and the elements of the matrices was also clarified:

> '. . . any constant of integration of the dynamical system . . . can be represented by a matrix whose elements are constants, there being one row and one column of the matrix corresponding to each eigenfunction ψ_n.'

Dirac's clear and concise exposition of the fundamentals of quantum mechanics provided a complete demonstration of the equivalence and merits of the different approaches of matrix and wave mechanics. He had no hesitation in adapting Schrödinger's approach to his way of thinking.

Dirac had a long-standing interest in developing the quantum mechanics of multiple-electron systems and in Sect. 3 of his paper concentrated upon the case of the helium atom and then the statistics of identical particles (Dirac, 1926f). The nub of his thinking is contained in the following quotation from his paper.

> 'Consider now a system that contains two or more similar particles, say, for definiteness, an atom with two electrons. Denote by (mn) that state of the atom in which one electron is in an orbit labelled m, and the other in the orbit n. The question arises whether the two states (mn) and (nm), which are physically indistinguishable as they differ only by the

interchange of the two electrons, are to be counted as two different states or as only one? If the first alternative is right, then the theory would enable one to calculate the intensities due to the two transitions $(mn) \rightarrow (m'n')$ and $(mn) \rightarrow (n'm')$ separately, as the amplitude corresponding to either would be given by a definite element in the matrix representing the total polarisation. The two transitions are, however, physically indistinguishable, and only the sum of the intensities for the two together could be determined experimentally. Hence, in order to keep the essential characteristic of the theory that it shall enable one to calculate only observable quantities, one must adopt the second alternative that (mn) and (nm) count as one state.'

This paragraph encapsulates the key concept in his development of what was to become Fermi–Dirac statistics.

This requirement of the indistinguishability of particles led to inconsistencies in a pure matrix mechanical scheme, but it had a straightforward interpretation in terms of the eigenfunction representation of the stationary states of the helium atom. Just as in the case of Heisenberg's analysis of the helium atom, if the interaction term between the electrons is ignored, the Hamiltonian (16.11) is separable so that the wavefunction of the two electrons can be written

$$\psi_m(x_1, y_1, z_1, t)\, \psi_n(x_2, y_2, z_2, t) = \psi_m(1)\, \psi_n(2)\,. \tag{16.27}$$

The eigenfunction $\psi_m(2)\psi_n(1)$, however, corresponds to exactly the same identical state. There must therefore be only one row and column of the matrices corresponding to both (mn) and (nm) and this is achieved by writing the wavefunction ψ_{mn} in the form

$$\psi_{mn} = a_{mn}\, \psi_m(1)\, \psi_n(2) + b_{mn}\, \psi_m(2)\, \psi_n(1)\,, \tag{16.28}$$

where the a_{mn}s and b_{nm}s are constants, the sets containing only one ψ_{mn} corresponding to both (mn) and (nm). Furthermore, the a_{mn}s and b_{nm}s must be chosen so that the matrix can represent any symmetric function A of the two electrons. Thus, it must be possible to expand the wavefunction ψ_{mn} in terms of the complete set of eigenfunctions so that

$$A\psi_{mn} = \sum_{m'n'} \psi_{m'n'}\, A_{m'n',mn}\,, \tag{16.29}$$

where the $A_{m'n',mn}$ are constants or only functions of time.

There are then two ways of choosing the set of functions ψ_{mn} to satisfy these conditions. Either $a_{mn} = b_{mn}$, resulting in ψ_{mn} being a symmetrical function of the two electrons and so all the eigenfunctions are symmetrical, or $a_{mn} = -b_{mn}$, so that ψ_{mn} is an antisymmetrical function of the two electrons and all the eigenfunctions are antisymmetrical. Dirac concludes

'An antisymmetrical eigenfunction vanishes identically when two of the electrons are in the same orbit. This means that in the solution of the problem with antisymmetric eigen-functions there can be no stationary states with two or more electrons in the same orbit, which is just Pauli's exclusion principle. The solution with symmetrical eigenfunctions, on the other hand, allows any number of electrons to be in the same orbit, so that this solution cannot be the correct one for the problem of electrons in an atom.'

In the next section of this very impressive paper, Dirac makes clear the application of the symmetrical and antisymmetrical wavefunctions to light and particles.

'The solution with symmetrical eigenfunctions must be the correct one when applied to light quanta, since it is known that the Einstein–Bose statistical mechanics leads to Planck's law of black-body radiation. The solution with antisymmetrical eigenfunctions, though, is probably the correct one for gas molecules, since it is known to be the correct one for electrons in an atom, and one would expect molecules to resemble electrons more closely than light-quanta.'

Dirac proceeds to determine the equation of state of the gas, under the assumption that there can only be one or zero electrons associated with a given wave, or eigenfunction. The waves are divided into a number of sets A_s, each of the same energy E_s. Then, the number of different ways in which N_s molecules can be distributed among the A_s waves is given by the standard relation

$$p_s = \frac{A_s!}{N_s!(A_s - N_s)!},$$
(16.30)

the factorials in the denominator removing duplicates of identical occupancy or vacancies. The total probability is then found by multiplying the probabilities of all possible states, subject to the constraints $N = \sum_s N_s$ and $E = \sum_s E_s N_s$. Dirac writes

$$W = \Pi_s \frac{A_s!}{N_s!(A_s - N_s)!}.$$
(16.31)

This has reduced the problem to the same standard calculation in the calculus of variations which was carried out by Fermi (1926a,b) and discussed in Sect. 16.3. The result is the standard Fermi–Dirac distribution, which in Dirac's notation is

$$N_s = \frac{A_s}{e^{\alpha + E_s/kT} + 1},$$
(16.32)

exactly the same as Fermi's expression (16.20). The expressions for the total numbers of particles and their pressure in volume V follow by the standard procedure for counting the numbers of states in phase space,

$$N = \sum N_s = \frac{2\pi V(2m)^{3/2}}{h^3} \int_0^\infty \frac{E_s^{1/2}\, dE_s}{e^{\alpha + E_s/kT} + 1},$$
(16.33)

$$E = \sum E_s N_s = \frac{2\pi V(2m)^{3/2}}{h^3} \int_0^\infty \frac{E_s^{3/2}\, dE_s}{e^{\alpha + E_s/kT} + 1}.$$
(16.34)

Dirac noted that, since $pV = \frac{2}{3}E$, by eliminating α between (16.33) and (16.34), the equation of state of the gas could be found. He noted also that the specific heat of the gas tended steadily to zero at zero temperature, as required by Nernst's heat theorem. The phenomenon of Bose–Einstein condensation does not occur.

Much later in an interview which took place in 1963, Dirac recalled that he had seen Fermi's paper, but had not paid much attention to it. Only when Fermi wrote to him about it, Dirac recalled the paper and fully acknowledged the fact that Fermi had independently discovered what is now referred to as the *Fermi–Dirac distribution*.[4] While there is no doubt about the priority of Fermi's pioneering paper, Dirac's had the greater impact since it addressed the issue of quantum statistics in greater generality and was quickly assimilated into the structure of quantum mechanics.

16.5 Building spin into quantum mechanics – Pauli spin matrices

Despite Heisenberg and Jordan's success in accounting for the anomalous Zeeman effect (Sect. 16.1), the analysis was only a quantum mechanical extension of the classical expression for a spinning electron in a magnetic field. The spin was represented by a spin vector s which was added vectorially to the orbital angular momentum L. Pauli fully appreciated that spin had not been properly incorporated into the scheme of quantum mechanics for a simple reason. Whilst called an 'angular momentum', spin could not be represented in the same way as orbital angular momentum because the orbital angular moment has three independent directions, corresponding to rotation about the x-, y- and z-axes, M_x, M_y and M_z, whereas the postulates of Uhlenbeck and Goudsmit required only two spin states associated with $s = +\frac{1}{2}$ and $s = -\frac{1}{2}$. The full significance of this difficulty was appreciated by Darwin (1927a) who wrote,

> 'When what is required is to double the number of states of the electron, it is at the least generous to introduce three extra degrees of freedom and then make an arbitrary (though not unnatural) assumption which cuts down the triple infinity to two. The electron is in fact given a complete outfit of Eulerian angles, even if it may not be necessary so to express the matter explicitly. Now we regard the electron as the most primitive thing in Nature, and it would therefore be much more satisfactory if the duality could be obtained without such great elaboration.'

Darwin elaborated his proposal to treat what Pauli called the electron's 'classically not-describable two-valuedness' as a vector quantity which involved two separate Schrödinger-type equations to describe the different polarisations (Darwin, 1927b,c). He doubled the number of equations so that the four new variables could be treated as the components of a four-vector in relativity. The scheme was complex and incomplete and was overtaken by Pauli's invention of spin matrices which could be incorporated into quantum mechanics, but this involved a quite different approach to the problem. The essence of what Pauli (1927b) did is pleasantly summarised in the exposition by Lindsay and Margenau (1957) who described the problem as follows:

> '... the operator corresponding to this new degree of freedom, which we shall metaphorically continue to call 'spin', must have *only two eigenvalues* corresponding to the values $\pm h/4\pi$ of angular momentum in any one direction in which the external field is imagined to be placed. ... To make a long story short, it is found very difficult to obtain a differential operator ... which, acting upon a function of a continuous variable, has only two eigenvalues.'

Pauli's approach was quite different from that previously used in quantum mechanics – he chose a spin variable which has finite values at only two points. The variable s measures the spin in, say, the z-direction. The associated spin state function is $\phi(s)$. Now we suppose the range of s is only the two points 1 and -1. It is convenient for visualisation purposes to think of s as the cosine of the angle between the electron's spin axis and some direction in space, so that $s = +1$ corresponds to alignment parallel to the chosen direction and $s = -1$

to the antiparallel alignment. The state function $\phi(s)$ can only have finite values a at $s + 1$ and b at $s = -1$ and so the most general function of $\phi(s)$ is

$$\phi(s) = a\,\delta_{s,+1} + b\,\delta_{s,-1}\,, \tag{16.35}$$

where the δs are delta functions. The state function has to be normalised to unity and so

$$\int \phi^*(s)\,\phi(s)\,\mathrm{d}s = \int (a^*a\,\delta_{s,+1}\delta_{s,+1} + a^*b\,\delta_{s,+1}\delta_{s,-1} + b^*a\,\delta_{s,-1}\delta_{s,+1} + b^*b\,\delta_{s,-1}\delta_{s,-1})\,\mathrm{d}s, \tag{16.36}$$

$$= a^*a + b^*b = 1\,. \tag{16.37}$$

Interpreting $\phi^*(s)\,\phi(s)$ as the probability that the system is found in the state s, we can find the state function, even without knowing the operator. If the eigenfunction describing the system as being certainly in state $s = +1$ is $\psi_+(s)$, then $\psi_+^*(s)\,\psi_+(s) = \delta_{s,+1}$. Now substituting this result into (16.35), it follows that $a = 1$, $b = 0$ and so

$$\psi_+(s) = \delta_{s,+1}\,. \tag{16.38}$$

Similarly, if the system is certainly in the state $s = -1$, the eigenfunction is

$$\psi_-(s) = \delta_{s,-1}\,. \tag{16.39}$$

For the purposes of the spin properties of the electron, (16.38) and (16.39) form a complete orthonormal set.

We can now determine directly the spin operators σ_x, σ_y, σ_z. Suppose we wish to determine the spin angular momentum about the z-axis σ_z. Then, by hypothesis, the eigenvalues are $\pm h/4\pi$ while the eigenfunctions are given by (16.38) and (16.39). If we write for economy

$$\sigma_{x,y,z} = \frac{h}{4\pi} S_{x,y,z}\,, \tag{16.40}$$

$S_{x,y,z}$ has eigenvalues ± 1. Then, the operator S_z must satisfy the equations

$$S_z\,\psi_+ = \psi_+\,, \quad S_z\,\psi_- = -\psi_-\,. \tag{16.41}$$

This is a key point in the argument. S_z cannot be a differential operator because the functions ψ are not differentiable. Rather (16.41) *defines* the operator S_z. When it operates on ψ_+, it leaves it unchanged; when it operates on ψ_-, it changes the sign of ψ_-.

It is convenient to regard S_z as a matrix. Then, ψ_+ and ψ_- can be written as a column vector ψ,

$$\psi = \begin{bmatrix} \psi_+ \\ \psi_- \end{bmatrix}\,, \tag{16.42}$$

so that we can write

$$S_z\psi = s_z\psi \quad \text{with} \quad S_z = \begin{bmatrix} 1 & 0 \\ 0 & -1 \end{bmatrix}\,. \tag{16.43}$$

To find the spin operators corresponding to S_x and S_y, we assume that the operator equivalents of the matrix equations for the relations between the components of the angular momentum (12.83) are applicable and so, for example,

$$\sigma_y \sigma_x - \sigma_x \sigma_y = \frac{h}{2\pi i} \sigma_z .$$ (16.44)

Using the notation (16.40), the relations between the spin operators become

$$\left. \begin{array}{l} S_x S_y - S_y S_x = 2iS_z \\ S_y S_z - S_z S_y = 2iS_x \\ S_z S_x - S_x S_z = 2iS_y . \end{array} \right\}$$ (16.45)

Given (16.43), these equations can be readily solved to find what are referred to as the *Pauli spin matrices*,

$$S_z = \begin{bmatrix} 1 & 0 \\ 0 & -1 \end{bmatrix} , \quad S_y = \begin{bmatrix} 0 & -i \\ i & 0 \end{bmatrix} , \quad S_x = \begin{bmatrix} 0 & 1 \\ -1 & 0 \end{bmatrix} .$$ (16.46)

Notice a number of important features of these matrices. Only S_z is diagonal and so determines the eigenvalues of the spin operator S_z. The others are not diagonal, as must be the case since diagonal matrices commute and (16.45) shows that the spin matrices do not commute. Notice that the spin matrices are simply a convenient shorthand for the following operator equations:

$$\left. \begin{array}{l} S_z \psi_+ = \psi_+ \\ S_z \psi_- = -\psi_- \end{array} \right\} \quad \left. \begin{array}{l} S_y \psi_+ = -i\psi_- \\ S_y \psi_- = i\psi_+ \end{array} \right\} \quad \left. \begin{array}{l} S_x \psi_+ = \psi_- \\ S_x \psi_- = \psi_+ . \end{array} \right\}$$ (16.47)

It is apparent that ψ_+ and ψ_- are eigenstates only for the operator S_z, while the operators S_y and S_x represent mixed states of the ψs.

The spin operators can now be used in Schrödinger's equation following the usual rules of replacing variables by operators. Consider, for example, the simplest case of an electron in a uniform magnetic field $\boldsymbol{B}$. The energy equation is $H\psi = E\psi$ and we replace the spin terms in the Hamiltonian H by spin operators. The energy term involving the interaction between the magnetic moment of the electron $\boldsymbol{\mu}$ and the magnetic field is

$$H = \mu_z B_z + \mu_y B_y + \mu_x B_x .$$ (16.48)

Using the vector relation $\boldsymbol{\mu} = (he/m_e c)\boldsymbol{s}$, the corresponding operator for the magnetic moment of the electron is $(he/m_e c)S$ and so Schrödinger's equation becomes

$$\frac{eh}{4\pi m_e c} \left(S_z B_z + S_y B_y + S_x B_x \right) \psi = E\psi .$$ (16.49)

In the simplest case in which the magnetic field is oriented in the z-direction, (16.49) becomes

$$\frac{eh}{4\pi m_e c} S_z B_z \psi = E\psi , \quad \text{or} \quad |\mu| B S_z \psi = E\psi .$$ (16.50)

ψ is given by the column vector (16.42) which then provides the two solutions for the magnetic moment being parallel and antiparallel to the magnetic field direction[5] with energies $+|\mu|B$ and $-|\mu|B$. Thus, the spin operators fit naturally into the scheme of the operator formalism, although they are not differential operators. Notice that the spin operator has the property that it converts Schrödinger's equation directly into an algebraic equation, rather than a differential equation.

The extension to incorporate spin into atomic physics follows by exactly the same formulation as in the case of the electron in a uniform magnetic field. For example, for the case of an electron in an atom in the presence of a uniform magnetic field, we can write $H\psi = E\psi$ with the Hamiltonian now consisting of two parts, H_0 associated with the operator which appears in Schrödinger's equation for the atom and a second term which represents the interaction of the electron with the magnetic field,

$$H = H_0 + |\mu| B S_z \,. \tag{16.51}$$

Suppose we represent the wavefunction by the product $\psi(q)\psi(s)$, where q stands for the three space coordinates and s for the spin coordinates. H_0 operates only on the space coordinates and $\psi(q)$ is the resulting standard solution of the wave equation. S_z operates only on the spin part of the wavefunction $\psi(s)$. Consequently, by separation of variables, the solution is found by solving separately the equations

$$H_0\psi(q) = E_q\psi(q) \,, \quad |\mu| B s_z \psi(s) = E_s \psi(s) \,, \tag{16.52}$$

with $E = E_q + E_s$. For any given stationary state, there are two solutions associated with the spin of the electron with energies $\pm|\mu|B$ relative to the case in which there is no magnetic field. There is another pleasant consequence of this result. If the magnetic field is decreased, the two spin states will tend towards degeneracy as $B \to 0$ and indeed the two spin states are degenerate in the limit of zero field. In fact, however, because of the effects of spin–orbit coupling, the electron experiences a magnetic field, as discussed in Sect. 16.1. Therefore, in practice, the degeneracy is lifted and this coupling accounts for the fine splitting of the energy levels of the atom.

The important result of these calculations was that Pauli had succeeded in incorporating spin into the scheme of quantum mechanics. This achievement had profound implications for future studies of the quantum properties of matter since the way was opened up for fully incorporating spin into many different types of quantum mechanical problem, in particular, for systems of more than one electron. These studies led to the development of a self-consistent scheme of spin in quantum mechanics and was able to explain a large number of phenomena in atomic and condensed matter physics. There was, however, no deeper understanding of the origin of the spin of the electron – what Pauli had achieved was to find the correct formalism for including spin in quantum calculations, once it is assumed that the electron is a spin-half particle. The solution came with Dirac's relativistic formulation of quantum mechanics.

16.6 The Dirac equation and the theory of the electron

From the outset, there were difficulties in incorporating relativity into quantum mechanics. Sommerfeld's *tour de force* in using relativity to account for the splitting of the hydrogen lines according to the old quantum theory was a remarkable feat, but it was recognised by the pioneers of quantum mechanics that building relativity into quantum and wave mechanics was non-trivial. We recall that Schrödinger had begun with a fully relativistic version of his wave equation which he had to abandon because it resulted in incorrect energy levels for the hydrogen atom. With the undoubted successes of matrix and wave mechanics and the incorporation of spin into the formalism of quantum mechanics, the need to find a route to a relativistic version of Schrödinger's equation was pressing. The hero of this story is undoubtedly Dirac who continued to pursue his own individual agenda for the development of a fully relativistic theory of quantum mechanics. He achieved this goal in a pair of remarkable papers published in 1928 in which he generalised Schrödinger's equation, resulting in the *linear Dirac equation* (Dirac, 1928a,b).

There is no question about Pauli's priority in the discovery of the spin matrices, which he had discussed with Heisenberg in November 1926. Dirac worked independently on the problem of electron spin and derived his own version of the spin matrices during the early months of 1927. He was certainly aware of Pauli's paper during the summer of 1928 when he prepared a set of *Lectures on Modern Quantum Mechanics* for delivery in the Michaelmas and Lent terms of the 1927–1928 Cambridge academic year. As discussed in Sect. 13.3, Dirac was also aware that the set of 2×2 matrices described by Baker in his *Principles of Geometry* (1922) were q-numbers and so would fit naturally into his approach to quantum mechanics.[6] According to Dirac's reminiscences, he was not interested in incorporating spin into a relativistic theory of quantum mechanics when he developed that theory in the autumn of 1927. In his words,

> 'I was not interested in bringing the spin of the electron into the wave equation, did not consider the question at all and did not make any use of Pauli's work. The reason for this is that my dominating interest was to get a relativistic theory agreeing with my general physical interpretation and transformation theory. I thought that this problem should first be solved in the simplest possible case, which was presumably a spinless particle, and only after that should one go on to consider how to bring in spin. It was a great surprise to me when I later on discovered that the simplest possible case did involve a spin.' (Dirac, 1977)

The key role played by Pauli's introduction of spin matrices is vividly portrayed by Kronig who wrote in the Pauli memorial volume (Kronig, 1960),

> 'In the formulation of quantum mechanics reported so far, the material particles are considered as characterised by position coordinates only. It now became an urgent task to incorporate the phenomenon indicated by the term 'spin' into the new scheme of things. The initial step was taken by Pauli himself [(1927b)], who proposed that the wave function of the electron, besides being determined by continuously variable position coordinates, should be considered as also depending on a spin variable, capable of two values only. . . . Thereby Pauli paved the way for the relativistic theory of the electron and of hydrogen-like atoms which we owe to Dirac (1928a).'

Pauli's advance was to incorporate the spin of the electron consistently into the scheme of quantum mechanics by introducing a state vector $\psi(s)$, which as demonstrated by (16.42), has two components. As van der Waerden remarked,

> 'The step from one to two components is large, whereas the step from two to four components is small.' (van der Waerden, 1960)

16.6.1 The Dirac equation of a free electron

Let us outline how the Dirac equation can be formulated without going into too many technical details. The objective is to find an operator formalism for the relativistic equivalent of Schrödinger's equation which reduces correctly to the standard equation in the non-relativistic limit. The requirement of the relativistic theory must be that the equations are invariant under Lorentz transformations. Schrödinger, Oskar Klein and Gordon had all carried what might be regarded as the 'natural extension' of the operator formalism of quantum mechanics to the relativistic Lagrangian of an electron in an electromagnetic field.

Let us consider first the case of a free particle, for which the energy equation is

$$\frac{p^2}{2m_e} = E \ . \tag{16.53}$$

To find Schrödinger's equation, we replace p and E by the differential operators

$$p \to -\frac{ih}{2\pi}\nabla \ , \quad E \to \frac{ih}{2\pi}\frac{\partial}{\partial t} \ . \tag{16.54}$$

Schrödinger's time-dependent wave equation follows immediately:

$$-\frac{h^2}{8\pi^2 m_e}\nabla^2\psi = \frac{ih}{2\pi}\frac{\partial\psi}{\partial t} \ . \tag{16.55}$$

In the case of a free relativistic particle, the relation between momentum and energy, found by equating the norms of the momentum four-vector of the electron in the external frame and the rest frame of the electron, is given by the expression

$$p^2c^2 + m_e^2c^4 = E^2 \ . \tag{16.56}$$

Using the definitions (16.54), this expression becomes

$$-\frac{h^2c^2}{4\pi^2}\nabla^2\psi + m_e^2c^4\psi = \frac{h^2}{4\pi^2}\frac{\partial^2\psi}{\partial t^2} \ . \tag{16.57}$$

This is the simplest form of the *Klein–Gordon equation*, derived independently by Schrödinger, Klein and Gordon. As is apparent, it can only be used for particles with zero spin. As Schrödinger had found in his earliest searches for a wave equation to accommodate de Broglie waves, the solutions give the wrong answers for the energy levels of the hydrogen atom. Notice also the issue that the equation involves the second derivative of the wavefunction with respect to time, whereas one of the great virtues of the standard Schrödinger equation is that it is linear, involving only the first derivative of the wavefunction with respect to time.

An alternative approach would be to express (16.56) in terms of E rather than E^2, in which case we would write

$$E = \sqrt{p^2c^2 + m_e^2c^4} \quad \text{and so} \quad \frac{h}{2\pi}\frac{\partial\psi}{\partial t} = \sqrt{-\frac{h^2c^2}{4\pi^2}\nabla^2\psi + m_e^2c^4\psi} \ . \tag{16.58}$$

This approach runs into difficulties because the spatial differential operators on the right-hand side of the equation are inside the square root. Dirac had profound misgivings about this approach to discovering the correct relativistic formulation of Schrödinger's equation. Specifically, he found that, although the Klein–Gordon equation could reproduce the same results for the position coordinates of a particle as in the non-relativistic case, this was not the case for dynamical variables such as momentum and angular momentum. Dirac was convinced that his transformation theory approach to fundamental quantum processes should underlie a proper relativistic formulation of quantum mechanics and this had consequences for how the translation from relativistic mechanics to relativistic quantum mechanics should be carried out. He stated in his great paper of 1928,

> 'The general interpretation of non-relativity quantum mechanics is based on transformation theory, and is made possible by the wave equation being of the form
>
> $$(H - W)\psi = 0 \,, \tag{16.59}$$
>
> that is, being linear in W or $\partial/\partial t$, so that the wave function at any time determines the wave function at any later time. The wave equation of the relativity theory must also be linear in W if the general interpretation is to be possible.' (Dirac, 1928a)

Dirac's approach to the laws of quantum mechanics through the use of q-numbers had the great advantage that he did not need to specify exactly what the q-numbers were – what was important was that, whatever they were, the non-commutative properties ensured that the non-commutation present at the heart of quantum mechanics would be preserved. Specifically, the wavefunctions involved need not be scalar functions of a single variable, but could be vectors, matrices, tensors, . . . Already, he and Pauli had shown that the spin matrices provided the solution to the problem of incorporating spin into non-relativistic quantum mechanics and so he was in the ideal position to extend these insights into the development of the relativistic theory of the electron.

In his own words, he found the solution by 'playing around with mathematics'. The specific insight came from his considerations of finding relativistically invariant quantities which could be built into his scheme of quantum mechanics. He needed the equivalent of four-vectors, for which the norm, or the sum of squares of the four components of the four-vector, are invariants. Again, in his words,

> 'It took me quite a while, studying over this dilemma, before I suddenly realised that there was no need to stick to the quantities σ [the spin matrices], which can be represented by matrices with just two rows and columns. . . . Why not go over to four rows and four columns? Mathematically, there was no objection to this at all. Replacing the σ-matrices by four-row and column matrices, one could easily take the square root of the sum of four squares, or even five squares if one wanted to.' (Dirac, 1977)

Dirac worked out the new relativistic theory of quantum mechanics of the electron over the 1927 Christmas break, using the insights which again must have been provoked by his close understanding of Baker's *Principles of Geometry* (1922). On page 69, Baker writes:

'Another symbol with the same laws of combination [non-commutation] ... may be represented by

$$
\begin{bmatrix}
\delta, & \gamma, & -\alpha, & -\beta \\
-\gamma, & \delta, & \beta, & -\alpha \\
\alpha, & -\beta, & \delta, & -\gamma \\
\beta, & \alpha, & \gamma, & \delta
\end{bmatrix}.
\tag{16.60}
$$

To illustrate what is involved in the derivation of the Dirac equation and its consequences, we follow the presentation by Lindsay and Margenau (1957). There is an alternative way of writing the energy of a free electron in special relativity. Let us write down the four-vectors associated with the momentum and velocity of the electron, $\boldsymbol{P}$ and $\boldsymbol{U}$ respectively:

$$
\boldsymbol{P} = \left[\frac{E}{c}, \ p_x, p_y, p_z \right]; \quad \boldsymbol{U} = \left[\gamma c, \ \gamma v_x, \ \gamma v_y, \ \gamma v_z \right],
\tag{16.61}
$$

where $\gamma = (1 - v^2/c^2)^{-1/2}$ is the Lorentz factor.[7] The scalar product of the four-vectors $\boldsymbol{P}$ and $\boldsymbol{U}$ is an invariant and so

$$
\boldsymbol{P} \cdot \boldsymbol{U} = \gamma E - \gamma (\boldsymbol{p} \cdot \boldsymbol{v}) = \text{constant} = m_e c^2,
\tag{16.62}
$$

since $\boldsymbol{P} \cdot \boldsymbol{U} = m_e c^2$ in the rest frame of the electron. Note that $\boldsymbol{p}$ and $\boldsymbol{v}$ are the relativistic three-momentum, $\boldsymbol{p} = \gamma m_e \boldsymbol{v}$, and three-velocity $\boldsymbol{v}$ of the electron. Therefore, since $H = E$, the Hamiltonian of the relativistic electron can also be written

$$
H = \boldsymbol{p} \cdot \boldsymbol{v} + \frac{m_e c^2}{\gamma} = \boldsymbol{p} \cdot \boldsymbol{v} + \sqrt{1 - \frac{v^2}{c^2}} m_e c^2.
\tag{16.63}
$$

This is now a first-order expression for H.

We now make the following translation into the language of operators:

$$
\boldsymbol{p} \rightarrow \frac{h}{2\pi i} \nabla; \ v_x \rightarrow c\alpha_x; \ v_y \rightarrow c\alpha_y; v_z \rightarrow c\alpha_z; \ \sqrt{1 - \frac{v^2}{c^2}} \rightarrow \alpha_4,
$$

$$
H \equiv \frac{h}{2\pi i} c \left(\alpha_x \frac{\partial}{\partial x} + \alpha_y \frac{\partial}{\partial y} + \alpha_z \frac{\partial}{\partial z} \right) + \alpha_4 m_e c^2.
\tag{16.64}
$$

The momentum operator is the same type of differential operator as before, but the nature of the α operators is undefined – it is convenient to call them 'symbols' using Baker's language, the nature of which have yet to be determined. Their algebraic properties can be found by applying the operator (16.64) twice and then comparing with the results of the operator substitutions which resulted in the Klein–Gordon equation (16.57). These are two different ways of describing H^2. Therefore, we require

$$
\left[\frac{h}{2\pi i} c \left(\alpha_x \frac{\partial}{\partial x} + \alpha_y \frac{\partial}{\partial y} + \alpha_z \frac{\partial}{\partial z} \right) + \alpha_4 m_e c^2 \right] \cdot \left[\frac{h}{2\pi i} c \left(\alpha_x \frac{\partial}{\partial x} + \alpha_y \frac{\partial}{\partial y} + \alpha_z \frac{\partial}{\partial z} \right) + \alpha_4 m_e c^2 \right]
$$

$$
= -\frac{h^2 c^2}{4\pi^2} \nabla^2 + m_e^2 c^4.
\tag{16.65}
$$

Now, the expectation is that the αs will correspond to something akin to the spin matrices, which are independent of the coordinates x, y, z, and so they will commute with

$\partial/\partial x$, $\partial/\partial y$, $\partial/\partial z$, so that, for example, $(\alpha_x\, \partial/\partial x - \partial/\partial x\, \alpha_x) = 0$. Then, multiplying out the scalar product in (16.65) and preserving the order in which the operators are applied, we find

$$- \frac{hc^2}{4\pi^2}\left[\alpha_x^2\frac{\partial^2}{\partial^2 x} + \alpha_y^2\frac{\partial^2}{\partial^2 y} + \alpha_z^2\frac{\partial^2}{\partial^2 z}\right]$$

$$- \frac{hc^2}{4\pi^2}\left[(\alpha_x\alpha_y + \alpha_y\alpha_x)\frac{\partial}{\partial x}\frac{\partial}{\partial y} + (\alpha_y\alpha_z + \alpha_z\alpha_y)\frac{\partial}{\partial y}\frac{\partial}{\partial z} + (\alpha_z\alpha_x + \alpha_x\alpha_z)\frac{\partial}{\partial z}\frac{\partial}{\partial x}\right]$$

(16.66)

$$+ \frac{hm_ec^2}{2\pi i}c\left[(\alpha_4\alpha_x + \alpha_x\alpha_4)\frac{\partial}{\partial x} + (\alpha_4\alpha_y + \alpha_y\alpha_4)\frac{\partial}{\partial y} + (\alpha_4\alpha_z + \alpha_z\alpha_4)\frac{\partial}{\partial z}\right] + \alpha_4^2\, m_e^2 c^4$$

$$= -\frac{h^2c^2}{4\pi^2}\nabla^2 + m_e^2 c^4\ .$$

It immediately follows that we can find the commutation relations which the α symbols must obey. Specifically, we require

$$\alpha_x^2 = \alpha_y^2 = \alpha_z^2 = \alpha_4^2 = 1\ ,$$

$$(\alpha_x\alpha_y + \alpha_y\alpha_x) = (\alpha_y\alpha_z + \alpha_z\alpha_y) = (\alpha_z\alpha_x + \alpha_x\alpha_z) = 0\ ,$$

(16.67)

$$(\alpha_4\alpha_x + \alpha_x\alpha_4) = (\alpha_4\alpha_y + \alpha_y\alpha_4) = (\alpha_4\alpha_z + \alpha_z\alpha_4) = 0\ .$$

These results tell us that, applying the individual α operators twice, the result is the identity operator, that is, simply multiplying by one. Secondly, notice that the operators α *anti-commute* – in other words, for commuting variables $ab - ba = 0$, for anti-commuting variables $ab + ab = 0$. Thus, returning to the Hamiltonian equation (16.63) and the translations (16.64), the energy equation for ψ becomes

$$\left[\frac{hc}{2\pi i}\left(\alpha_x\frac{\partial}{\partial x} + \alpha_y\frac{\partial}{\partial y} + \alpha_z\frac{\partial}{\partial z}\right) + \alpha_4 m_e c^2\right]\psi = E\psi\ ,$$

(16.68)

where the properties of the symbols $\alpha_x, \alpha_y, \alpha_z, \alpha_4$ are determined by the commutation relations (16.67). Equation (16.68) is *Dirac's equation* for a free electron.

Pauli's spin matrices are 2×2 matrices which describe the two spin states of the electron. Now we have four αs with the commutation properties (16.67). The simplest way of accommodating these properties is by 4×4 matrices which were derived by Dirac. It will not prove necessary to write these out for our present purposes. What is important is that the wavefunction ψ must also be a matrix and the simplest form it can have is as a column matrix with four elements,

$$\psi = \begin{bmatrix} \psi_1 \\ \psi_2 \\ \psi_3 \\ \psi_4 \end{bmatrix}.$$

(16.69)

Thus, the Dirac equation (16.68) is actually four equations, one for each component of ψ. To normalise the wavefunctions, we introduce the row vector ψ^*, with components which are the complex conjugates of ψ. Thus, $\psi^* = [\psi_1^*, \psi_2^*, \psi_3^*, \psi_4^*]$ and so the normalisation condition is

$$\int \left(\psi_1^* \psi_1 + \psi_2^* \psi_2 + \psi_3^* \psi_3 + \psi_4^* \psi_4 \right) d\tau = 1 \,. \qquad (16.70)$$

Let us now work out the energies of the eigenstates of a free electron according to Dirac's equation (16.68). For simplicity consider only motion in the x-direction. Then, (16.68) reduces to

$$(E - \alpha_4 m_e c^2)\psi - \frac{hc}{2\pi i} \alpha_x \frac{d\psi}{dx} = 0 \,. \qquad (16.71)$$

We note that (16.71) is a matrix equation and so E means E times the unit matrix, α_4 is a 4×4 matrix and the components of $d\psi/dx$ are $(d\psi_1/dx, d\psi_2/dx, d\psi_3/dx, d\psi_4/dx)$. The solution of (16.71) is an exponential function which we can write

$$\psi = A \exp\left(\frac{2\pi i}{h} px \right) , \qquad (16.72)$$

where p is a number which will turn out to be the relativistic three-momentum of the electron and A is a column matrix with elements (A_1, A_2, A_3, A_4). Now substituting this solution back into (16.71), we find

$$(E - m_e c^2 \alpha_4 - cp\alpha_x) A = 0 \,. \qquad (16.73)$$

Now, we can operate on both sides of this equation with the same operator to the left of A so that

$$(E - m_e c^2 \alpha_4 - cp\alpha_x) \cdot (E - m_e c^2 \alpha_4 - cp\alpha_x) A = 0 \,. \qquad (16.74)$$

Recalling that we need to preserve the order of the operators and that E is a diagonal matrix multiplied by E, we find

$$\left[E^2 + m_e^2 c^4 \alpha_4^2 + c^2 p^2 \alpha_x^2 - 2E(mc^2 \alpha_4 + cp\alpha_x) + pmc^3 (\alpha_4 \alpha_x + \alpha_x \alpha_4) \right] A = 0 \,. \qquad (16.75)$$

Now, the last term in round brackets $\alpha_4 \alpha_x + \alpha_x \alpha_4$ in the left-hand side of this expression is zero because of the commutation relations and the second term in round brackets is E because of (16.73). Therefore, since $\alpha_x^2 = \alpha_4^2 = 1$,

$$(-E^2 + m_e^2 c^4 + c^2 p^2) A = 0 \,. \qquad (16.76)$$

Notice that what has happened is that the operators associated with each of the terms is a unit matrix and so the terms in round brackets form an algebraic expression. Since A cannot vanish, it follows that

$$E^2 = m_e^2 c^4 + c^2 p^2 \,, \qquad (16.77)$$

and so the energy eigenvalues associated with the Dirac equation are

$$E = \pm \sqrt{m_e^2 c^4 + c^2 p^2} \,. \qquad (16.78)$$

We can demonstrate that p is the average momentum of the electron using Schrödinger's prescription and the solution (16.72) of the Dirac equation. The momentum operator is

$(h/2\pi i) \, \partial/\partial x$ and so,

$$\frac{\int \psi^* \dfrac{h}{2\pi i} \dfrac{\partial}{\partial x} \psi \, dx}{\int \psi^* \psi \, dx} = p \frac{\int A^* A \, dx}{\int A^* A \, dx} = p \,. \tag{16.79}$$

The energies of the eigenstates of the electron according to the Dirac equation are one of its most remarkable features. According to classical physics, the negative energy solution can be discarded as meaningless. But this is not the case in quantum mechanics which allows discontinuous transitions between the discrete energy levels which are the solutions of the wave equation. These negative energy solutions were a paradox, to which we will return.

Our presentation contains the elements of what Dirac set out in the second section of his paper. In Sect. 3, he demonstrated that the formalism is fully Lorentz invariant. The real triumph of the theory came, however, in Sect. 4 of the paper in which the motion of an electron in electric and magnetic fields is discussed.

16.6.2 The Dirac equation for an arbitrary electromagnetic field

In compliance with Dirac's dictum that the equations of classical mechanics are not at fault, simply the interpretation of the algebra, the natural extension of the scheme is to use the classical expression for the Hamiltonian in electromagnetic fields, but now armed with the new precepts of relativistic quantum mechanics. According to classical Hamiltonian mechanics we need to make the replacements

$$\boldsymbol{p} \to \boldsymbol{p} + \frac{e}{c} \boldsymbol{A} \quad ; \quad H \to H + e\phi \,, \tag{16.80}$$

where A is the vector potential and ϕ the scalar potential. It will be recalled that $[\phi/c, A_x, A_y, A_z]$ forms the electromagnetic four-potential. Then, (16.68) becomes

$$H\psi = \left\{ c \left[\alpha_x \left(\frac{h}{2\pi i} \frac{\partial}{\partial x} + \frac{e}{c} A_x \right) + \alpha_y \left(\frac{h}{2\pi i} \frac{\partial}{\partial y} + \frac{e}{c} A_y \right) \right. \right.$$
$$\left. \left. + \alpha_z \left(\frac{h}{2\pi i} \frac{\partial}{\partial z} + \frac{e}{c} A_z \right) \right] + \alpha_4 m_e c^2 \psi - e\phi \right\} \psi$$
$$= E\psi \,. \tag{16.81}$$

As, before, (16.81) is a set of four differential equations for ψ_1, ψ_2, ψ_3 and ψ_4. In the case of the hydrogen atom, the potential is given by the Coulomb formula, $\phi = Ze/4\pi\epsilon_0 r$, and there is no magnetic field, $A = 0$. The solutions of the equations for the hydrogen atom agreed precisely with the experimentally measured values, including the details of the fine structure of the energy levels, as demonstrated in the fourth edition of Sommerfeld's *Atombau und Spektrallinien*, which included additional material on Schrödinger's wave mechanics and Dirac's papers of 1928 (Sommerfeld, 1929). Note that because the 4×4 matrices automatically take account of the spin of the electron and the new formulation is fully relativistic, the fine-structure calculations include all the effects of spin and relativity. Sommerfeld's fine-structure constant appears naturally in these calculations.

Let us simplify the notation slightly. We can write

$$\boldsymbol{p}' = \boldsymbol{p} + \frac{e}{c}\boldsymbol{A} \quad ; \quad H' = H + e\phi , \qquad (16.82)$$

and then (16.81) can be written in more compact form as

$$\left(c\boldsymbol{\alpha} \cdot \boldsymbol{p}' + \alpha_4 mc^2 - e\phi\right)\psi = E\psi . \qquad (16.83)$$

This has been written like a vector equation, but it will be understood that the components of the 'vector' $\boldsymbol{\alpha}$ are the 4×4 α matrices and the equation is in fact a matrix equation. If we add $e\phi$ to both sides of (16.81), we obtain

$$H'\psi = E'\psi , \qquad (16.84)$$

where $E' = E + \phi$ – notice that this is no longer an eigenfunction equation since E' is no longer a constant.

We now need to carry out some straightforward manipulation of the equations to reduce the energy equation to a form like that of Schrödinger's equation. First, we pre-multiply (16.84) by E' so that,

$$E'H'\psi = E'^2\psi . \qquad (16.85)$$

We can rewrite the left-hand side of (16.85) as $H'E'\psi + (E'H' - H'E')\psi$ and then, because of (16.84), the first term of this expression is H'^2. Therefore,

$$[H'^2 + (E'H' - H'E')]\psi = E'^2\psi . \qquad (16.86)$$

Notice that the operator H' is just the expression in the first equality of (16.81) or (16.83) without the term $-e\phi$. The objective is now to evaluate the terms in square brackets in (16.86).

Let us deal first with the term $H'^2\psi$. We can write this term as

$$H'^2\psi = c^2[\alpha_x p'_x + \alpha_y p'_y + \alpha_z p'_z + \alpha_4 mc^2][\alpha_x p'_x + \alpha_y p'_y + \alpha_z p'_z + \alpha_4 mc^2] . \quad (16.87)$$

We use the commutation properties of the α matrices (16.67) and the $\boldsymbol{p}'$ operators to reduce (16.87) to the expression

$$H'^2\psi = \left\{c^2\left[p'^2 + \alpha_x\alpha_y(p'_x p'_y - p'_y p'_x) + \alpha_y\alpha_z(p'_y p'_z - p'_z p'_y) + \alpha_z\alpha_x(p'_z p'_x - p'_x p'_z)\right] + m_e c^2\right\}\psi . \qquad (16.88)$$

Next, we need to evaluate the terms such as $(p'_x p'_y - p'_y p'_x)\psi$ in (16.88) by expanding the p' operators as $p' = p + (e/c)A$.

$$(p'_x p'_y - p'_y p'_x)\psi = \left(\frac{h}{2\pi i}\frac{\partial}{\partial x} + \frac{e}{c}A_x\right)\left(\frac{h}{2\pi i}\frac{\partial}{\partial y} + \frac{e}{c}A_y\right)\psi$$
$$- \left(\frac{h}{2\pi i}\frac{\partial}{\partial y} + \frac{e}{c}A_y\right)\left(\frac{h}{2\pi i}\frac{\partial}{\partial x} + \frac{e}{c}A_x\right)\psi$$
$$= \frac{he}{2\pi ic}\left[\frac{\partial}{\partial x}(A_y\psi) + A_x\frac{\partial\psi}{\partial y} - \frac{\partial}{\partial y}(A_x\psi) - A_y\frac{\partial\psi}{\partial x}\right]$$
$$= \frac{he}{2\pi ic}\left(\frac{\partial A_y}{\partial x} - \frac{\partial A_x}{\partial y}\right) = \frac{h}{2\pi i}(\nabla \times A)_z\,\psi\,. \tag{16.89}$$

But the magnetic flux density $B = \nabla \times A$ and so

$$(p'_x p'_y - p'_y p'_x)\psi = \frac{he}{2\pi ic}B_z\,\psi\,. \tag{16.90}$$

The corresponding expressions for $(p'_y p'_z - p'_z p'_y)$ and $(p'_z p'_x - p'_x p'_z)$ are found by cyclic permutation of the indices x, y, z.

To complete the reduction of (16.88), let us introduce a new set of matrices defined by

$$\sigma_z = -i\alpha_x\alpha_y\,, \quad \sigma_y = -i\alpha_z\alpha_x\,, \quad \sigma_x = -i\alpha_y\alpha_z\,. \tag{16.91}$$

Then, (16.88) can be written in the simple form

$$H'^2\psi = \left[c^2\left(p'^2 + \frac{he}{2\pi c}\boldsymbol{\sigma}\cdot\boldsymbol{B}\right) + m_e^2 c^4\right]\psi\,. \tag{16.92}$$

Next, we have to tackle the term $(E'H' - H'E')\psi$ in (16.86). We use (16.82), (16.83) and the relation $E' = E + e\phi$ to write this term as

$$(E'H' - H'E')\psi = e(\phi H' - H'\phi)\psi = ce(\phi\boldsymbol{\alpha}\cdot\boldsymbol{p}' - \boldsymbol{\alpha}\cdot\boldsymbol{p}'\phi)\psi\,,$$
$$= \frac{hce}{2\pi i}(\phi\boldsymbol{\alpha}\cdot\nabla\psi - \boldsymbol{\alpha}\cdot\nabla(\phi\psi))\,,$$
$$= -\frac{hce}{2\pi i}\boldsymbol{\alpha}\cdot(\nabla\phi)\psi = \frac{hce}{2\pi i}\boldsymbol{\alpha}\cdot\boldsymbol{E}\,\psi\,. \tag{16.93}$$

E is the electric field strength which is assumed to be given by $\boldsymbol{E} = -\nabla\phi$, in other words, there is no induced electric field and so $\dot{A} = 0$. Now substituting (16.92) and (16.93) into (16.86), we obtain,

$$\left(c^2 p'^2 + m^2 c^4 + \frac{hec}{2\pi}\boldsymbol{\sigma}\cdot\boldsymbol{B} + \frac{hce}{2\pi i}\boldsymbol{\alpha}\cdot\boldsymbol{E}\right)\psi = E'^2\psi\,. \tag{16.94}$$

To make a comparison with the non-relativistic Schrödinger equation, we use the expressions,

$$p' = \frac{h}{2\pi i}\nabla + \frac{e}{c}A \quad; \quad E' = E + e\phi = m_e c^2 + W + e\phi\,. \tag{16.95}$$

The reason for writing the expression for the energy of the electron as $m_e c^2 + W$ is that, throughout the analysis of this section, E has been the total energy of the electron and so to compare with the Schrödinger equation, we need to separate out the rest mass energy. W represents the energy level of the electron according to non-relativistic physics. Carrying out these substitutions and dividing through by $2m_e c^2$, we find

$$\left[\left(-\frac{h^2}{8\pi^2 m_e}\nabla^2 - e\phi\right) + \left(\frac{he}{2\pi i c m_e}\boldsymbol{A}\cdot\nabla + \frac{e^2}{2m_e c^2}A^2\right)\right.$$
$$\left. + \left(\frac{he}{4\pi m_e c}\boldsymbol{\sigma}\cdot\boldsymbol{B} + \frac{he}{4\pi i m_e c}\boldsymbol{\alpha}\cdot\boldsymbol{E}\right)\right]\psi = \left[W + \frac{(W+e\phi)^2}{2m_e c^2}\right]\psi . \quad (16.96)$$

It is convenient to compare this equation with Schrödinger's non-relativistic wave equation (14.37),

$$\left(-\frac{h^2}{8\pi^2 m_e}\nabla^2\psi - e\phi\right) = W , \quad (16.97)$$

where we have written $\phi = e/4\pi\epsilon_0 r$ and $E = W$. If we remove all terms in $1/c$ from (16.96), we see that it is identical with Schrödinger's equation (16.97). In this case, the four components of ψ, $[\psi_1, \psi_2, \psi_3, \psi_4]$, all satisfy the same Schrödinger equation, and so Dirac's equation reduces correctly to the non-relativistic case.

The additional terms which appear in (16.96) are purely relativistic effects. The vector potential terms in the second round bracket on the left of the equation are the usual terms which appear when terms in $1/c$ are included and the relativistic correction on the right-hand side, $(W + e\phi)^2/2m_e c^2$, is a relativistic correction to the energy W. The remarkable feature is the presence of the terms in the third round bracket on the left-hand side of (16.96),

$$\left(\frac{he}{4\pi m_e c}\boldsymbol{\sigma}\cdot\boldsymbol{B} + \frac{he}{4\pi i m_e c}\boldsymbol{\alpha}\cdot\boldsymbol{E}\right). \quad (16.98)$$

Suppose the electron had magnetic dipole moment $\boldsymbol{\mu}$ and electric dipole moment $\boldsymbol{\mu}_e$. Then, classically, the electron would have additional energy contributions associated with the interaction with a magnetic and electric field which would be

$$\Delta W = \boldsymbol{\mu}\cdot\boldsymbol{B} + \boldsymbol{\mu}_e\cdot\boldsymbol{E} . \quad (16.99)$$

Thus, the terms in (16.98) correspond to the electron having magnetic and electric moments

$$\boldsymbol{\mu} = \frac{he}{4\pi m_e c}\boldsymbol{\sigma} \quad ; \quad \boldsymbol{\mu}_e = \frac{he}{4\pi i m_e c}\boldsymbol{\alpha} . \quad (16.100)$$

Notice that what have been written as products of vectors in (16.98) are in fact 4×4 matrices. They are, however, closely related to the Pauli spin matrices. It is straightforward to derive the forms of the matrices $\boldsymbol{\sigma}$ and $\boldsymbol{\alpha}$ from Dirac's paper and these are given in the endnotes.[8] The spin matrices $\boldsymbol{\sigma}$ are of particular interest. As an example, it is shown in endnote 8 that the σ_z spin matrix is

$$\sigma_z = -i\alpha_x\alpha_y = \begin{bmatrix} 0 & 1 & 0 & 0 \\ 1 & 0 & 0 & 0 \\ 0 & 0 & 0 & 1 \\ 0 & 0 & 1 & 0 \end{bmatrix} . \quad (16.101)$$

This is just an extension of the 2×2 spin matrix S_z of (16.46) to a corresponding 4×4 matrix which Dirac created by duplicating the 2×2 matrix and filling up the remaining elements with zeros. In fact, in his analysis of his equation, Dirac recognised that the use of 4×4 matrices duplicated the solutions for the spin states of the electron.

The result (16.100) is the remarkable result of Dirac's great paper. It shows that, in the relativistic formulation of quantum mechanics, there is necessarily a magnetic moment associated with the spin of the electron and that its magnitude is $eh/4\pi m_e c$. This is exactly the result derived empirically by Uhlenbeck and Goudsmit that the magnetic moment associated with the spin angular momentum had to be twice that associated with orbital angular momentum.

In addition, (16.100) shows that there is also an electric dipole moment associated with the electron, but that it is imaginary. Dirac appreciated the problem, but expressed his view as follows:

> 'This magnetic moment is just that assumed in the spinning electron model. The electric moment, being a pure imaginary, we should not expect to appear in the model. It is doubtful whether the electric moment has any physical meaning, since the Hamiltonian in [(16.81)] that we started from is real and the imaginary part only appeared when we multiplied it up in an artificial way in order to make it resemble the Hamiltonian of previous theories.' (Dirac, 1928a)

It has been worth all the effort to produce these remarkable results which formally demonstrate the origin of the spin of the electron in quantum mechanics. The words of Lindsay and Margenau reveal just how remarkable it is:

> 'Dirac's theory, therefore, produces the spin properties without a special postulate, and this is its major achievement. ... But equation [(16.96)] also warns us not to take the electron spin too literally. The equation merely indicates in a formal way that the electron, if placed in a field, has an added energy part of which may be interpreted by saying that the electron spins. ... Furthermore, there is no term in [(16.96)] which could be interpreted as energy due to mechanical rotation. On the whole, then, the situation is more complex than would be in accord with the simple classical statement: the electron spins.' (Lindsay and Margenau, 1957)

16.7 The discovery of the positron

16.7.1 Dirac's prediction of the anti-electron and antimatter

The existence of negative energy states with $E = -m_e c^2$ according to Dirac's theory was a major concern. According to quantum mechanics, these states could not be ignored. Indeed, they had to be included if Dirac's relativistic quantum theory was to reduce correctly to the classical results. In a letter to Pauli of 31 July 1928, Heisenberg showed that it was necessary to include these negative energy terms if the correct expression for the dispersion formula was to be obtained. Furthermore, Oskar Klein and Yoshio Nishina in their derivation of the relativistic quantum theory of the scattering of high energy radiation by electrons found that it was essential to include the negative energy states in order to account for the Compton scattering properties of electrons at energies $h\nu \gtrsim m_e c^2$ (Klein and Nishina, 1928, 1929).

Dirac proposed an ingenious solution to the problem by invoking Pauli's exclusion principle. He came up with a picture in which the Universe is so densely packed with electrons in the negative energy states that they are all filled and so the electrons cannot make transitions from the positive to the negative energy states. The proposal was that the 'vacuum' was filled with electrons, what became known as the 'Dirac sea'. But there might well be vacancies, or 'holes' in the sea, a situation similar to that which gives rise to X-ray lines when an electron is removed from, say, the K-shell of the atom. On 26 November 1929, Dirac wrote to Bohr,

> 'Such a hole ... would appear experimentally as a thing with positive energy, since to make the hole disappear (that is, to fill it up) one would have to put negative energy into it. Further one can easily see that such a hole would move in an electromagnetic field as though it had positive charge. These holes I believe to be protons. When an electron of positive energy drops into a hole and fills it up, we have an electron and a proton disappearing simultaneously and emitting radiation in the form of radiation.'

These ideas were published by Dirac in a paper entitled *A theory of electrons and protons* in the 1 January 1930 edition of the *Proceedings of the Royal Society of London* (Dirac, 1930b). The theory was immediately challenged on a number of grounds, the most significant being the conclusion demonstrated independently by Igor Tamm, Robert Oppenheimer and Hermann Weyl that Dirac's theory predicted that the electrons and holes should have the same mass. Ultimately, Dirac withdrew his proposal and replaced it by the idea that the holes would be 'anti-electrons'. In his words,

> 'It appears that we must abandon the identification of the holes with protons and must find some other interpretation for them. Following Oppenheimer (1930), we must assume that in the world as we know it, all, and not nearly all, of the negative-energy states for the electrons are occupied. A hole, if there were one, would be a new kind of particle, unknown to experimental physics, having the same mass and opposite charge to an electron. We may call such a particle an anti-electron.' (Dirac, 1931)

Dirac went further and stated that the protons must be unconnected with the electrons and that both electrons and protons should have negative energy states, thus introducing the concept of both anti-electrons and antiprotons (Dirac, 1931). This represented the introduction of the concept of antimatter into physics.

This famous prediction was only the prelude to the principal concern of his paper which was entitled *Quantized singularities in the electromagnetic field* and concerned the theoretical possibility of the existence of magnetic monopoles (Dirac, 1931). He was well aware of the fact that this concept involved 'a symmetry between electricity and magnetism quite foreign to current views'. His predicted value for the elementary magnetic pole strength was $\mu = hc/4\pi e$. Despite many investigations, the magnetic monopole has never been detected.

16.7.2 The discovery of positive electrons – the positrons

The late 1920s and the 1930s were periods of unprecedented discovery in atomic and nuclear physics. These will be summarised in Chapter 18. Suffice to say that the discoveries of the positron, the neutron, artificially induced nuclear interactions, the discovery of the mesotron and so on were all rapidly built into the pantheon of the new physics with quantum mechanics at its core. Here we are only concerned with the discovery of the positron.

We need to retrace our steps somewhat to take up the story of the discovery of cosmic rays which were to provide a natural source of very high energy particles.[9] In the early 1900s, it was known that there is a small amount of residual ionisation in the atmosphere and, close to the surface of the Earth, this could be attributed to the effects of natural radioactivity in rocks. In their pioneering experiments, Victor Hess and Werner Kolhörster made high altitude balloon observations of the degree of ionisation of the atmosphere and showed that above about 2 km altitude, the ionisation increases with increasing altitude. This phenomenon was attributed to some form of *cosmic radiation* which originated from above the Earth's atmosphere (Hess, 1913; Kolhörster, 1913). They demonstrated that the increase was exponential with increasing height, with $n(l) \propto \exp(\alpha l)$ and $\alpha \sim 10^{-3} m^{-1}$. This increasing ionisation corresponded to a path length of much more penetrating radiation than was found for the most penetrating γ-rays observed in radioactive decays. As Hess expressed it in his paper,

> 'The results of the present observations seem to be most readily explained by the assumption that a radiation of very high penetrating power enters our atmosphere from above, and still produces in the lower layers a part of the ionisation observed in closed vessels.'

Originally it was assumed that the *cosmic rays*, as they were named by Millikan in 1925, were high energy γ-rays with greater penetrating power than those observed in natural radioactivity. In 1929, Dmitri Skobeltsyn, working in his father's laboratory in Leningrad, constructed a cloud chamber which was placed in the jaws of a strong magnet so that the curvature of the tracks of the charged particles could be measured. Among the tracks, he noted some which were hardly deflected at all and had all the appearance of being electrons with energies greater than 15 MeV. He identified them with secondary electrons produced by the 'Hess ultra γ-radiation'.

A key technical development for these studies was the invention of the *Geiger–Müller detector* by Hendrik (Hans) Geiger and Walther Müller in 1928. This enabled individual cosmic rays to be detected and the times of their arrival measured precisely (Geiger and Müller, 1928, 1929). In 1929, Böthe and Kolhörster carried out one of the key experiments in cosmic ray physics in which they introduced the concept of *coincidence counting* to eliminate spurious background events (Bothe and Kolhörster, 1929). By using two counters, one placed above the other, they found that simultaneous discharges of the two detectors occurred very frequently, even when a strong absorber was placed between the detectors. In one of the key experiments, slabs of lead and gold 4 cm thick were placed between the counters and the mass absorption coefficient was found to agree very closely with that of the attenuation of the cosmic radiation in the atmosphere. The

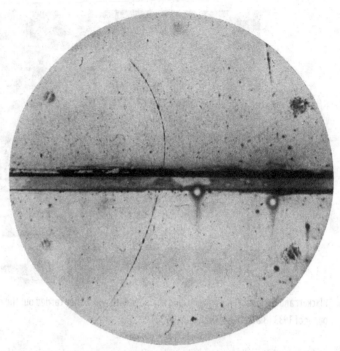

One of the discovery records of the positron. This cloud chamber photograph shows a 63 MeV positron passing through a 6 mm lead plate and emerging as a 23 MeV positron. According to Anderson, the length of the latter path is at least 10 times greater than the possible length of a proton path of this curvature (Anderson, 1933).

experiment demonstrated that the cosmic radiation consisted of highly energetic charged particles.

From the 1930s until about 1960, the cosmic radiation provided a natural source of very high energy particles, of very much greater energies than those produced in radioactive decays. In 1930, Millikan and Anderson used an electromagnet 10 times stronger than that used by Skobeltsyn to study the tracks of particles passing through the cloud chamber. Anderson (1932) observed curved tracks identical to those of electrons, but with positive electric charges (Fig. 16.2).

This discovery was confirmed by Patrick Blackett and Guiseppe Occhialini in 1933 using an improved technique in which the cloud chamber was only triggered after it was certain that a cosmic ray had passed through the supersaturated vapour within the chamber (Fig. 16.3) (Blackett and Occhialini, 1933). They obtained many excellent photographs of the positive electrons, on many occasions showers containing equal numbers of positive and negative electrons created by cosmic-ray interactions within the body of the apparatus being observed. Blackett and Occhialini's analysis went considerably further than that of Anderson in that they interpreted the positive and negative electrons as being produced simultaneously in the interaction of the incoming cosmic ray particle with the material of the chamber. They stated:

Fig. 16.3   Blackett and Occhialini's automatic cloud chamber with which they carried out the experiments described in their paper of 1933 (Blackett and Occhialini, 1933).

'In this way one can imagine that negative and positive electrons may be born in pairs during the disintegration of light nuclei. If the mass of the positive electron is the same as that of the negative electron, such a twin birth requires an energy of $2m_ec^2 \sim 1$ million [electron] volts, that is much less than the translatory energy with which they appear in general in the showers.' (Blackett and Occhialini, 1933)

They used Dirac's calculations to show that the annihilation of the positron would occur very rapidly resulting in the formation of a pair of high energy photons. Thus, their paper introduced the concept of pair-production and annihilation of electrons and positrons. These experiments were conclusive evidence for the positron as predicted by Dirac's theory of the electron and the first example of the existence of antimatter.

The interpretation of quantum mechanics

In completing the story of spin, we have run far ahead of the continued development of the understanding of the matrix, operator and wave mechanical approaches to quantum mechanics. The reconciliation of these approaches was described in Chap. 15, but there remained the issue of the interpretation of the wavefunction and the deeper implications of the theory. The understanding came gradually with Born's interpretation of the wavefunction, Ehrenfest's demonstration of the equivalence of the classical and quantum pictures and Heisenberg's enunciation of the uncertainty principle. These led to what became known as the Copenhagen interpretation of quantum mechanics. At the same time, the formal mathematical foundations of the different approaches to quantum phenomena were set on a secure foundation thanks to the efforts of Hilbert and many others. These developments resulted in what may be referred to as the completion of quantum mechanics, in the sense that it laid the foundations for all the future development of physics at the atomic and subatomic level – some of these achievements are summarised in Chap. 18.

17.1 Schrödinger's interpretation (1926)

Schrödinger regarded wave mechanics as superior to the matrix mechanical approach to quantum physics, not only because it was based upon the well-known eigenfunction techniques of classical physics, but also because it was much more visualisable. His first attempt at interpreting the wavefunction appeared in the final Sect. 7 of the fourth part of his great series of papers (Schrödinger, 1926f) and was entitled *On the physical significance of the field scalar*. There he identified the quantity $\psi \psi^*$ as the 'weight function' of the distribution of charge so that $\rho_e = e\psi\psi^*$ is the electric charge density. In support of this picture, he carried out an analysis using the time-dependent wave equations (14.102) to evaluate the rate of change of charge density according to this prescription. The rate of change of charge is then

$$\frac{\partial}{\partial t} \int e\psi\psi^*\rho\, \mathrm{d}V = e \int \left(\psi \frac{\partial \psi^*}{\partial t} + \psi^* \frac{\partial \psi}{\partial t} \right) \rho\, \mathrm{d}V . \tag{17.1}$$

Using the Schrödinger equations (14.102) for $\partial\psi/\partial t$ and $\partial\psi^*/\partial t$,

$$\frac{\partial}{\partial t} \int e\psi\psi^*\rho\, \mathrm{d}V = \int \frac{he\rho}{4\pi \mathrm{i} m_e} \left(\psi^*\nabla^2\psi - \psi\nabla^2\psi^* \right) \mathrm{d}V . \tag{17.2}$$

We now use Green's theorem to convert the integral on the right-hand side into a surface integral:

$$\frac{\partial}{\partial t} \int e\psi\psi^* \rho \, \mathrm{d}V = -\int_A \frac{he\rho}{4\pi \mathrm{i} m_e} \left(\psi^* \nabla \psi - \psi \nabla \psi^*\right) \cdot \mathrm{d}A \,, \qquad (17.3)$$

where $\mathrm{d}A$ is an element of surface area. Next, we define the vector $\boldsymbol{S}$ as

$$\boldsymbol{S} = \frac{he}{4\pi \mathrm{i} m_e} \left(\psi^* \nabla \psi - \psi \nabla \psi^*\right) \,, \qquad (17.4)$$

and then convert the surface integral back into a volume integral using the divergence theorem,

$$\frac{\partial}{\partial t} \int e\psi\psi^* \rho \, \mathrm{d}V = -\int \rho \,\mathrm{div}\, \boldsymbol{S} \, \mathrm{d}V \,. \qquad (17.5)$$

Taking the time differential inside the integral, (17.4) is the equation of continuity for the conservation of electric charge in classical electrodynamics,

$$\frac{\partial \rho_e}{\partial t} + \mathrm{div}\, \boldsymbol{S} = 0 \,, \qquad (17.6)$$

recalling that the density of charge is defined as $\rho_e = e\psi\psi^*$. $\boldsymbol{S}$ is the electric current density, corresponding to the quantity $\boldsymbol{J}$ which appears in Maxwell's equations of classical electrodynamics.

Schrödinger's interpretation was reinforced by his analysis of the harmonic oscillator according to wave mechanics which was presented in Sect. 14.6 and illustrated in Fig. 14.3. His interpretation was that the particles are represented by wave groups composed of the superposition of an infinite series of wavefunctions and that Fig. 14.3 represents the oscillation of the electric charge in a harmonic oscillator. Therefore, just as in classical physics, the oscillating charge emits dipole radiation at the frequency of the oscillator. Schrödinger believed that the motion of the electron in the hydrogen atom could be interpreted in the same way:

> 'We can definitely foresee that, in a similar way, wave groups can be constructed which move round highly quantised Kepler ellipses and are the representation by wave mechanics of the hydrogen electron. But the technical difficulties in the calculation are greater than in the especially simple case which we have treated here.' (Schrödinger, 1926a)

But this interpretation could not be correct as was pointed out by Heisenberg and Born. In general, the wave-packets do spread out in space. Heisenberg showed that the case of the harmonic oscillator was rather special because the successive energy levels are spaced by equal amounts. Furthermore, in general, the wavefunction ψ is a function in a multi-dimensional space and so the interpretation of the motion of the electron as a wave-packet in three-dimensional space is not viable. Heisenberg was particularly opposed to Schrödinger's interpretation which seemed to omit many of the very considerable achievements of quantum physics. As he wrote to Pauli in June 1926,

> 'The more I ponder about the physical part of Schrödinger's theory, the more horrible I find it. One should imagine the rotating electron, whose charge is distributed over the entire space and which has an axis in a fourth and fifth dimension. What Schrödinger writes about the visualizability of his theory . . . I find rubbish.'

Soon after, he complained that, while he admired the power of Schrödinger's equation in simplifying the evaluation of matrix elements in quantum mechanics, Schrödinger's interpretation

> 'throws overboard everything which is 'quantum theoretical': namely, the photoelectric effect, the Franck[–Hertz] collisions, the Stern–Gerlach effect, . . . '

In addition, there was the problem of understanding the electron diffraction experiments in crystals and collision phenomena involving electrons. In interpreting these experiments, it had to be assumed that the waves dispersed, the classical analogue being Huygens' construction for the interference of light waves, and so how could the stability of the particle as a discrete entity be explained by Schrödinger's picture? The new interpretation came from Born's study of the scattering of electrons by atoms.

17.2 Born's probabilistic interpretation of the wavefunction ψ (1926)

In his Nobel Prize speech of 1954, Born explained that he was opposed to Schrödinger's interpretation:

> 'On this point, I could not follow him. This was connected with the fact that my Institute and that of James Franck were housed in the same building of the Göttingen University. Every experiment by Franck and his assistants on electron collisions (of the first and second kind) appeared to me as a new proof of the corpuscular nature of the electron.' (Born, 1961a)

Born set about carrying out a quantum mechanical calculation of the scattering of charged particles such as α-particles or electrons by atoms using wave mechanical techniques. As he remarked,

> '. . . among the various forms of the theory, only Schrödinger's formalism proved itself appropriate for this purpose; for this reason I am inclined to regard it as the most profound formulation of the quantum laws.' (Born, 1926a)

These remarks were made in a preliminary report of his calculations which were explained in much more detail in two further papers (Born, 1926b,d). The analysis involved what became known as the *Born approximation*, in which the incoming particle is represented by plane-wave functions incident upon the scattering centre from the positive z-direction. The scattered outgoing waves are represented by plane waves at infinity. Born treated the scattered wave as a first-order perturbation of the combined unperturbed wavefunctions of the scattering centre and the incoming particle. If the unperturbed wavefunction of the atom is $\psi_n^0(q)$ and the energy of the incoming electron $E = p^2/2m_e = h^2/2m_e\lambda^2$, he took the eigenfunction of the unperturbed system to be

$$\psi_{nE}^0(q, z) = \psi_n^0(q) \sin(2\pi z/\lambda) \,, \tag{17.7}$$

where n labels the nth wavefunction and q the spatial coordinates relative to the scattering centre. Then, if $V(x, y, z, q)$ is the potential energy of interaction between the charged particle and the atom, he could apply the techniques of perturbation theory to evaluate the amplitude of the scattered plane wave at infinity,

$$\psi_{nE}^{(1)}(x, y, z, q) = \sum_m \iint d\omega \, \psi_{nm}^{(E)}(\alpha, \beta, \gamma) \, \sin k_{nm}^{(E)}(\alpha x + \beta y + \gamma z + \delta) \, \psi_m^0(q) \, . \quad (17.8)$$

This equation has the following meaning. The superscript (1) means the first-order perturbation solution for the scattered wave. The double integral is over the solid angle $d\omega$, which is the element of solid angle in the direction of the unit vector, the components of which are α, β and γ; δ is an additional scalar phase factor. $\psi_{nm}^{(E)}(\alpha, \beta, \gamma)$ is a wavefunction which determines what is now referred to as the *differential cross-section* for scattering in the (α, β, γ) direction. Since all the experiments on electron scattering indicated that the scattered electrons had a 'corpuscular nature', Born inferred that the only possible interpretation of the expression $|\psi_{nm}^{(E)}(\alpha, \beta, \gamma)|^2$ was that it represented the *probability* that the electron, approaching along the z-axis, is scattered in the α, β, γ direction. The wavefunction $\psi_{nE}^{(1)}(x, y, z, q)$ was then related to the total cross-section for the scattering of the electron by the atom.

The implications of Born's calculations were profound. He concluded that quantum mechanics does not answer the question, 'What, precisely, is the state of the system after the collision?', but rather the question 'What is the probability of a particular state after the collision?' Thus, Born introduced the concept that the wavefunction ψ and the square of its amplitude $|\psi^2|$ determine the probabilities of events occurring in quantum mechanics. Born's thinking was strongly influenced by Einstein's interpretation of the relation between electromagnetic waves and light quanta, the very heart of the wave–particle duality. According to Born, Einstein interpreted the electromagnetic field $E(x, y, z, t)$ as a 'phantom' or 'ghost' field, a '*Gespensterfeld*', which served to guide the light quanta. The intensities, and hence the density of light quanta, were determined by the square of the amplitude of the electromagnetic field. In Born's interpretation, the strict equivalence of the wavefunction and the properties of electromagnetic waves was reinforced by the comparison of the wavefunction of a particle of energy E and momentum p and the expression for the amplitude of an electromagnetic wave,

$$\underbrace{\exp\left[2\pi i \nu \left(t - \frac{x}{c}\right)\right]}_{\text{Electromagnetic wave}} \equiv \underbrace{\exp\left[\frac{2\pi i}{h}(Et - px)\right]}_{\text{de Broglie wave}}, \quad (17.9)$$

where $\nu = E/h$ and $\lambda = h/p$. The first expression is proportional to the amplitude of the electromagnetic wave and the second to the amplitude of the de Broglie wave associated with the electron. Since the energy density of the light waves depends upon the square of the amplitude of the waves, Born translated this into the statement that

'...it was almost self-understood to regard $|\psi|^2$ as the probability density of particles.'

It was soon appreciated that these probabilities are different from those used in classical statistical mechanics and the theory of Gaussian statistics. Einstein had already understood the difference between the statistics of waves and particles in his great paper of 1909 which was analysed in Sect. 3.6[1] (Einstein, 1909) . According to classical statistics, if p_1 and p_2 are the probabilities of outcomes 1 and 2 taking place in some experiment, the combined probability that one or other of them occurring is $p_1 + p_2$. Translating this into Born's interpretation of the wavefunctions, this would mean that the probabilities would correspond to $|\psi_1|^2 + |\psi_2|^2$, where ψ_1 and ψ_2 are the wavefunctions associated with p_1 and p_2. But this is not the correct rule for the superposition of waves. We need first to form the sum of the two wavefunctions $\psi_1 + \psi_2$ and then the probability is given by the square of the modulus of the joint wavefunction,

$$p_{12} = |\psi_1 + \psi_2|^2 = |\psi_1|^2 + |\psi_2|^2 + \psi_1 \psi_2^* + \psi_2 \psi_1^* . \tag{17.10}$$

The last two terms are the 'interference terms' which give rise to the phenomena of electron diffraction and the 'wave' aspects of the wave–particle duality. Note that these are similar in form to the terms which appear in the statistical properties of electromagnetic waves (see Sect. 3.6).

Jammer's commentary on what Born had achieved is revealing:

'For Einstein the notion of probability, even as he applied it to reconcile his light-quantum hypothesis with Maxwell's theory of electromagnetic waves, was the traditional concept of classical physics, a mathematical objectivisation of the human deficiency of complete or exact knowledge but ultimately a creation of the human mind ... For Born probability, as far as it was related to the wave function, was not merely a mathematical fiction but something endowed with physical reality, for it evolved with time and propagated in space in accordance with Schrödinger's equation. It differed, however, from ordinary physical agents in one fundamental aspect: it did not transmit energy or momentum. Since in classical physics, whether Newtonian mechanics or Maxwellian electrodynamics, only what transfers energy or momentum (or both) is regarded as physically 'real', the ontological status of ψ had to be considered as something intermediate.' (Jammer, 1989)

Born appreciated that the wavefunction ψ could be expressed as an expansion in a complete, orthonormal set of eigenfunctions which are solutions of the appropriate Schrödinger equation,

$$\psi = \sum_n c_n \psi_n , \tag{17.11}$$

with a completeness relation which results from the orthogonality relations for the eigenfunctions,

$$\int |\psi(q)|^2 \, \mathrm{d}q = \sum_n |c_n|^2 . \tag{17.12}$$

Born then interpreted $\int |\psi(q)|^2 \, \mathrm{d}q$ as the total number of particles and the $|c_n|^2$ as the statistical frequency of occurrence of the nth eigenstate of the solutions of Schrödinger's equation.

This interpretation had a number of immediate successes. Wentzel (1926b) used Born's wave mechanical approach to derive the Rutherford scattering formula, while Faxén and Holtsmark (1927), Bethe (1930) and Mott (1928) used Born's approach to study the passage of fast and slow particles through matter. Among the successes of these papers was the interpretation of the Ramsauer–Townsend effect, the minimum in the cross-section for the scattering of low-energy electrons in the noble gases, argon, krypton and xenon (Ramsauer, 1921; Townsend and Bailey, 1922) – this phenomenon had no explanation according to classical physics, but is found naturally in the quantum theory of electron scattering.

Born (1926c) next tackled the interpretation of the time-dependence of the wavefunction using the time-dependent Schrödinger wave equation,

$$\nabla^2 \psi - \frac{8\pi^2 m_e}{h^2} U(x)\psi - \frac{4\pi i m_e}{h} \frac{\partial \psi}{\partial t} = 0 \,. \tag{17.13}$$

Assuming the wavefunctions $\psi_n(x)$ are normalised, the general solution was taken to be

$$\psi(x, t) = \sum_n c_n \psi_n(x) \exp\left(\frac{2\pi i}{h} W_n t\right) \,, \tag{17.14}$$

where W_n is the energy of the nth eigenstate. Therefore, at time $t = 0$, the wavefunction is

$$\psi(x, 0) = \sum_n c_n \psi_n(x) \,. \tag{17.15}$$

Born considered the result of applying a force $F(x, t)$ to the system which acts only during the time interval $0 \le t \le T$. He treated the action of this force as a small perturbation to the potential $U(x)$ so that $U(x)$ was replaced by $U(x) + x F(x, t)$ where the factor x is later used as a small expansion parameter. Using the perturbation techniques described by Schrödinger in Part 4 of his series (Schrödinger, 1926f), Born demonstrated that, for the simple case in which $\psi(x, t)$ consists of only a single eigenfunction $\psi_n(x)$, the solution of the time-dependent problem for $t \ge T$ is

$$\psi_n(x, t) = \sum_m b_{nm} \psi_m(x) \exp\left(\frac{2\pi i}{h} W_m t\right) \,, \tag{17.16}$$

the coefficients b_{nm} being determined by the action of the force $F(x, t)$ during the interval $0 \le t \le T$. In the spirit of the probability interpretation of the wavefunction, he immediately interpreted the quantity $|b_{nm}|^2$ as the probability that the system changed from the initial state n to the final state m, in other words, the quantities $|b_{nm}|^2$ represent the *transition probabilities* between the states n and m.

Having dealt with a single wavefunction $\psi_n(x)$, Born could then generalise to the case in which the initial state is given by (17.15) and so, for $t \ge T$,

$$\psi(x, t) = \sum_n c_n \psi_n(x, t) \,. \tag{17.17}$$

He now worked out the total transition probability from the state n by writing

$$\psi(x, t) = \sum_n C_n \psi_n(x) \,, \tag{17.18}$$

for times $t \geq T$. Using the orthogonality properties of the wavefunctions $\psi_n(x)$, he derived the key expression

$$C_n = \int \psi(x, T)\, \psi^*(x)\, \mathrm{d}x = \sum_m b_{mn}\, c_m \exp\left(\frac{2\pi \mathrm{i}}{h} W_m\, T\right), \qquad (17.19)$$

and so

$$|C_n|^2 = \left| \sum_m c_m b_{mn} \right|^2. \qquad (17.20)$$

In carrying out this calculation, Born had uncovered the rules for the determination of probabilities in quantum mechanics which differ crucially from their classical counterparts. Thus, classically, if the transition probability from the state m to n is $P_1 = |b_{nm}|^2$ and the probability of that state m occurring is $P_2 = |c_m|^2$, the joint probability would be $P = P_1 P_2 = |b_{nm}|^2 |c_m|^2$, which is quite different from the quantum rule for the addition of amplitudes of the wavefunctions to generate the probabilities (17.20) – note the same differences between classical and quantum probabilities as described by (17.10). As expressed by Jammer (1989),

> '... Born advanced two theorems which were destined to play a fundamental role in the further development of quantum theory, its interpretation, and its theory of measurement:
>
> 1. the theorem of spectral decomposition according to which there corresponds a possible state of motion to every component ψ_n in the expansion or superposition of ψ;
> 2. the theorem of interference of probabilities according to which the phases of the expansion coefficients, and not only their absolute values, are physically significant.'

These are profound insights into the physical meaning of quantum mechanical calculations. The amplitudes *and* phases of the wavefunctions are both crucial in describing quantum phenomena and consequently complex numbers, which incorporate both amplitude and phase, are the natural language of quantum mechanics. These results were independently discovered by Dirac in his important paper *On the theory of quantum mechanics* using his rather different approach to quantum mechanics (Dirac, 1926f).

17.3 Dirac–Jordan transformation theory

Born's probabilistic interpretation of the wavefunction was a key advance which was to be deepened as the formalism of quantum mechanics developed. At the same time, while the matrix and wave mechanical approaches had been shown to be equivalent, a unifying mathematical theory was not available at the time Born published his paper in 1926. We took the story of Dirac's formulation of quantum mechanics as far as his solution of the problem of the hydrogen atom in Chap. 13 and then jumped ahead to his discovery of the relativistic version of the Schrödinger equation, the Dirac equation, the magnetic moment of the electron and the prediction of the positron and antimatter in Chap. 16.

We pick up the story from Sect. 13.3 where the elements of what became known as *transformation theory* were described in the context of his introduction of q-numbers and their algebra. There, we encountered the rules for the transformation of the canonical variables Q_r and P_r, which are functions of q-numbers, obeying the following rules in terms of their Poisson brackets,

$$[Q_r, P_s] = \delta_{rs} , \qquad [Q_r, Q_s] = [P_r, P_s] = 0 . \tag{17.21}$$

Then, Q_r and P_r can be transformed to another set of canonical variables q_r and q_s by relations of the form

$$Q_r = bq_rb^{-1} , \qquad P_r = bp_rb^{-1} , \tag{17.22}$$

where b is a q-number. We observed the strong similarity to the canonical transformations (12.64) introduced by Born and his colleagues in the matrix treatment of quantum algebra. As Born remarked,

'A function **S** is to be determined, such that when

$$\mathbf{p} = \mathbf{S}\mathbf{p}_0\mathbf{S}^{-1} , \quad \mathbf{q} = \mathbf{S}\mathbf{q}_0\mathbf{S}^{-1} , \tag{17.23}$$

the function

$$\mathbf{H}(\mathbf{p}, \mathbf{q}) = \mathbf{S}\mathbf{H}(\mathbf{p}_0, \mathbf{q}_0)\mathbf{S}^{-1} = \mathbf{W} , \tag{17.24}$$

becomes a diagonal matrix.' (Born *et al.*, 1926)

The rigorous proof of this result was given by Jordan (1926) who developed the theory for action and angle variables. Despite these advances, the theory was of limited usefulness since, in general, it proved difficult to find the reciprocal matrices $\mathbf{S}^{-1}$. Furthermore, the matrix mechanical approach could not deal with the case of constant p. With Schrödinger's demonstration of the equivalence of the matrix and wave mechanical approaches to quantum phenomena, London (1926) carried over the concepts of matrix mechanics into the framework of Schrödinger's wave mechanics. It was only after this work was completed that it was realised that London had formulated the problem using closely analogous procedures to those employed in the use of linear operators in functional spaces, which were discussed in detail in Chap. 15. Jammer (1989) summarises London's achievements and what Dirac published a few weeks later as follows:

'[The paper by London] started by applying canonical transformations to the wave mechanics of discrete eigenvalue problems and ended up with discrete transformation matrices. A few weeks later, Dirac (1926g) published a paper which began by applying canonical transformations to continuous or discrete matrices in Dirac's matrix mechanics of continuous and discrete eigenvalue problems. Dirac's work thus complemented London's in two respects: it showed, so to speak, the reversibility of the conceptual process under discussion and generalised it to continuous transformations.'

Dirac had no doubt about the central importance of the transformation theory of quantum mechanics. As he wrote in the preface to the first edition of his classic text *The Principles of Quantum Mechanics* (Dirac, 1930a),

'... the growth of the use of transformation theory ... is the essence of the new method in theoretical physics.'

Dirac's paper contained a number of innovations which were inspired by his appreciation of Lanczos's rewriting of matrix mechanics in terms of integral equations (Lanczos, 1926) (see Sect. 15.2). Dirac appreciated that the elements of the matrices were now continuous functions and the matrices were *continuous matrices*.

17.3.1 Discrete and continuous matrices

We can illustrate the similarities and differences between discrete and continuous matrices by comparing Fourier series and Fourier integrals (Bohm, 1951). In the case of a Fourier series expansion, a wavefunction $\psi(x)$ can be written as a Fourier series

$$\psi(x) = \sum_n a_n \psi_n(x),$$ (17.25)

where $\psi_n(x)$ may be taken to be, for example, the complete set of orthonormal harmonic functions $\exp(2\pi i n x / L)$, where $0 \leq n \leq \infty$. Thus, if the functions $\psi_n(x)$ form a complete orthonormal set, any wavefunction can be expressed as the sum (17.25) over all the discrete values of n. Now consider a new wavefunction $\phi_m(x)$ obtained by applying the operator A to $\psi_m(x)$,

$$A\psi_m(x) = \phi_m(x).$$ (17.26)

Since the $\psi_n(x)$ form a complete orthonormal series, $\phi_m(x)$ can be expressed as a sum over all the components of the series so that

$$A\psi_m(x) = \sum_n a_{nm} \psi_n(x).$$ (17.27)

The values of a_{nm} can be found using the orthonormal properties of the wavefunctions ψ_n in the usual way

$$a_{nm} = \int \psi_n^*(x) \, A \, \psi_m(x) \, dx.$$ (17.28)

Now the effect of operating on any function $\psi(x)$ can be found by writing

$$A\psi(x) = A \sum_m C_m \psi_m(x) = \sum_m C_m \, A \, \psi_m(x) = \sum_n \sum_m C_m \, a_{nm} \, \psi_n(x).$$ (17.29)

In the above exposition, the matrix elements a_{nm} are associated with the *discrete* eigenfunctions ψ.

In the case of Fourier transforms, the discrete functions $\phi_n(x)$ are replaced by continuous functions. Thus, representing the function ψ by a Fourier integral

$$\psi(x) = \frac{1}{\sqrt{2\pi}} \int \phi(k) \, e^{ik \cdot x} \, dk,$$ (17.30)

the orthonormal functions are now the continuous set of functions $e^{ik \cdot x}$ and $\phi(k)$ are the corresponding continuous expansion coefficients. The matrix elements $a_{kk'}$ are now found by analogy with (17.28):

$$a_{kk'} = \frac{1}{2\pi} \int e^{-ik \cdot x} A\, e^{ik' \cdot x}\, dx \,, \tag{17.31}$$

with the key distinction that k and k' are now continuous functions, rather than the discrete functions associated with the Fourier components n and m. Thus, for any set of continuous matrices ψ_p, we may define the matrix element $a_{pp'}$

$$a_{pp'} = \int \psi_p^* A\, \psi_{p'}\, dx \,. \tag{17.32}$$

The following expressions, similar to those found for discrete matrices, follow naturally. Any $\psi(x)$ can be represented by

$$\psi(x) = \int C_p \psi_p\, dp \,, \tag{17.33}$$

and so if the operator A acts upon $\psi(x)$,

$$A\, \psi(x) = \iint C_p a_{pp'} \psi_{p'}(x)\, dp'\, dp \,. \tag{17.34}$$

The product rule for continuous matrices becomes

$$(AB)_{pp'} = \int a_{pp''} b_{p''p'}\, dp'' \,. \tag{17.35}$$

17.3.2 Dirac's interpretation of quantum mechanics

Dirac now translated the canonical transformation (17.22) of the dynamical variable g into G into the language of continuous matrices (Dirac, 1926f). The transformation

$$G = bgb^{-1} \tag{17.36}$$

was written in terms of integrals over continuous matrices as

$$g(\xi'\xi'') = \iint \left(\frac{\xi'}{\alpha'}\right) d\alpha'\, g(\alpha'\alpha'')\, d\alpha'' \left(\frac{\alpha''}{\xi''}\right) \,, \tag{17.37}$$

following the rules described in Sect. 17.3.1. The primed and double primed quantities are continuous parameters, c-numbers, which number the rows and columns of the matrix elements; (ξ'/α') and (α''/ξ'') represent the transformation functions $b(\xi'/\alpha')$ and $b^{-1}(\alpha''/\xi'')$ respectively; $g(\xi'\xi'')$ and $g(\alpha'\alpha'')$ are dynamical variables and are q-numbers.

One of the objectives of Dirac's paper was to discover the means of evaluating the transformation functions $b(\xi'/\alpha')$ and $b^{-1}(\alpha''/\xi'')$. In the course of his analysis, he introduced the famous *Dirac δ-function*. His training in electrical engineering proved to be invaluable since, as he said,

> 'All electrical engineers are familiar with the idea of a pulse, and the δ-function is just a way of expressing a pulse mathematically.'

The δ-function had been introduced by Kirchhoff and was used extensively by Oliver Heaviside in his pioneering studies of electromagnetic theory. The δ-function was defined in the usual way as

$$\delta(x) = 0 \quad \text{for all } x \neq 0 \quad \text{and} \quad \int \delta(x)\,dx = 1 \,. \tag{17.38}$$

Dirac was well aware of the fact that the δ-function is what Jammer calls a 'convenient mathematical artifice' and was candid about its mathematical status:

> 'Strictly, of course, $\delta(x)$ is not a proper function of x, but can only be regarded as a limit of a certain sequence of functions. All the same one can use $\delta(x)$ as though it were a proper function for practically all purposes of quantum mechanics without getting incorrect results. One can also use the differential coefficients of $\delta(x)$, namely $\dot{\delta}(x)$, $\ddot{\delta}(x)$, ..., which are even more discontinuous and less "proper" than $\delta(x)$ itself.'

Dirac went on to show that the n-th derivative of $\delta(x)$ could be written

$$\int_{-\infty}^{\infty} f(x)\,\delta^{(n)}(a - x)\,dx = f^{(n)}(a) \,. \tag{17.39}$$

This result was needed to define the elements of the unit continuous diagonal matrix $I(\alpha', \alpha'')$ labelled by the continuous parameters α' and α'' as

$$I(\alpha', \alpha'') = \delta(\alpha' - \alpha'') \,. \tag{17.40}$$

Therefore, the elements of the generalised continuous diagonal matrix could be written

$$f(\alpha', \alpha'') = f(\alpha')\,\delta(\alpha' - \alpha'') \,. \tag{17.41}$$

These results were needed in the succeeding analysis in his paper.

The spectacular result of Dirac's analysis was his demonstration that the function (ξ'/α') was just the appropriate solution of Schrödinger's equation, with the substitutions $\xi \to q$, $\eta \to p$, $(\xi'/\alpha') \to \psi_E(q)$ and $f(\alpha') \to E$. In Dirac's own words

> 'The eigenfunctions of Schrödinger's wave equation are just the transformation functions (or the elements of the transformation matrix previously denoted b) that enable one to transform from the (q) scheme of matrix representation to a scheme in which the Hamiltonian is diagonal.'

In fact, Dirac's analysis was a generalisation of Schrödinger's wave equation and he proceeded to provide a generalisation of Born's interpretation of the wavefunction. Born's result could be written

$$\psi(q, t) = \sum_n c_n(t)\,\psi_n(q) \,, \tag{17.42}$$

where $|c_n(t)|^2$ was interpreted as the probability that the transition to the state n took place. Now, Dirac generalised this result to the continuous energy ranges described by his new formalism so that the equivalent of (17.42) became

$$\psi(q, t) = \int c(E, t)\,dE\,\psi_E(q) \,, \tag{17.43}$$

where $|c(E, t)|^2 \, dE$ was to be interpreted as the transition probability to a state with energy in the range E to $E + dE$. Reformulating Born's treatment of the scattering of electrons by atoms to his new formulism, he found perfect agreement, provided

> 'the coefficients that enable one to transform from the one set of matrices to the other are just those that determine the transition probabilities.'

Born's statistical interpretation was also generalised by Pauli in a footnote to his paper on Fermi statistics (Pauli, 1927a). He stated that the probability of finding the position coordinates $q_1, q_2, \ldots, q_f$ of a system of N particles in the volume element $dq_1 dq_2 \ldots dq_f$ of the configuration space was given by $|\psi(q_1, q_2, \ldots, q_f)|^2 \, dq_1 \, dq_2 \ldots dq_f$, if the system is in a state characterised by ψ. As Pauli wrote to Heisenberg,

> 'Born's interpretation may be viewed as a special case of a more general interpretation. Thus, for example, $|\psi(p)|^2 \, dp$ may be interpreted as the probability that the particle has momentum between p and $p + dp$.'

He further asserted that, for every pair of quantum mechanical quantities q and β, a function $\phi(q, \beta)$ exists, the 'probability amplitude', such that $|\phi(q_0, \beta)|^2 \, dq$ is the probability that q lies between the values q_0 and $q_0 + dq$, if β has a fixed value β_0. These insights were to be used by Jordan to expound the first axiomatic approach to the statistical transformation theory.

17.3.3 Jordan's axiomatic synthesis of the statistical transformation theory

Inspired by Pauli's concept of probability amplitudes, Jordan developed an axiomatic approach to the transformation theory which was independent of London's and Dirac's formulations (Jordan, 1927). According to Jammer, the three major axioms of his approach were:

(1) '[the probability amplitude] $\phi(q, \beta)$ is independent of the mechanical nature (Hamiltonian function) of the system and depends only on the kinematic relation between q and β.

(2) the probability (density) that for a fixed value β_0 of β the quantum-mechanical quantity q has the value q_0 is the same as the probability (density) that, for a fixed q_0 of q, β has the value β_0.

(3) probabilities are combined by superposition, that is, if $\phi(x, y)$ is the probability amplitude for the value x of q at a fixed value y of β, and $\chi(x, y)$ is the probability amplitude for the value x of Q at the fixed value y of q, then the probability amplitude for x of Q at a fixed value y of β is given by

$$\Phi(x, y) = \int \chi(x, z) \, \phi(z, y) \, dz . \tag{17.44}$$

In the particular case $Q = \beta$, Jordan's $\Phi(x, y)$ becomes Dirac's $\delta(x - y)$.'

In Jordan's axiomatic approach, p is defined as the momentum canonically conjugate to q if the probability amplitude $\rho(x, y)$, for every possible value x of p at a fixed y of q, is given by

$$\rho(x, y) = \exp\left(\frac{2\pi xy}{ih}\right) . \tag{17.45}$$

Jordan inferred that for a fixed value of q, all possible values of p are equally probable and *vice versa*.

Jordan's somewhat formal and mathematical approach was not readily accessible, except to experts in transformational theory, but it had some remarkable features. It will be noted that (17.45) is a solution of the time-independent Schrödinger wave equation for a particle moving with constant momentum. Hence, if the momentum is precisely known, all values of q are equally likely, a precursor of the Heisenberg uncertainty principle. Jordan's great achievement was that, by adopting the full apparatus of Hermitian operators, he was able to demonstrate that his theory encompassed not only Schrödinger's wave equation and Heisenberg's matrix mechanics, but also the Born–Weiner operator calculus and Dirac's q-number calculus. The synthesis of all these approaches was to find its ultimate expression in the application of functional analysis to the formalism of quantum mechanics.

17.3.4 The new perspective

It is worthwhile reviewing what had been achieved as a result of Born's probabilistic interpretation of the wavefunction and the Dirac–Jordan statistical transformation theory, following Jammer's careful exposition. The achievements of Schrödinger and Heisenberg may be summarised as follows. Schrödinger's discovery of his wave equation was the route for determining the energy eigenvalues of quantum mechanical systems. Similarly, in Heisenberg's matrix mechanics, the solutions for p and q were found for the diagonal matrices for which the diagonal terms were the energy eigenvalues. The off-diagonal terms were interpreted as transition probabilities. The observables were the energies of the stationary states and the transition probabilities. But matrix mechanics could not deal with the motion of a free electron, nor was the position variable accorded any status in the scheme, being disguised though the process of taking Fourier transforms.

In contrast, the transformation theory, to quote Jammer (1989),

'introduced the experimentally required generalisations by postulating that, in principle, any Hermitian matrix A represents an observable quantity a, on a par with the energy, and that the eigenvalues of A are possible results of measuring a. Through Dirac's introduction of continuous matrices and Born's probabilistic interpretation of Schrödinger's wave equation, the notion of position was retrieved.'

17.4 The mathematical completion of quantum mechanics

In 1926, Heisenberg appealed to the mathematicians to take up the challenge of coming up with the mathematics which would underpin the different approaches to quantum mechanics. He was fortunate that the leader of this attack was David Hilbert who was already well versed in the mathematical issues of quantum mechanics and was a close neighbour of Born's at Heidelberg. In late 1926, he began a systematic study of the mathematical foundations of quantum mechanics supported by his assistants Lothar Nordheim and John von Neumann. During the winter of 1926–1927, Hilbert gave a two-hour lecture every Monday and Thursday morning on the mathematics underlying quantum mechanics and a summary of these was published in 1927 (Hilbert *et al.*, 1927).

It would take us too far into pure mathematics to describe exactly what Hilbert and his colleagues achieved, but suffice to say that, as might be expected of the mathematician who had axiomatised the fundamentals of geometry, he set about coming up with a self-consistent and mathematically rigorous axiomatic basis for the application of probability amplitudes in quantum mechanics. They established six axioms which the amplitudes had to fulfil. Hilbert associated an operator with every dynamical variable and the resulting operator calculus was to be the mathematics of the probability amplitudes associated with each operator. Hilbert recognised, however, that strict mathematical rigour was not able alone to encompass the needs of quantum physics. As he stated,

> 'It is difficult to understand such a theory if the formalism and its physical interpretation are not strictly kept apart. Such a separation shall be adhered to even though at the present stage of the development of the theory no complete axiomatisation has as yet been achieved. However, what is definite by now is the analytical apparatus which will not admit any alternations in its purely mathematical aspects. What can, and probably will, be modified is its physical interpretation for it allows a certain freedom of choice.'

Hilbert adopted the integral equation approach to the operator formalism which he had already pioneered in his paper of 1912 (Hilbert, 1912). Summarising the results of their analysis, Hilbert and his colleagues found that the condition that the relative probability density is real and non-negative is that the operators had to be Hermitian. Furthermore, by introducing Dirac's δ-function, they were able to derive both the time-independent and time-dependent Schrödinger wave equations for both the energy and position coordinates. Their transformation theory was fully consistent with Born's probabilistic interpretation of the wavefunction, that is, that the function $|\psi_n(x)|^2$ is the probability that the atom is found in the nth state and that it is located at x while in that state. The Hilbert–Neumann–Nordheim reformulation of the transformation theory of Dirac and Jordan included both wave and matrix mechanics and laid the rigorous mathematical foundations for quantum mechanics.

There remained, however, the awkward properties of Dirac's δ-function which appeared in the reduction of the formal operator description in terms of integral equations into Schrödinger's wave equation. The legitimising of Dirac's δ-function was only to take place very much later.[2] Von Neumann therefore adopted a different approach involving concepts developed by Hilbert on linear equations and used these to provide a new mathematical framework for quantum mechanics (von Neumann, 1927). This provided the most suitable formalism for the elaboration of quantum mechanics and its subsequent extensions into relativistic quantum mechanics and quantum field theory. This involved the introduction of what von Neumann called *Hilbert space*, an infinite-dimensional complete separable linear space with a positive definite metric. Within this space, he developed a theory of linear operators. In one realisation, the operators become functionals as studied in functional analysis. Likewise, adjoint and Hermitian operators appeared naturally in the theory and the most general description of Born's statistical and probabilistic interpretation of quantum mechanics was described.

The formal elaboration of the technical details of von Neumann's scheme was a formidable achievement and, although the overall scheme was laid out in his paper of 1927, it was not until 1929 that he had solved all the formal issues involved (von Neumann, 1929). The modern axiomatic exposition of quantum mechanics is ultimately founded upon von Neumann's ground-breaking papers. These cannot be appreciated without considerable effort and this qualitative summary does scant justice to his remarkable achievement. Jammer provides more of the mathematical details, but even he has to refer the interested reader to the original papers for a full appreciation of their mathematical content.

17.5 Heisenberg's uncertainty principle

While the tools of quantum mechanics were approaching completion, the interpretation of the quantum mechanical operators and variables was still unclear. Heisenberg's initial reaction had been to reject the concept of position and velocity inside atoms as having no meaning since they could not be observed – for Heisenberg, the only observables at the atomic level were the emission and absorption properties of atoms, their frequencies, intensities and polarisations of the radiation. Certainly, the concepts of position, velocity and momentum could not have their classical significance at the atomic level in view of the fundamental quantum non-commutability relation $pq - qp = h/2\pi i$. And yet, the formalism of quantum mechanics had its roots in classical physics which certainly worked splendidly for macroscopic bodies. Born's probabilistic interpretation of the wavefunction offered a compelling way forward, indicating that the outcome of experiments at the atomic level have an intrinsic indeterminacy. This feature of quantum mechanics was certainly appreciated by Dirac (1926g) who remarked that

> 'One cannot answer any question on the quantum theory, which refers to numerical values for both the p and the q. One would expect, however, to be able to answer questions in which only the q or only the p are given numerical values ...'

Jordan (1927) likewise stated that 'for a given value of q all values of p are equally possible'. Heisenberg and his colleagues fully appreciated that, according to quantum mechanics, it was

> ... 'meaningless to speak of the place of a particle with a definite velocity ... But if one does not take it too seriously with the accuracy in using the notions of velocity and position, then it may well make sense.' (Heisenberg, 1960)

In a letter to Pauli of 28 October 1926, Heisenberg asserted that it made no sense to talk about a monochromatic wave at a definite instant or extremely short period of time.

After four months of deliberation on these issues, he came up with a self-consistent solution to reconcile the classical and quantum interpretations of non-commuting variables, in particular, in defining the range of applicability of the classical concepts of position and momentum. These are embodied in what became known as *Heisenberg's uncertainty principle*. The concepts were contained in a 14-page letter to Pauli who reacted positively and enthusiastically to its contents. The contents of that letter constituted the bulk of Heisenberg's famous paper on the uncertainty principle which was submitted to the *Zeitschrift für Physik* at the end of March 1927 (Heisenberg, 1927).

Heisenberg used the newly developed statistical transformation theory of Dirac and Jordan to define precisely the theoretically permissible values of non-commuting variables such as q and p in terms of the statistical distributions of their p and q values. As discussed in Sect. 17.3, this theory had the advantage of being able to deal with matrix elements which were continuous functions. Heisenberg used results derived by Jordan in his version of the Dirac–Jordan transformational theory (Jordan, 1927). Jordan took the probability amplitude for q to be of the following form

$$S(\eta, q) \propto \exp\left[-\frac{(q - q')^2}{2q_1^2} - \frac{2\pi i p'(q - q')}{h}\right] . \tag{17.46}$$

This expression has the following meaning. η is some fixed parameter which will not enter into the argument. $S(\eta, q)$ is the probability amplitude that the electron will be at position q if the mean value of the position is q' with uncertainty q_1. The probability is found by taking the square of the modulus of the probability amplitude and so

$$|S(\eta, q)|^2 = SS^* \propto \exp\left[-\frac{(q - q')^2}{q_1^2}\right] . \tag{17.47}$$

This formulation has the advantage that it results in a Gaussian probability distribution of possible values of q with 'uncertainty' q_1.

Next, Heisenberg used the rules of the transformation theory to write down the corresponding probability amplitude for p by using the relation

$$S(\eta, p) = \int S(\eta, q) S(q, p) \, dq . \tag{17.48}$$

The function $S(q, p)$ is given by (17.45) and so, carrying out the integration, Heisenberg found that the probability amplitude for p is

$$S(\eta, p) \propto \exp\left[-\frac{(p - p')^2}{2p_1^2} + \frac{2\pi i q'(p - p')}{h}\right] , \tag{17.49}$$

and the corresponding probability distribution for p

$$|S(\eta, p)|^2 = SS^* \propto \exp\left[-\frac{(p - p')^2}{p_1^2}\right] , \tag{17.50}$$

where

$$p_1 q_1 = \frac{h}{2\pi} . \tag{17.51}$$

This is Heisenberg's uncertainty principle describing the intrinsic indeterminacy with which p and q can be determined and provides a statistical interpretation of the basic non-commutability relation $pq - qp = h/2\pi i$. As Heisenberg wrote,

'The more accurately the position is determined, the less accurately the momentum is known and conversely.'

In fact, the essence of Heisenberg's calculation is most easily appreciated from the properties of the integral Fourier transforms of a Gaussian distribution, as was demonstrated by Darwin (1927d) – a simple analysis using that approach is presented in the endnote to this chapter.[3] It is striking that the Fourier transform approach indicates clearly the origin of the complex terms in the probability amplitudes (17.46) and (17.49) in Jordan's and Heisenberg's analyses.

Let us convert (17.51) into conventional notation by rewriting the Gaussian distributions (17.49) and (17.50) in proper normalised form. In order that the distributions be standard Gaussians, their *standard deviations* Δp and Δq are related to p_1 and q_1 by $\Delta p = p_1/\sqrt{2}$ and $\Delta q = q_1/\sqrt{2}$. Hence, (17.51) may be written

$$\Delta p \, \Delta q = \frac{h}{4\pi} \, . \tag{17.52}$$

Ditchburn (1930) demonstrated that in fact, because of the choice of Gaussian distributions for p and q, the uncertainly relation (17.52) represents the minimum indeterminacy of p and q, the inequality applying for all non-Gaussian distributions,

$$\Delta p \, \Delta q \geq \frac{h}{4\pi} \, . \tag{17.53}$$

Heisenberg's great paper on the uncertainty principle was of central importance for many different aspects of the understanding of quantum physics and for physics in general. Let us highlight just a few of these consequences.

1. At the most elementary level, the principle tells us the scales on which the classical and quantum theories are applicable. For example, applying the principle to an electron in the Bohr model of the atom, its velocity in the ground state, $n = 1$, is 2.2×10^6 m s^{-1} and hence its momentum is $p = m_e v = 2 \times 10^{-24}$ kg m s^{-1}. Therefore, taking $\Delta p = p$ and setting $\Delta p \Delta x = h/4\pi$, we find $\Delta x = h/4\pi \Delta p = 0.3 \times 10^{-10}$ m, roughly the size of the first Bohr orbit – this is no accident. What this calculation is telling us is that, on the scale of atoms, we cannot know precisely where the electron is at any moment – we can only describe very precisely where it is likely to be.

2. More generally, the principle tells us that, *at any time*, we cannot define the state of any system at the microscopic level absolutely precisely. Thus, the concept of setting up a system with a perfectly defined set of initial conditions and then following precisely the future evolution of the system is not feasible.

3. Another way of expressing this same concern is that *causality* at the microscopic level becomes meaningless since the intrinsic uncertainty means that we cannot predict exactly the outcome of any process. We can make accurate predictions about the various possible

outcomes of the experiment, but we cannot state with absolute certainty which will actually occur.

4. The principle has profound implications for the *theory of measurement* and this topic became a major preoccupation in the theory of quantum processes.

5. The principle had a major impact upon philosophy, in many ways demolishing many of the basic tenets of classical philosophy and logic. For example, the way in which probability amplitudes and probabilities are evaluated in quantum mechanics tells us that events which might have happened, but didn't, can affect the outcome of an experiment. There is no scope for such constructs in classical physics, but they are an integral part of the Universe we live in.

6. Perhaps most remarkable of all is the fact that statistical concepts were not explicitly incorporated into the basic postulates and structure of the theory, and yet the predictions are all of probability amplitudes and probabilities. Heisenberg should have the last word on this subject

> 'We have not assumed that the quantum theory, unlike classical physics, is essentially a statistical theory in the sense that from exact data only statistical data can be inferred. For such an assumption is refuted, for example, by the well-known experiments of Bothe and Geiger. However, in the strong formulation of the causal law, "If we know exactly the present, we can predict the future' it is not the conclusion but rather the premise which is false. We *cannot* know, as a matter of principle, the present in all its details."

It is not perhaps surprising that it took some time before the full implications of Heisenberg's calculations were fully appreciated since they ran contrary to many of the most cherished tenets of classical physics. Perhaps the most important conclusion of these calculations was the clear distinction between the realms of applicability of classical and quantum physics.

17.6 Ehrenfest's theorem

An important link between the classical and quantum pictures was provided by Ehrenfest who carried out a 'short elementary calculation without approximations' (Ehrenfest, 1927). In modern language, he showed that, according to quantum mechanics, the expectation value of the time derivative of the momentum is equal to the expectation value of the negative gradient of the potential function, the quantum equivalent of Newton's second law of motion. In only a page and a half, Ehrenfest quoted the result of his calculations, without giving a mathematical derivation, which he regarded as 'elementary'. The argument goes as follows.

For simplicity, consider the one-dimensional, time-dependent Schrödinger wave equation and its complex conjugate, as discussed in relation to (14.102):

$$-\frac{h^2}{8\pi^2 m_e}\frac{\partial^2 \psi}{\partial x^2} + V(x)\psi = \frac{ih}{2\pi}\frac{\partial \psi}{\partial t} \; ; \tag{17.54}$$

$$-\frac{h^2}{8\pi^2 m_e}\frac{\partial^2 \psi^*}{\partial x^2} + V(x)\psi^* = -\frac{ih}{2\pi}\frac{\partial \psi^*}{\partial t} \; . \tag{17.55}$$

Following Schrödinger's prescription, we define the mean values of the position and momentum of the particle, $\langle x \rangle$ and $\langle p \rangle$, what are now referred to as their *expectation values*, by the relations

$$\langle x \rangle = \int_{-\infty}^{\infty} \psi^* x \psi \, dx \,, \tag{17.56}$$

$$\langle p \rangle = \int_{-\infty}^{\infty} \psi^* p \psi \, dx = -\frac{ih}{2\pi} \int_{-\infty}^{\infty} \psi^* \frac{\partial \psi}{\partial x} \, dx \,, \tag{17.57}$$

where x is the scalar position operator and p the momentum operator $-(ih/2\pi)\,\partial/\partial x$. We now find the derivative of $\langle x \rangle$ with respect to time.

$$\frac{d}{dt} \int_{-\infty}^{\infty} \psi^* x \psi \, dx = \int_{-\infty}^{\infty} \left(\frac{\partial \psi^*}{\partial t} (x\psi) + \psi^* x \frac{\partial \psi}{\partial t} \right) dx \,. \tag{17.58}$$

Using the expressions (17.54) and (17.55) for $\partial \psi/\partial t$ and $\partial \psi^*/\partial t$ respectively, we find

$$\frac{d}{dt} \int_{-\infty}^{\infty} \psi^* x \psi \, dx = \frac{ih}{4\pi m_e} \int_{-\infty}^{\infty} \left[\psi^* x \frac{\partial^2 \psi}{\partial x^2} + \frac{\partial^2 \psi^*}{\partial x^2} (x\psi) \right] dx \,. \tag{17.59}$$

Now,

$$\frac{\partial^2 (x\psi)}{\partial x^2} = 2 \frac{\partial \psi}{\partial x} + x \frac{\partial \psi^2}{\partial x^2} \,, \tag{17.60}$$

and so, substituting for $x\,\partial^2 \psi/\partial x^2$ in (17.59),

$$\frac{d}{dt} \int_{-\infty}^{\infty} \psi^* x \psi \, dx = \frac{ih}{4\pi m_e} \int_{-\infty}^{\infty} \left[\psi^* \frac{\partial^2 (x\psi)}{\partial x^2} + \frac{\partial^2 \psi^*}{\partial x^2} (x\psi) \right] dx - \frac{ih}{2\pi m_e} \int_{-\infty}^{\infty} \psi^* \frac{\partial \psi}{\partial x} \, dx \,, \tag{17.61}$$

$$= \frac{ih}{4\pi m_e} \int_{-\infty}^{\infty} \frac{\partial}{\partial x} \left[\psi^* \frac{\partial (x\psi)}{\partial x} + \frac{\partial \psi^*}{\partial x} (x\psi) \right] dx + \frac{1}{m_e} \int_{-\infty}^{\infty} \psi^* \left(-\frac{ih}{2\pi} \right) \frac{\partial \psi}{\partial x} \, dx \,. \tag{17.62}$$

The first integral on the right-hand side of (17.62) becomes

$$\frac{ih}{4\pi m_e} \left[\psi^* \frac{\partial (x\psi)}{\partial x} + \frac{\partial \psi^*}{\partial x} (x\psi) \right]_{-\infty}^{\infty} \tag{17.63}$$

and this must be zero since the wavefunctions ψ and ψ^* are zero at $\pm\infty$. The term $-(ih/2\pi)\,\partial/\partial x$ in the last integral of (17.62) is the momentum operator and so from (17.56) and (17.57)

$$m_e \frac{d\langle x \rangle}{dt} = \langle p \rangle \,. \tag{17.64}$$

Thus, quantum mechanically, the mean, or expectation, value of the momentum is equal to the product of the mass of the electron times the mean, or expectation, value of the velocity. This is the exact equivalent of the definition of momentum in classic mechanics. Notice that Planck's constant has disappeared from this expression, despite the fact that the momentum is defined purely quantum mechanically.

Let us now take the next time derivative of the mean value of x with respect to time. We follow exactly the same procedure as above. From (17.57), we find,

$$m_e \frac{d^2 \langle x \rangle}{dt^2} = -\frac{ih}{2\pi} \int_{-\infty}^{\infty} \frac{\partial}{\partial t} \left[\psi^* \frac{\partial \psi}{\partial x} \right] dx \ . \tag{17.65}$$

Carrying out the partial differentiation and then substituting for the terms in $\partial/\partial t$ using the pair of Schrödinger equations (17.54) and (17.55), we find

$$m_e \frac{d^2 \langle x \rangle}{dt^2} = \left(-\frac{h^2}{8\pi^2 m_e} \right) \int_{-\infty}^{\infty} \left[\frac{\partial^2 \psi^*}{\partial x^2} \frac{\partial \psi}{\partial x} - \psi^* \frac{\partial}{\partial x} \left(\frac{\partial^2 \psi}{\partial x^2} \right) \right] dx - \int_{-\infty}^{\infty} \psi^* \frac{\partial V(x)}{\partial x} \psi \, dx \ . \tag{17.66}$$

Using the fact that ψ and ψ^* tend to zero at $\pm\infty$, after a bit of manipulation of the partial derivatives, the first integral on the right-hand side of (17.66) is zero. The second term is just the expectation value of the gradient of the potential $V(x)$, which is the expectation value of the force $\langle f \rangle$. Hence,

$$m_e \frac{d^2 \langle x \rangle}{dt^2} = \frac{d\langle p \rangle}{dt} = \int_{-\infty}^{\infty} \psi^* \left(-\frac{\partial V(x)}{\partial x} \right) \psi \, dx = \langle f \rangle \ . \tag{17.67}$$

This is *Ehrenfest's theorem* which states that the rate of change of the expectation value of the momentum is equal to the expectation value of the applied force. This is exactly the same statement as Newton's second law of motion, but derived purely from the rules of quantum mechanics. Notice again that Planck's constant has disappeared from the expression. Ehrenfest's theorem provided physicists with a natural continuity between the quantum description of the action of forces and the world of classical physics.

Ehrenfest's brief paper, in which only the results of his calculations were presented, was important in furthering the cause of quantum mechanics among practising physicists. The fact that the equivalent of Newton's second law of motion could be derived from a purely quantum mechanical set of operations made the theory much more acceptable to physicists, despite the fact that the classical and quantum pictures are built on completely different foundations.

17.7 The Copenhagen interpretation of quantum mechanics

By 1927, most of the elements of non-relativistic quantum mechanics were in place, but their interpretation was the subject of hot debate among the principal contributors to the new discipline. At the centre of these debates was Bohr, who continued to ponder deeply about the meaning of the new vistas emerging from the brilliant analyses of Heisenberg, Jordan, Born, Schrödinger, Pauli, Wiener, von Neumann and many others. The invitation to visit Bohr in Copenhagen was a singular honour but also a test of the character of the individual to match Bohr's inexhaustibility in maintaining an intellectual argument over many hours or days. Schrödinger was invited by Bohr to visit Copenhagen in September 1926 to discuss his brilliant papers on wave mechanics. The discussion would often last whole days, leaving Schrödinger in a state of exhaustion. The debates concerned the rationalisation of Bohr's insistence upon the concept of 'quantum jumps' and Schrödinger's conviction that these were unphysical and should be replaced by his continuous wavefunctions. At one point, in exasperation, Schrödinger is reported to have stated,

> 'If one has to stick to this damned quantum jumping, then I regret ever having become involved in this thing.'

In conciliatory mood, Bohr responded

> 'But we others are very grateful to you that you were, since your work did so much to promote this theory.'

There is a large literature on the vicissitudes associated with the interpretation of quantum mechanics, many of the key protagonists holding different views. Einstein was particularly aggressive in his opposition to the probability interpretation of the wavefunction, his conviction of the incompleteness of the new quantum mechanics being expressed by his well-known remark in a letter to Born of 4 December 1926,

> 'I, at any rate, am convinced that He (God) does not throw dice.'

17.7.1 Bohr and complementarity

Bohr remained the godfather of the new discipline of quantum mechanics and agonised over the interpretation of the new scheme of things. He had held out against the reality of the wave–particle duality, particularly Einstein's concept of light quanta, but eventually he conceded, following the decisive results of the Bothe–Geiger experiments and the consequent rejection of the Bohr–Kramers–Slater picture (see Sect. 10.2). For Bohr, the wave–particle duality for radiation was at the heart of the new conceptions of quantum mechanics – how was it possible for radiation to possess simultaneously both wave properties, as exhibited by the phenomena of interference and diffraction, and the particle properties found in the photoelectric and Compton effects? To resolve this apparent paradox, he introduced the concept of *complementarity*, a conception which was designed as what Jammer calls a 'new logical instrument' for the interpretation of classical and quantum phenomena. The words of Jammer give as close to a definition of complementarity as will be found in the literature:

> '[Bohr] called it "complementarity", denoting thereby the logical relation between two descriptions or sets of concepts which, though mutually exclusive, are nevertheless both necessary for an exhaustive description of the situation. In Heisenberg's reciprocal uncertainty relations he saw a mathematical expression which defines the extent to which complementary notions may overlap, that is, may be applied simultaneously, but, of course, not rigorously. The uncertainty relations, Bohr contended, tell us the price we have to pay for violating the rigorous exclusion of notions, the price for applying to the description of a physical phenomenon two categories of notions which, strictly speaking, are contradictory to each other.' (Jammer, 1989)

Bohr went on to relate the notion of complementarity to the issue of measurement in quantum mechanics. In his interpretation, it was now impossible to separate out what was being observed from the means by which it was observed. He contended that measuring instruments produce results which are expressed in classical terms and that by observing the system in different ways, complementary variables may be determined which however can only be determined and reconciled according to the limitations imposed by Heisenberg's uncertainty principle.

After a number of years in which he had published little on quantum mechanics, Bohr first described his new understanding at the 1927 International Congress of Physics held in Como to celebrate the centenary of the death of Alessandro Volta, who had been born and died there (Bohr, 1928). In his address *The quantum postulate and the recent development of quantum theory*, Bohr set out his concepts for the first time and these might be considered a primitive form of what was later to be known as the *Copenhagen interpretation* of quantum mechanics. Bohr's arguments did not make a strong immediate impression upon his audience, but the emphasis upon the role of experiment in defining what was measurable at the elementary level was of lasting importance.

Bohr gave no exact definition of the principle of complementary in his various expositions of the concept, and indeed later used the flexibility of its definition to include areas outside physics, as described in Pais's biography of Bohr (Pais, 1991). Pauli sharpened up the concept to give a precise operational meaning to the principle (Pauli, 1933). He stated

> '[two classical concepts – and not two modes of description – are] complementary if the applicability of one (for example, position coordinate) stands in the relation of exclusion to that of the other (for example, momentum).'

There ensued a debate among theorists, such as von Weizsäcker and Feyerabend about the exact meaning of Bohr's complementarity, largely engendered by the vagueness of Bohr's definition of the conception. For some theorists its very vagueness was a virtue in allowing flexible interpretations of the relation of non-commuting variables to measurement and observation.

17.7.2 Dirac notation

The first complete exposition of the formal foundations of quantum mechanics was given by Dirac in his classic text *The Principles of Quantum Mechanics*, the first edition of which was published in 1930 (Dirac, 1930a). Dirac's book was an extraordinary achievement setting out in detail the mathematical foundations of quantum mechanics axiomatically. The remarkable feature of his exposition is the constant interplay between the formal structure of the statistical transformation theory and the need to account for observed physical phenomena. In the first edition, Dirac used the same notation he had adopted in his papers from 1925 to 1930. This exposition of the principles of quantum mechanics used the language and techniques of statistical transformation theory which he developed from his pioneering geometric-algebraic approach discussed in Chap. 14 and developed in Sect. 17.3. The various manipulations of operators and wavefunctions developed over the last eight chapters were the tools used in Dirac's exposition of 1930.

In 1939, Dirac invented the bra and ket notation which resulted in a significantly more elegant notation for the underlying algebraic structure of the theory (Dirac, 1939). This new formalism was incorporated into the third edition (Dirac, 1947) and remained essentially unchanged in the subsequent fourth and definitive edition (Dirac, 1958). Let us summarise the translation from the notation of 1930 to the post-1939 version. This also completed the transformation of the theory from a more-or-less self-consistent set of assumptions and rules into the modern axiomatic exposition of quantum mechanics. Dirac's notation works as follows:

- The wavefunction $\Psi(x, t)$ incorporates all knowledge about the evolution of the system. In general, it is defined in an infinite dimensional function space, what is referred to as *Hilbert space*. $\Psi(x, t)$ can be thought of as a *vector* in this space. $\Psi(x, t)$ may consist of functions of the spatial coordinates and time as well as spin coordinates. It is referred to as the *state* of the system, or the *state vector* and is written $|\Psi\rangle$, meaning 'the state with state function $\Psi(x, t)$'. In the time-independent case, lower case Greek symbols are used so that $|\psi\rangle$ means 'the time-independent state with state function $\psi(x)$'.
- In carrying out the evaluation of the integrals over complete sets of eigenfunctions, we need the complex conjugates of the wavefunctions in order to form integrals such as

$$\int_{-\infty}^{\infty} \phi^*(x)\,\psi(x)\,\mathrm{d}x \, . \tag{17.68}$$

Dirac introduced the notation $\langle\phi|$ to rewrite the integral (17.68) as

$$\int_{-\infty}^{\infty} \phi^*(x)\,\psi(x)\,\mathrm{d}x \,=\, \langle\phi|\psi\rangle \, , \tag{17.69}$$

so that $|\psi\rangle$ and $\langle\phi|$ are shorthand notation for the wavefunction ψ and the complex conjugate of the wavefunction ϕ respectively and the combination of the two, written as $\langle\phi|\psi\rangle$, is an abbreviation for the integral (17.69). If the wavefunction ψ is normalised, $\langle\psi|\psi\rangle = 1$. In Dirac's language of 1939,

- ○ $|\psi\rangle$ is called a *ket vector*,
- ○ $\langle\psi|$ is called a *bra vector*, the complex conjugate of ψ,
- ○ $\langle\phi|\psi\rangle$ is called a Dirac bra(c)ket, or *inner product* of the state vectors.

- It is conventional to write operators with a 'hat' so that the operator A is written $\widehat{A}$. Then, the state found by operating upon ψ with $\widehat{A}$ is denoted $|\widehat{A}\psi\rangle$. We can also define the bra vector corresponding to this state by analogy with the relation $\phi^*(x) \to \langle\phi|$ so that

$$(\widehat{A}\,\psi)^* = \langle \widehat{A}\,\psi| \, . \tag{17.70}$$

The integral corresponding to

$$\int_{-\infty}^{\infty} \phi^* \,\widehat{A}\,\psi \,\mathrm{d}x \tag{17.71}$$

can therefore be written in Dirac notation as

$$\int_{-\infty}^{\infty} \phi^* \,\widehat{A}\,\psi \,\mathrm{d}x = \langle\phi|\widehat{A}|\psi\rangle \, . \tag{17.72}$$

The quantity $\langle\phi|\widehat{A}|\psi\rangle$ is called the *matrix element* associated with the state vectors $|\psi\rangle$ and $|\phi\rangle$. Because of the rules for forming the complex conjugates of the states and the adjoint properties of $\widehat{A}$,

$$\langle\phi|\widehat{A}|\psi\rangle^* = \langle\psi|\widehat{A}^\dagger|\phi\rangle \tag{17.73}$$

where $\widehat{A}^\dagger$ is the adjoint operator associated with $\widehat{A}$. It follows that $\widehat{A}^\dagger\widehat{A} \equiv \widehat{I}$, where $\widehat{I}$ is the identity operator.

- The expectation value for an observable quantity represented by an operator $\widehat{A}$ with the system in the state $|\psi\rangle$ can be written

$$\langle A\rangle = \int_\infty^\infty \psi^*(x)\widehat{A}\psi(x)\,dx = \langle\psi|\widehat{A}|\psi\rangle . \tag{17.74}$$

An integral part of the formalism is that the operator $\widehat{A}$ should be Hermitian, or self-adjoint, so that the eigenvalues associated with the eigenfunctions are real. For Hermitian operators, $\widehat{A} = \widehat{A}^\dagger$ and so

$$\langle A\rangle^* = \langle\psi|\widehat{A}|\psi\rangle^* = \langle\psi|\widehat{A}^\dagger|\psi\rangle = \langle\psi|\widehat{A}|\psi\rangle = \langle A\rangle . \tag{17.75}$$

- In the Dirac notation, the eigenvalue equation is written

$$|\widehat{A}|\psi\rangle = a|\psi\rangle . \tag{17.76}$$

Thus, $|\psi\rangle$ is the *eigenstate* corresponding to the operator $\widehat{A}$ and a is the associated *eigenvalue*; $\psi(x)$ is the corresponding eigenfunction of $\widehat{A}$. As examples of operators in the Dirac notation, the momentum operator is $\widehat{p} \equiv (h/2\pi i)\,\partial/\partial x$ in one dimension and the Hamiltonian operator is

$$\widehat{H} = \widehat{T} + \widehat{V} = \frac{h^2}{8\pi^2 m_e}\nabla^2 + V(x) , \tag{17.77}$$

where the two terms correspond to the kinetic energy operator $\widehat{T}$ and the scalar potential energy operator $\widehat{V}$. Thus, the eigenvalue equation for the stationary states of a system is given by the solution of the eigenfunction equation

$$\widehat{H}\psi = \widehat{T}\psi + \widehat{V}\psi = \frac{h^2}{8\pi^2 m_e}\nabla^2\psi + V(x)\psi = E\psi , \tag{17.78}$$

which is the time-independent Schrödinger equation with energy eigenvalues E.

Dirac's great book repays close reading. The various complex pathways by which the new understandings were obtained are bypassed and the structure no longer depended upon concepts such as the correspondence principle. Just as Newtonian mechanics needed first the development of differential and integral calculus and then the Lagrangian and Hamiltonian approaches to perfect the mathematical structures, so Dirac's book is an account of the non-commutative algebra of operators. On first encounter, it appears to be a somewhat formal mathematical approach to quantum mechanics, but close reading shows how Dirac's thought was strongly governed at each stage by the need to account for a relatively small number of phenomena which contain the essence of what the algebra had to encompass. It is one of the great books of physics.

17.7.3 The Copenhagen interpretation

In due course, the rules of quantum mechanics were hammered out into a self-consistent set of postulates which form the basis of what may be called the *axiomatic approach* to quantum mechanics. It is often referred to as the *Copenhagen interpretation* and is based upon the results of the many different analyses described in the last seven chapters. It has to be said that exactly what constitutes the Copenhagen interpretation is a matter of debate, but the following sets of rules, are fully consistent with what Bohr and his collaborators eventually agreed upon.[4]

- The most complete knowledge we can have of a system is represented by the state vector $|\Psi\rangle$.
- To every observable A there corresponds a Hermitian operator $\widehat{A}$. The result of a measurement of A must be one of the eigenvalues of $\widehat{A}$.
- If the eigenvalue a corresponds to the eigenstate $|\phi\rangle$, then the probability of obtaining the result a when the system is in the state Ψ is $|\langle\phi|\Psi\rangle|^2$.
- As a result of the measurement of A in which the result a is obtained, the state of the system is changed to the corresponding eigenstate $|\phi\rangle$.
- Between measurements, the state vector $|\Psi\rangle$ evolves with time according to the time-dependent Schrödinger equation

$$\frac{ih}{2\pi}\frac{\partial}{\partial t}|\Psi\rangle = \widehat{H}|\Psi\rangle .\qquad (17.79)$$

The consequences of these rules are profound. For example, we described in Sect. 17.2.2 how the *expectation value* of any measurement is found for the observable A which has state vector $|\psi\rangle = \sum_j c_j|\phi_j\rangle$ in the time-independent case. The result of any particular measurement is that the system is found in the eigenstate $|\phi_j\rangle$ with eigenvalue a_j and that the probability of that value being obtained is $|c_j|^2 = |\langle\phi_j|\psi\rangle|^2$. Hence, the expectation value, in the sense of the mean value of a large number of experiments, is $\langle A\rangle = \sum_j a_j|c_j|^2$. But, once the measurement is made, the system is in one particular state $|\phi\rangle_j$. The effect of the measurement of the observable A is to force the system into one of the eigenstates of $\widehat{A}$, with the result that c_j is now 1 and all the other c_is for which $i \neq j$ are zero. This is a key feature of the way in which quantum mechanics works. This process of forcing the system into one of the possible state functions is referred to as the *collapse* or *reduction* of the wavefunction.

This is one of the most remarkable results of the formalism of quantum mechanics. In terms of amplitudes, the theory is linear – it is only when an experiment is made that the system is forced into one of the eigenstates and the *a priori* probabilities of this happening are given by the square of the modulus of the probability amplitudes.

18 The aftermath

We have reached the goal I set myself when this journey began. From about 1927 onwards, the quantum theory in its modern guise for non-relativistic quantum mechanics was essentially complete, although there remained problems of interpretation which took a number of years to unravel – some of them are still hotly debated. But much of the apparatus was already in place and the subsequent developments changed completely the face of physics. Jammer (1989) summarises the achievement as follows:

> 'Since 1927, the development of quantum mechanics and its applications to molecular physics, to the solid state of matter, to liquids and gases, to statistical mechanics, as well as to nuclear physics, demonstrated the overwhelming generality of its methods and results. In fact, never has a physical theory given a key to the explanation and calculation of such a heterogeneous group of phenomena and reached such a perfect agreement with experience as has quantum mechanics.'

As noted by Mehra and Rechenberg (2001), the 1930s also saw the beginning of the compartmentalisation of physics into separate quantum disciplines. Thus, during the 1930s, with the general acceptance and success of quantum mechanics, the quantum physicists began to specialise in disciplines such as atomic physics, molecular physics, solid state physics, including metal and semiconductor physics, condensed matter physics and low temperature physics, while at high energies, nuclear, particle and cosmic ray physics developed as disciplines in their own right. Whereas the pioneers of quantum mechanics regarded the whole province of quantum physics as their domain, the various branches of quantum physics became fragmented into these specialisms, not so different from those encountered in any physics department today.

The years up to the outbreak of the Second World War were a golden age of theoretical and experimental physics. Let us outline briefly some of the major achievements of theory, experiment, observation and the interpretation of these data. These topics are now part of the basic infrastructure of modern physics and the details of the experiments and calculations can be found in the standard textbooks.

18.1 The development of theory

Theoretical quantum mechanics developed into the various subdisciplines mentioned above, but a key priority for theory and experiment was the development of a formalism which could encompass the quantisation of the electromagnetic field as well as the mechanics and dynamics of particles. In fact, the first attempt to include the quantisation of the electromagnetic field into the scheme of matrix mechanics appeared in the last sections of the pioneering papers by Born and Jordan (1925b) and in the Three-Man Paper (Born *et al.*, 1926). In this approach, the electromagnetic waves were treated as a system of harmonic oscillators. The interaction of radiation and matter next appeared in Dirac's treatment of Compton scattering using a relativistic version of his q-number approach to quantum mechanics (Dirac, 1926c). In his next assault on the quantum theory of radiation, Dirac extended the theory of the emission and absorption of radiation by requiring the q-numbers which characterise the radiation to be the Fourier components of the energy E_r and its conjugate 'phase' θ_r and that they should satisfy the commutation relations

$$\theta_r E_r - E_r \theta_r = \frac{ih}{2\pi} \ . \tag{18.1}$$

His achievement in this paper was his demonstration that this procedure led to the determination of the values of Einstein's A and B coefficients (Dirac, 1927). According to Jammer (1989), Dirac's procedure for quantising the electromagnetic field marked the beginning of quantum electrodynamics. The derivation of the Dirac equation and the predictions of the magnetic moment of the electron and positrons were spectacular achievements (Dirac, Dirac (1928a,b), see Sect. 16.6) but the electromagnetic field itself still had to be built into a consistent theory of relativistic quantum mechanics. In the succeeding years, Dirac, Heisenberg, Jordan, Klein and Pauli made major inroads into the proper formulation of quantum electrodynamics, step-by-step incorporating the features now familiar in their modern contexts.

To mention only a few of the steps along the way, in 1928, Jordan and Pauli published their paper on the quantum electrodynamics of charge-free fields in which they extended Dirac's formalism by introducing non-commuting q-numbers for the electromagnetic field and casting them in relativistically invariant form (Jordan and Pauli, 1928). Heisenberg and Pauli returned to the attack in early 1929. They had already introduced a relativistic Lagrangian in which the quantum field variables are given in a form similar to that used in classical mechanics. The problem they faced was that the formalism led to the conjugate momentum of the fields being identically zero. This was remedied by a 'trick' discovered by Heisenberg of adding an extra term to the Lagrangian which was multiplied by the small quantity ε. This resolved the problem of the vanishing conjugate momentum and led to sensible results in the limit $\varepsilon \to 0$. The result of these endeavours was a long and important paper by Heisenberg and Pauli (1929), which begins with the words,

> 'So far it has not been possible to connect in quantum theory the mechanical and electrodynamic laws, electrostatic and magnetostatic interactions on the one hand, and interactions mediated by radiation on the other hand, in a unified point of view. In particular, one has not proceeded to take into account correctly the finite velocity of propagation of the actions due to electromagnetic forces.'

Their paper set out to provide this unified view and take account of retardation effects in radiation problems. The three chapters describe the 'general methods', the derivation of the 'fundamental equations of the theory for electromagnetic and matter fields' and 'approximation methods for the integration of the equations and physical applications'. The new relativistic quantum field theory contained both Fermi–Dirac and Bose–Einstein statistics.

Pauli and Heisenberg sought to continue along the lines of their paper of 1929 and were stimulated by the publication of Weyl's *Gruppentheorie und Quantenmechanik* (1928) to generalise the group-theoretical methods of Weyl to the case of wave fields. This had the effect of simplifying the presentation of relativistic quantum field theory and of eliminating the 'trick' of the vanishing ε which Heisenberg had introduced in their first paper (Heisenberg and Pauli, 1930).

In classical mechanics, there is a close relation between the symmetry properties of a mechanical system and conservation laws.[1] A much more general approach to the problems of symmetry under different types of mathematical operation is provided by *group theory*. Already in his paper of June 1926, Heisenberg had made use of group theoretical results in his study of the many-body problem and resonances in quantum mechanics (Heisenberg, 1926a). These ideas were to be taken very much further by the young Eugene Wigner. In 1927, he published an ambitious paper entitled *Some consequences from Schrödinger's theory for the term structures* (Wigner, 1927) in which group-theoretical concepts were applied to Schrödinger's wave equation. From these considerations, Wigner was able to derive the selection rules for azimuthal quantum numbers as well as the term structure and selection rules when atoms are subject to electric and magnetic fields. He concluded triumphantly that he been able to

'demonstrate that one can, by rather simple symmetry considerations with the Schrödinger equation, already explain an essential part of the purely qualitative spectroscopic experience.'

In his subsequent development of the full power of group theory as applied to quantum mechanical problems, he was joined by his colleague John von Neumann. The full significance of Wigner's innovations can be appreciated from the remark of van den Waerden, himself a specialist in group theory, that

'Wigner's paper seems to be the first in which group theory was applied to [quantum] physics. In this paper, some rules of term zoology were deduced, but the spin was left out of account. In the papers of von Neumann and Wigner [Neumann and Wigner (1928a,b,c)], the whole apparatus of group characters and representations was put into action and the complete system of term zoology, including selection rules, intensity formulae [and] Stark effect was developed.' (van der Waerden, 1960)

For example, in the third paper by von Neumann and Wigner, the expression for the Landé *g*-factor was derived purely from group theory, specifically from the group-theoretical multiplication formula, which resulted in

$$g = 1 + \frac{j(j+1) - l(l+1) + s(s+1)}{2j(j+1)}. \tag{18.2}$$

The works of Wigner and Neumann were incorporated into Weyl's *Gruppentheorie und Quantenmechanik* (1928), Weyl stating that he had independently derived many of the key results.

The group-theoretical approach opened up new ways of tackling the problem of the origin of the chemical bond. Hund had made use of symmetry arguments derived from the group-theoretical aspects of Heisenberg's and Wigner's papers of 1927 to interpret the binding of atoms in molecules (Hund, 1927c). These arguments provided a more visualisable approach to the combination of wavefunctions for the atoms of a molecule. A pure group-theoretical approach was adopted by Walter Heitler and Fritz London who investigated the forces between two neutral atoms using Heisenberg's exchange integral. The result was an explanation for the homopolar covalent bonds found in molecular chemistry (Heitler and London, 1927).

Following this advance, Heitler went on in his next paper to explore the full capabilities of group theory for quantum chemistry (Heitler, 1927). What he did is best explained in his own words.

> 'London did not join me in that: he thought is was too complicated. Wigner's paper had appeared by that time, and I saw immediately that it could be useful for further development of the theory of chemical bonds, but London wanted to go on in his own more intuitive way . . .
>
> So I started to study group theory. I first read the book by Speiser . . . Later on I read the papers by Schur and other people. The very nice thing was that the mathematicians had prepared group theory so well for the use of the physicists without knowing it that sometimes I could just copy, word for word, pages from a group-theory paper and use them for my purposes.' (Heitler interview: the *Archives for the History of Quantum Physics*, 1963)

Among the achievements of this paper was his demonstration that the closed shells of an atom can result in no new splittings of the energy levels, but only a shift in the levels. As he remarked,

> '[This conclusion is] solely responsible for the existence of the periodic system with *homologous sequences*, in which the elements exhibit *homologous spectroscopic* and *chemical* properties.'

These events reflect the fact that the theory of quantum mechanics was becoming increasingly mathematically sophisticated and involved mathematics which was often beyond the reach of the average physicist. Slater was among those disaffected by these developments, referring to the group theory approach as the '*Gruppenpest*', the pest of group theory (Slater, 1975). But Slater did not simply complain, but converted the formalism of many-electron atoms into a viable scheme for the determination of the energy levels and structure of atoms. He realised that the atomic properties of atoms including their spin properties could be written entirely in terms of antisymmetric wavefunctions, thus explicitly including the Pauli exclusion principle (Slater, 1929). Combining these concepts with Douglas Hartree's 'self-consistent field' approach to the determination of atomic structure (Hartree, 1927a,b) led to many important results in the understanding of atoms and their spectra. For example, he was able to recover Heitler's result that filled shells make no contribution to the splitting of the energy levels of atoms. With some pride Slater recalled,

> 'As soon as this paper became known, it was obvious that a great many other physicists were as disgusted as I had been with the group-theoretical approach to the problem. As I heard later, there were remarks made such as "Slater has slain the 'Gruppenpest' ". I believe that no other piece of work I had done was so universally popular.' (Slater, 1975)

Despite these somewhat reactionary remarks, group theory was there to stay in the armoury of weapons to be used in tackling problems in quantum mechanics.

To complete the story of the understanding of spin up to 1940, the final piece of the jigsaw was the relation of the spin of the particles to their statistical properties. This connection was discovered in the studies of Markus Fierz who worked as an assistant to Pauli from 1936 to 1940. Fierz analysed the properties of quantum field theories and, using the spinor calculus of van der Waerden (1932), found that

> 'Particles with integral spin must always satisfy Bose statistics and particles with half-integral spin Fermi statistics.' (Fierz, 1939)

These results were consolidated in joint papers with Pauli in 1939 and 1940 (Fierz and Pauli, 1939; Pauli and Fierz, 1940). An even more general approach which gave the same result was developed by Frederick Joseph Belinfante who introduced *undons*. These objects could describe both integral and half-integral spin particles (Pauli and Belinfante, 1940).

18.2 The theory of quantum tunnelling

One of the earliest triumphs of quantum mechanics was the theory of quantum mechanical tunnelling as applied to molecules and to the emission of α-particles in radioactive decays.[2] Friedrich Hund's objective was to understand the nature of the stationary states of diatomic molecules. He modelled the potential well experienced by the electrons in molecules by a double-well potential (Fig. 18.1) in order to work out the eigenstates of molecules (Hund, 1927a,b,d). He found that the superposition of the even ground state and the odd first excited state shown in Fig. 18.1*b* resulted in a non-stationary state in which the electron oscillated back and forth between the two potential wells, the period T of the oscillation being

$$\frac{T}{\tau} \approx \sqrt{\frac{h\nu}{V}} \exp \frac{V}{h\nu} , \qquad (18.3)$$

where $\tau = 1/\nu$ is the period of oscillation of the electron in one of the potential wells when they are widely separated and V is the height of the potential barrier. Notice the exponential dependence of the oscillation frequency upon the height of the potential barrier.

In the same year, Lothar Nordheim published a paper on barrier penetration associated with the thermionic emission of electrons from a heated metal surface (Nordheim, 1927). In his calculations, he introduced the concept of a rectangular potential barrier, a picture now familiar in every quantum mechanics textbook (Fig. 18.2).

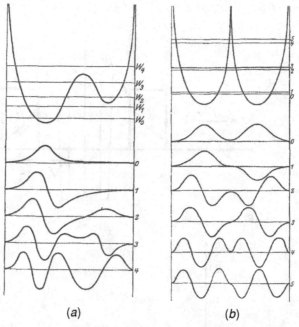

(a)   (b)

Fig. 18.1   Examples of the double potential wells considered by Hund in his determination of the eigenstates of diatomic molecules. (*a*) An asymmetric double potential showing the first five eigenfunctions. For states W_2 and W_3, transitions through the potential barrier are not allowed according to classical physics. (*b*) A symmetric double potential well with a finite potential barrier, also showing the first six eigenfunctions. For all the states shown, transitions through the potential barrier are not allowed classically (Hund, 1927a).

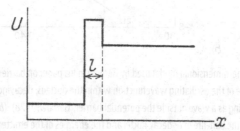

Fig. 18.2   The rectangular potential barrier used by Nordheim to estimate the barrier penetration rate of hot electrons within a metal surface (Nordheim, 1927).

Perhaps the most famous example of the application of barrier penetration was to the process of α decay, the first application of quantum mechanics to the atomic nucleus. When George Gamow arrived in Göttingen from the Soviet Union in 1927, he read Rutherford's paper of 1927 on the problem of understanding the α decay in thorium C′, or polonium-212 (^{212}Pu). Geiger's α-scattering experiments had shown that the height of the electrostatic

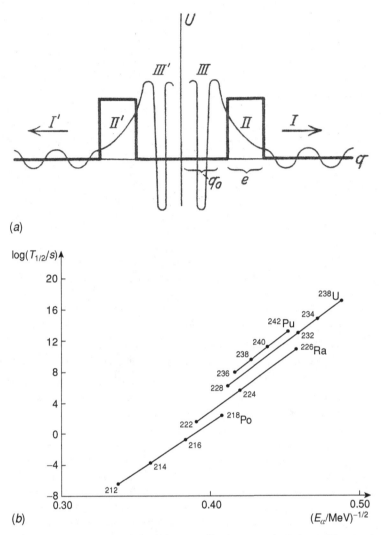

Fig. 18.3 (*a*) The one-dimensional model used by Gamow in his paper on barrier penetration by α-particles, showing the amplitude of the oscillating wavefunction within the nucleus, decaying through the potential barrier and then propagating as a wave outside the potential barrier (Gamow, 1928). (*b*) The Geiger–Nuttall law showing the relation between the half-life of α decay nuclei and the energies of the emitted α-particles. Note the enormous range of half-lives shown on a logarithmic scale on the ordinate and the small range of energies shown on a linear scale on the abscissa. The measurements follow closely the expectation of barrier penetration theory, specifically the expression (18.4) (Courtesy of Creative Commons).

potential barrier within which the nucleons were confined was at least 8.57 MeV, and yet the energies of the α-particles observed in the α decay of thorium C′ were less than half this value, 4.2 MeV. Gamow realised that this was an example of barrier penetration in quantum mechanics. The nuclear potential could be modelled as a deep rectangular potential well, as illustrated in Fig. 18.3*a*. Then, according to the quantum calculation for barrier penetration, although the barrier is impenetrable according to classical physics, there is a

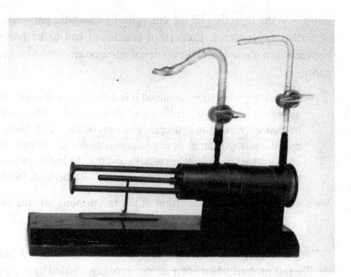

Fig. 18.4 The apparatus with which Rutherford demonstrated the nuclear disintegration of nitrogen atoms (Rutherford, 1919, 1920).

finite probability that the α-particles can reach the other side because of the wave properties of the particle. This is illustrated by the diagram from Gamow's paper (Fig. 18.3a) which shows the amplitude of the wavefunction on both sides of the barrier and within it, another diagram which appears in all the standard textbooks. In fact, Gamow and independently Ronald Gurney and Edward Condon almost simultaneously solved Schrödinger's equation for the nuclear potential shown in Fig. 18.3a and derived a relationship between the decay constant λ of the nucleus against α-particle decay and the energy of the α-particle (Gamow, 1928; Gurney and Condon, 1928, 1929). This theory of α decay could account naturally for the very narrow range of energies of the α-particle and the enormous range of decay constants, as found by Hans Geiger and John Nuttall in 1911. The law can be written

$$\ln \lambda = -A \frac{Z}{\sqrt{E}} + B \, , \qquad (18.4)$$

where $\lambda = \ln 2/(\text{half-life})$, Z is the atomic number, E the total kinetic energy of the α-particle and the residual nucleus, and A and B are constants (Geiger and Nuttall, 1911). A modern version of the Geiger–Nuttall law is shown in Fig. 18.3b.

18.3 The splitting of the atom and the Cockcroft and Walton experiment

During the last years of his tenure of the Chair of Physics at Manchester University, Rutherford carried out a key experiment in which he bombarded nitrogen atoms with the α-particles produced in the radioactive decays of radium-C (Fig. 18.4). The surprising result was the detection of the emission of energetic particles, which, after a careful set of further experiments, Rutherford concluded had to be protons liberated in collisions between the α-particle and the nuclei of nitrogen atoms. As he wrote in the conclusion of his paper,

'From the results so far obtained it is difficult to avoid the conclusion that the long-range atoms arising from the collision of alpha particles with nitrogen are not nitrogen atoms but probably atoms of hydrogen, or atoms of mass 2. If this be the case, we must conclude that the nitrogen atom is disintegrated under the intense forces developed in a close collision with a swift alpha particle, and that the hydrogen atom which is liberated formed a constituent part of the nitrogen nucleus.' (Rutherford, 1919)

We now understand that the origin of the fast protons was the nuclear interaction

$$^{14}\text{N} + \alpha \rightarrow {}^{17}\text{O} + \text{p}. \tag{18.5}$$

Further experiments were undertaken by Rutherford and James Chadwick who found similar fast protons when other atoms were bombarded by α-particles. A problem was that the energies of the α-particles were limited to those available from naturally occurring radioactive nuclides. Ideally, a more controlled beam of incident particles was required. The problem was that, as can be seen from Fig. 18.3b, the energies of the particles would have to be in the MeV energy range and it was a serious technological challenge to create such electrostatic potentials in the laboratory.

Gamow realised that his theory of barrier penetration could account for the penetration of α-particles into nitrogen nuclei in Rutherford's experiments. He explained the theory in a manuscript sent to Rutherford and his colleague John Cockcroft in December 1929. Cockcroft repeated Gamow's calculations and showed that, because of the process of barrier penetration, protons accelerated to only 300 keV could penetrate a boron nucleus with about 0.6% probability. Cockcroft inferred that an accelerating electric potential of only 300 keV would be sufficient to penetrate the boron nucleus and induce nuclear transmutations. By 1932, after a great deal of effort, Cockcroft and Ernest Walton had developed the technology of electrostatic particle acceleration to the extent that potentials of 700 keV could be sustained to accelerate protons to these energies (Fig. 18.5). They succeeded in inducing the first artificial nuclear disintegrations by bombarding lithium nuclei with high energy protons (Cockcroft and Walton, 1932). The process involved was

$$^{7}\text{Li} + \text{p} \rightarrow {}^{4}\text{He} + {}^{4}\text{He}. \tag{18.6}$$

The energies of the accelerated protons were precisely known, as were the rest masses of the lithium and helium atoms. The kinetic energies of the helium nuclei ejected in the interaction (18.6) could be measured. As a result, this nuclear interaction provided the first direct experimental test of Einstein's mass–energy relation $E = m_e c^2$ – precise agreement was found with Einstein's prediction. This experiment marked the beginning of experimental high energy physics in which particles are accelerated to high energies and used as probes of the structure of the nucleus and as tools for the discovery of new particles.[3]

Fig. 18.5 The apparatus with which Cockcroft and Walton artificially disintegrated lithium nuclei (Cockcroft and Walton, 1932). Walton is sitting inside the little tent, observing the decay products on a luminescent screen. Cockcroft is on the left.

18.4 Discovery of the neutron

Rutherford's Bakerian lecture to the Royal Society of London in 1920 provides a vivid picture of the state of knowledge of the properties of the atomic nucleus at that time (Rutherford, 1920). Atomic nuclei have masses about two or more times that which can be attributed to positively charged protons. The commonly held explanation was that the nucleus was composed of electrons and protons, the 'inner' electrons neutralising the extra protons. The fact that certain nuclei ejected electrons in radioactive β decays supported this point of view. Rutherford speculated in his review that the neutral mass in the nucleus might be in the form of some new type of particle, similar to the proton but with no electric charge. As he wrote,

> '... it may be possible for an electron to combine much more closely with the H-nucleus [than is the case in the ordinary hydrogen atom] ... It is the intention of the writer to test [this idea] ... The existence of such atoms seems almost necessary to explain the building up of the heavy elements.' (Rutherford, 1920)

During the 1920s Rutherford and his colleagues, particularly James Chadwick, made a number of unsuccessful attempts to find evidence for these particles, which became known as *neutrons*. Little attention was paid to Rutherford's proposal.

The apparatus with which Chadwick discovered the neutron (Chadwick, 1932).

In 1930, Walther Bothe and Herbert Becker discovered that very penetrating radiation was emitted when light elements such as beryllium were bombarded by α-particles (Bothe and Becker, 1930). Because the penetrating particles did not cause ionisation, they postulated that the neutral particles were high energy γ-rays. In 1932, Irène Joliot-Curie and her husband Frédéric Joliot in France carried out a similar series of experiments in which the penetrating neutral radiation hit a block of paraffin wax, which was found to emit energetic protons. If the energetic protons were produced by Compton scattering, the γ-rays would have had to be of very high energy, ~ 50 MeV (Curie and Joliot, 1932). Chadwick guessed that the penetrating radiation was rather a flux of the elusive neutrons. Within a matter of weeks, he performed the key experiment in which α-particles bombarded a beryllium target, releasing neutrons which then collided with a block of paraffin wax (Fig. 18.6). Energetic protons were emitted in collisions between the neutrons and the protons in the paraffin wax which were detected in an ionisation chamber, enabling the mass of the invisible neutron to be estimated. It was inferred that the neutral particles had masses roughly the same as that of the proton (Chadwick, 1932). He interpreted the nuclear interaction as the following process,

$$\alpha + {}^{9}\text{Be} \rightarrow {}^{12}\text{C} + \text{n}. \tag{18.7}$$

This was the discovery of the neutron.

18.5 Discovery of nuclear fission

The discovery of the neutron had immediate implications for experimental nuclear physics. Unlike the electron or the α-particle, the neutron is electrically neutral and so could penetrate the Coulomb barrier of the nucleus. The discipline of nuclear physics was transformed since heavy nuclei such as uranium could be bombarded with neutrons, resulting in the formation of new isotopes. Such experiments were initiated by Fermi and his colleagues in Rome in 1934. They believed that they had demonstrated the formation of a new element with atomic number 94 (Fermi *et al.*, 1934), but the result was viewed with some scepticism.

The experiments were repeated by Otto Hahn, Lise Meitner and Fritz Strassmann. In 1938, following the Anschluss in which she lost her citizenship, Meitner fled to Sweden and continued her collaboration with Hahn by mail. During this correspondence, Hahn informed Meitner of his discovery of traces of barium when uranium was bombarded with neutrons. This came as a complete surprise since barium has only 40% the atomic weight of uranium. Meitner soon convinced herself and Hahn that the barium resulted from what became known as the *nuclear fission* of the uranium nuclei. The results were published by Hahn and Strassmann (1939). Meitner and her nephew Otto Frisch, who was also working in Sweden, published the results of their calculations that a new type of nuclear reaction had been observed in Hahn and Strassmann's experiments (Meitner and Frisch, 1939).

Leo Szilard was well aware of the significance of these experiments for nuclear energy generation. In 1933, following Chadwick's discovery of the neutron, Szilard had realised that a self-sustaining nuclear chain reaction would be possible if the neutrons liberated in the types of interaction involved in the Cockcroft and Walton experiment could be used to initiate further nuclear interactions. He filed patents for this concept and also carried out unsuccessful experiments in which light elements were bombarded with neutrons to demonstrate the effect. As soon as the results of Hahn and Strassmann's experiments were published, he immediately realised that this provided a route to a nuclear chain reaction, both for the generation of nuclear power and for the creation of nuclear weapons. He urged restraint in the publication of these results because of the impending war, but Joliot and his colleagues in Paris did not hesitate. The theory of nuclear chain reactions was published by both groups in 1939 (von Halban *et al.*, 1939; Szilard and Zinn, 1939). The details of this story and the subsequent development of nuclear weapons is vividly told in Rhodes' classic book *The Making of the Atomic Bomb* (1986).

18.6 Pauli, the neutrino and Fermi's theory of weak interactions

Unlike the process of α decay in which the decay of a radioactive nuclide results in a well-defined energy for the emitted α-particle, the β decay process results in a broad spectrum of electron energies. An example of the continuous energy spectrum of electrons found in the decay of radium-E (^{210}Bi) is shown in Fig. 18.7 (Neary, 1940). There is an upper limit to the energies of the emitted electrons of just over 1 MeV, but the spread of energies extends to less than 4% of this value, the maximum occurring at just less than 300 keV and the average energy amounting to 390 keV. During the 1920s, there was an ongoing debate about whether or not the broad continuum electron energy spectrum could be attributed to what was termed 'ordinary' processes, meaning that the electrons were created with a single energy which was then redistributed by 'ordinary' processes such a Compton scattering.

After two years of challenging experiments, Charles Ellis and William Alfred Wooster completed calorimetric experiments in which they showed that the average energy deposited in their calorimeter was about 350 keV per disintegration, rather than about 1 MeV as

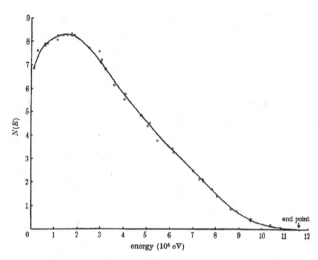

Fig. 18.7 An example of the energy spectrum of electrons emitted in the β-decay process. This spectrum shows the continuous electron energy spectrum found in the radioactive decay of radium-E (^{210}Bi) (Neary, 1940).

might be expected if all the energy was injected with maximum energy of 1 MeV and then dissipated by 'ordinary' processes (Ellis and Wooster, 1927). The experiment indicated that the measured electron energy spectrum was indeed the intrinsic energy spectrum of the β-decay process. The problem was that the process seemed to violate conservation of energy. Furthermore, with the understanding of the quantum mechanical rules for the addition of angular momentum, the hyperfine splitting of atomic lines could be used to determine the magnetic moments and hence the spin of nuclei. The process of β decay seemed to involve a violation of the law of conservation of angular momentum at the nuclear level. Thus, the conservation of both energy and angular momentum were in jeopardy.

For some time, Bohr returned to his old concern about the validity of the law of conservation of energy at the atomic level. In 1930, in desperation, Pauli suggested that the problem might be solved by invoking the existence of a neutral particle which he called a 'neutron'. Note that at this date, the only known 'subatomic particles' were the proton, the electron and the photon. Pauli's radical proposal was contained in an impassioned letter to his expert colleagues working on radioactivity at their meeting in Tübingen.[4]

> 'Dear Radioactive Ladies and Gentlemen,
>
> I have come to a desperate way out regarding the "wrong" statistics of the N- and ^{6}Li nuclei, as well as the continuous β-spectrum, in order save the "alternation law" of statistics and the energy law. To wit, the possibility that there could exist in the nucleus electrically neutral particles, which I will call neutrons, which have spin 1/2 and satisfy the exclusion principle and which are further distinct from light-quanta in that they do not move with light velocity. The mass of the neutrons should be of the same order of magnitude as the electron mass and in any case not larger than 0.01 times the proton mass. ... The continuous β-spectrum would then become understandable from the assumption that in β-decay a neutron is emitted along with the electron in such a way that the sum of the energies of the neutron and the electron is constant. ...

For the time being I dare not publish anything about this idea and address myself confidentially to you, dear radioactive ones, with the question how it would be with the experimental proof of such a neutron, if it were to have a penetrating power equal to or about ten times larger than a γ-ray.

I admit that my way out may not seem very probable *a priori* since one would probably have seen the neutrons a long time ago if they exist. But only who dares wins and the seriousness of the situation concerning the continuous β-spectrum, is illuminated by my honoured predecessor, Mr. Debye, who recently said to me in Brussels: "Oh, it is best not to think about this at all, as with new taxes". One must therefore discuss seriously every road to salvation. Thus, dear radioactive ones, examine and judge.

Unfortunately, I cannot appear personally in Tübingen since a ball which takes place in Zürich the night of sixth to seventh of December makes my presence here indispensable. . . .

Your most humble servant,

W. Pauli'

In 1932, Chadwick's discovery of the neutron, meaning the neutral partner of the proton, changed the picture. In the following year, Fermi suggested that Pauli's 'neutron' might be better called a *neutrino* and that usage was established from then on. In the following year, Fermi published his theory of weak interactions and β decay (Fermi, 1934). In his famous paper, he treated the process by analogy with the process of the emission of radiation according to the rapidly developing theory of quantum electrodynamics. Neutrinos have a very small cross-section for interaction with matter and it was not until 1956 that they were detected experimentally by Frederick Reines and Clyde Cowan who used a fission reactor as a source of neutrinos (Reines and Cowan, 1956). The discovery was made only two and a half years before Pauli's death. His response on receiving the news was sent in a congratulatory telegram:

'Thanks for message. Everything comes to him who knows how to wait. Pauli.'

18.7 Cosmic rays and the discovery of elementary particles

From the 1930s until the early 1950s, the cosmic radiation provided a natural source of very high energy particles which were energetic enough to penetrate the nucleus. This was the principal technique by which new particles were discovered until the early 1950s. As already recounted in Sect. 16.7.2, in 1930 Millikan and Anderson used an electromagnet 10 times stronger than that used by Skobeltsyn to study the tracks of particles passing through a cloud chamber. Anderson observed curved tracks identical to those of electrons but corresponding to particles with positive electric charge (Anderson, 1932). The discovery of the positron was confirmed by Patrick Blackett and Giuseppe Occhialini in 1933 using an improved technique in which the cloud chamber was only triggered after it was certain that a cosmic ray had passed through (Blackett and Occhialini, 1933). They obtained many excellent photographs of the positive electrons, on many occasions showers containing equal numbers of positive and negative electrons created by cosmic ray interactions.

There were more surprises in store, however. Anderson noted that there were often much more penetrating positive and negative particle tracks in the cloud chamber pictures. These particles displayed little evidence of interaction with the gas in the chamber. By 1936, Anderson and Seth Neddermeyer were sufficiently confident of their results to announce the discovery of particles with mass intermediate between that of the electron and the proton (Anderson and Neddermeyer, 1936). These *mesotrons* had mass between about 50 and 400 times the mass of the electron. This discovery coincided rather conveniently with the theoretical prediction by Hideki Yukawa concerning the nature of the strong force which binds neutrons and protons together in the nucleus. According to Yukawa's theory, the strong short-range force could be understood in terms of the exchange of particles about 250 times as massive as the electron (Yukawa, 1935). In fact, the particles discovered by Anderson and Neddermeyer, nowadays known as *muons*, are not the particles which bind nuclei together. The identification was somewhat unsatisfactory because the mesotrons showed little interaction with nuclei in the chamber, whereas the exchange particle was expected to show a strong interaction with nuclei.

The same procedures were used immediately after the Second World War by George Rochester and Clifford Butler who constructed a new cloud chamber to use with a large electromagnet obtained by Blackett before the War. In 1947 they reported the discovery of two cases of particle tracks in the form of 'V's with apparently no incoming particle (Rochester and Bulter, 1947). They correctly suggested that the Vs resulted from the spontaneous decay of an unknown particle, the mass of which could be estimated from the decay products. Both had mass about half that of the proton. To obtain higher fluxes of cosmic radiation, the experiments were repeated at much higher altitudes. Two years later, the experiments were carried out by Blackett's group working at the Pic du Midi Observatory in the Pyrenees and by Anderson and Cowan on White Mountain in California. Many more examples of Vs were found and this class of particle became known as *strange particles*. Both neutral and charged strange particles were discovered. Most of them had mass about half that of the proton and are what are now referred to as *charged* and *neutral kaons* (K^+, K^-, K^0). There were a few examples, however, of neutral particles with mass greater than the mass of the proton – these are now known as *lambda particles* (Λ). What puzzled the physicists was their long lifetimes – 10^{-8} and 10^{-10} s, which is many orders of magnitude greater than the time-scale associated with the strong interactions.

Meanwhile another powerful tool for the study of particle collisions and interactions had been developed by Cecil Powell at Bristol University. Photographic plates had played a key role in the discovery of X-rays and radioactivity in the 1890s. Powell, in collaboration with the Ilford company, developed special 'nuclear' emulsions which were sufficiently sensitive to register the tracks of protons, electrons and all the other types of charged particle which had been discovered. Powell and his colleagues mastered the techniques of producing thick layers of emulsion by stacking layer upon layer of emulsion, resulting in a three-dimensional picture of the interactions taking place in the emulsion. Among the first discoveries using this high precision technique was that of the *pion* (π) in 1947, which was the particle predicted by Yukawa in 1936 (Lattes *et al.*, 1947).

By 1953, accelerator technology had developed to the point where energies comparable to those available in the cosmic rays could be produced in the laboratory with known energies and directed precisely onto the chosen target. After about 1953, the future of high energy physics lay in the accelerator laboratory rather than in the use of cosmic rays.

18.8 Astrophysical applications[5]

The solution of the problem of energy generation in the Sun was one of the first fruits of the discovery of quantum mechanics. In his early papers Eddington had advocated the annihilation of matter as an inexhaustible source of energy for the stars but in 1920 he realised that, although there was no known mechanism by which nuclear energy could be released, at least energetically this provided an attractive means of powering the stars. In a remarkably prescient paragraph of his Presidential Address to the Mathematical and Physics Section of the British Association for the Advancement of Science at its Annual Meeting, held in Cardiff, he stated (Eddington, 1920):

> 'Certain physical investigations in the past year...make it probable to my mind that some portion of this sub-atomic energy is actually being set free in the stars. F. W. Aston's experiments seem to leave no room for doubt that all the elements are constituted out of hydrogen atoms bound together with negative electrons. The nucleus of the helium atom, for example, consists of 4 hydrogen atoms bound with two electrons. But Aston has further shown conclusively that the mass of the helium atom is less than the sum of the masses of the 4 hydrogen atoms which enter into it; and in this at any rate the chemists agree with him. There is a loss of mass in the synthesis amounting to about 1 part in 120, the atomic weight of hydrogen being 1.008 and that of helium 4. ... Now mass cannot be annihilated, and the deficit can only represent the mass of the electrical energy set free in the transmutation. We can therefore at once calculate the quantity of energy liberated when helium is made out of hydrogen. If 5 per cent of the star's mass consists initially of hydrogen atoms, which are gradually being combined to form more complex elements, the total heat liberated will more than suffice for our demands, and we need look no further for the source of a star's energy.'

Eddington had the good fortune to be working at the Observatories at Cambridge University, only a twenty minute walk from the Cavendish Laboratory where Francis Aston was carrying out his precise measurements of atomic and isotopic masses. At that time, nuclear energy generation could be no more than a hypothesis, but Eddington had indeed hit upon the correct solution for the energy source of the Sun. The beauty of Eddington's argument was that it did not depend upon the precise nature of the nucleus, but only upon the conservation of energy and the mass–energy relation $E = mc^2$.

The problem was that, even at the high temperatures of stellar interiors, the Coulomb repulsion between protons and nuclei is so great that, according to classical physics, protons could not penetrate the nucleus and so this energy source could not be tapped. The solution of this problem had to await the theory of quantum mechanical tunnelling (Gamow, 1928; Gurney and Condon, 1928). One year later, Robert Atkinson and Fritz Houtermans applied

Gamow's theory to the physics of nuclear reactions in the hot central regions of stars (Atkinson and Houtermans, 1929). By considering the process of barrier penetration by a Maxwellian distribution of protons, they established two key features of the process of nuclear energy generation in stars. First, the most effective energy sources involve interactions with nuclei of small electric charge since the Coulomb barriers are lower than for nuclei with large charges. Second, the particles which can penetrate the Coulomb barriers are those few particles in the high energy tail of the Maxwellian distribution. As a result, nuclear reactions can take place at temperatures which are considerably lower than might have been expected. These ideas also suggested why the luminosity of the stars should be a sensitive function of temperature. As the temperature increases, the rate of barrier penetration increases exponentially and so hotter, more massive stars should be more luminous than less massive stars.

Atkinson's objective was to account for the origin of the chemical elements by the successive addition of protons to nuclei. He argued that the process of forming helium by the combination of four protons was very unlikely and proposed instead that helium could be formed by the successive addition of protons to heavier nuclei which, when they became too massive for nuclear stability, would eject α-particles and so create helium (Atkinson, 1931a,b). This proposal was the precursor of the carbon-nitrogen-oxygen (CNO) cycle, which was discovered independently by Carl von Weizsäcker and Hans Bethe in 1938 (Weizsäcker, 1937, 1938; Bethe, 1939). In this cycle, carbon acts as a catalyst for the formation of helium through the successive addition of protons accompanied by two β^+ decays as follows:

$$^{12}C + p \rightarrow {}^{13}N + \gamma \; ; \; {}^{13}N \rightarrow {}^{13}C + e^+ + \nu_e \; ; \; {}^{13}C + p \rightarrow {}^{14}N + \gamma$$
$$^{14}N + p \rightarrow {}^{15}O + \gamma \; ; \; {}^{15}O \rightarrow {}^{15}N + e^+ + \nu_e \; ; \; {}^{15}N + p \rightarrow {}^4He + {}^{12}C.$$

In the meantime, it had become possible to make estimates of the reaction rates for the simplest nuclear reaction, the combination of pairs of protons to form deuterium nuclei which can then combine with other deuterons to form ^3He and ^4He. The first calculations were carried out by Atkinson in 1936 (Atkinson, 1936) and were much refined in 1938 by Bethe and Critchfield who combined Fermi's theory of weak interactions with Gamow's theory of barrier penetration (Bethe and Critchfield, 1938). The principal series of reactions in the proton-proton (or p-p) chain are as follows:

$$p + p \rightarrow {}^2H + e^+ + \nu_e \qquad {}^2H + p \rightarrow {}^3He + \gamma$$
$$^3He + {}^3He \rightarrow {}^4He + 2p.$$

The crucial first reaction in the chain involves a weak interaction in which a positron and neutrino are released in what may be thought of as the transformation of one of the protons into a neutron. This reaction accounts for most of the energy release in the p-p chain but it has never been measured experimentally at the energies of interest for nucleosynthesis in the Sun. Bethe and Critchfield showed that this series of reactions could account for the luminosity of the Sun. In addition, they found that the rate of energy production ε of the p-p chain depends upon the central temperature of the star as $\varepsilon \propto T^4$. In 1939, Bethe worked out the corresponding energy production rate for the CNO cycle and found a very much stronger dependence, $\varepsilon \propto T^{17}$ (Bethe, 1939). He concluded that the CNO cycle was dominant in massive stars while the p-p chain was the principal energy source for stars with mass less than roughly the mass of the Sun, $M_\odot$.

The theory of white dwarfs was one of the first triumphs of the new quantum theory of statistical mechanics as applied to astrophysics. In 1926, Ralph Fowler used Fermi–Dirac statistics to derive the equation of state of a cold degenerate electron gas (Fowler, 1926) and found the important result

$$p = \frac{(3\pi^2)^{2/3}}{5} \frac{\hbar^2}{m_e} \left(\frac{\rho}{\mu_e m_u} \right)^{5/3}, \tag{18.8}$$

where μ_e is the mean molecular weight of the material of the star per electron and m_u is the unified atomic mass constant. The important aspect of this equation of state is that it is independent of temperature and so the structure of white dwarfs can be derived directly from the Lane–Emden equation of stellar structure.[6] Unlike main sequence stars, in which pressure support is provided by the thermal pressure of hot gas, the white dwarfs are supported by electron degeneracy pressure. The source of their luminosity is the internal thermal energy with which they were endowed on formation. According to Fowler's picture, the white dwarfs simply radiate away their internal thermal energies and end up as inert cold stars with all the nuclei and electrons in their ground states.

In 1929, Wilhelm Anderson showed that the degenerate electrons in the centres of white dwarfs with mass roughly that of the Sun become relativistic (Anderson, 1929). In the extreme relativistic limit, the equation of state of the degenerate electron gas becomes

$$p = \frac{(3\pi^2)^{1/3} \hbar c}{4} \left(\frac{\rho}{\mu_e m_u} \right)^{4/3}. \tag{18.9}$$

Once again, the result is independent of temperature but the change in the dependence of pressure upon density from $p \propto \rho^{5/3}$ to $p \propto \rho^{4/3}$ has profound implications. Anderson and Edmund Stoner realised that the consequence was that there do not exist equilibrium configurations for degenerate stars with mass greater than about the mass of the Sun (Anderson, 1929; Stoner, 1929). The most famous analysis of this result was carried out by Subrahmanyan Chandrasekhar, who had begun working on this problem before he arrived to take up a fellowship at Trinity College, Cambridge in 1930. He found the crucial result that, in the extreme relativistic limit, there is an upper limit to the mass of stable white dwarfs,

$$M_{Ch} = \frac{(3\pi)^{3/2}}{2} \left(\frac{\hbar c}{G} \right)^{3/2} \times \frac{2.01824}{(\mu_e m_u)^2} = \frac{5.836}{\mu_e^2} M_\odot. \tag{18.10}$$

This mass is known as the *Chandrasekhar mass* (Chandrasekhar, 1931). The critical mass depends upon the chemical composition of the material of the star through the value of μ_e, the mean molecular weight of the stellar material per electron. Other than that, the Chandrasekhar mass only depends upon fundamental constants. Since $\mu_e \approx 2$ for the material of compact stars, the Chandrasekhar mass is usually quoted as $M_{Ch} = 1.46 M_\odot$. The cause of the instability is that, in the extreme relativistic limit, both the internal thermal energy U_{th} and the gravitational potential energy U_{grav} of the star depend upon the radius in the same way, $U_{th} = (1/2)U_{grav} \propto R^{-1}$. Now, the gravitational potential energy is proportional to M^2 whereas the thermal energy is proportional to the mass of the star and so, for massive enough stars, the gravitational energy term dominates causing collapse, which cannot be stabilised by the pressure of the degenerate gas since the two energies always depend upon the radius in the same way. The inference is that there is nothing to prevent degenerate stars more massive than M_{Ch} from collapsing to very high densities indeed and possibly to a state of complete gravitational collapse.

Quite independently in 1932, Lev Landau had come to the conclusion that gravitational collapse to a singularity should be taken seriously (Landau, 1932) and in 1938 Robert Oppenheimer and Hartland Snyder gave the first general relativistic analysis of what would be observed in the final stages of the gravitational collapse of a pressureless sphere (Oppenheimer and Snyder, 1939). In their paper, they described the key observed features of what are now termed *black holes*.

Following Chadwick's discovery of the neutron in 1932 (Chadwick, 1932), the first mention of the possibility of neutron stars appears as the famous 'Additional Remark' to a paper by Walter Baade and Fritz Zwicky of 1934 (Baade and Zwicky, 1934b). In that year, they published two papers on the energetics of what they termed 'super-novae'. In their first paper, Baade and Zwicky proposed that the population of novae consists of two types, the ordinary novae, which are relatively common phenomena and which had been used by Lundmark as distance indicators for spiral nebulae, and the super-novae which are very rare but very energetic indeed (Baade and Zwicky, 1934a). In their second paper, they suggested that such events might be the sources of the cosmic rays, discovered by Victor Hess in 1912 (Hess, 1913). Both proposals are remarkably close to the truth. As an addendum to the second paper (Baade and Zwicky, 1934b), they wrote

> 'With all reserve we advance the view that a super-nova represents the transition of an ordinary star into a *neutron star*, consisting mainly of neutrons. Such a star may possess a very small radius and an extremely high density. As neutrons can be packed much more closely than ordinary nuclei and electrons, the "gravitational packing" energy in a *cold* neutron star may become very large, and under certain circumstances, may far exceed the ordinary nuclear packing fractions. A neutron star would therefore represent the most stable configuration of matter as such. The consequences of this hypothesis will be developed in another place, where also will be mentioned some observations that tend to support the idea of stellar bodies made up mainly of neutrons.'

It is best to allow Zwicky to describe how these ideas were received in a quotation from the extraordinary preface to his *Catalogue of Selected Compact Galaxies and of Post-Eruptive Galaxies* of 1968 (Zwicky, 1968).

> 'In the *Los Angeles Times* of January 19, 1934, there appeared an insert in one of the comic strips, entitled "Be Scientific with Ol'Doc Dabble" quoting me as having stated "Cosmic rays are caused by exploding stars which burn with a fire equal to 100 million suns and then shrivel from $\frac{1}{2}$ million miles diameter to little spheres 14 miles thick", Says Prof. Fritz Zwicky, Swiss Physicist. This, in all modesty, I claim to be one of the most concise triple predictions ever made in science. More than 30 years were to pass before this statement was proved to be true in every respect.'

In the meantime, Gamow showed in 1937 that a gas of neutrons could be compressed to a much higher density than a gas of nuclei and electrons and estimated the probable densities of such a star to be about 10^{17} kg m^{-3} (Gamow, 1937, 1939). The issue of the maximum mass of neutron stars was discussed by Landau in 1938 (Landau, 1938) and in much greater detail by Oppenheimer, Robert Serber and George Volkoff (Oppenheimer and Serber, 1938; Oppenheimer and Volkoff, 1939). The physics is the same as in the case of the white dwarfs but now neutron degeneracy pressure holds up the star. Complications arise because it is necessary to take into account the details of the equation of state of neutron matter at nuclear densities and the effects of general relativity can no longer be neglected. They found an upper mass limit of about 0.7 $M_\odot$. This result is not so different from the best modern estimates which correspond to about 2–3 $M_\odot$.

This work created some theoretical interest but little enthusiasm from the observers. The radii of typical neutron stars were expected to be about 10 km and so there was no prospect of detecting significant fluxes of thermal radiation from such tiny stars. The idea of Baade and Zwicky that a neutron star might be the remnant left behind after a supernova explosion was proved correct 33 years later with the discovery of pulsars by Antony Hewish, Jocelyn Bell(-Burnell) and their colleagues in 1987 (Hewish et al., 1968).

Epilogue

As expected when I started out on this project, this has proved to be a complex and, at times, difficult story. After all, what was involved was tearing up the foundations of classical physics, which had been extraordinarily successful in explaining the macroscopic world about us, and replacing it by something radically different and non-intuitive in terms of our everyday experience. But the effort involved has been more than repaid by the very much deeper appreciation I have gained of the extraordinary works of the pioneers of quantum mechanics, both the theorists and the experimenters. If the brilliant theoretical researches of Planck, Einstein, Bohr, Heisenberg, Born, Jordan, Schrödinger, Pauli, Dirac and many others form the central core of this story, it should be remembered that their researches were inspired by the equally brilliant achievements of experimental physics. Another huge bonus has been a deepened understanding of quantum mechanics itself – if only I had these insights more than 50 years ago when I first encountered the subject.

There is a great deal more that could be said. I must reiterate that I have presented a somewhat streamlined version of the story in order to ensure that there is some continuous pathway, however tortuous, to the way in which the new understandings came about. For a full appreciation of the complexity of the story and the numerous blind alleys and diversions which took place, there is no substitute for in-depth absorption in Mehra and Rechenberg's magisterial exposition of the history of quantum theory. Likewise, Jammer's wonderful survey of the conceptual and philosophical development of the subject is indispensable. I make no claim even to begin to approach the depth and thoroughness of these expositions. Building on these authors' achievements, my much more modest aim has been to demonstrate at an accessible level exactly what the great pioneers actually did and to enable readers to reconstruct for themselves the extraordinary efforts of the imagination involved in coming to these new understandings. In Chandrasekhar's words, the appreciation of the subject cannot be achieved with 'modest effort'.

The story has an immediacy which corresponds exactly with my own experience as a research scientist. It reveals how experimental and theoretical physicists go about discovering new laws of nature and the way in which these then shape our world. I have concentrated upon the intellectual, theoretical and experimental side of the story, but there is an equally absorbing human story which is the subject of the numerous excellent biographies of the protagonists of this story. My admiration for all the founders of quantum mechanics has increased enormously through the detailed study of their original works. I personally find it a dramatic and awe-inspiring adventure which changed forever our understanding of the nature of the physical world we live in.

Notes

Chapter 1

1. Pippard's review provides an illuminating account of the profession of the experimental and theoretical physicist in 1900 (Pippard, 1995). Equally important are the statistics in the essay by Heilbron (1977) concerning the central role of European physicists, particularly those in Germany, at the turn of the century.
2. Many more details of these issues and problems are contained in the various case studies described in the second edition of my book *Theoretical Concepts in Physics* (*TCP2*) (Longair, 2003). The various topics will only be summarised here with references to the relevant sections of *TCP2*.
3. I have enjoyed revisiting my old undergraduate chemistry textbook *General and Inorganic Chemistry* by P. J. Durrant (1952). I have used the formulation of the laws as expressed by Durrant.
4. In the modern SI system, the carbon-12 atom is used as the standard for atomic weights. One mole is defined as the mass of a substance which contains the same number of chemical units (atoms or molecules) as exactly 12 grams of carbon-12. For example, since the molecular weight of oxygen is 31.9988, one mole of oxygen has mass 31.9988 gram.
5. See *TCP2*, pp. 257–263. The outline of Maxwell's derivation is as follows. The total number of molecules is N and the x, y and z components of their velocities are v_x, v_y and v_z. Maxwell supposed that the velocity distribution in the three orthogonal directions must be the same after a great number of collisions, that is,

$$Nf(v_x)\,dv_x = Nf(v_y)\,dv_y = Nf(v_z)\,dv_z,$$

where f is the same function. The three perpendicular components of the velocity are entirely independent and hence the number of molecules with velocities in the range v_x to $v_x + dv_x$, v_y to $v_y + dv_y$ and v_z to $v_z + dv_z$ is

$$Nf(v_x)f(v_y)f(v_z)\,dv_x\,dv_y\,dv_z$$

But, the total velocity of any molecule v is $v^2 = v_x^2 + v_y^2 + v_z^2$. Because large numbers of collisions have taken place, the probability distribution for the total velocity v, $\phi(v)$, must be isotropic and depend only on v, that is,

$$f(v_x)f(v_y)f(v_z) = \phi(v) = \phi[(v_x^2 + v_y^2 + v_z^2)^{1/2}]. \tag{1}$$

where we have normalised the function $\phi(v)$ so that

$$\int_{-\infty}^{\infty}\int_{-\infty}^{\infty}\int_{-\infty}^{\infty} \phi(v)\,dv_x\,dv_y\,dv_z = 1.$$

Equation (1) is a *functional equation*. By inspection, a suitable solution is

$$f(v_x) = C\,e^{Av_x^2}, \quad f(v_y) = C\,e^{Av_y^2}, \quad f(v_z) = C\,e^{Av_z^2}.$$

where C and A are constants. The distribution must converge as $v \to \infty$ and hence A must be negative. After normalisation (1.1) is obtained.
6. *TCP2* Sect. 10.5.

7. See *TCP2*, p. 266, where the origin of the term 'Maxwell's demon' is explained.

8. *TCP2*, Chap. 4.

9. For the sake of clarity of exposition, I adopt SI units throughout this text. In Maxwell's great papers of 1861 and 1865, the equations were written out in a somewhat lengthier form than (1.5)–(1.8) and involved seven equations. Maxwell's equations were reorganised into their more familiar form by Heaviside, Helmholtz and Hertz in the subsequent years.

10. *TCP2*, pp. 289–297.

11. *TCP2*, pp. 400–403.

12. The reasons for the neglect of Voigt's analysis are discussed by Ernst and Hsu (2001) who point out that the general scale-invariant transformations are

$$\begin{cases} t' = \kappa\gamma\left(t - \dfrac{Vx}{c^2}\right), \\ x' = \kappa\gamma(x - ct), \qquad \gamma = \left(1 - \dfrac{V^2}{c^2}\right)^{-1/2}. \\ y' = \kappa y, \\ z' = \kappa z, \end{cases} \tag{2}$$

Voigt set the constant $\kappa = \gamma^{-1}$. Much later, long after the papers by Lorentz and Einstein, Lorentz, Weichert, Minkowski, Born and Sommerfeld all acknowledged that Voigt had found the form of the Lorentz transforms in 1887. Not only is the equation scale invariant, it is also conformally invariant.

13. This is demonstrated in my book *High Energy Astrophysics*, third edition, pp. 149–151 and Fig. 5.4.

14. *TCP2*, pp. 404–405.

15. *TCP2*, Sect. 11.2.

16. Balmer's paper of 1885 published in the *Annalen der Physik und Chemie* was a synthesis of two papers originally published in the *Verhandlungen der Naturforschenden Gesellschaft in Basel* **7**, 548–560 and 750–752.

17. *TCP2*, Sect. A9.2. Maxwell's relations provide partial differential relations between the four thermodynamic coordinates p, V, S and T.

18. This result is derived in *TCP2*, pp. 297–300.

Chapter 2

1. In his excellent book *Inward Bound*, Pais (1985) provides a detailed account of the development of experimental physics and its discoveries through the whole period covered by this book.

2. As noted in Sect. 1.7.2, at the microscopic level, the system comes into thermal equilibrium, whatever the properties of the walls of the enclosure as a result of the principle of detailed balance.

3. Quoted by M. J. Klein (1967) p. 3.

4. The argument also works for perfectly absorbing walls since, according to Kirchhoff's theorem, a perfect absorber is also a perfect emitter of radiation. Planck uses the case of perfectly reflecting walls so that the argument is purely electrodynamic.

5. Notice that this intensity is the total power per unit area from 4π steradians. The usual definition of intensity is in terms of W m^{-2} Hz^{-1} sr^{-1}. Correspondingly, in (2.19), the relation between $I(\omega)$ and $u(\omega)$ is $I(\omega) = u(\omega)c$ rather than the usual relation $I(\omega) = u(\omega)c/4\pi$.

6. See note 5 above about the relation between intensity and energy density.

7. I have given a detailed analysis of Rayleigh's paper in *TCP2*, Sect. 12.6 and only summarise the key results here.

8. See *TCP2*, Sect. 12.4. Wien's law (2.29) can be written

$$u(\nu) = \frac{8\pi\alpha}{c^3}\nu^3 e^{-\beta\nu/T} \,.$$

Combining this with the relation (2.26) between $u(\nu)$ and E, $u(\nu) = (8\pi\nu^2/c^3)E$,

$$E = \alpha\nu\, e^{-\beta\nu/T} \,.$$

The thermodynamic relation between the entropy S, internal energy U, the pressure p and the volume V is

$$T\, dS = dU + p\, dV \,.$$

Therefore, placing the oscillator in a fixed volume V,

$$\left(\frac{\partial S}{\partial U}\right)_V = \frac{1}{T} \,. \tag{3}$$

U and S are additive functions of state and hence the above relation refers to the properties of an individual oscillator as well as to an ensemble of them. Therefore,

$$\frac{1}{T} = \left(\frac{\partial S}{\partial E}\right)_V = -\frac{1}{\beta\nu}\ln\left(\frac{E}{\alpha\nu}\right) \,. \tag{4}$$

Integrating with respect to E, we obtain Planck's definition of the entropy of the oscillator

$$S = -\frac{E}{\beta\nu}\ln\frac{E}{\alpha\nu e} \,. \tag{5}$$

9. See *TCP2*, Sect. 12.5.
10. See *TCP2*, Sect. 13.3.
11. See *TCP2*, Sect. 10.7.
12. The simplest way of deriving this relation is to run ahead of our story and consider the radiation within the enclosure to consist of particles travelling in random directions at the speed of light. Then, according to classical kinetic theory, the number arriving at unit area of the enclosure per unit time is $\frac{1}{4}nc$. The total energy of particles arriving at unit area is therefore $\frac{1}{4}nc\varepsilon = \frac{1}{4}uc$. In thermodynamic equilibrium, this must also equal the radiated energy per unit surface area.
13. See Klein (1967), p. 17.

Chapter 3

1. These papers are conveniently available in English translations in *Einstein's Miraculous Year*, edited and introduced by John Stachel (1998). The translations are taken from *The Collected Papers of Albert Einstein: Vol. 2. The Swiss Years: Writings, 1900–1909* (Stachel and Cassidy, 1999), which can also be strongly recommended.
2. See *TCP2*, Sect. 14.4.

Chapter 4

1. More details of the contributions of Zeeman and Lorentz are given by Kox (1997).
2. See the rules for electric dipole radiation in Sect. 2.3.1.
3. See Sect. 9.2.1 of *High Energy Astrophysics* (Longair, 2011).
4. Thomson's formula is similar to that derived for the interaction of cosmic ray electrons with thermal electrons. The formula is derived in Sect. 6.1 of *High Energy Astrophysics* (Longair, 2011).

5. A derivation of the formulae for Rutherford scattering is contained in the Appendix to Chapter 2 of *TCP2* (Longair, 2003).

Chapter 5

1. Different authors user different notation for the quantum numbers. For consistency, I will endeavour to use a consistent notation throughout, but it will be different from those appearing in the primary literature. As we will find, the quantum numbers have slightly different meanings in the final version of quantum mechanics and so using n_ϕ, n_r and n_θ in the old quantum theory should not cause problems and should clarify the differences between the old and new theories.
2. I have preserved the italic and bold fonts of the original text.
3. See, for example, Goldstein (1950).
4. There are four transformations between the p_i, q_i, P_i and Q_i. Two of them are given in (5.75) and (5.76). The other two are:

$$q_i = -\frac{\partial S}{\partial p_i}, \quad P_i = -\frac{\partial S}{\partial Q_i} \quad \text{and} \quad K = H + \frac{\partial S}{\partial t}, \tag{6}$$

and

$$q_i = -\frac{\partial S}{\partial p_i}, \quad Q_i = \frac{\partial S}{\partial P_i} \quad \text{with} \quad K = H + \frac{\partial S}{\partial t}. \tag{7}$$

5. This follows the notation of Goldstein (1950) who uses S in the time-dependent version of the Hamilton–Jacobi equations and W in the time-independent version. W is referred to as *Hamilton's characteristic function*.
6. In the quantum mechanics of Heisenberg and Schrödinger, the principal quantum number is denoted by n, rather than $(n_r + n_\theta + m)$. As will become apparent, the significance of n in defining the stationary states in the final theory of quantum mechanics is different from that in the Bohr–Sommerfeld picture. The other quantum numbers have corresponding equivalences which will be described in due course.

Chapter 6

1. The term 'photon' meaning quantum of light was introduced by Gilbert N. Lewis in 1926 (Lewis, 1926).
2. See, for example, *High Energy Astrophysics*, Sect. 9.2.2 (Longair, 2011).
3. Note that, in Bohr's usage, ω is the frequency of orbital motion, not the angular frequency. This usage is adopted in (6.25) and (6.26).
4. Note that the quantum defects in (1.18) and (1.19)–(1.21) are written with a plus sign. The difference between these and (6.63) is explained by adopting different initial values for the principal quantum number n.

Chapter 7

1. Epstein was a pupil of Sommerfeld's and, because he was a Polish citizen, he was interned as an enemy alien in Munich, during which period of internment, he carried out these pioneering calculations.
2. Schwarzschild's paper was published on the day he died, 11 May 1916, from the illness pemphigus contracted while he was on military service on the eastern front in Russia.
3. The details of the development of these ideas are described by Mehra and Rechenberg (Mehra and Rechenberg, 1982a) (Vol. 1, Sect. IV.4) and summarised by Jammer (Jammer, 1989).

Chapter 8

1. Details of Niels Bohr's life and times are contained in the books *Niels Bohr: A Centenary Volume* edited by A. P. French and P. S. Kennedy (1985) and *Niels Bohr's Times, in Physics, Philosophy and Polity* by A. Pais (1991)
2. See Mehra and Rechenberg (1982a), **1**, 230.
3. The Wolfskehl Prize was awarded to Andrew Wiles on 28 June 1997 almost a century after Wolfskehl's death. Wiles completed his proof in 1995 but it took a further two years of expert analysis to confirm that the theorem had at last been proved.
4. A priority dispute ensued between the French and Danish physicists about the discovery of hafnium. This caused Bohr considerable distress (see, for example, the discussion by Kragh (1985)).
5. The history of X-ray spectroscopy is presented in the book *Fifty Years of X-ray Diffraction* (ed. P. P. Ewald) (1962). This book is available on-line in pdf form from the International Union of Crystallography at http://www.iucr.org/publ/50yearsofxraydiffraction.
6. Van der Waerden has made a careful study of why Pauli did not press ahead and propose the spin and magnetic moment of the electron when he was so very close to their discovery (van der Waerden, 1960).
7. More details of this episode and its context are given in Sect. 16.1.

Chapter 9

1. See Mehra and Rechenberg (1982a), p. 511.
2. These results are derived in Chapter 9 of my book *High Energy Astrophysics* (Longair, 2011).
3. The differences were to find an explanation in quantum mechanics and are associated with the symmetries of the wavefunctions for particles of different spins (see Chap. 16 and Sect. 18.1).
4. An excellent concise treatment of the derivation of the Bose–Einstein distribution is given by Huang in his book *Introduction to Statistical Physics* (Huang, 2001).
5. In terms of quantum mechanics, the explanation for this distinction is neatly summarised by Huang who remarks: 'The classical way of counting in effect accepts all wave functions regardless of their symmetry properties under the interchange of coordinates. The set of acceptable wave functions is far greater than the union of the two quantum cases [the Fermi–Dirac and Bose–Einstein cases].'
6. In his paper, de Broglie uses the principle of least action in the form of *Maupertuis' principle* which can be applied in the case of a particle moving in a stationary potential (de Broglie, 1924b). This is discussed in more detail in Sect. 14.5.1.
7. For a demonstration of this result, see *Theoretical Concepts in Physics*, Sect. 7.3 (Longair, 2003).

Chapter 10

1. The term 'quantum mechanics' first appeared in the paper by Born (1924) which is discussed later in this chapter. Jammer (1989) discusses the origin of this term and its significance.
2. See, for example, Sect. 5.3.2 of *Theoretical Concepts in Physics* (Longair, 2003).
3. This result is demonstrated in, for example, Sect. 6.11 of *Theoretical Concepts in Physics* (Longair, 2003).
4. Mehra and Rechenberg (1982a) provide a detailed account of the history of appointments to the chairs of mathematics and physics at Göttingen (see pp. 262–313). Their discussion tracks the various routes through the academic profession from doctoral student, to Habilitation, to Privatdozent, to extraordinary professor and finally to the most prestigious position as ordinary professor. Because of a prohibition on appointing professorships by internal promotion from Privatdozent, the winning of a professorship normally meant moving from one University to another. This accounts for the somewhat complex movements of physicists and mathematicians as they made their way up through the university profession.
5. Mehra and Rechenberg, *op. cit.*, Vol. 1, Part 1, p. 275.

Chapter 11

1. Fermi's problems with Heisenberg's paper are described by van der Waerden (1967). Weinberg's concerns are expressed in his book *Dreams of a Final Theory* (Weinberg, 1992). He writes

   'If the reader is mystified at what Heisenberg was doing, he or she is not alone. I have tried several times to read the paper that Heisenberg wrote on returning from Heligoland, and, although I think I understand quantum mechanics, I have never understood Heisenberg's motivation for the mathematical steps in his paper. Theoretical physicists in their most successful work tend to play one of two roles: they are either *sages* or *magicians* ... It is usually not difficult to understand the papers of the sage-physicists, but the papers of the magician-physicists are often incomprehensible. In that sense, Heisenberg's 1925 paper was pure magic.'

2. The issues of the notation used by Heisenberg are discussed by van der Waerden (1967) who shows how they are related to the notation used in the paper by Kramers and Heisenberg (1925) and to his correspondence with Kronig and Pauli.
3. We obtain exact agreement if we set $|x_0| = |x_0|/2$ (see Mehra and Rechenberg (1982b), p. 301). The justification for this step is given in Sect. 11.5, equation (11.70).
4. As pointed out by Aitchison *et al.* (2004), Heisenberg confusingly uses the terminology $a(n, n - \tau)$ in both (11.40) and (11.70) without explaining why the factor of 4 disappears from the latter expression.
5. The expression (11.104) is identical to the result found in the quantum theory of rotation, the n being replaced by the angular momentum quantum number j. Therefore,

$$W = \frac{h^2}{8\pi^2 I} \left[j(j + 1) + \tfrac{1}{2} \right] . \tag{8}$$

Chapter 12

1. I enjoyed revisiting the elementary text book by A.C. Aitken *Determinants and Matrices* (1959) from which I learned matrix algebra.
2. The term *libration* means periodic perturbations to the elliptical orbits of planets in celestial mechanics and of electrons according to the old quantum theory.
3. I have given an elementary proof of this relation in Appendix A4 of my text *Theoretical Concepts in Physics* (Longair, 2003).
4. In Sect. IV.5 of Volume 3 of their series, *The Historical Development of Quantum Theory. The Formulation of Matrix Mechanics and its Modifications* 1925–1926, Mehra and Rechenberg (1982c) provide a clear summary of the contents of Pauli's important paper.

Chapter 13

1. Poisson's theorem is the statement that, for arbitrary functions F_1, F_2 and F_3,

$$[[F_1, F_2], F_3] + [[F_2, F_3], F_1] + [[F_3, F_1], F_2] = 0 , \tag{9}$$

   where the square brackets represent Poisson brackets.
2. I use the convention that Poisson brackets are contained within square brackets. In the literature, they are often enclosed in round brackets.
3. In Dirac's paper, he uses h to mean what we would now call $\hbar = h/2\pi$. Likewise, when he uses the word frequency, he means what we would call angular frequency ω. I have made these translations in the usage employed throughout this book and also converted to SI units.

4. Pauli's paper was received by *Zeitschrift für Physik* on 17 January 1926 and published on 27 March 1926. Dirac's paper was received by *Philosophical Transactions of the Royal Society* on 22 January 1926 and published on 1 March 1926.

Chapter 14

1. These papers and others were published in English in the book *Collected Papers on Wave Mechanics* (Schrödinger, 1928).
2. The origin of the Bose–Einstein condensation can be understood from the following arguments. The Bose–Einstein distribution can be written

$$n_k = \frac{g_k}{e^{\alpha + \beta \varepsilon_k} - 1} \,, \tag{10}$$

where the constants α and β are to be determined. On thermodynamics grounds, $\beta = 1/kT$ and α is determined by fixing the number of photons, atoms or molecules. For black-body radiation, the number of photons is unconstrained and so $\alpha = 0$. The number of photons is matched to the total energy in the Planck distribution – all the properties of the radiation are determined by the temperature T alone.

In the case of the atoms of a monatomic gas, as shown in Sect. 14.2, $g_k = 4\pi p^2 \, dp$ and $E = p^2/2m$. Therefore, converting to a continuous energy distribution,

$$N(E) \, dE = \frac{g_k}{(Be^{E/kT} - 1)} = \frac{4\pi V (2m^3 E)^{1/2}}{h^3 (Be^{E/kT} - 1)} \, dE \,, \tag{11}$$

where $B = e^\alpha$. The constant B cannot be less than 1 or else the number density of particles could be negative. Therefore, $B \geq 1$, $\alpha \geq 0$. This expression can be rewritten as follows,

$$N(E) \, dE = \frac{V}{4\pi^2} \left(\frac{2m}{\hbar^2} \right)^{3/2} \frac{E^{1/2}}{(Be^{E/kT} - 1)} \, dE \,, \tag{12}$$

where $\hbar = h/2\pi$. Therefore, the total number of particles is given by

$$N = \frac{V}{4\pi^2} \left(\frac{2m}{\hbar^2} \right)^{3/2} \int_0^\infty \frac{E^{1/2}}{(Be^{E/kT} - 1)} \, dE = \frac{V}{4\pi^2} \left(\frac{2mkT}{\hbar^2} \right)^{3/2} \int_0^\infty \frac{y^{1/2} \, dy}{(Be^y - 1)}$$

$$N = V \left(\frac{mkT}{2\pi \hbar^2} \right)^{3/2} F(B) \,, \tag{13}$$

where $y = E/kT$ and

$$F(B) = \frac{2}{\sqrt{\pi}} \int_0^\infty \frac{y^{1/2} \, dy}{(Be^y - 1)} \,. \tag{14}$$

A plot of $F(B)$ as a function of B is shown in Fig. 1. For large values of B, the function $F(B)$ tends to $1/B$ and the particle distribution to the Boltzmann distribution. The figure also shows that $F(B)$ tends to a finite limit as $B \to 1$. For $B = 1$, the integral for $F(B)$ becomes

$$F(B) = \frac{2}{\sqrt{\pi}} \int_0^\infty \frac{y^{1/2} \, dy}{(e^y - 1)} = \zeta(3/2) = 2.612 \,, \tag{15}$$

where ζ is the zeta function. There is therefore a critical temperature T_B at which $F(B)$ reaches this finite value given by

$$T_B = \frac{2\pi \hbar^2}{mk} \left(\frac{N}{2.612 V} \right)^{2/3} \,. \tag{16}$$

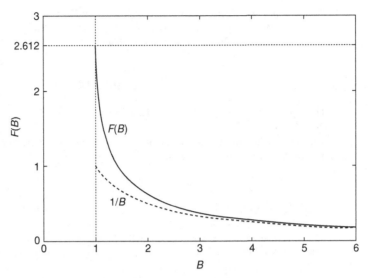

Fig. 1   The function $F(B)$ as a function of B.

Einstein appreciated that this result led to a strange behaviour of the particle distribution at temperatures below T_B. According to (11), if N is fixed, we would expect that, as the temperature decreases, $F(B)$ would increase, but Bose–Einstein statistics show that this increase does not continue below T_B. If $F(B)$ reaches the limiting value of 2.612, the number of particles N must decrease at temperatures lower than T_B.

Einstein realised that what had gone wrong was that the continuum approximation whereby we replaced g_i by $N(E)\,dE$ does not take account of the fact that the quantised zero energy state can contain a finite number of particles. The solution was that, at temperatures below T_B, the particles accumulate in the zero energy state and that at low enough temperatures, all the particles would be condensed into that state. This was the discovery of the *Bose–Einstein condensation*. Although it was not realised at the time, this was the first demonstration of a phase transition using the techniques of statistical mechanics of identical particles.

3. In 1927, Willem Keesom and Mieczyslaw Wolfke discovered the strange variation of the specific heat capacity of helium at the *lambda point* at 2.19 K which was identified as a phase transition between the phases known as He I and He II (Keesom and Wolfke, 1928). In 1938, Fritz London proposed interpreting this phase transition as the onset of the Bose–Einstein condensation (London, 1938). Taking the number density of liquid helium at the λ-point to be $2.18 \times 10^{28}\ \mathrm{m}^{-3}$, $T_B = 3.1$ K. There were however concerns about this identification since the expression (14) for T_B was derived for an ideal gas, whereas the phase transition is observed in liquid helium.

4. Throughout this chapter, I have converted Schrödinger's notation of quantum numbers to modern usage. Thus, in Schrödinger's notation, the angular quantum number l is written as n.

5. We need to evaluate

$$\delta J = \delta \iiint dx\,dy\,dz \left[\left(\frac{\partial \psi}{\partial x} \right)^2 + \left(\frac{\partial \psi}{\partial y} \right)^2 + \left(\frac{\partial \psi}{\partial z} \right)^2 - \frac{2m}{K^2} \left(E + \frac{Ze^2}{4\pi\epsilon_0 r} \right) \psi^2 \right] = 0 \quad (17)$$

where the integral is taken over all space. In Sect. 5.4.2, we derived the procedures for finding the stationary values of the Lagrangian $\mathcal{L}$ with respect to time. We can reformulate these procedures for the present problem for a single variable q_i by replacing $\mathcal{L}(\dot{q}_i, q_i, t)$ by $J(\dot{\psi}, \psi, q_i)$, η by $\delta\psi$ and carrying out the variation with respect to spatial coordinate q_i. Making this replacement in (5.57),

$$S = \int_{q_1}^{q_2} J[\dot{\psi}(q_i), \psi(q_i), q_i] \, dq_i + \int_{q_1}^{q_2} \left[\frac{\partial J}{\partial \dot{\psi}(q_i)} \delta \dot{\psi} + \frac{\partial J}{\partial \psi} \delta \psi \right] dq_i \,, \tag{18}$$

where $\dot{\psi}(q_i)$ means $\partial \psi / \partial q_i$. As in Sect. 5.4.2, we can set the first integral equal to S_0 and integrate the second term by parts. Then,

$$S = S_0 + \left[\frac{\partial J}{\partial \dot{\psi}(q_i)} \delta \psi \right]_{q_i(1)}^{q_i(2)} - \int_{\psi_1}^{\psi_2} \left[\frac{d}{dq_i} \left(\frac{\partial J}{\partial \dot{\psi}(q_i)} \right) \delta \psi - \frac{\partial J}{\partial \psi} \delta \psi \right] dq_i \,. \tag{19}$$

In the case discussed in Sect. 5.4.2, the first term in square brackets disappears because the endpoints were assumed to be fixed. In the case of the hydrogen atom, this becomes a condition on the convergence properties of the integral as $r \to \infty$. Let us continue with the integration in the x-direction. The first term of (19) can be evaluated in the x-direction from the terms inside the triple integral in (17) as follows:

$$\left[\frac{\partial J}{\partial \dot{\psi}(x)} \delta \psi \right]_{x_1}^{x_2} = 2 \iint dx \, dy \left(\frac{\partial \psi}{\partial x} \right) \delta \psi = 2 \iint_{y,z} dA \left(\frac{\partial \psi}{\partial n} \right) \delta \psi \,, \tag{20}$$

where dn is the differential element of distance perpendicular to the element of area dA. Extending this calculation to three dimensions, this term becomes

$$2 \oint dA \left(\frac{\partial \psi}{\partial n} \right) \delta \psi. \tag{21}$$

Extending the last integral of (19) to three dimensions,

$$\iiint dx \, dy \, dz \, \delta \psi \left[\frac{\partial}{\partial x} \left(\frac{\partial J}{\partial \dot{\psi}(x)} \right) + \frac{\partial}{\partial y} \left(\frac{\partial J}{\partial \dot{\psi}(y)} \right) + \frac{\partial}{\partial z} \left(\frac{\partial J}{\partial \dot{\psi}(z)} \right) - \frac{\partial J}{\partial \psi} \right]. \tag{22}$$

Therefore the stationary function $\psi(x, y, z)$ is found from the function which minimises the expression

$$\frac{1}{2} \delta S = \oint dA \left(\frac{\partial \psi}{\partial n} \right) \delta \psi$$
$$+ \iiint dx \, dy \, dz \, \delta \psi \left[\frac{\partial}{\partial x} \left(\frac{\partial J}{\partial \dot{\psi}(x)} \right) + \frac{\partial}{\partial y} \left(\frac{\partial J}{\partial \dot{\psi}(y)} \right) + \frac{\partial}{\partial z} \left(\frac{\partial J}{\partial \dot{\psi}(z)} \right) - \frac{\partial J}{\partial \psi} \right]. \tag{23}$$

The two terms must be separately zero, the first being the integral over a sphere as $r \to \infty$ and the second must be true for all small changes $\delta \psi$ about the stationary solution. Therefore, these conditions reduce to the requirements

$$\frac{\partial J}{\partial \psi} = \frac{\partial}{\partial x} \left(\frac{\partial J}{\partial \dot{\psi}(x)} \right) + \frac{\partial}{\partial y} \left(\frac{\partial J}{\partial \dot{\psi}(y)} \right) + \frac{\partial}{\partial z} \left(\frac{\partial J}{\partial \dot{\psi}(x)} \right) \tag{24}$$

and

$$\oint dA \left(\frac{\partial \psi}{\partial n} \right) \delta \psi = 0 \,. \tag{25}$$

Inserting the expression for J from (17) into (24), we immediately find

$$\frac{\partial^2 \psi}{\partial x^2} + \frac{\partial^2 \psi}{\partial y^2} + \frac{\partial^2 \psi}{\partial z^2} + \frac{2m}{K^2} \left(E + \frac{e^2}{4\pi \epsilon_0 r} \right) \psi = 0 \,, \tag{26}$$

or

$$\nabla^2 \psi + \frac{2m}{K^2} \left(E + \frac{e^2}{4\pi \epsilon_0 r} \right) \psi = 0 \,. \tag{27}$$

The expressions (25) and (27) are those derived by Schrödinger in his first paper on wave mechanics (Schrödinger, 1926b).

6. In Schrödinger's paper, he uses the wave equation in the form (14.37) until he derived the energy levels of the hydrogen atom and then infers $K = h/2\pi$. He certainly knew this relation had to be true from the outset of the paper.

7. A simple demonstration of the origin of Maupertuis' principle in one dimension can be obtained from the Euler–Lagrange equation (5.59) which is derived by precisely the same procedure for finding the stationary function between two fixed points. Writing $2T$ for $\mathcal{L}$ in (5.59) and considering only motion in the x-direction,

$$\frac{\partial T}{\partial x} - \frac{\mathrm{d}}{\mathrm{d}t}\left(\frac{\partial T}{\partial \dot{x}}\right) = 0 \,. \tag{28}$$

In a conservative field of force, $T = \frac{1}{2}m\dot{x}^2 = E - U(x)$ and E is the constant total energy of the particle. Inserting these relations into (14.44), we find immediately

$$m\frac{\mathrm{d}^2 x}{\mathrm{d}t^2} = -\frac{\mathrm{d}U(x)}{\mathrm{d}x} = f_x \,, \tag{29}$$

which is Newton's law of motion and hence the proof of Maupertuis' principle.

8. Hamilton's action function is defined to be $S = \int_{t_0}^{t} \mathcal{L} \, \mathrm{d}t$, where $\mathcal{L} = (T - U)$ is the Lagrangian, and defines an *action surface* $S(x, y, z, t) = \text{constant}$. The properties of the action surface are directly related to the dynamics of the particles of the system. These can be most simply appreciated from the exposition by Landau and Lifshitz (1976). Using their notation, the expression for δS (5.58), can be rewritten

$$\delta S = \left[\frac{\partial \mathcal{L}}{\partial \dot{q}_i} \, \mathrm{d}q\right]_{t_1}^{t_2} - \int_{t_1}^{t_2} \left[\frac{\mathrm{d}}{\mathrm{d}t}\left(\frac{\partial \mathcal{L}}{\partial \dot{q}}\right) - \frac{\partial \mathcal{L}}{\partial q}\right] \mathrm{d}q \, \mathrm{d}t \,. \tag{30}$$

Because of Lagrange's equations of motion (5.59), the integral on the right-hand side of (22) is zero and so

$$\delta S = \left[\frac{\partial \mathcal{L}}{\partial \dot{q}_i} \, \mathrm{d}q\right]_{t_1}^{t_2} \,. \tag{31}$$

We now set $\delta q(t_1) = 0$ and let $\delta q(t_2) = \delta q$. Then, since $\partial \mathcal{L}/\partial \dot{q}_i = p$, we find

$$\delta S = p \, \delta q \,, \tag{32}$$

or, summing over all the degrees of freedom,

$$\delta S = \sum_i p_i \, \delta q_i \,. \tag{33}$$

It follows that

$$p_i = \frac{\delta S}{\delta q_i} \,. \tag{34}$$

Thus, the components of the momentum are the gradients of the action surface S in different directions. In general we can write

$$\boldsymbol{p} = \nabla S \,. \tag{35}$$

From the definition of the action given at the beginning of this endnote, the total time derivative is

$$\frac{\mathrm{d}S}{\mathrm{d}t} = \mathcal{L} \,. \tag{36}$$

We can now relate the total derivative of S to the partial derivatives with respect to time and the coordinates q_i by the relations

$$\frac{\mathrm{d}S}{\mathrm{d}t} = \frac{\partial S}{\partial t} + \sum_i \frac{\partial S}{\partial q_i}\dot{q}_i = \frac{\partial S}{\partial t} + \sum_i p_i\dot{q}_i = \mathcal{L}, \tag{37}$$

from (34). Hence,

$$\frac{\partial S}{\partial t} = \mathcal{L} - \sum_i p_i\dot{q}_i, \tag{38}$$

or,

$$\frac{\partial S}{\partial t} = -H. \tag{39}$$

Rewriting this expression in the notation used in (14.48) for a single particle, we find

$$\frac{\partial S}{\partial t} = \mathcal{L} - \boldsymbol{p}\cdot\boldsymbol{v} = -E. \tag{40}$$

Equations (35) and (40) are those used by Hamilton in comparing particle dynamics and the paths of light rays.

9. The simple way of deriving this result, with a great deal of hindsight, is to recognise that the quantisation of angular momentum results in the expression $J^2 = l(l+1)\hbar^2$. Using the fact that the expression for the energy of a stationary state is the same in quantum mechanics as in classical mechanics,

$$E = \frac{J^2}{2I} = \frac{l(l+1)\hbar^2}{2I} = \frac{l(l+1)h^2}{8\pi^2 I}. \tag{41}$$

10. In the mathematical description of the propagation of a wave-packet in a dispersive medium, the pulse shape $A(x)$ is modulated by a 'carrier wave' with wavevector K. The profile of the wave-packet is therefore $f(x, t = 0) = A(x)\,\mathrm{e}^{\mathrm{i}Kx}$. Taking the Fourier transform of the wave-packet, we can write

$$A(x)\,\mathrm{e}^{\mathrm{i}Kx} = \left(\int_{-\infty}^{\infty} B(q)\,\mathrm{e}^{\mathrm{i}qx}\,\mathrm{d}q\right)\mathrm{e}^{\mathrm{i}Kx}. \tag{42}$$

If the dispersion relation for the waves is $\omega(k)$, it is a straightforward calculation to show that, if $q \ll K$, at a later time $t > 0$, the wave-packet has the form

$$\left[\int_{-\infty}^{\infty} \underbrace{B(q)\,\mathrm{e}^{\mathrm{i}q(x-\omega' t)}}_{(ii)}\ \underbrace{\exp(-\tfrac{1}{2}q^2\omega'' t) + \cdots)}_{(iii)}\,\mathrm{d}q\right]\underbrace{\mathrm{e}^{\mathrm{i}(Kx-\omega t)}}_{(i)}, \tag{43}$$

where $\omega' = \mathrm{d}\omega/\mathrm{d}k$ is the group velocity and $\omega'' = \mathrm{d}^2\omega/\mathrm{d}k^2$ is the second derivative of the angular frequency with respect to wavenumber k.

Let us consider the representation of the motion of a uniformly moving particle by a wave-packet. The envelope of the wave-packet is modulated by the carrier frequency which moves at the phase velocity $v_{\mathrm{ph}} = \omega/k$ in the positive x-direction, term (i) in (43). Term (ii) in (43) shows that all the Fourier components of the waveform move at the *group velocity*, $v_{\mathrm{gr}} = \mathrm{d}\omega/\mathrm{d}k$ in the positive x-direction. For a free particle, it is shown in endnote 11 that the dispersion relation is $\omega = \hbar k^2/2m$ and so $v_{\mathrm{gr}} = \mathrm{d}\omega/\mathrm{d}k = \hbar k/m = p/m = v$, where v is the constant speed of the particle. Thus, the wave-packet propagates with the speed of the particle it is to represent.

The term (iii) represents the dispersion of the wave because generally, in classical physics, the group velocity is slightly different at the leading and trailing edges of the wave-packet. For a particle moving at constant speed according to de Broglie's hypothesis, we have shown, however, that $\mathrm{d}\omega/\mathrm{d}k = v = $ constant and so $\mathrm{d}^2\omega/\mathrm{d}k^2 = 0$. In other words, there is no dispersion of the wave-packet which represents a particle moving at constant speed in the x-direction.

11. De Broglie postulated that the wavevector k associated with the momentum of a particle should be $p = \hbar k$ and that the relation between the energy of the particle E and angular frequency ω should be similar to that for photons, $E = \hbar\omega$. For a free particle, the energy is the kinetic energy $E = \frac{1}{2}mv^2 = p^2/2m$. Therefore,

$$E = \hbar\omega = \frac{p^2}{2m} = \frac{\hbar^2 k^2}{2m} , \quad \text{or} \quad \omega = \frac{\hbar k^2}{2m} . \tag{44}$$

This dispersion relationship must be the solution of our quantum mechanical wave equation for a free particle. The expression $\omega = \hbar k^2/2m$ is significantly different from the dispersion relation for light waves, $\omega = ck$. However, the *group velocity*, $v_g = \mathrm{d}\omega/\mathrm{d}k$, is

$$v_g = \frac{\mathrm{d}\omega}{\mathrm{d}k} = \frac{\hbar k}{m} = \frac{p}{m} = v ,$$

the speed of the particle, as discussed in Sect. 14.5.1.

Suppose the wavefunction for a free particle is a sine wave, $\psi(x,t) = A\sin(kx - \omega t)$. Taking the second derivative of ψ with respect to x, the expression for $\sin(kx - \omega t)$ is multiplied by k^2:

$$\frac{\partial^2\psi(x,t)}{\partial x^2} = -Ak^2\sin(kx - \omega t) . \tag{45}$$

To obtain a factor containing ω, we take the first derivative with respect to t

$$\frac{\partial\psi(x)}{\partial t} = -\omega A\cos(kx - \omega t) . \tag{46}$$

We cannot find an equation involving a second derivative with respect to x and a first derivative with respect to t if the solution is to be simply a sine wave of the form $\sin(kx - \omega t)$. If instead the solution for a free particle is to be a complex wave of the form $\psi(x,t) = A\exp[i(kx - \omega t)]$, then, taking partial derivatives with respect to x twice and once with respect to t, we find

$$\frac{\partial^2\psi}{\partial x^2} = -k^2 A\mathrm{e}^{i(kx-\omega t)} = -\frac{p^2}{\hbar^2}\psi , \quad \frac{\partial\psi}{\partial t} = -i\omega A\mathrm{e}^{i(kx-\omega t)} = -\frac{iE}{\hbar}\psi . \tag{47}$$

We can also write $E\psi = p^2/2m\,\psi$ and so

$$-\frac{\hbar^2}{2m}\frac{\partial^2\psi}{\partial x^2} = i\hbar\frac{\partial\psi}{\partial t} , \quad E\psi = i\hbar\frac{\partial\psi}{\partial t} . \tag{48}$$

These relations are of exactly the same form as (14.102) for the time dependence of the wavefunction.

Chapter 15

1. I use the term *Heisenberg approach* to refer to matrix mechanics as developed by Born, Heisenberg and Jordan in their seminal papers of 1925 and 1926 (Born and Jordan, 1925b; Born *et al.*, 1926).

2. Note that Schrödinger does not use a consistent set of suffices for the matrix elements used in this development. I have largely maintained the usage in Schrödinger's paper.

3. This remark was made as part of a recorded interview with Born on 17 October 1962. The interview is preserved in the *Archive for the History of Quantum Physics* (see Jammer (1989)).

4. This letter was translated by van der Waerden and appears with a commentary and analysis in his paper *From matrix mechanics and wave mechanics to unified quantum mechanics* (van der Waerden, 1973).

5. The method had been previously known to mathematical physicists, prominent among them being Harold Jeffreys. For this reason, a 'J' is often added to WKB and the order of the initials may vary from country to country.

Chapter 16

1. See Mehra and Rechenberg (1987), p. 736.
2. I have given this simple physical derivation in Sect. 13.2.1 of *High Energy Astrophysics* (Longair, 2011).
3. Note that in his papers Dirac takes the Planck's constant to be what is now designated $\hbar$. I have translated his notation into standard contemporary usage.
4. Dirac's remark was made as part of a recorded interview carried out on 7 May 1963. The interview is preserved in the *Archive for the History of Quantum Physics*. See also, Mehra and Rechenberg (1987), p. 767.
5. In his paper, Pauli (1927b) writes the non-relativistic Schrödinger equation as a pair of equations, one for each spin state

$$\left.\begin{array}{l} H\left(\dfrac{h}{2\pi\mathrm{i}}\dfrac{\partial}{\partial q_k}, q_k, \sigma_x, \sigma_y, \sigma_z\right)\psi_{E,\alpha} = E\psi_\alpha\,, \\[2mm] H\left(\dfrac{h}{2\pi\mathrm{i}}\dfrac{\partial}{\partial q_k}, q_k, \sigma_x, \sigma_y, \sigma_z\right)\psi_{E,\beta} = E\psi_\beta\,, \end{array}\right\} \tag{49}$$

the ψ_α and ψ_β corresponding to our ψ_+ and ψ_-.
6. Specifically, on page 69 of Baker's *Principles of Geometry* (1922), a set of four 2×2 matrices is given as an example which could form the basis of a self-consistent non-commutative geometry. To quote Baker, 'Using 0, 1, i, ... let us consider then the four symbols

$$U = \begin{bmatrix} 1 & 0 \\ 0 & 1 \end{bmatrix}, \quad I = \begin{bmatrix} 0 & -1 \\ 1 & 0 \end{bmatrix}, \quad J = \begin{bmatrix} 0 & \mathrm{i} \\ \mathrm{i} & 0 \end{bmatrix}, \quad K = \begin{bmatrix} -\mathrm{i} & 0 \\ 0 & \mathrm{i} \end{bmatrix}; \tag{50}$$

it is then easy to compute that

$$JK = -KJ = I, \; KI = -IK = J, \; IJ = -JI = K\,, \tag{51}$$

$$I^2 = J^2 = K^2 = -U\,.' \tag{52}$$

Evidently, Dirac had to find corresponding matrices which could represent the three orthogonal spin orientations, σ_x, σ_y and σ_z.
7. I am using the conventions used by Rindler (2001) in which the Lorentz transformations are written:

$$\left.\begin{array}{l} ct' = \gamma(ct - Vx/c)\,, \\ x' = \gamma(x - Vt)\,, \\ y' = y\,, \\ z' = z\,, \end{array}\right\} \tag{53}$$

where $\gamma = (1 - v^2/c^2)^{-1/2}$ is the Lorentz factor. The prototype displacement four-vector is $R = [ct, x, y, z]$ and the norm of the displacement four-vector is $\mathrm{norm}(R) = c^2t^2 - x^2 - y^2 - z^2$.
8. The forms of the 4×4 α matrices can be readily derived from Dirac's paper (Dirac, 1928a). They are:

$$\alpha_1 = \alpha_x = \begin{bmatrix} 0 & 0 & 0 & 1 \\ 0 & 0 & 1 & 0 \\ 0 & 1 & 0 & 0 \\ 1 & 0 & 0 & 0 \end{bmatrix}, \quad \alpha_2 = \alpha_y = \begin{bmatrix} 0 & 0 & 0 & -\mathrm{i} \\ 0 & 0 & \mathrm{i} & 0 \\ 0 & -\mathrm{i} & 0 & 0 \\ \mathrm{i} & 0 & 0 & 0 \end{bmatrix},$$

$$\alpha_3 = \alpha_z = \begin{bmatrix} 0 & 0 & 1 & 0 \\ 0 & 0 & 0 & -1 \\ 1 & 0 & 0 & 0 \\ 0 & -1 & 0 & 0 \end{bmatrix}, \quad \alpha_4 = \begin{bmatrix} 1 & 0 & 0 & 0 \\ 0 & 1 & 0 & 0 \\ 0 & 0 & -1 & 0 \\ 0 & 0 & 0 & -1 \end{bmatrix}.$$

The spin matrices introduced in (16.91) immediately follow from the above definitions,

$$\sigma_z = -\mathrm{i}\alpha_x\alpha_y \begin{bmatrix} 0 & 1 & 0 & 0 \\ 1 & 0 & 0 & 0 \\ 0 & 0 & 0 & 1 \\ 0 & 0 & 1 & 0 \end{bmatrix}, \sigma_y = -\mathrm{i}\alpha_z\alpha_x = \begin{bmatrix} 0 & -\mathrm{i} & 0 & 0 \\ \mathrm{i} & 0 & 0 & 0 \\ 0 & 0 & 0 & -\mathrm{i} \\ 0 & 0 & \mathrm{i} & 0 \end{bmatrix},$$

$$\sigma_x = -\mathrm{i}\alpha_y\alpha_z = \begin{bmatrix} 1 & 0 & 0 & 0 \\ 0 & -1 & 0 & 0 \\ 0 & 0 & 1 & 0 \\ 0 & 0 & 0 & -1 \end{bmatrix}.$$

9. More details of the history of cosmic rays can be found in my book *The Cosmic Century*, Sect. 7.2 (Longair, 2006).

Chapter 17

1. This topic is dealt with in much more detail in Sect. 15.3 of my book *Theoretical Concepts in Physics* (Longair, 2003).
2. See the discussion by Jammer (1989), pp. 328–329.
3. The essence of Heisenberg's analysis can be appreciated using Schrödinger's representation of a particle by a wave-packet and then using de Broglie's relations between momentum and wavelength. A *localised wave-packet* can be represented by a superposition of an infinite number of waves. Following Jordan and Heisenberg, let us choose a continuous distribution of waves which has a Gaussian distribution of amplitudes about some central wavenumber k',

$$A(k) = \exp\left[-\frac{(k-k')^2}{2k_1^2}\right]. \tag{54}$$

k_1 is the standard deviation of the amplitudes of the waves of different values of k about the mean value k'. When we dealt with discrete values of k_n, we described the function as a summation of sine or cosine waves. Let us consider the cosine waves for convenience,

$$\psi(x) = \sum_n A_n(k_n)\cos k_n x. \tag{55}$$

Converting this into a continuous distribution of wavenumbers, ψ becomes

$$\psi(x) = \int_{-\infty}^{+\infty} A(k)\cos kx \, \mathrm{d}k. \tag{56}$$

We carry out this integral for the above Gaussian distribution $A(k)$. The integral becomes

$$\int_{-\infty}^{\infty} \exp\left[-\frac{(k-k')^2}{2k_1^2}\right]\cos kx \, \mathrm{d}k. \tag{57}$$

We write $\cos kx = \Re(\mathrm{e}^{\mathrm{i}kx})$ where the symbol $\Re$ means 'take the real part of'. Therefore, the integral becomes

$$\Re\left\{\int_{-\infty}^{\infty} \exp\left[-\frac{(k-k')^2}{2k_1^2} + \mathrm{i}kx\right] \mathrm{d}k\right\}. \tag{58}$$

Note that this expression has exactly the same form as (17.49) and explains the origin of the complex terms in Jordan's expression for the probability amplitude. Now, let us deal with the expression inside the exponential. To integrate the expression, we need to complete the square and so we write

$$-\frac{(k-k')^2}{2k_1^2} + ikx = -\frac{k^2 - 2kk' + k'^2 - 2ikxk_1^2}{2k_1^2}$$

$$= -\frac{k^2 - 2k[k' + ixk_1^2] + [k' + ixk_1^2]^2}{2k_1^2} + \frac{2ik'xk_1^2 - x^2k_1^4}{2k_1^2}$$

$$= -\frac{[k-k'-ixk^2]^2}{2k'^2} + ik'x - \frac{x^2k'^2}{2}. \tag{59}$$

Therefore, we have to integrate

$$\int_{-\infty}^{\infty} \exp\left\{-\frac{[k-k'-ixk_1]^2}{2k_1}\right\} \times \exp(ik') \exp\left[-\frac{x^2k_1^2}{2}\right] dk. \tag{60}$$

Notice that the terms after the 'times' sign do not contain the variable k and so these terms can be taken outside the integral. Now, we change variables to

$$y = \frac{k - k' - ixk_1^2}{\sqrt{2}k_1}, \quad dk = \sqrt{2}k_1\, dy. \tag{61}$$

Thus, the integral reduces to

$$\exp(ik'x) \exp\left[-\frac{x^2k_1^2}{2}\right] \sqrt{2}k_1 \int_{-\infty}^{\infty} e^{-y^2}\, dy. \tag{62}$$

But, the integral

$$\int_{-\infty}^{\infty} e^{-y^2}\, dy = \sqrt{\pi}.$$

Therefore, taking the real part of the expression in front of the definite integral, we find that the answer is

$$\psi(x) = \Re\left\{\exp(ik'x)\exp\left[-\frac{x^2k_1^2}{2}\right]\sqrt{2\pi}k_1\right\} \tag{63}$$

$$= A\cos k'x\, \exp\left[-\frac{x^2}{2x_1^2}\right], \tag{64}$$

where $x_1 = k_1^{-1}$. Thus, the function $\psi(x)$ has a Gaussian envelope with standard deviation x_1 which is related to the spread in wavenumbers k_1 by the relation $x_1 = k_1^{-1}$. According to de Broglie's relation $k = 2\pi/\lambda = 2\pi h/p_1$ and so

$$x_1 p_1 = \frac{h}{2\pi}, \tag{65}$$

exactly Heisenberg's relation (17.51).

4. These are the rules which we teach our students in their first serious introduction to quantum mechanics. I am grateful to my colleagues Prof. Michael Payne and Dr. Howard Hughes for permission to reproduce these statements which were at the heart of their lecture courses on quantum mechanics.

Chapter 18

1. I give a number of simple examples of symmetry and conservation laws in Sect. 7.5 of my book *Theoretical Concepts in Physics* (Longair, 2003) starting from the Euler–Lagrange equations.
2. An excellent summary of the early history of quantum tunnelling is given by Merzbacher (2002).

3. A delightful account of the history of the Cockcroft and Walton experiment is contained in the book by Brian Cathcart *The Fly in the Cathedral: How a Small Group of Cambridge Scientists Won the Race to Split the Atom* (Cathcart, 2005)

4. This translation is taken from Pais's book *Inward Bound* in which many more details of this complex story are recounted (Pais, 1985).

5. The section is an abbreviated version of sections of Chap. 3 and 4 of my book *The Cosmic Century* (2006).

6. This equation is derived in Sect. 13.2.2 in my book *High Energy Astrophysics* (Longair, 2011) where it is used to derive the expression (18.10).

References

Abraham, M. (1903). Principien der Dynamik des Elektron, *Annalen der Physik*, **10**, 105–179.

Aitchison, I. J. R., MacManus, D. A., and Snyder, T. M. (2004). Understanding Heisenberg's 'magical' paper of July 1925: A new look at the calculational details, *American Journal of Physics*, **72**, 1370–1379.

Aitkin, A. (1959). *Determinants and Matrices*, ninth edition. Edinburgh and London: Oliver and Boyd Limited. The first edition was published in 1939.

Allen, H. S. and Maxwell, R. S. (1952). *A Text-book of Heat. Vol. II.* London: MacMillan and Co.

Anderson, C. and Neddermeyer, S. (1936). Cloud chamber observations of cosmic rays at 4300 metres elevation and near sea-level, *Physical Review*, **50**, 263–271.

Anderson, C. D. (1932). The apparent existence of easily deflected positives, *Science*, **76**, 238–239.

Anderson, C. D. (1933). The positive electron, *Physical Review*, **43**, 491–494.

Anderson, W. (1929). Gewöhnliche Materie und Strahlende Energie als Verschiedene "Phasen" eines und Desselben Grundstoffes (Ordinary matter and radiation energy as different phases of the same underlying matter), *Zeitschrift für Physik*, **54**, 433–444.

Andrade, E. N. d. C. (1964). *Rutherford and the Nature of the Atom.* New York: Doubleday. The quotation is on page 111.

Arvidsson, G. (1920). Eine Untersuchung über die Ampéreschen Molecularströme nach der Methode von A. Einstein und W. J. de Haas, *Physikalische Zeitschrifte*, **21**, 88–91.

Atkinson, R. d. (1931a). Atomic synthesis and stellar energy I, *Astrophysical Journal*, **73**, 250–295.

Atkinson, R. d. (1931b). Atomic synthesis and stellar energy II, *Astrophysical Journal*, **73**, 308–347.

Atkinson, R. d. (1936). Atomic synthesis and stellar energy III, *Astrophysical Journal*, **84**, 73–84.

Atkinson, R. d. and Houtermans, F. (1929). Zur Frage der Aufbaumöglichkeit der Elemente in Sternen (On the possible synthesis of the elements in stars), *Zeitschrift für Physik*, **54**, 656–665.

Baade, W. and Zwicky, F. (1934a). On super-novae, *Proceedings of the National Academy of Sciences*, **20**, 254–259.

Baade, W. and Zwicky, F. (1934b). Cosmic rays from super-novae, *Proceedings of the National Academy of Sciences*, **20**, 259–263.

Baker, H. F. (1922). *Principles of Geometry. Vol. 1. Foundations.* Cambridge: Cambridge University Press.

Balmer, J. J. (1885). Note on the spectral lines of hydrogen, *Annalen der Physik und Chemie,* **25**, 80–87.

Barkla, C. G. (1906). Polarisation of secondary Röntgen radiation, *Proceedings of the Royal Society of London,* **A77**, 247–255.

Barkla, C. G. (1911a). Note on the energy of scattered X-radiation, *Philosophical Magazine (6),* **21**, 648–652.

Barkla, C. G. (1911b). The spectra of the fluorescent Röntgen radiations, *Philosophical Magazine (6),* **22**, 396–412.

Barkla, C. G. and Ayres, T. (1911). The distribution of secondary X-rays and the electro-magnetic pulse theory, *Philosophical Magazine (6),* **21**, 275–278.

Barkla, C. G. and Sadler, C. A. (1908). Homogeneous secondary Röntgen radiations, *Philosophical Magazine (6),* **16**, 550–584.

Barnett, S. J. (1915). Magnetization by rotation, *Physical Review (2),* **6**, 239–270.

Barnett, S. J. and Barnett, L. H. (1922). Improved experiments on magnetization by rotation, *Physical Review (2),* **20**, 90–91.

Bates, L. F. (1969). Edmund Clifton Stoner, *Biographical Memoirs of Fellows of the Royal Society,* **15**, 201–237.

Beck, E. (1919a). Zum experimentellen Nachweis der Ampèreschen Molekularströme, *Annalen der Physik (4),* **60**, 109–148.

Beck, E. (1919b). Zum experimentellen Nachweis der Ampèreschen Molekularströme, *Physikalische Zeitschrift,* **20**, 490–491.

Becquerel, H. (1896). Sur les Radiations Invisibles émises par les corps phosphorescents (On the invisible radiation emitted by phosphorescent bodies), *Comptes Rendus de l'Academie des Sciences,* **122**, 501–503.

Bethe, H. (1930). Zur Theorie des Durchgangs schneller Korpuskularstrahlen durch Materie, *Annalen der Physik,* **5**, 375–400.

Bethe, H. (1939). Energy production in stars, *Physical Review,* **55**, 434–456.

Bethe, H. and Critchfield, C. (1938). The formation of deuterons by proton combination, *Physical Review,* **54**, 248–254.

Blackett, P. M. S. and Occhialini, G. P. S. (1933). Some photographs of the tracks of penetrating radiation, *Proceedings of the Royal Society of London,* **A139**, 699–722.

Blatt, F. (1992). *Modern Physics.* New York: McGraw-Hill.

Bloch, F. (1976). Reminiscences of Heisenberg and the early days of quantum mechanics, *Physics Today,* **29**, 23–27.

Bôcher, M. (1911). *Introduction to Higher Algebra.* New York: Macmillan Publishing Company. The English version was published in 1907, the German translation being published in 1911: Leipzig: B.G. Teubner.

Bohm, D. (1951). *Quantum Theory.* New York: Prentice-Hall. The book was republished in 1989 by Dover Publications.

Bohr, N. (1912). Unpublished memorandum for Ernest Rutherford, in *On the Constitution of Atoms and Molecules,* 1963, ed. Rosenfeld, L. Copenhagen: Munksgaard.

Bohr, N. (1913a). On the constitution of atoms and molecules (Part I), *Philosophical Magazine (6)*, **26**, 1–25.

Bohr, N. (1913b). On the constitution of atoms and molecules. Part II. Systems containing only a single electron, *Philosophical Magazine (6)*, **26**, 476–502.

Bohr, N. (1913c). On the constitution of atoms and molecules. Part III. Systems containing several nuclei, *Philosophical Magazine (6)*, **26**, 857–875.

Bohr, N. (1913d). On the spectra of helium and hydrogen, *Nature*, **92**, 231–232.

Bohr, N. (1914). On the effect of electric and magnetic fields on spectral lines, *Philosophical Magazine (6)*, **27**, 506–524.

Bohr, N. (1915). On the quantum theory of radiation and the structure of the atom, *Philosophical Magazine (6)*, **30**, 394–415.

Bohr, N. (1918a). On the quantum theory of line spectra. Part I. On the general theory, *Mathematisk-Fysiske Meddelelser, Det Kgl. Danske Videnskabernes Selskab: Skrifter 8*, **4.1**, 1–36. Reprinted in *Collected Works*, **3**, 67–102.

Bohr, N. (1918b). On the quantum theory of line spectra. Part II. On the hydrogen spectrum, *Mathematisk-Fysiske Meddelelser, Det Kgl. Danske Videnskabernes Selskab: Skrifter 8*, **4.1**, 37–100. Reprinted in *Collected Works*, **3**, 103–166.

Bohr, N. (1921a). Atomic structure, *Nature*, **107**, 104–107.

Bohr, N. (1921b). Atomic structure, *Nature*, **108**, 208–209.

Bohr, N. (1922). The structure of the atom, in *Nobel Lectures 1922–1941*, pp. 7–43. Amsterdam: Elsevier Publishing Company. This volume was published in 1965.

Bohr, N. (1928). The quantum postulate and the recent development of quantum theory, in *Atti del Congresso Internazionale dei Fisica, Como–Pavia–Roma*, **2**, pp. 565–588. Bologna: Zanichelli. The essence of the paper was also published in *Nature*, **121**, 580–590, 1928.

Bohr, N. (1963). Introduction by L. Rosenfeld, in *On the Constitution of Atoms and Molecules*, ed. Rosenfeld, L. Copenhagen: Munksgaard.

Bohr, N. (1977). Wolfskehl lectures, in *Niels Bohr Collected Works, Vol. 4. The Periodic System (1920–1923)*, ed. Nielsen, J. R., pp. 341–419. Amsterdam: North Holland Publishing Company.

Bohr, N., Kramers, H. A., and Slater, J. C. (1924). The quantum theory of radiation, *Philosophical Magazine (6)*, **47**, 785–822.

Boltzmann, L. (1884). Ableitung des Stefan'schen Gesetzes, betreffend die Abhängigkeit der Wärmestrahlung von der Temperatur aus der electromagnetischen Lichttheorie, *Annalen der Physik (3)*, **22**, 291–294.

Born, M. (1923). Atomtheorie des festen Zustandes (Dynamik der Kristallgitter), *Encyklopädie der mathematischen Wissenschaften mit Einschluss ihrer Anwendungen*, **3**, 527–781.

Born, M. (1924). Über quantenmechanik (On quantum mechanics), *Zeitschrift für Physik*, **26**, 379–395.

Born, M. (1926a). Zur quantenmechanik der Stossvorgänge, *Zeitschrift für Physik*, **37**, 863–867.

Born, M. (1926b). Quantenmechanik der Stossvorgänge, *Zeitschrift für Physik*, **38**, 803–827.

Born, M. (1926c). Das Adiabatenprincip in der Quantenmechanik, *Zeitschrift für Physik*, **40**, 167–192.

Born, M. (1926d). Zur Wellenmechanik der Stossvorgänge, *Göttinger Nachrichten*, pp. 146–160.

Born, M. (1961a). Bemerkungen zur statistischen Deutung der Quantenmechanik, in *Werner Heisenberg und die Physik seiner Zeit*, ed. Hermann, A., pp. 103–108. Stuttgart: Deutsche Verlags-Anstalt. This book was published in 1977 but contains a reprint of Born's remarks of 1961.

Born, M. (1961b). Erwin Schrödinger, *Physikalische Blatter*, **17**, 85–86.

Born, M. (1978). *My Life: Recollections of a Nobel Laureate*. London: Taylor and Francis; New York: Charles Scribner's Sons.

Born, M. and Heisenberg, W. (1923a). Über Phazenbeziehungen bei den Bohrschen Modellen von Atomen und Molekeln, *Zeitschrift für Physik*, **14**, 44–55.

Born, M. and Heisenberg, W. (1923b). Die Elektronenbahnen im angeregten Heliumatom, *Zeitschrift für Physik*, **16**, 229–243.

Born, M., Heisenberg, W., and Jordan, P. (1926). Zur Quantenmechanik. II, *Zeitschrift für Physik*, **35**, 557–615.

Born, M. and Jordan, P. (1925a). Zur Quantentheorie aperiodischer Vorgänge, *Zeitschrift für Physik*, **33**, 479–505.

Born, M. and Jordan, P. (1925b). Zur Quantenmechanik, *Zeitschrift für Physik*, **34**, 858–888.

Born, M. and von Kármán, T. (1912). Über Schwingungen von Raumgittern, *Physikalische Zeitschrift*, **13**, 297–309.

Born, M. and Wiener, N. (1926). Eine neue Formulierung der Quantumgesetze für periodische und nichtperiodische Vorgänge, *Zeitschrift für Physik*, **36**, 174–187. The paper was simultaneously published in English as *A new formulation of the laws of quantization of periodic and aperiodic phenomena* in the *MIT Journal of Mathematics and Physics*, **5**, 84–98.

Bose, S. N. (1924). Planck's Gesetz und Lichtquantenhypothese (Planck's law and the hypothesis of light quanta), *Zeitschrift für Physik*, **26**, 178–181.

Bothe, W. and Becker, H. (1930). Künstliche Erregungen von Kern γ-Strahlen, *Zeitschrift für Physik*, **66**, 289–306.

Bothe, W. and Geiger, H. (1924). Ein Weg zur experimentellen Nachprürung der Theorie von Bohr, Kramers und Slater, *Zeitschrift für Physik*, **26**, 44–44.

Bothe, W. and Kolhörster, W. (1929). The nature of the high-altitude radiation, *Zeitschrift für Physik*, **56**, 751–777.

Bragg, W. H. and Bragg, W. L. (1913a). The reflection of X-rays from crystals, *Proceedings of the Royal Society of London (A)*, **88**, 428–438.

Bragg, W. H. and Bragg, W. L. (1913b). The reflection of X-rays from crystals, *Proceedings of the Royal Society of London (A)*, **89**, 246–248.

Brillouin, L. (1926). La méchanique ondulatoire de Schrödinger; une méthod générale de résolution par approximations successives, *Comptes Rendus*, **183**, 24–26.

Burgers, J. M. (1916). Adiabatische invarienten bij mechanische systemen (Adiabatic invariants of mechanical systems), *Verslag van de gewone Vergadering der wis-en natuurdige Afdeeling, Koniklijle Akadamie van Wetenschappen te Amsterdam*, **25**, 849–857,

918–922, 1055–1061. English translation published in *Proceedings of the Amsterdam Academy*, **20**, 149–157, 158–162, 163–169.

Campbell, N. R. (1920). Atomic structure, *Nature*, **106**, 408–409.

Cathcart, B. (2005). *The Fly in the Cathedral: How a Small Group of Cambridge Scientists Won the Race to Split the Atom*. London: Viking.

Chadwick, J. (1932). The existence of a neutron, *Proceedings of the Royal Society of London*, **A136**, 692–708.

Chandrasekhar, S. (1931). The maximum mass of ideal white dwarfs, *Astrophysical Journal*, **74**, 81–82.

Charlier, C. V. L. (1902). *Die Mechanik des Himmels*. Liepzig: Veit.

Clausius, R. (1857). Über die Art der Bewegung, die wir Wärme nennen" (On the nature of the motion, which we call heat), *Annalen der Physik*, **176**, 353–380.

Clausius, R. (1858). Ueber die mittlere Länge der Wege, welche bei der Molecularbewegung gasförmiger Körper von den einzelnen Molecülen zurückgelegt werden; nebst einigen anderen Bemerkungen über die mechanische Wärmetheorie. (On the average length of paths which are traversed by single molecules in the molecular motion of gaseous bodies), *Annalen der Physik*, **181**, 239–258.

Cockcroft, J. D. and Walton, E. T. S. (1932). Experiments with high velocity positive ions. II. The disintegration of elements by high velocity protons, *Proceedings of the Royal Society of London*, **A137**, 229–242.

Compton, A. H. (1922). The spectrum of secondary rays, *Physical Review (2)*, **19**, 267–268.

Compton, A. H. (1923). A quantum theory of the scattering of X-rays by light elements, *Physical Review (2)*, **21**, 483–502.

Compton, A. H. (1961). The scattering of X rays as particles, *American Journal of Physics*, **29**, 817–820.

Compton, A. H. and Simon, A. W. (1925). Directed quanta of scattered X-rays, *Physical Review (2)*, **26**, 289–299.

Cornu, A. (1898). Sur queques résultats nouveaux relatifs au phénomène découvert par M. le Dr. Zeeman, *Comptes Rendus (Paris)*, **126**, 181–186.

Courant, R. and Hilbert, D. (1924). *The Methods of Mathematical Physics, Vol. 1*. Berlin: Springer-Verlag.

Curie, I. and Joliot, F. (1932). The emission of high energy photons from hydrogenous substances irradiated with very penetrating alpha rays, *Comptes Rendus (Paris)*, **194**, 273–275.

Curie, M. P. and Skłodowska-Curie, M. (1898). On a new radioactive substance contained in pitchblende, *Comptes Rendus (Paris)*, **127**, 175–178.

Curie, M. P., Skłodowska-Curie, M., and Bémont, G. (1898). On a new, strongly radioactive substance, contained in pitchblende, *Comptes Rendus (Paris)*, **127**, 1215–1217.

Dalton, J. (1808). *A New System of Chemical Philosophy*. Manchester: R. Bickerstaff.

Darwin, C. G. (1922). A quantum theory of optical dispersion, *Nature*, **110**, 841–842.

Darwin, C. G. (1927a). The electron as a vector wave, *Nature*, **119**, 282–284.

Darwin, C. G. (1927b). The Zeeman effect and spherical harmonics, *Proceedings of the Royal Society of London*, **A115**, 1–19.

Darwin, C. G. (1927c). The electron as a vector wave, *Proceedings of the Royal Society of London*, **A116**, 227–233.

Darwin, C. G. (1927d). Free motion in wave mechanics, *Proceedings of the Royal Society of London*, **A117**, 258–293.

Davisson, C. and Germer, L. H. (1927). Diffraction of electrons by a crystal of nickel, *Physical Review*, **30**, 705–740.

Davisson, C. and Kunsman, C. H. (1921). The scattering of electrons by nickel, *Science*, **54**, 522–524.

de Broglie, L. (1923a). Ondes et Quanta, *Comptes Rendus (Paris)*, **177**, 507–510.

de Broglie, L. (1923b). Quanta de lumière, diffraction et interférence, *Comptes Rendus (Paris)*, **177**, 548–550.

de Broglie, L. (1923c). Les quanta, la théorie cinétique de gaz et la principe de Fermat, *Comptes Rendus (Paris)*, **177**, 630–632.

de Broglie, L. (1924a). *Recherches sur la théorie des quanta: doctoral dissertation*. Paris: Masson et Cie.

de Broglie, L. (1924b). A tentative theory of light quanta, *Philosophical Magazine*, **47**, 446–458.

Debye, P. (1912). Zur Theorie der spezifischen Wärme (On the theory of specific heats), *Annalen der Physik (4)*, **39**, 789–839. English translation: *Collected Papers of Peter J. W. Debye*, 1954, pp. 650–696. New York: Interscience Publishers.

Debye, P. (1916). Quantenhypothese und Zeeman-Effekt, *Physikalische Zeitschrift*, **17**, 507–512.

Debye, P. (1923). Zerstreuung von Röntgenstrahlen und Quantentheorie, *Physikalische Zeitschrift*, **24**, 161–166.

Delaunay, C.-E. (1860). Théorie du mouvement de la lune, *Mémoires de l'Académie des Sciences (Paris)*, **28**.

Delaunay, C.-E. (1867). Théorie du mouvement de la lune, *Mémoires de l'Académie des Sciences (Paris)*, **29**.

Dirac, P. A. M. (1925). The fundamental equations of quantum mechanics, *Proceedings of the Royal Society of London*, **A109**, 642–653.

Dirac, P. A. M. (1926a). *Quantum mechanics*. Cambridge: PhD dissertation.

Dirac, P. A. M. (1926b). On quantum algebra, *Mathematical Proceedings of the Cambridge Philosophical Society*, **23**, 412–418.

Dirac, P. A. M. (1926c). Relativity quantum mechanics with an application to Compton scattering, *Proceedings of the Royal Society of London*, **A110**, 405–423.

Dirac, P. A. M. (1926d). Quantum mechanics and a preliminary investigation of the hydrogen atom, *Proceedings of the Royal Society of London*, **A110**, 561–579.

Dirac, P. A. M. (1926e). The elimination of nodes in quantum mechanics, *Proceedings of the Royal Society of London*, **A111**, 281–305.

Dirac, P. A. M. (1926f). On the theory of quantum mechanics, *Proceedings of the Royal Society of London*, **A112**, 661–677.

Dirac, P. A. M. (1926g). The physical interpretation of the quantum dynamics, *Proceedings of the Royal Society of London*, **A113**, 621–641.

Dirac, P. A. M. (1927). The quantum theory of emission and absorption of radiation, *Proceedings of the Royal Society of London*, **A114**, 243–265.

Dirac, P. A. M. (1928a). The quantum theory of the electron, *Proceedings of the Royal Society of London*, **A117**, 610–624.

Dirac, P. A. M. (1928b). The quantum theory of the electron, *Proceedings of the Royal Society of London*, **A118**, 351–361.

Dirac, P. A. M. (1930a). *Principles of Quantum Mechanics*. Oxford: Clarendon Press.

Dirac, P. A. M. (1930b). A theory of electrons and protons, *Proceedings of the Royal Society of London*, **A126**, 360–375.

Dirac, P. A. M. (1931). Quantized singularities in the electromagnetic field, *Proceedings of the Royal Society of London*, **A133**, 60–72.

Dirac, P. A. M. (1939). A new notation for quantum mechanics, *Proceedings of the Cambridge Philosophical Society*, **35**, 416–418.

Dirac, P. A. M. (1947). *Principles of Quantum Mechanics*, third edition. Oxford: Clarendon Press.

Dirac, P. A. M. (1958). *Principles of Quantum Mechanics*, fourth edition. Oxford: Clarendon Press.

Dirac, P. A. M. (1977). Recollections of an exciting era, in *History of Twentieth Century Physics: 57th Varenna International School of Physics, 'Enrico Fermi'*, ed. Weiner, C., pp. 109–146. New York and London: Academic Press.

Ditchburn, R. W. (1930). The uncertainty principle in quantum mechanics, *Proceedings of the Royal Irish Academy*, **39**, 73–80.

Drude, P. (1900). Zur Geschichte der elektromagnetischen Dispersionsgleichungen, *Annalen der Physik (4)*, **1**, 437–440.

Durrant, P. J. (1952). *General and Inorganic Chemistry*. London: Longman, Green and Co.

Dyson, F. (1999). Why is Maxwell's theory so hard to understand?, in *James Clerk Maxwell Commemorative Booklet*, pp. 8–13. Edinburgh: James Clerk Maxwell Foundation.

Eckart, C. (1926a). Operator calculus and the solution of the equations of quantum dynamics, *Physical Review (2)*, **28**, 711–726.

Eckart, C. (1926b). The solution of the problem of the simple oscillator by a combination of the Schrödinger and the Lanczos theories, *Proceedings of the National Academy of Sciences of the USA*, **12**, 473–476.

Eddington, A. (1920). The internal constitution of the stars, *Observatory*, **43**, 341–358.

Eddington, A. S. (1924). *The Mathematical Theory of Relativity*. Cambridge: Cambridge University Press.

Ehrenfest, P. (1913). Een mechanische theorema van Boltzmann en zinje betrekking tot de quanta theorie (A mechanical theorem of Boltzmann and its relation to the theory of quanta), *Verslag van de gewone Vergadering der wis-en natuurdige Afdeeling, Koniklijle Akadamie van Wetenschappen te Amsterdam*.

Ehrenfest, P. (1916). Over adiabatische veranderingen van stelsel in verband met de theorie der quanta (On adiabatic changes of a system in connection with the quantum theory), *Verslag van de gewone Vergadering der wis-en natuurdige Afdeeling, Koniklijle Akadamie van Wetenschappen te Amsterdam*.

Ehrenfest, P. (1927). Bemerkung über die angenäherte Gültigkeit der klassischen Mechanik innerhalb der Quantenmechanik, *Zeitschrift für Physik*, **45**, 455–457.

Einstein, A. (1905a). Über einen die Erzeugung und Verwandlung des Lichtes betreffenden heuristischen Gesichtspunkt (On a heuristic point of view concerning the production and transformation of light), *Annalen der Physik (4)*, **17**, 132–148.

Einstein, A. (1905b). Über die von der molekularkinetischen Theorie der Wärme geforderte Bewegung von in ruhenden Flüssigkeiten suspendierten Teilchen (On the motion of small particles suspended in stationary liquids required by the molecular-kinetic theory of heat), *Annalen der Physik (4)*, **17**, 549–560.

Einstein, A. (1905c). Zur Elektrodynamik bewegter Körper (On the electrodynamics of moving bodies), *Annalen der Physik (4)*, **17**, 891–921.

Einstein, A. (1906a). Zur Theorie der Brownschen Bewegung (On the theory of Brownian motion), *Annalen der Physik (4)*, **19**, 371–381.

Einstein, A. (1906b). Zur Theorie der Lichterzeugung und Lichtabsorption (On the theory of light emission and absorption), *Annalen der Physik (4)*, **20**, 199–206.

Einstein, A. (1906c). Die Plancksche Theorie der Strahlung und die Theorie der spezifischen Wärme, *Annalen der Physik (4)*, **22**, 180–190.

Einstein, A. (1909). Zum gegenwärtigen Stand des Strahlungsproblems, *Physikalische Zeitschrift*, **10**, 185–193.

Einstein, A. (1910). Letter to Laub of 16 March 1910, in Kuhn, T. S., *Black-Body Theory and the Quantum Discontinuity 1894–1912*, Oxford: Clarendon Press. See also Einstein Archives: Princeton.

Einstein, A. (1912). L'Etat Acutal du Probléme des Chaleurs Spécifiques, in *The Theory of Radiation and Quanta: Proceedings of the First Solvay Conference*, eds Langevin, P. and de Broglie, M., pp. 407–437. Paris: Gautier-Villars. The interchange between Einstein and Lorentz is recorded in the discussion of Einstein's paper on page 450.

Einstein, A. (1916). Quantentheorie der Strahlung, *Mitteilungen der Physikalischen Gesellschaft Zürich*, **16**, 47–62. Also published in *Physikalische Zeitschrift*, **18**, 121–128, 1917.

Einstein, A. (1924). Quantentheorie des einatomigen idealen Gases, *Sitzungberichte der (Kgl.) Preussischen Akademie der Wissenschaften (Berlin)*, pp. 261–267.

Einstein, A. (1925). Quantentheorie des einatomigen idealen Gases, *Sitzungberichte der (Kgl.) Preussischen Akademie der Wissenschaften (Berlin)*, pp. 3–14.

Einstein, A. (1949). Autobiographisches–Autobiographical Notes, in *Albert Einstein: Philosopher Scientist*, ed. Schilpp, P. A., pp. 1–95. New York: Tudor Publishing Company. The quotation is on pp. 45–47.

Einstein, A. (1979). *Autobiographical Notes*. La Salle, Illinois; Open Court. This is the fourth reprint of Einstein' autobiographical notes edited by P. A. Schilpp.

Einstein, A. (1993). Letter to Conrad Habicht of May 1905: item No. 27, in *The Collected Papers of Albert Einstein: Vol. 5. The Swiss Years: Correspondence, 1902–1914 (English translation supplement)*, eds Klein, M. J., Kox, A. J., and Schulman, R. Princeton: Princeton University Press.

Einstein, A. and de Haas, W. J. (1915). Experimenteller Nachweis der Ampéreschen Molekularströme, *Verhandlungen der Deutschen Physikalischen Gesellschaft (2)*, **17**, 152–170.

Einstein, A. and Ehrenfest, P. (1922). Quantentheoretische Bemerkungen zum Experiment von Stern und Gerlach, *Zeitschrift für Physik*, **11**, 31–34.

Ellis, C. D. and Wooster, W. A. (1927). The average energy of disintegration of radium E, *Proceedings of the Royal Society*, **A117**, 109–123.

Elsasser, W. (1925). Bemerkungen zur Quantenmechanik freier Elektronen, *Die Naturwissenschaften*, **13**, 711.

Epstein, P. S. (1916a). Zur Theorie des Starkeffektes (On the theory of the Stark effect), *Annalen der Physik (4)*, **50**, 489–520.

Epstein, P. S. (1916b). Zur Theorie des Starkeffektes (On the theory of the Stark effect), *Physikalische Zeitschrift*, **17**, 148–150.

Ernst, A. and Hsu, J. (2001). First proposal of the universal speed of light by Voigt in 1887, *Chinese Journal of Physics*, **39**, 211–230.

Everitt, C. W. F. (1975). *James Clerk Maxwell: Physicist and Natural Philosopher*. New York: Charles Scribner's Sons. See also, *Dictionary of Scientific Biography*, Vol. 9, 198–230. New York: Charles Scribner's Sons.

Ewald, P. P. (1962). *Fifty Years of X-ray Diffraction*. Utrecht, The Netherlands: N. V. A. Oosthoek's Uitgeversmaatschappij. This volume was edited by Ewald. It includes articles by and biographies of many of the pioneers of X-ray spectroscopy.

Farmelo, G. (2009). *The Strangest Man: The Hidden Life of Paul Dirac, Quantum Genius*. London: Faber and Faber.

Faxén, H. and Holtsmark, J. (1927). Beitrag zur Theorie des ganges langsamer Elektronen durch Gase, *Zeitschrift für Physik*, **45**, 307–324.

Fermi, E. (1926a). Sulla quantizzazione del gas perfetto monatomico, *Rendiconti del Reale Accademia Lincei (3)*, **3**, 145–149.

Fermi, E. (1926b). Zur Quantelung des idealen einatomigen Gases, *Zeitschrift für Physik*, **36**, 902–912.

Fermi, E. (1934). Versuch einer Theorie der β-Strahlen. I, *Zeitschrift für Physik*, **88**, 161–177.

Fermi, E., Amaldi, E., D'Agostino, O., Rasetti, F., and Segré, E. (1934). Radioattività provocata da bombardamento di neutroni III, *La Recherca Scientifica*, **5**, 452–453.

Fierz, M. (1939). Über die relativistische Theorie kräftefreier Teilchen mit beliebigen Spin, *Helvetica Physica Acta*, **12**, 3–37.

Fierz, M. and Pauli, W. (1939). On relativistic wave equations for particles of arbitrary spin in an electromagnetic field, *Proceedings of the Royal Society of London*, **A173**, 211–232.

Fitzgerald, G. F. (1889). The aether and the Earth's atmosphere, *Science*, **13**, 390.

Fitzgerald, G. F. (1902). On Ostward's energetics [1896], in *Scientific Writings*, ed. Larmor, J., p. 388. Dublin: Hodges, Figges; London: Longman; Green.

Foster, J. S. (1930). Some leading features of the Stark effect, *Journal of the Franklin Institute*, **209**, 585–588.

Foucault, L. (1849). Lumière électrique (Electric light), *L'Institut, Journal Universal des Sciences*, **17**, 44–46.

Fowler, A. (1912). Observations of the principal and other series in the spectrum of hydrogen, *Monthly Notices of the Royal Astronomical Society*, **73**, 62–71.

Fowler, A. (1913a). The spectra of hydrogen and helium, *Nature*, **92**, 95–96.

Fowler, A. (1913b). The spectra of hydrogen and helium, *Nature*, **92**, 232–233.

Fowler, R. (1926). On dense matter, *Monthly Notices of the Royal Astronomical Society*, **87**, 114–122.

Franck, J. and Hertz, G. (1914). Über Zussamenstösse zwischen Elektronen und den Molekülen des Quecksilberdampfes und die Ionisierungsspannung desselben, *Verhandlungen der Deutschen Physikalischen Gesellschaft (2)*, **16**, 457–467.

Fraunhofer, J. (1817a). Bestimmung des Brechungs- und des Farbenzerstreuungs-Vermögens Verschiedener Glasarten, in Bezug auf die Vervollkommnung Achromatischer Fernröhre (On the refractive and dispersive power of different species of glass in reference to the improvement of achromatic telescopes, with an account of the lines or streaks which cross the spectrum), *Denkschriften der königlichen Akademie der Wissenschaften zu München*, **5**, 193–226. Translation: *Edinburgh Philosophical Journal*, **9**, pp. 288–299, 1823; **10**, pp. 26–40, 1824.

Fraunhofer, J. (1817b). Bestimmung des Brechungs- und des Farbenzerstreuungs-Vermögens Verschiedener Glasarten, in Bezug auf die Vervollkommnung Achromatischer Fernröhre (On the refractive and dispersive power of different species of glass in reference to the improvement of achromatic telescopes, with an account of the lines or streaks which cross the spectrum), *Gilberts Annalen der Physik*, **56**, 264–313.

French, A. P. and Kennedy, P. J., eds (1985). *Niels Bohr: A Centenary Volume*. Cambridge, Mass: Harvard University Press.

Friedrich, W., Knipping, P. and Laue, M. v. (1912). Interferenz-Erscheinungen bei Röntgenstrahlen (Interference effects with Röntgen rays), *Sitzberichte der Königlich Bayerischen Akademie der Wissenschaften*, pp. 303–312.

Gamow, G. (1928). Zur Quantentheorie des Atomkernes, *Zeitschrift für Physik*, **51**, 204–212.

Gamow, G. (1937). *Atomic Nuclei and Nuclear Transformations*. Oxford: Oxford University Press.

Gamow, G. (1939). Physical possibilities of stellar evolution, *Physical Review*, **55**, 718–725.

Geiger, H. and Marsden, E. (1913). The laws of deflexion of α-particles through large angles, *Philosophical Magazine*, **25**, 604–623.

Geiger, H. and Müller, W. (1928). Das Electronenzählrohr (The electron-counting tube), *Physikalische Zeitschrift*, **29**, 839–841.

Geiger, H. and Müller, W. (1929). Technische Bermerkungen zum Electronenzählrohr (Technical remarks on the electron-counting tube), *Physikalische Zeitschrift*, **30**, 489–493.

Geiger, H. and Nuttall, J. M. (1911). The ranges of the α-particles from various radioactive substances and a relation between range and period of transformation, *Philosophical Magazine*, **22**, 613–621.

Gerlach, W. (1969). Zur Enddeckung des "Stern-Gerlach-Effektes", *Physikalische Blätter*, **25**, 472–472.

Gerlach, W. and Stern, O. (1922a). Der experimentelle Nachweis der Richtungsquantelung im Magnetfeld, *Zeitschrift für Physik*, **9**, 349–352.

Gerlach, W. and Stern, O. (1922b). Das magnetische Moment des Silberatoms, *Zeitschrift für Physik*, **9**, 353–355.

Gibbs, J. W. (1902). *Elementary Principles of Statistical Mechanics*. New York: Charles Scribner's Sons.

Goldstein, H. (1950). *Classical Mechanics*. Reading, Mass: Addison-Wesley.

Goudsmit, S. and Kronig, R. (1925). Dir Intenstität der Zeemancomponenten, *Die Natur-wissenschaften*, **12**, 90.

Gray, J. A. (1913). The scattering and the absorption of the gamma rays of radium, *Philosophical Magazine*, **26**, 611–623.

Gray, J. A. (1920). The scattering X-rays and γ-rays, *Journal of the Frankin Institute*, **190**, 633–655.

Gurney, R. W. and Condon, E. U. (1928). Quantum mechanics and radioactive disintegration, *Nature*, **122**, 439.

Gurney, R. W. and Condon, E. U. (1929). Quantum mechanics and radioactive disintegration, *Physical Review*, **33**, 127–140.

Haas, A. E. (1910a). Der Zusammenhang des Planckschen elementaren Wirkungsquantums mit den Grundgrössen der Elektronentheorie, *Jahrbuch d. Radioaktivität und Elektronik*, **7**, 261–268.

Haas, A. E. (1910b). Über die electrodynamische bedeutung des Planck'schen Strahlungs-gesetzes und über eine neue Bestimmung des elektrischen Elementarquantums unde der Dimensionen des Wasserstoffatoms, *Sitzberichte der Kaiserlichen Akademie der Wissenschaften (Wien). Abteilung II*, **119**, 119–144.

Haas, A. E. (1910c). Über eine neue theoretische Methode zur Bestimmung des elek-trischen Elementarquantums unde des Halbmessers des Wasserstoffatoms, *Physikalische Zeitschrift*, **11**, 537–538.

Hahn, O. and Strassmann, F. (1939). Über den Nachweis und das Verhalten der bei der Bestrahlung des Urans mittels Neutronen entstehenden Erdalkalimetalle, *Natur-wissenschaften*, **27**, 11–15.

Hamilton, W. (1833). On a general method of expressing the path of light, and of the planets, by the coefficients of a characteristic function, *Dublin University Review 1833*, pp. 795–826.

Hamilton, W. R. (1931). Essay on the theory of the systems of rays, in *The Mathematical Papers of Sir William Rowan Hamilton*, eds Conway, A. W. and Synge, J. L., pp. 1–294. These four papers were first published in the *Transactions of the Royal Irish Academy*, **15**, 69–174 (1828), **16**, 1–61 (1830), **16**, 93–125 (1931), **17**, 1–144 (1837).

Hartree, D. R. (1927a). The wave mechanics of the atom with a non-Coulomb central field. Part I. Theory and methods, *Mathematical Proceedings of the Cambridge Philosophical Society*, **24**, 89–110.

Hartree, D. R. (1927b). The wave mechanics of the atom with a non-Coulomb central field. Part II. Some results and discussion, *Mathematical Proceedings of the Cambridge Philosophical Society*, **24**, 111–132.

Heilbron, J. (1977). Lectures on the history of atomic physics 1900–1922, in *History of Twentieth Century Physics: 57th Varenna International School of Physics, 'Enrico Fermi'*, ed. Weiner, C., pp. 40–108. New York and London: Academic Press.

Heisenberg, W. (1925). Über die quantentheoretische Umdeutung kinematischer und mech-anischer Beziehungen (Quantum-theoretical re-interpretation of kinematic and mechan-ical relations), *Zeitschrift für Physik*, **33**, 879–893.

Heisenberg, W. (1926a). Mehrkörperproblem und Resonanz in der Quantenmechanik, *Zeitschrift für Physik*, **38**, 411–426.

Heisenberg, W. (1926b). Über die Spektra von Atomsystemen mit zwei Electronen, *Zeitschrift für Physik*, **39**, 499–518.

Heisenberg, W. (1927). Über den anschaulichen Inhalt der quantentheoretischen Kinematik und Mechanik, *Zeitschrift für Physik*, **43**, 172–198. An English translation is included in the volume *Quantum Theory and Measurement*, eds J. A. Wheeler and W. H. Zurek, 62–84, Princeton: Princeton University Press, 1983.

Heisenberg, W. (1928). Zur Theorie des Ferromagnetismus, *Zeitschrift für Physik*, **49**, 619–636.

Heisenberg, W. (1929). Die Entwicklung der Quantentheorie 1918–1928 (The development of the quantum theory 1918–1928), *Naturwissenschaften*, **17**, 490–496.

Heisenberg, W. (1960). Erinnerungen an die Zeit der Entwicklung der Quantenmechanik, in *Theoretical Physics in the Twentieth Century*, eds Fierz, M. and Weisskopf, V. F., pp. 40–47. New York: Interscience Publishers.

Heisenberg, W. (1963). Archive for the History of Quantum Physics Interview with Heisenberg. Archived interview. This interview was conducted on 27 February 1963.

Heisenberg, W. and Jordan, P. (1926). Anwendung der Quantenmechanik auf das Problem der anomalen Zeemaneffekte, *Zeitschrift für Physik*, **37**, 263–277.

Heisenberg, W. and Pauli, W. (1929). Zur Quantendynamik der Wellenfelder, *Zeitschrift für Physik*, **56**, 1–61.

Heisenberg, W. and Pauli, W. (1930). Zur Quantentheorie der Wellenfelder. II, *Zeitschrift für Physik*, **59**, 168–190.

Heitler, W. (1927). Störungsenergie und Austausch beim Mehrkörperproblem, *Zeitschrift für Physik*, **46**, 47–72.

Heitler, W. and London, F. (1927). Wechselwirkung neutraler Atome und homöopolare Bindung nach der Quantenmechanik, *Zeitschrift für Physik*, **44**, 455–472.

Hellinger, E. (1909). Neue Begründung der Theorie quadratischer Formen von unendlichvielen Veränderlichen, *Journal für die reine und angewandte Mathematik (Crelle's Journal)*, **136**, 210–271.

Hertz, H. (1893). *Electric Waves*. London: Macmillan and Company. The original book, *Untersuchungen über die Ausbreitung der elektrischen Kraft*, was published by Johann Ambrosius Barth in Leipzig in 1892.

Herzberg, G. (1944). *Atomic Spectra and Atomic Structure*. New York: Dover Publications.

Hess, V. F. (1913). Über Beobachtungen der durchdringenden Strahlung bei sieben Freiballonfahrten (Concerning observations of penetrating radiation on seven free balloon flights), *Physikalische Zeitschrift*, **13**, 1084–1091.

Hewish, A., Bell, S., Pilkington, J., Scott, P., and Collins, R. (1968). Observations of a rapidly pulsating radio source, *Nature*, **217**, 709–713.

Hilbert, D. (1900). Mathematische Probleme, *Nachrichten von der Königl. Gesellschaft der Wissenschaften zu Göttingen*, pp. 253–297. Hilbert's address is translated into English in *Bulletin of the American Mathematical Society*, **8**, 437–479, 1902.

Hilbert, D. (1912). *Grundzüge einer allgemeinen Theorie der linearen Integralgleichungen*. Leipzig: Teubner.

Hilbert, D. (1915). Die Grundlagen der Physik (The Foundations of Physics), *Nachrichten von der Königl. Gesellschaft der Wissenschaften zu Göttingen*, pp. 395–407.

Hilbert, D., von Neumann, J. and Nordheim, L. (1927). Über die Grundlagen der Quantenmechanik, *Mathematische Annalen*, **98**, 1–30.

Honl, H. (1925). Die Intensitäten der Zeemancomponenten, *Zeitschrift für Physik*, **34**, 340–354.

Huang, K. (2001). *Introduction to Statistical Physics*. London: Macmillan and Company.

Hund, F. (1927a). Zur Deutung der Molekelspektren. I, *Zeitschrift für Physik*, **40**, 742–764.

Hund, F. (1927b). Zur Deutung der Molekelspektren. II, *Zeitschrift für Physik*, **42**, 93–120.

Hund, F. (1927c). Symmetriecharaktere von Termen bei Systemen mit gleichen Partikeln in der Quantenmechanik, *Zeitschrift für Physik*, **43**, 788–804.

Hund, F. (1927d). Zur Deutung der Molekelspektren. III, *Zeitschrift für Physik*, **43**, 805–826.

Jacobi, G. J. (1841). Sur un théorème de Poisson, *Comptes rendus (Paris)*, **11**, 529.

Jammer, M. (1989). *The Conceptual Development of Quantum Mechanics*, second edition. New York: American Institute of Physics and Tomash Publishers. The first edition of Jammer's important book was published in 1966 by the McGraw-Hill Book Company in the International Series of Pure and Applied Physics. The 1989 edition is an enlarged and revised version of the first edition and I use this as the primary reference to Jammer's history.

Jones, B. (1870). *The Life and Letters of Faraday (in two volumes)*. London: Longmans, Green and Company. This two-volume work has been reprinted as a single volume by Cambridge University Press (2010).

Jordan, P. (1926). Über kanonische Transformationen in der Quantenmechanik, *Zeitschrift für Physik*, **37**, 383–386. See also *Zeitschrift für Physik*, **38**, 513–317.

Jordan, P. (1927). Über eine neue Bergründung der Quantenmechanik, *Zeitschrift für Physik*, **40**, 809–838.

Jordan, P. and Pauli, W. (1928). Zur Quantenelektrodynamik ladungsfreier Felder, *Zeitschrift für Physik*, **47**, 151–173.

Kaufmann, W. (1902). Die Elektromagnetische Masse des Elektrons, *Physikalische Zeitschift*, **4**, 54–56.

Keesom, W. H. and Wolfke, M. (1928). Two different liquid states of helium, *Proceedings of the Koniklijle Akadamie van Wetenschappen te Amsterdam*, **30**, 90–94.

Kellner, G. W. (1927). Die Ionisierungsspannung des Heliums nach der Schrödingerschen Theorie, *Zeitschrift für Physik*, **44**, 91–109.

Kirchhoff, G. (1859). Ueber den Zusammenhang zwischen Emission und Absorption von Licht und Wärme (On the connection between emission and absorption of light and heat), *Berlin Monatsberichte*, pp. 783–787.

Kirchhoff, G. (1861). Untersuchungen über das Sonnenspektrum und die Spectren der Chemischen Elemente (Investigations of the solar spectrum and the spectra of the chemical elements), Part 1, *Abhandlungen der königlich Preussischen Akademie der Wissenschaften zu Berlin*, pp. 62–95.

Kirchhoff, G. (1862). Untersuchungen über das Sonnenspektrum und die Spectren der Chemischen Elemente (Investigations of the solar spectrum and the spectra of the chemical elements), Part 1 (continued), *Abhandlungen der königlich Preussischen Akademie der Wissenschaften zu Berlin*, pp. 227–240.

Kirchhoff, G. (1863). Untersuchungen über das Sonnenspektrum und die Spectren der Chemischen Elemente (Investigations of the solar spectrum and the spectra of the chemical elements), Part 2, *Abhandlungen der königlich Preussischen Akademie der Wissenschaften zu Berlin*, pp. 225–240.

Klein, M. J. (1967). The beginnings of the quantum theory, in *History of Twentieth Century Physics: 57th Varenna International School of Physics, 'Enrico Fermi'*, ed. Weiner, C., volume 57, pp. 1–39. New York and London: Academic Press.

Klein, M. J. (1970). *Paul Ehrenfest: The Making of a Theoretical Physicist*. Amsterdam-London: North-Holland.

Klein, O. and Nishina, Y. (1928). The scattered light of free electrons according to Dirac's new relativistic dynamics, *Nature*, **122**, 398–399.

Klein, O. and Nishina, Y. (1929). Über die Streuung von Strahlung durch freie Elektronen nach der neuen relativischen Quantendynamik von Dirac, *Zeitschrift für Physik*, **52**, 853–868.

Kolhörster, W. (1913). Messungen der Durchdringenden Strahlung im Freiballon in Grösseren Höhen (Measurements of penetrating radiation in free balloon flights at great altitudes), *Physikalische Zeitschrift*, **14**, 1153–1156.

Kossel, W. (1914). Bemerkung zur Absorption homogener Röntgenstrahlen, *Verhandlungen der Deutschen Physikalischen Gesellschaft (2)*, **16**, 898–909, 953–963.

Kossel, W. (1916). Bemerkung zum Seriencharakter der Röntgenspektren, *Verhandlungen der Deutschen Physikalischen Gesellschaft (2)*, **18**, 339–359.

Kox, A. J. (1997). The discovery of the electron: II. The Zeeman effect, *European Journal Physics*, **18**, 139–144.

Kragh, H. (1985). Bohr and the Periodic Table, in *Niels Bohr: A Centenary Volume*, eds French, A. P. and Kennedy, P. J., pp. 50–67. Cambridge, Mass. and London: Harvard University Press.

Kramers, H. A. (1924). The law of dispersion and Bohr's theory of spectra, *Nature*, **113**, 673–674. Also published in van der Waerden, *Sources of Quantum Mechanics*, 1967, *op. cit.*, 177–180.

Kramers, H. A. (1926). Wellenmechanik und halbzahlige Quantisierung, *Zeitschrift für Physik*, **39**, 828–840.

Kramers, H. A. and Heisenberg, W. (1925). Über die Streuung von Strahlung durch Atome (On the dispersion of radiation by atoms), *Zeitschrift für Physik*, **31**, 681–708. English translation published in van der Waerden, *Sources of Quantum Mechanics*, 1967, *op. cit.*, 223–252.

Kramers, H. A. and Holst, H. (1923). *The Atom and the Bohr Theory of its Structure: An Elementary Presentation*. London and Copenhagen: Gyldendal.

Kramers, H. A. and Ittmann, G. P. (1929). Zur Quantelung des symmetrischen Kreisels II, *Zeitschrift für Physik*, **58**, 217–231.

Kratzer, B, A. (1922). Störungen und Kombinationsprinzip in System der violetten Cyanbanden, *Sitzungsberichtle der (Kgl.) Bayerischen Akademie der Wissenschaften (München), Mathematisch-physikalische Klasse*, pp. 107–118.

Kronig, R. d. (1925). Spinning electrons and the structure of spectra, *Nature*, **117**, 555–555.

Kronig, R. d. (1960). The turning point, in *Theoretical Physics in the Twentieth Century*, eds Fierz, M. and Weisskopf, V. F., pp. 5–39. New York: Interscience Publishers.

Kuhn, T. S. (1978). *Black-Body Theory and the Quantum Discontinuity 1894–1912*. Oxford: Clarendon Press.

Kuhn, W. (1925). Über die Gesamtstärke der von einem Zustande ausgehenden Absorptionslinien (On the total intensity of absorption lines emanating from a given state), *Zeitschrift für Physik*, **33**, 408–412. English translation in van der Waerden 1967 (*op. cit.*), pp. 253–257.

Ladenburg, R. (1921). Der quantentheoretische Deutung der Zahl der Dispersionselectronen (The quantum-theoretical interpretation of the number of dispersion electrons), *Zeitschrift für Physik*, **4**, 451–468.

Lanczos, C. (1926). Über eine feldmässige Darstellung der neuen Quantenmechanik, *Zeitschrift für Physik*, **35**, 812–830.

Landau, L. (1932). On the theory of stars, *Physicalische Zeitschrift der Sowjetunion*, **1**, 285–288.

Landau, L. (1938). Origin of stellar energy, *Nature*, **141**, 333–334.

Landau, L. D. and Lifshitz, E. (1976). *Mechanics, Course of Theoretical Physics, Vol. 1*. third edition. Amsterdam: Elsevier.

Landé, A. (1919). Eine Quantenregel für die räumliche Orientierung von Elektron-ringen, *Verhandlungen der Deutschen Physikalischen Gesellschaft*, **21**, 585–588.

Landé, A. (1921a). Über den anomalen Zeemaneffekt, *Zeitschrift für Physik*, **5**, 231–241.

Landé, A. (1921b). Über den anomalen Zeemaneffekt, *Zeitschrift für Physik*, **7**, 398–405.

Landé, A. (1922). Zur Theorie der anomalen Zeeman- und magnetomechanischen Effekte, *Zeitschrift für Physik*, **11**, 353–363.

Landé, A. (1923a). Termstruktur und Zeemaneffekt der Multipletts, *Zeitschrift für Physik*, **15**, 189–205.

Landé, A. (1923b). Zur Theorie der Röntgenspektren, *Zeitschrift für Physik*, **16**, 391–395.

Landé, A. (1923c). Termstruktur und Zeemaneffekt der Multipletts, *Zeitrschrift für Physik*, **19**, 112–123.

Langevin, P. and De Broglie, M. (1912). *La Théorie du rayonnement et les quanta: Rapports et discussions de la réunion tenue à Bruxelles, du 30 octobre au 3 novembre 1911*. Paris: Gautier-Villars.

Larmor, J. (1897). On the theory of the magnetic influence on spectra; and on the radiation of moving ions, *Philosophical Magazine (5)*, **44**, 503–512.

Lattes, C., Occhialini, G., and Powell, C. (1947). Observations on the tracks of slow mesons in photographic emulsions, *Nature*, **160**, 453–456.

Laue, M. v. (1912). Eine quantative Prüfung der Theorie für die InterferenzErscheinungen bei Röntgenstrahlung (A quantitative test of the theory of X-ray interference phenomena), *Sitzberichte der Königlich Bayerischen Akademie der Wissenschaften*, pp. 363–373.

Lénárd, P. (1902). Über die lichtelektrische Wirkung (The photoelectric effect), *Annalen der Physik (4)*, **8**, 149–198.

Lewis, G. N. (1926). The conservation of photons, *Nature*, **118**, 874–875.

Lindsay, R. B. and Margenau, H. (1957). *Foundations of Physics*. New York: Dover Publications.

Lo Surdo, A. (1913). Sul fenomeno analogo a quello di Zeeman nel campo elettrico, *Atti del Accademia Reale dei Lincei (5)*, **22**, 664–666.

London, F. (1926). Winkelvariable und kanonische Transformationen in der Undulationsmechanik, *Zeitschrift für Physik*, **40**, 193–210.

London, F. (1938). The *lambda*-phenomenon of liquid helium and the Bose–Einstein degeneracy, *Nature*, **141**, 643–644.

Longair, M. S. (2003). *Theoretical Concepts in Physics: An Alternative View of Theoretical Reasoning in Physics*. Cambridge: Cambridge University Press.

Longair, M. S. (2006). *The Cosmic Century: A History of Astrophysics and Cosmology*. Cambridge: Cambridge University Press.

Longair, M. S. (2011). *High Energy Astrophysics*, third edition. Cambridge: Cambridge University Press.

Lorentz, H. A. (1892a). De relatieve beweging van de aarde en den aether (The relative motion of earth and aether), *Verslag van de gewone Vergadering der wis- en natuurdige Afdeeling, Koniklijle Akadamie van Wetenschappen te Amsterdam*, **1**, 74–79. English translation in Lorentz's *Collected Papers*, eds Zeeman, P. and Fokker, A. D., Vol. 4, 219–223. The Hague: Martinus Nijhoff.

Lorentz, H. A. (1892b). La théorie électromagnétique de Maxwell et son application aux corps mouvants (The electromagnetic theory of Maxwell and its application to moving bodies), *Archives néerlandaises des Sciences exactes et naturelles*, **25**, 363–551.

Lorentz, H. A. (1904). Electromagnetic phenomena in a system moving with any velocity less than that of light. *Proceedings of the Academy of Science of Amsterdam*, **6**, 809–831.

Lorentz, H. A. (1909). Letter to W. Wien of 12 April 1909, in *The Genesis of Quantum Theory (1899–1913)*, ed. A. Hermann, p. 56. Cambridge, Mass: MIT Press (1971).

Lyman, T. (1914). An extension of the line spectra to the extreme violet, *Physical Review (2)*, **3**, 504–505.

Maxwell, J. C. (1860a). Illustrations of the dynamical theory of gases. Part 1. On the motions and collisions of perfectly elastic spheres, *Philosophical Magazine*, **19**, 19–32. Also published in *The Scientific Papers of James Clerk Maxwell*, ed. Niven, W. D. 1890. Vol. 1, Pages 377–391. Cambridge: Cambridge University Press.

Maxwell, J. C. (1860b). Illustrations of the dynamical theory of gases. Part 2. On the process of diffusion of two or more kinds of moving particles among one another., *Philosophical Magazine*, **20**, 21–33. Also published in *The Scientific Papers of James Clerk Maxwell*, ed. Niven, W. D. 1890. Vol. 1, Pages 392–405. Cambridge: Cambridge University Press.

Maxwell, J. C. (1860c). Illustrations of the dynamical theory of gases. Part 3. On the collision of perfectly elastic bodies of any form, *Philosophical Magazine*, **20**, 33–37. Also published in *The Scientific Papers of James Clerk Maxwell*, ed. Niven, W. D. 1890. Vol. 1, Pages 405–409. Cambridge: Cambridge University Press.

Maxwell, J. C. (1861a). On physical lines of force. I. The theory of molecular vortices applied to magnetic phenomena, *Philosophical Magazine*, **21**, 161–175. Also published in *The Scientific Papers of James Clerk Maxwell*, ed. Niven, W. D. 1890. Vol. 1, Pages 451–466. Cambridge: Cambridge University Press.

Maxwell, J. C. (1861b). On physical lines of force. II. The theory of molecular vortices applied to electric currents, *Philosophical Magazine*, **21**, 281–291, + plate, 338–348. Also published in *The Scientific Papers of James Clerk Maxwell*, ed. Niven, W. D. 1890. Vol. 1, Pages 467–488. Cambridge: Cambridge University Press.

Maxwell, J. C. (1862a). On physical lines of force. III. The theory of molecular vortices applied to statical electricity, *Philosophical Magazine*, **23**, 12–24. Also published in *The Scientific Papers of James Clerk Maxwell*, ed. Niven, W. D. 1890. Vol. 1, Pages 489–502. Cambridge: Cambridge University Press.

Maxwell, J. C. (1862b). On physical lines of force. IV. The theory of molecular vortices applied to the action of magnetism on polarised light, *Philosophical Magazine*, **23**, 85–95. Also published in *The Scientific Papers of James Clerk Maxwell*, ed. Niven, W. D. 1890. Vol. 1, Pages 502–512. Cambridge: Cambridge University Press.

Maxwell, J. C. (1865). A dynamical theory of the electromagnetic field, *Philosophical Transactions of the Royal Society of London*, **155**, 459–512. Also published in *The Scientific Papers of James Clerk Maxwell*, ed. Niven, W. D. 1890. Vol. 1, Pages 526–597. Cambridge: Cambridge University Press.

Maxwell, J. C. (1867). On the dynamical theory of gases, *Philosophical Transactions of the Royal Society of London*, **157**, 49–88. Also published in *The Scientific Papers of James Clerk Maxwell*, ed. Niven, W. D. 1890. Vol. 2, Pages 26–78. Cambridge: Cambridge University Press.

Maxwell, J. C. (1873). *A Treatise on Electricity and Magnetism*. Two volumes. Oxford: Clarendon Press.

Maxwell, J. C. (1890). Introductory lectures on experimental physics, *Scientific Papers*, **2**, 241–255. This is Maxwell's inaugural lecture as first Cavendish Professor of Experimental Physics delivered in October 1871. The quotation is on page 244.

Mehra, J. and Rechenberg, H. (1982a). *The Historical Development of Quantum Theory. Vol. 1, The Quantum Theory of Planck, Einstein, Bohr and Sommerfeld: Its Foundation and the Rise of its Difficulties*. Berlin: Springer-Verlag. Volume 1 consists of two separate books, Part 1 and Part 2.

Mehra, J. and Rechenberg, H. (1982b). *The Historical Development of Quantum Theory. Vol. 2, The Discovery of Quantum Mechanics*. Berlin: Springer-Verlag.

Mehra, J. and Rechenberg, H. (1982c). *The Historical Development of Quantum Theory. Vol. 3, The Formulation of Matrix Mechanics and its Modifications 1925–1926*. Berlin: Springer-Verlag.

Mehra, J. and Rechenberg, H. (1982d). *The Historical Development of Quantum Theory. Vol. 4 (Part 1), The Fundamental Equations of Quantum Mechanics 1925–1926. (Part 2) The Reception of the New Quantum Mechanics*. Berlin: Springer-Verlag.

Mehra, J. and Rechenberg, H. (1987). *The Historical Development of Quantum Theory. Vol. 5 (Parts 1 and 2), Erwin Schrödinger and the Rise of Wave Mechanics*. Berlin: Springer-Verlag. This volume was published in two separate parts. The first is entitled

Schrödinger in Vienna and Zurich, 1887–1925 and the second *The Creation of Wave Mechanics; Early Response and Applications 1925–1926*.

Mehra, J. and Rechenberg, H. (2000). *The Historical Development of Quantum Theory. Vol. 6 (Part 1), The Completion of Quantum Mechanics 1926–1941*. Berlin: Springer-Verlag. This first part is entitled *The Probability Interpretation and Statistical Transformation Theory, the Physical Interpretation and the Empirical and Mathematical Foundations of Quantum Mechanics*.

Mehra, J. and Rechenberg, H. (2001). *The Historical Development of Quantum Theory. Vol. 6 (Part 2), The Completion of Quantum Mechanics 1926–1941*. Berlin: Springer-Verlag. This second part is entitled *The Conceptual Completion and the Extensions of Quantum Mechanics 1932–1941* and *Epilogue: Aspects of the Further Development of Quantum Theory 1942–1999*.

Meitner, L. and Frisch, O. R. (1939). Disintegration of uranium by neutrons: a new type of nuclear reaction, *Nature*, **143**, 239–240.

Mendeleyev, D. I. (1869). On the relationship of the properties of the elements to their atomic weights, *Zhurnal Russkoe Fiziko-Khimicheskoe Obshchestvo*, **1**, 60–77.

Mensing, L. (1926). Die Rotations-Schwingungsbanden nach der Quantummechanik, *Zeitschrift für Physik*, **36**, 814–823.

Merzbacher, E. (2002). The early history of quantum tunneling, *Physics Today*, **55**(8), 080000–50.

Michelson, A. A. (1891). On the application of interference-methods to spectroscopic measurements – I, *Philosophical Magazine (4)*, **31**, 338–346.

Michelson, A. A. (1892). On the application of interference-methods to spectroscopic measurements – II, *Philosophical Magazine (4)*, **34**, 280–299.

Michelson, A. A. (1897). Radiation in a magnetic field, *Philosophical Magazine (4)*, **44**, 109–115. See also: *Astrophysical Journal*, **7**, 131–138.

Michelson, A. A. (1903). *Light Waves and their Uses*. Chicago: Chicago University Press. This quotation appears on page 24.

Michelson, A. A. (1927). *Studies in Optics*. Chicago: Chicago University Press.

Michelson, A. A. and Morley, E. W. (1887). On the relative motion of the Earth and the luminiferous ether, *American Journal of Science*, **34**, 333–345.

Millikan, R. A. (1916). A direct photoelectric determination of Planck's *h*, *Physical Review*, **7**, 355–388.

Moseley, H. G. J. (1913). The high frequency spectra of the elements, *Philosophical Magazine (6)*, **26**, 1024–1034.

Moseley, H. G. J. (1914). The high frequency spectra of the elements. Part II, *Philosophical Magazine (6)*, **27**, 703–713.

Mott, N. F. (1928). The solution of the wave equation for the scattering of particles by a Coulombian centre of force, *Proceedings of the Royal Society of London*, **A118**, 542–549.

Nagaoka, H. (1904a). Kinematics of a system of particles illustrating the line and band spectrum and the phenomenon of radioactivity, *Philosophical Magazine*, **7**, 445–455.

Nagaoka, H. (1904b). Kinematics of a system of particles illustrating the line and band spectrum and the phenomenon of radioactivity, *Nature*, **69**, 392–393.

Neary, G. J. (1940). The β-ray spectrum of radium-E, *Proceedings of the Royal Society of London*, **A175**, 71–87.

Nicholson, J. W. (1911). A structural theory of chemical elements, *Philosophical Magazine (6)*, **22**, 864–889.

Nicholson, J. W. (1912). The spectrum of nebulium, *Monthly Notices of the Royal Astronomical Society*, **72**, 49–64.

Niessen, K. F. (1928). Über die annähernden komplexen Lösunden der Schrödingerschen Differentialgleichung für den harmonischen Oszillator, *Annalen der Physik*, **85**, 497–514.

Noether, E. (1918). Invariante Variationsprobleme (Invariant variation problems), *Nachrichten von der Königl. Gesellschaft der Wissenschaften zu Göttingen*, pp. 235–257.

Nordheim, L. (1927). Zur Theorie der thermische Emission und der Reflexion von Elektronen an Metallen, *Zeitschrift für Physik*, **46**, 833–855.

Oppenheimer, J. and Snyder, H. (1939). On continued gravitational contraction, *Physical Review*, **56**, 455–459.

Oppenheimer, J. and Volkoff, G. (1939). On massive neutron cores, *Physical Review*, **55**, 374–381.

Oppenheimer, J. R. (1930). On the theory of electrons and protons, *Physical Review (2)*, **35**, 562–563.

Oppenheimer, J. R. and Serber, R. (1938). On the stability of stellar neutron cores, *Physical Review*, **54**, 540–540.

Pais, A. (1985). *Inward Bound*. Oxford: Clarendon Press.

Pais, A. (1991). *Niels Bohr's Times, in Physics, Philosophy and Polity*. Oxford: Clarendon Press.

Paschen, F. (1908). Zur Kenntnis ultraroter Linienspektra (Normalwellenlängen bis 27 000 Å-E), *Annalen der Physik (4)*, **27**, 537–570.

Paschen, F. and Back, E. (1912). Normale und anomale Zeemaneffekte, *Annalen der Physik (4)*, **39**, 897–932.

Pauli, W. (1925). Über den Zussamenhang des Abschlusses der Elektronengruppen im Atom mit der Komplexstrucktur der Spektren, *Zeitschrift für Physik*, **31**, 765–785.

Pauli, W. (1926). Quantentheorie, *Handbuch der Physik*, **23**, 1–278.

Pauli, W. (1927a). Über Gasentartung und Paramagnetismus, *Zeitschrift für Physik*, **41**, 81–102.

Pauli, W. (1927b). Zur Quantenmechanik des magnetischen Elektrons, *Zeitschrift für Physik*, **43**, 601–623.

Pauli, W. (1933). Die allgemeinen Prinzipien der WellenMechanik, in *Handbuch der Physik*, eds Geiger, H. and Scheel, K., volume 24, pp. 83–272. Berlin: Springer-Verlag.

Pauli, W. (1964). *Wolfgang Pauli: Collected Scientific Papers, Vols. 1 and 2*, eds Kronig, R. and Weisskopf, V. F. New York: Interscience Publishers.

Pauli, W. and Belinfante, F. J. (1940). On the statistical behaviour of known and unknown elementary particles, *Physica*, **7**, 177–192.

Pauli, W. and Fierz, M. (1940). Über relativistische Feldgreichlungnen von Teilchen mit beliebigen Spin, *Helvetica Physica Acta*, **12**, 297–301.

Perrin, J. B. (1901). Les hypothèses moléculaires, *Revue Scientifique*, **15**, 449–461.

Perrin, J. B. (1909). Mouvement brownien et réalité moléculaire, *Annales de Chemie et de Physique (VIII)*, **18**, 5–114.

Perrin, J. B. (1910). *Brownian Movement and Molecular Reality*. London: Taylor and Francis. Perrin's book was translated by F. Soddy.

Persico, E. (1938). Dimostrazione elementare del metodo di Wentzel e Brillouin, *Il Nuovo Cimento*, **15**, 133–138.

Pickering, E. C. (1896). Stars having peculiar spectra. New variable stars in Crux and Cygnus, *Astrophysical Journal*, **4**, 369–370. This note is also Harvard College Observatory Circular No. 12.

Pincherle, A. (1906). Funktionaloperationen und -gleichungen, *Encyclopädie der mathematischen Wissenschaften*, **II/I**, 761–817.

Pippard, A. B. (1995). Physics in 1900, in *Twentieth Century Physics*, eds Brown L. M., Pais A., and Pippard A. B., pp. 1–41. Bristol: Institute of Physics Publishing; New York: American Institute of Physics.

Planck, M. (1899). Über irreversible Strahlungsvorgänge. 5. Mitteilung, *Sitzungberichte der (Kgl.) Preussischen Akademie der Wissenschaften (Berlin)*, pp. 440–480. Also published in Planck's collected papers: *Physikalische Abhandlungen und Vortrage*, **1**, pp. 560-600, Braunschweig: Friedr. Vieweg und Sohn.

Planck, M. (1900a). Entropie und Temperatur strahlender Wärme, *Annalen der Physik (4)*, **1**, 719–737. Also published in Planck's collected papers: *Physikalische Abhandlungen und Vortrage*, **1**, pp. 668–686, Braunschweig: Friedr. Vieweg und Sohn.

Planck, M. (1900b). Über eine Verbesserung der Wien'schen Spektralgleichung (On an improvement of Wien's spectral distribution), *Verhandlungen der Deutschen Physikalischen Gesellschaft*, **2**, 202–204. Also published in Planck's collected papers: *Physikalische Abhandlungen und Vortrage*, **1**, pp. 687–689, Braunschweig: Friedr. Vieweg und Sohn. English Translation: *Planck's Original Papers in Quantum Physics*, (1972), annotated by H. Kangro, pp. 38–45. London: Taylor and Francis.

Planck, M. (1900c). Zur Theorie des Gesetzes der Energieverteilung im Normalspektrum, *Verhandlungen der Deutschen Physikalischen Gesellschaft*, **2**, 237–245. Also published in Planck's collected papers: *Physikalische Abhandlungen und Vortrage*, **1**, pp. 698–706, Braunschweig: Friedr. Vieweg und Sohn. English translation: Hermann, A. (1971). *The Genesis of Quantum Theory (1899–1913)*, p. 10. Cambridge, Mass: MIT Press.

Planck, M. (1902). Über die Natur des weisen Lichtes, *Annalen der Physik (4)*, **7**, 390–400. Also published in Planck's collected papers: *Physikalische Abhandlungen und Vortrage*, **1**, pp. 763–773, Braunschweig: Friedr. Vieweg und Sohn.

Planck, M. (1906). *Vörlesungen über die Theorie der Wärmstrahlung (Lectures on the Theory of Thermal Radiation)*. Leipzig: Barth.

Planck, M. (1907). Letter to Einstein of 6 July 1907, in *The Genesis of Quantum Theory (1899–1913)*, ed. Hermann A., p. 56. Cambridge, Mass: MIT Press (1971). This letter is in the Einstein Archives, Princeton.

Planck, M. (1910). Letter to W. Nernst of 11 July 1910, in *Black-Body Theory and the Quantum Discontinuity 1894–1912*, ed. Kuhn, T., p. 230. Oxford: Clarendon Press (1978).

Planck, M. (1925). *A Survey of Physics*. London: Methuen and Company.

Planck, M. (1931a). Letter to R. W. Wood of 7 October 1931, in *The Genesis of Quantum Theory (1899–1913)*, ed. Hermann, A., pp. 37–38. Cambridge, Mass: MIT Press. Hermann's book was published in 1971.

Planck, M. (1931b). *The Universe in the Light of Modern Physics*. New York: Norton. The remark is on page 30.

Planck, M. (1950). *Scientific Autobiography and Other Papers*. London: Williams and Norgate.

Poincaré, H. (1912). Sur la théorie des quanta (On the theory of quanta), *Journal de physique théorique et appliquée (Paris)(5)*, **2**, 5–34.

Preston, T. (1898). Radiative phenomena in a strong magnetic field, *Scientific Transactions of the Royal Dublin Society*, **6**, 385–391.

Ramsauer, C. (1921). Über den Wirkungsquerschnitt der Gasmoleküle gegenüber langsamen Elektronen, *Annalen der Physik (4)*, **64**, 513–540.

Rayleigh, J. W. (1900). Remarks upon the law of complete radiation, *Philosophical Magazine*, **49**, 539–540. Rayleigh's paper is also contained in *Scientific Papers by John William Strutt, Baron Rayleigh, 1892–1901. Vol. 4*. Cambridge: Cambridge University Press.

Rayleigh, J. W. (1902). *Scientific Papers by John William Strutt, Baron Rayleigh, 1892–1901*, Vol. 4. Cambridge: Cambridge University Press.

Rechenberg, H. (1995). Quanta and quantum mechanics, in *Twentieth Century Physics*, eds Brown, L. M., Pais A., and Pippard A. B., pp. 143–248. Bristol: Institute of Physics Publishing; New York: American Institute of Physics.

Reines, F. and Cowan, C. L. (1956). The neutrino, *Nature*, **178**, 446–449.

Rhodes, R. (1986). *The Making of the Atomic Bomb*. New York: Simon and Schuster.

Rindler, W. (2001). *Relativity: Special, General and Cosmological*. Oxford: Oxford University Press.

Ritz, W. (1908). Magnetische Atomfelder und Serienspektren, *Annalen der Physik (4)*, **25**, 660–696. A shorter paper 'On a new law of series spectra' by Ritz was published in 1908 in *Astrophysical Journal*, **28**, 237–243.

Rochester, G. and Bulter, C. (1947). Evidence for the existence of new unstable elementary particles, *Nature*, **160**, 855–857.

Röntgen, W. C. (1895). Über eine neue Art von Strahlen. (On a new type of ray. Preliminary communication), *Erste Mittheilung: Sitzungsberichte der Physikalisch-Medizinische Gesellschaft, Würzburg*, p. 137. Röntgen's paper was published in December 1895. It was also published in English in 1896, *Nature*, **53**, 274.

Roschdestwensky, D. (1920). Das innere Magnetfeld des Atoms erzeugt die Dublette und Triplette der Spektralserien, *Transactions of the Optical Insitute in Petrograd*, **1**, 1–20.

Rubens, H. and Kurlbaum, F. (1901). Über die Emission langwelliger Lichtwellen durch den schwarzen Körper bei verschiedenen Temperaturen, *Annalen der Physik*, **4**, 649–666. This paper was presented at the meeting of the Prussian Academy of Sciences on 25 October 1900 by F. Kohlrausch.

Rubinowicz, A. (1918a). Bohrsche Frequenzbedingnug und Erhaltung des Impulsmomentes I. Tiel, *Physikalische Zeitschrift*, **19**, 441–445.

Rubinowicz, A. (1918b). Bohrsche Frequenzbedingnug und Erhaltung des Impulsmomentes II. Tiel, *Physikalische Zeitschrift*, **19**, 465–474.

Rubinowicz, A. (1921). Zur Polarisation der Bohrschen Strahlung, *Zeitschrift für Physik*, **4**, 343–346.

Runge, C. (1908). Über die Zerlegung von Spektrallinien im magneteschen Felde, *Physikalische Zeitschrift*, **8**, 232–237.

Runge, C. and Paschen, F. (1900). Studium des Zeeman-Effekts im Quecksilberspektrum, *Physikalische Zeitschrift*, **1**, 480–481.

Rutherford, E. (1899). Uranium radiation and the electrical conduction produced by it, *Philosophical Magazine (5)*, **47**, 109–163.

Rutherford, E. (1903). The electric and magnetic deviation of the easily absorbed rays from radium, *Philosophical Magazine (5)*, **5**, 177–187.

Rutherford, E. (1911). The scattering of α and β particles by matter and the structure of the atom, *Philosophical Magazine (6)*, **21**, 669–688.

Rutherford, E. (1913). The structure of the atom, *Nature*, **92**, 423.

Rutherford, E. (1919). Collisions of α particles with light atoms, IV. An anomalous effect in nitrogen, *Philosophical Magazine, Series 6*, **37**, 581–587.

Rutherford, E. (1920). Bakerian Lecture. Nuclear constitution of atoms, *Proceedings of the Royal Society of London*, **A97**, 374–400.

Rutherford, E. and Andrade, E. N. d. C. (1913). The reflection of γ-rays from crystals, *Nature*, **92**, 267.

Rutherford, E. and Royds, T. (1909). The nature of the α particle from radioactive substances, *Philosophical Magazine, Series 6*, **15**, 281–286.

Schlesinger, L. (1900). *Einführung in die Theorie der Differentialgleichungen mit einer unabhängigen Variablen*. Leipzig: G. J. Göschensche Verlagshandlung.

Schmidt, G. C. (1898). Ueber die von den Thorvebindungen und einigen anderen Substanzen ausgehende Strahlung, *Annalen der Physik und Chemie (Wiedemanns Annalen)*, **65**, 141–151.

Schrödinger, E. (1914). Zur Dynamik elastisch gekoppelter Puncktsysteme, *Annalen der Physik (4)*, **44**, 961–934.

Schrödinger, E. (1919). Der Energieinhalt der Festkörper im Lichte der neueren Forschung, *Physikalische Zeitschrift*, **20**, 420–428, 450–455, 474–480, 497–503, 523–526.

Schrödinger, E. (1921). Versuch zur modelmässigen Deutung des Terms der scharfen Nebenserien, *Zeitschrift für Physik*, **4**, 347–354.

Schrödinger, E. (1926a). The continuous transition from micro- to macro-mechanics, *Die Naturwissenschaften*, **28**, 664–666.

Schrödinger, E. (1926b). Quantisierung als Eigenwertproblem. (Erster Mitteilung) (Quantisation as an eigenvalue problem (Part 1)), *Annalen der Physik (4)*, **79**, 361–376.

Schrödinger, E. (1926c). Quantisierung als Eigenwertproblem. (Zweite Mitteilung) (Quantisation as an eigenvalue problem (Part 2)), *Annalen der Physik (4)*, **79**, 489–527.

Schrödinger, E. (1926d). Über das Verhältnis der Heisenberg–Born–Jordanschen Quantenmechanik zu der meinen, *Annalen der Physik (4)*, **79**, 734–756.

Schrödinger, E. (1926e). Quantisierung als Eigenwertproblem. (Dritte Mitteilung: Störungstheorie, mit Anwendung auf den Starkeffekt der Balmerlinien) (Quantisation as an eigenvalue problem (Part 3). Perturbation theory, with application to the Stark effect in the Balmer lines), *Annalen der Physik (4)*, **80**, 437–490.

Schrödinger, E. (1926f). Quantisierung als Eigenwertproblem. (Vierte Mitteilung) (Quantisation as an eigenvalue problem (Part 4)), *Annalen der Physik (4)*, **81**, 109–139.

Schrödinger, E. (1928). *Collected Papers on Wave Mechanics*. London: Blackie and Sons.

Schrödinger, E. (1935). Biographical Notes, in *Les Prix Nobel en 1933*, pp. 361–363. Stockholm: Imprimérie Royale P.A. Norstedt & Soener.

Schwarzschild, K. (1916). Zur QuantenHypothese, *Sitzungberichte der (Kgl.) Preussischen Akademie der Wissenschaften (Berlin)*, pp. 548–568.

Semat, H. (1962). *Introduction to Atomic and Nuclear Physics*. London: Chapman and Hall.

Siegbahn, M. (1962). X-ray spectroscopy, in *Fifty Years of X-ray Diffraction*, ed. Ewald, P. P., pp. 265–276. Utrecht, The Netherlands: N.V. A. Oosthoek's Uitgeversmaatschappij.

Slater, J. C. (1924). Radiation and atoms, *Nature*, **113**, 307–308.

Slater, J. C. (1929). The theory of complex spectra, *Physical Review*, **34**, 1293–1322.

Slater, J. C. (1975). *Solid State and Molecular Theory. A Scientific Biography*. New York: John Wiley.

Sommerfeld, A. (1915a). Zur Theorie der Balmerschen Serie, *Münchener Berichte*, 425–458.

Sommerfeld, A. (1915b). Die Feinstruktur der Wassensttoff- und Wasserstoff-ähnlichen Linien, *Münchener Berichte*, 459–500.

Sommerfeld, A. (1916a). Zur Quantentheorie der Spektrallinien, *Annalen der Physik*, **51**, 1–94, 125–167.

Sommerfeld, A. (1916b). Zur Theorie des Zeeman-Effekts der Wasserstofflinien, mit einem Anhang über den Stark-Effekt, *Physikalische Zeitschrift*, **17**, 491–507.

Sommerfeld, A. (1919). *Atombau und Spektrallinien*. Braunschweig: Vieweg. The third edition was published in English as *Atomic Spectra and Spectral Lines*, trans. Brose, H. L., 1923. London: Methuen.

Sommerfeld, A. (1920a). Allgemeine spektroskopische Gesetze, insbesonderenein magnetooptischer Zerlegungssatz, *Annalen der Physik*, **63**, 221–263.

Sommerfeld, A. (1920b). Ein Zahlenmysterium in der Theorie des Zeeman-Effekts, *Naturwissenschaften*, **8**, 61–64.

Sommerfeld, A. (1923). Über die Deutung verwickelter Spektren (Manga, Chrom, usw) nach der Methode der inner Quantenzahlen, *Annalen der Physik*, **70**, 32–62.

Sommerfeld, A. (1924). Zur Theorie der Multipletts und ihre Zeemaneffekte, *Annalen der Physik*, **73**, 209–227.

Sommerfeld, A. (1925). Zur Theorie des periodischen Systems, *Physikalische Zeitschrift*, **26**, 70–74.

Sommerfeld, A. (1929). *Atombau und Spektrallinien, Wellenmechanischer Ergäzungsband*. Braunschweig: Vieweg. This is the fourth edition of Sommerfeld's famous text.

Stachel, J. (1998). *Einstein's Miraculous Year: Five Papers That Changed the Face of Physics*. Princeton: Princeton University Press.

Stachel, J. and Cassidy, D. C. (1999). *The Collected Papers of Albert Einstein: Vol. 2. The Swiss Years: Writings, 1900–1909 (English translation supplement)*. Princeton: Princeton University Press.

Stark, J. (1913). Beobachtungen über den Effekt des elektrischen Feldes auf Spektrallinien, *Sitzungberichte der (Kgl.) Preussischen Akademie der Wissenschaften (Berlin)*, pp. 932–946. Also published in *Annalen der Physik* (4), **43**, 1914.

Stark, J. (1914). Betrachtungen über den Effekt des elektrischen Feldes auf Spektrallinien. V. Feinzerlegung der Wasserstoffserie, *Nachrichten von der Kgl. Gesellschaft der Wissenschaften zu Göttingen*, pp. 427–444.

Stefan, J. (1879). Über die Beziehung zwischen der Wärmestrahlung und der Temperatur, *Sitzungberichte der Kaiserlichen Akademie der Wissenschaften (Wien)*, **79**, 391–428.

Stern, O. (1920). Eine direkte Messung der thermischen Molekulargeschwindigkeit, *Zeitschrift für Physik*, **2**, 49–56.

Stern, O. (1921). Ein Weg zur experimentellen Prüfung der Richtungsquantelung im Magnetfeld, *Zeitschrift für Physik*, **7**, 249–253.

Stoner, E. (1929). The limiting density in white dwarf stars, *Philosophical Magazine*, **7**, 63–70.

Stoner, E. C. (1924). The distribution of electrons among atomic levels, *Philosophical Magazine (6)*, **48**, 719–736.

Stoney, G. J. (1891). On the cause of double lines and of equidistant satellites in the spectra of gases, *Scientific Transactions of the Royal Dublin Society*, **4**, 563–608. The reference to the term *electron* appears on page 583.

Strutt (Lord Rayleigh), J. W. (1894). *The Theory of Sound*, two volumes. London: MacMillan.

Szilard, L. and Zinn, W. H. (1939). Instantaneous emission of fast neutrons in the interaction of slow neutrons with uranium, *Physical Review*, **55**, 799–800.

ter Haar, D. (1967). *The Old Quantum Theory*. Oxford: Pergamon Press.

Thomas, L. H. (1926). The motion of the spinning electron, *Nature*, **117**, 514.

Thomas, W. (1925). Über die Zahl der Dispersionselectronen, die einam stationären Zustande zugeordnet sind, *Naturwissenschaften*, **13**, 627.

Thomsen, J. (1895a). Classification des corps simple, *Oversigt over det Kongelige Danske Videnskabernes Selskaps Forhandlingar*, 132–136.

Thomsen, J. (1895b). Om den sandsynlige Forekomst af en Gruppe uvirksomme Grundstoffer (On the probability of the existence of a group of inactive elements), *Oversigt over det Kgl. Danske Videnskabernes Selskaps Forhandlingar*, 137–43.

Thomson, G. P. (1928). Experiments on the diffraction of cathode rays, *Proceedings of the Royal Society of London (A)*, **117**, 600–609.

Thomson, G. P. and Reid, A. (1927). Diffraction of cathode rays by thin films, *Nature*, **119**, 890.

Thomson, J. J. (1897). Cathode rays, *Philosophical Magazine*, **44**, 293–316.

Thomson, J. J. (1906). On the number of corpuscles in an atom, *Philosophical Magazine*, **11**, 769–781.

Thomson, J. J. (1907). *Conduction of Electricity through Gases*. Cambridge: Cambridge University Press.

Thorpe, G. (1910). *History of Chemistry, Vol. 2. From 1850 to 1910*. London: G. P. Putnam's Sons.

Townsend, J. S. and Bailey, V. A. (1922). The motion of electrons in argon, *Philosophical Magazine*, **43**, 593–600.

Uhlenbeck, G. E. and Goudsmit, S. (1925a). Ersetzung der Hypothese vom unmechanischen Zwang durch eine Forderung bezüglich des inneren Verhaltens jedes einzelnen Elektrons, *Die Naturwissenschaften*, **13**, 953–954.

Uhlenbeck, G. E. and Goudsmit, S. (1925b). Spinning electrons and the structure of spectra, *Nature*, **117**, 264–265.

van den Broek, A. J. (1913). Die Radioelemente, das Periodische System und die Konstitution der Atome, *Physikalische Zeitschrift*, **14**, 32–41.

van der Waerden, B. L. (1932). *Die gruppentheoretische Methode in der Quantenmechanik*. Berlin: Springer-Verlag.

van der Waerden, B. L. (1960). Exclusion principle and spin, in *Theoretical Physics in the Twentieth Century*, eds Fierz, M. and Weisskopf, V. F., pp. 199–244. New York: Interscience.

van der Waerden, B. L. (1967). *Sources of Quantum Mechanics*. Amsterdam: North-Holland Publishing Company. This book was republished by Dover Publications, New York in 1968 by a special arrangement with the North-Holland Publishing Company.

van der Waerden, B. L. (1973). From matrix mechanics and wave mechanics to unified quantum mechanics, in *The Physicist's Conception of Nature*, ed. Mehra, J., pp. 276–293. Dordrecht: D. Reidel.

van Vleck, J. H. (1924). The absorption of radiation by multiply periodic orbits and its relation to the correspondence principle and the Rayliegh–Jeans law, *Physical Review*, **24**, 330–365.

Villard, P. (1900a). Sur la Réflection et la Réfraction des Rayons Cathodique et les Rayons Déviables de Radium (On the reflection and refraction of cathode rays and the deviable rays of radium), *Comptes Rendus de L'Academie des Sciences*, **130**, 1010–1012.

Villard, P. (1900b). Sur le Rayonnement du Radium (On the radiation of radium), *Comptes Rendus de L'Academie des Sciences*, **130**, 1178–1179.

Voigt, W. (1887). Über das Dopplersche Princip (On Doppler's principle), *Göttingen Nachrichten*, **7**, 41–51. This paper was reprinted with additional comments by Voigt in *Physikalische Zeitschrift*, **16**, 381–386, 1915.

Voigt, W. (1901). Über das Elektrische Analogon des Zeemaneffectes, *Annalen der Physik (4)*, **4**, 197–208.

Voigt, W. and Hansen, H. M. (1912). Das neue Gitterspektroskop des Göttinger Institutes und seine Verwendung zur Beobachtung der manetischen Doppelbrechung im Gebiete der Absorptionslinien, *Physikalische Zeitschrift*, **13**, 217–224.

von Halban, H., Joliot, F., and Kowarski, L. (1939). Number of neutrons liberated in the nuclear fission of uranium, *Nature*, **143**, 680.

von Neumann, J. (1927). Mathematische Begründung der Quantenmechanik, *Göttinger Nachrichten*, pp. 1–57.

von Neumann, J. (1929). Allgemeine Eigenwerttheorie Hermitischer Funktionaloperatoren, *Mathematische Annalen*, **102**, 49–131.

von Neumann, J. and Wigner, E. (1928a). Zur Erklärung einiger Eigenschaften der Spektren aus der Quantenmechanik des Drehelektrons. Erster Teil, *Zeitschrift fur Physik*, **47**, 203–220.

von Neumann, J. and Wigner, E. (1928b). Zur Erklärung einiger Eigenschaften der Spektren aus der Quantenmechanik des Drehelektrons. Zweiter Teil, *Zeitschrift fur Physik*, **49**, 73–97.

von Neumann, J. and Wigner, E. (1928c). Zur Erklärung einiger Eigenschaften der Spektren aus der Quantenmechanik des Drehelektrons. Drittern Teil, *Zeitschrift fur Physik*, **51**, 844–858.

Warburg, E. (1913). Bemerkung zu der Aufspaltung der Spektallinien im electrischen Feld, *Verhandlungen der Deutschen Physikalischen Gesellschaft (2)*, **15**, 1259–1266.

Weinberg, S. (1992). *Dreams of a Final Theory*. New York: Pantheon.

Weizsäcker, C. v. (1937). Element transformation inside stars. I, *Physikalische Zeitschrift*, **38**, 176–191.

Weizsäcker, C. v. (1938). Element transformation inside stars. II, *Physikalische Zeitschrift*, **39**, 633–646.

Wentzel, G. (1926a). Eine Verallgemeinerung der Quantenbedingungen für die Zwerke der Wellemmechanik, *Zeitschrift für Physik*, **38**, 518–529.

Wentzel, G. (1926b). Zwei Bemerkungen über die Zerstreuung korpuscularer Strahlung als Beugungserscheinung, *Zeitschrift für Physik*, **40**, 590–593.

Weyl, H. (1928). *Gruppentheorie und Quantenmechanik*. Leipzig: S. Hirzel.

Whittaker, E. T. (1917). *A Treatise on the Analytic Dyanamics of Particles and Rigid Bodies*. Cambridge: Cambridge University Press. The first edition was published in 1904, the second in 1917, the third in 1927 and the fourth in 1939.

Wien, W. (1894). Temperatur und Entropie der Strahlung, *Annalen der Physik*, **52**, 132–165.

Wien, W. (1896). Über die Energieverteilung im Emissionspektrum eines scharzen Körpers, *Annalen der Physik*, **58**, 662–669.

Wiener, N. (1926). The operational calculus, *Mathematische Annalen*, **95**, 557–584.

Wiener, N. (1956). *I am a Mathematician*. New York: Doubleday. The quotation is on page 108.

Wigner, E. (1927). Einige Folgerungen aus der Schrödingerschen Theorie für die Termstrukturen, *Zeitschrift fur Physik*, **43**, 624–652.

Wollaston, W. H. (1802). A method of examining refractive and dispersive powers, by prismatic reflection, *Philosophical Transactions of the Royal Society*, **92**, 365–380.

Young, T. (1802). On the theory of light and colours, *Philosophical Transactions of the Royal Society*, **92**, 12–48.

Yukawa, H. (1935). On the interaction of elementary particles. I, *Proceedings of the Physical-Mathematical Society of Japan*, **17**, 48–57.

Zeeman, P. (1896a). Over den invloed eener magnetisatie op den aard van het door eenstof uitgezonden licht (On the influence of magnetism on the nature of light emitted by a substance), *Verslag van de gewone Vergadering der wis- en natuurdige Afdeeling, Koniklijle Akadamie van Wetenschappen te Amsterdam*, **5**, 181–185. English translations in *Philosophical Magazine (5)*, **43**, 226–239, 1897 and *Astrophysical Journal*, **5**, 332–347, 1897.

Zeeman, P. (1896b). Over den invloed eener magnetisatie op den aard van het door eenstof uitgezonden licht (On the influence of magnetism on the nature of light emitted by a substance), *Verslag van de gewone Vergadering der wis- en natuurdige Afdeeling, Koniklijle Akadamie van Wetenschappen te Amsterdam*, **5**, 242–248. English translations in *Philosophical Magazine (5)*, **43**, 226–239, 1897 and *Astrophysical Journal*, **5**, 332–347, 1897.

Zeeman, P. (1897). Over doubletten en tripletten in het spectrum, tweeggebracht door uitwendige magnetische krachten (Doublets and triplets in the spectrum produced by external magentic fields), *Verslag van de gewone Vergadering der wis- en natuurdige Afdeeling, Koniklijle Akadamie van Wetenschappen te Amsterdam*, **6**, 13–18, 99–102, 260–262. English translations in *Philosophical Magazine (5)*, **44**, 55–60 and 255–259, 1897.

Zwicky, F. (1968). *Catalogue of Selected Compact Galaxies and of Post-Eruptive Galaxies*. Guemlingen, Switzerland: F. Zwicky.

Name index

Subject index

Printed in the United States
By Bookmasters